elementary FLUID MECHANICS

4th edition

elementary FLUID MECHANICS

JOHN K. VENNARD
Professor of Fluid Mechanics, Stanford University

New York · London, John Wiley & Sons, Inc.

FOURTH EDITION

15 14 13 12 11

Library of Congress Catalog Card Number: 61-11174
Printed in the United States of America

ISBN 0 471 90585 2

To my devoted wife
DOROTHY W. VENNARD
for her assistance, patience, and understanding
through four editions

Preface

Fluid mechanics is the study of fluids under all conditions of rest and motion. Its approach is analytical and mathematical rather than empirical; it is concerned with the basic principles which allow the solution of the numerous and diversified problems encountered in many fields of engineering irrespective of the physical properties of the fluids involved.

This textbook is intended for the beginner who has completed his differential and integral calculus and his engineering courses in statics and dynamics, but has not yet encountered a course in engineering thermodynamics and has had no prior experience with fluid phenomena except that obtained in the basic physics courses. The limited experience of the beginner and the tremendous scope of fluid mechanics have led me to define *elementary fluid mechanics* as that portion of the subject which may feasibly be taken up in an introductory course; with more extensive training in mathematics and thermodynamics, the beginner in fluid mechanics could be expected to cover considerably more subject matter, especially in flowfields and compressible fluid motion, than I have presented here. Considerable experience in the teaching of fluid mechanics to the American undergraduate has led me to the belief that his difficulties are more of a conceptual than a mathematical nature; for this reason the emphasis in this book is on physical concepts rather than on mathematical manipulations. In this process I have sacrificed some mathematical rigor, but feel that some such oversimplification is necessary in introducing a subject to the beginner.

Since the publication of the first edition of the text some twenty years ago, the climate of engineering education has drastically changed, especially during the last five years, to a stronger emphasis on basic science; this in turn has attracted students more capable of a higher level of scientific attainment. The effect of these trends on fluid mechanics is an increased demand for mathematical rigor, emphasis on broader concepts, and greater expectations for speedier and deeper understanding by the student. This edition of the text reflects these trends and changes but does not depart radically from the scheme of the preceding editions. There is, however, considerably more emphasis on flowfield concepts necessitating use of the partial derivative, a somewhat deeper treatment of compressible flow, and a new chapter on some elementary concepts of mathematical hydrodynamics. However, the general philosophy of this edition remains the same as in the preceding ones, viz., that application of principles to specific useful problems is the optimum method for motivating engineering beginners in a new subject.

My assumption that the beginner in fluid mechanics is inexperienced in partial differentiation and in thermodynamics (yet needs both of these for treatment of certain areas of fluid mechanics) has led to some compromises, since it is not feasible to develop these separate subjects along with the fluid mechanics; frequently definitions and terminology have been stated and used as needed but without rigorous proof.

Like other courses in mechanics, fluid mechanics includes disciplinary features as well as factual information: the beginner must follow theoretical developments, stretch his imagination in visualizing new physical phenomena, and be forced to think his way through problems of theory and application. The text attempts to attain these objectives in the following ways: omission of subsidiary conclusions will encourage the student to come to some conclusions by himself; application of principles to specific problems develops his ingenuity; illustrative problems assist in overcoming numerical difficulties; challenging problems for the student to solve provide practice in analysis and show useful applications as well.

The text is written for the beginning student, rather than the mature professional, with the hope that by itself it will prove an understandable introduction to the various topics, allowing precious classroom time to be spent on amplifications and extensions of the material. It is my conviction that the use of a textbook in this manner provides vital training for later professional life when new subjects must be assimilated without teachers to introduce them.

Presentation of the subject begins with a discussion of fundamentals and fluid properties followed by fluid statics, at which point the student first encounters the concept of a fluid field and the use of the partial derivative. A chapter on kinematics takes up velocity, acceleration, vorticity, circulation, and the equation of continuity. After this come chapters on the flow of the incompressible and compressible ideal fluid, which include brief treatments of two-dimensional flowfields; the impulse-momentum principle is then developed in a separate chapter including applications to incompressible and compressible flow. Discussion of frictional processes and a chapter on similitude and dimensional analysis complete the presentation of the primary tools and lead to applications to pipes, open channels, and measuring devices. A chapter on elementary hydrodynamics then precedes the final chapter on flow about immersed objects. References have been provided to guide the inquiring reader to more exhaustive treatment of the various topics. More than a thousand problems will be found at the ends of the chapters, and most quantitative articles of the text are accompanied by numerical illustrative problems.

I should like to take this opportunity to say to the student of fluid mechanics that you will be confronted with certain special difficulties not encountered in parallel courses in the mechanics of solids. In the latter subjects, analytical treatment of a problem frequently leads to an engineering answer of satisfactory accuracy with the use of a minimum of experimentally determined quantities. In engineering problems of fluid flow, however, this is usually not the case and a variety of experimental factors often must be employed before an engineering answer can be obtained; this means that, along with your analysis of problems, you must also develop some ingenuity in handling and interpreting experimental results. Another difficulty, inherent in fluid mechanics, springs from your lack of casual experience with fluid phenomena, which may be almost nil in comparison with similar experience with simple structures and with solid objects in motion; this deficiency will require deeper concentration on descriptive material and new terminology, along with the analytical developments. Finally, do not forget that your studies of engineering mechanics have the twofold objective of sharpening analytical ability and acquiring new knowledge; attempting to sidestep the first of these aims by "formula-plugging" will not only prove fruitless and uninteresting but will also arrest the development of your analytical ability, which over the long run is more important than the content of any single course!

For many useful ideas, comments, and criticisms for the improvement of the text and problems, I am deeply indebted to many students and teachers, too numerous to mention here; I am especially indebted to Stanford colleagues Joseph B. Franzini, Byrne Perry, Peter O. Wolf (visiting), and Leonard B. Baldwin for their assistance and encouragement. The excellent work of Mrs. Thor H. Sjostrand and Mrs. Richard E. Platt in preparation of the manuscript is also greatly appreciated.

JOHN K. VENNARD

Palo Alto, California
April 1961

Contents

chapter 1

Fundamentals

1. Development of Fluid Mechanics. Man's desire for knowledge of fluid phenomena began with his problems of water supply, irrigation, navigation, and water power. With only a rudimentary appreciation for the physics of fluid flow he dug wells, constructed canals, operated crude water wheels and pumping devices, and, as his cities increased in size, constructed ever larger aqueducts, which reached their greatest size and grandeur in the City of Rome. However, with the exception of the thoughts of Archimedes (287–212 B.C.) on the principles of buoyancy, little of the scant knowledge of the ancients appears in modern fluid mechanics. After the fall of the Roman Empire (A.D. 476) there is no record of progress in fluid mechanics until the time of Leonardo da Vinci (1452–1519). This great genius designed and built the first chambered canal lock near Milan and ushered in a new era in hydraulic engineering; he also studied the flight of birds and developed some ideas on the origin of the forces which support them. However, down through da Vinci's time, concepts of fluid motion must be considered to be more art than science.

After the time of da Vinci, the accumulation of hydraulic knowledge rapidly gained momentum, the contributions of Galileo, Torricelli, Mariotte, Pascal, Newton, Pitot, Bernoulli, Euler, and d'Alembert to the fundamentals of the science being outstanding. Although the theories proposed by these scientists were in general confirmed by crude experiments, divergences between theory and fact led d'Alembert to observe in 1744, "The theory of fluids must necessarily be based upon experiment." D'Alembert showed that there is no resistance to motion when a body moves through an ideal (nonviscous or *inviscid*) fluid, yet obviously this conclusion is not valid for bodies moving through real fluids. This discrepancy between theory and experiment is called the *d'Alembert paradox*; it demonstrates the limitations of theory alone in solving fluid problems.

Because of the conflict between theory and experiment, two schools of thought arose in the treatment of fluid mechanics, one dealing with the theoretical and the other with the practical aspects of fluid flow. In a sense, these two schools of thought have persisted down to the present day, resulting in the mathematical field of *hydrodynamics* and the practical science of *hydraulics*. Before the beginning of the twentieth century notable contributions to theoretical hydrodynamics had been made by Euler, d'Alembert, Navier, Coriolis, Lagrange, Saint-Venant, Stokes, Helmholtz, Kirchhoff, Rayleigh, Rankine, Kelvin, and Lamb. In a broad sense, experimental hydraulics became a study of the flow phenomena occurring in orifices, pipes, and open channels. Among the many early pioneers who devoted their energies to this field were Chezy, Bossut, Borda, Du Buat, Coulomb, Venturi, de Prony, Eytelwein, Bidone, Belanger, Hagen, and Poiseuille.

Near the middle of the last century, Navier and Stokes succeeded in modifying the general equations for ideal fluid motion to fit those of a viscous fluid and in so doing showed the possibilities of adjusting the differences between hydraulics and hydrodynamics. About the same time, theoretical and experimental work on vortex motion by Helmholtz and Kirchhoff was aiding in explaining many of the divergent results of theory and experiment.

Meanwhile, hydraulic research went on apace, and large quantities of excellent data were collected or formulas proposed by Darcy, Bazin, Weisbach, Ganguillet, Kutter, Manning, Francis, and others. Unfortunately, researches led frequently to empirical formulas obtained by fitting curves to experimental data or by merely presenting the results in tabular form, and in many instances the relationship between physical facts and the resulting formula was not apparent.

Toward the end of the last century, new industries arose which demanded data on the flow of fluids other than water; this fact and many significant advances in knowledge tended to arrest the empiricism of hydraulics. These advances were: (1) the theoretical and experimental researches of Reynolds; (2) the development of dimensional analysis by Rayleigh; (3) the use of models by Froude, Reynolds, Vernon-Harcourt, Fargue, and Engels in the solution of fluid problems; and (4) the rapid progress of theoretical and experimental aeronautics in the work of Lanchester, Lilienthal, Kutta, Joukowsky,

Betz, and Prandtl. These advances supplied new tools for the solution of fluid problems and gave birth to modern fluid mechanics.

Since the beginning of the present century, empiricism has waned and fluid problems have been solved by increasingly rational methods; these methods have produced so many fruitful results and have aided so effectively in increasing knowledge of the details of fluid phenomena that the trend is certain to continue.

2. Physical Characteristics of the Fluid State. Matter exists in two states—the solid and the fluid, the fluid state being commonly divided into the liquid and gaseous states. Solids differ from liquids and liquids from gases in the spacing and latitude of motion of their molecules, these variables being large in a gas, smaller in a liquid, and extremely small in a solid. Thus it follows that intermolecular cohesive forces are large in a solid, smaller in a liquid, and extremely small in a gas. These fundamental facts account for the familiar compactness and rigidity of form possessed by solids, the ability of liquid molecules to move freely within a liquid mass, and the capacity of gases to fill completely the containers in which they are placed. In spite of the mobility and spacing of its molecules, a fluid is considered (for mechanical analysis) to be a *continuum* in which there can be no voids or holes; this assumption proves entirely satisfactory for most engineering problems.[1]

A more fruitful and rigorous mechanical definition of the solid and fluid states may be made on the basis of their actions under the various types of stress. Application of tension, compression, or shear stresses to a solid results first in elastic deformation and later, if these stresses exceed the elastic limits, in permanent distortion of the material. Fluids, however, possess elastic properties only under direct compression or tension, application of infinitesimal shear stress resulting in continual and permanent distortion. The inability of fluids to resist shearing stress gives them their characteristic ability to change their shape or to *flow*; their inability to support tension stress is an engineering assumption rather than a scientific fact, but it is a well-justified assumption because such stresses, which depend upon intermolecular cohesion, are usually extremely small.[2]

Because fluids cannot *support* shearing stresses, it does not follow that such stresses are nonexistent in fluids. During the flow of real fluids, the shearing (frictional) stresses assume an important role, and

[1] An important exception is a gas at very low pressure, where intermolecular spacing is very large and the "fluid" is an aggregation of widely separated particles.

[2] However, tests on some very pure liquids have shown them to be capable of supporting tension stresses up to a few thousands of pounds per square inch, but such liquids are seldom encountered in engineering practice.

their computations and prediction are a vital part of engineering work. Without flow, however, shearing stresses cannot exist, and compression stress or *pressure* is the only stress to be considered.

Since fluids at rest cannot contain shearing stresses, no component of stress can exist in a static fluid tangent to a solid boundary or tangent to an arbitrary section passed through the fluid. This means that pressures must be transmitted to solid boundaries or across arbitrary sections *normal* to these boundaries or sections at every

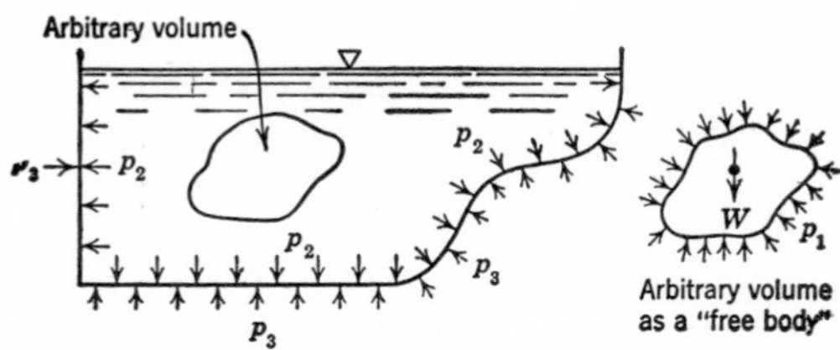

FIG. 1

point. Furthermore, if a *free body* of fluid is isolated as in Fig. 1, pressure must be shown acting *inward* (p_1) upon the free body (according to the usual conventions of mechanics for compression stress). Pressures exerted by the fluid on the container (p_2) will of course act *outward*, but their reactions (p_3) will act inward on the fluid as before. Another property of fluid pressure is that at a point in a fluid at rest it has the same magnitude in all directions; this may

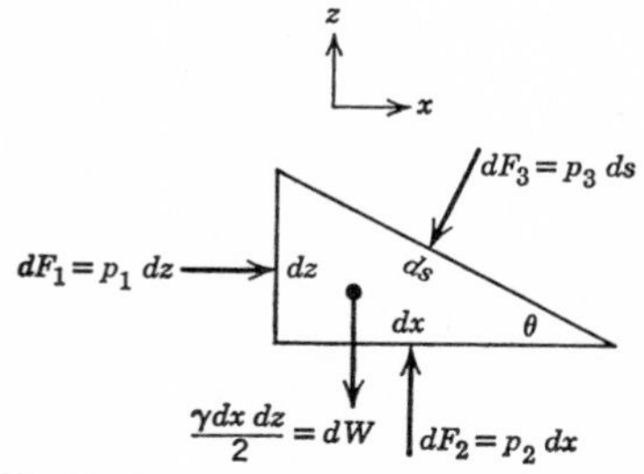

FIG. 2

be proved by considering a convenient two-dimensional [3] free body of fluid (Fig. 2) having unit width normal to the plane of the paper. Taking p_1, p_2, and p_3 to be the mean pressures on the respective

[3] A more general proof using a three-dimensional element will yield the same result.

surfaces of the element, γ the specific weight[4] of the fluid, and writing the equations of static equilibrium:

$$\Sigma F_x = p_1\, dz - p_3\, ds \sin\theta = 0$$

$$\Sigma F_z = p_2\, dx - \gamma\, dx\, dz/2 - p_3\, ds \cos\theta = 0$$

However, from geometry, $dz = ds \sin\theta$ and $dx = ds \cos\theta$. Substituting the first of these relations into the first equation above, $p_1 = p_3$ whatever the size of dz; substituting the second relation into the second equation, $p_2 = p_3 + \gamma\, dz/2$ whatever the size of dx. From these equations it is seen that p_1 and p_3 approach p_2 as dz approaches zero. Accordingly it may be concluded that at a point ($dx = dz = 0$) in a static[5] fluid $p_1 = p_2 = p_3$ and the pressure there is the same in all directions.

With the pressure at a point the same in all directions, it follows that pressure by itself can have no vector sense and hence is a *scalar* quantity; however, the differential forces produced by the action of pressures on differential areas will of course be vectors, since they will have directions normal to the areas; hence the resultant forces obtained by the integration of the differential forces will also be expected to be vector quantities.

Another well-known aspect of fluid pressure (which needs no formal proof) is that pressures imposed on a fluid at rest will be transmitted undiminished to all other points in the fluid; this follows directly from the static equilibrium of adjacent elements and the fact that the fluid mass is a continuum.

The reader may be uneasy concerning the treatment of liquids and gases by the same principles in view of their obvious differences of compressibility. In problems where compressibility is of small importance (and there are many of these in engineering) liquids and gases may be treated similarly, but where the effects of compressibility are predominant (as in high speed gas flow) the behavior of liquids and gases is quite dissimilar and is governed by very different physical laws. Usually, when compressibility is unimportant, fluid problems may be solved successfully with the principles of mechanics; when compressibility predominates, thermodynamics and heat transfer concepts must be used as well.

[4] A summary of symbols and their dimensions will be found in Appendix I.

[5] For the flow of an ideal (inviscid) fluid the same result may be proved, but for viscous fluid motion the pressure at a point is generally *not* the same in all directions owing to the action of shearing stress; however, this is of small consequence in most engineering problems, where shear stress is usually small compared to pressure.

3. Density, Specific Weight, Specific Volume, Specific Gravity. Density [6] is the *mass* of fluid contained in a unit of volume; specific weight,[6] the *weight* of fluid contained in a unit of volume. Both these terms are fundamentally measures of the number of molecules per unit of volume. Since molecular activity and spacing increase with temperature, fewer molecules exist in a given volume of fluid as temperature rises; thus density and specific weight decrease with increasing temperature.[7] Since a larger number of molecules can be forced into a given volume by application of pressure, it is to be expected that density and specific weight will increase with increasing pressure.

Density, ρ, will be expressed in the mass-length-time system of dimensions and will have the dimensions of mass units (slugs) per cubic foot (slugs/ft^3).

Specific weight, γ, will be expressed in the force-length-time system of dimensions and will have dimensions of pounds (weight) per cubic foot (lb/ft^3).

Since a weight, W, is related to its mass, M, by the equation

$$W = Mg$$

in which g is the acceleration due to gravity, density and specific weight (the mass and weight of a unit volume of fluid) will be related by a similar equation,

$$\gamma = \rho g \tag{1}$$

Using the fact that physical equations are dimensionally homogeneous, the foot-pound-second dimensions of ρ (which are equivalent to slugs per cubic foot) may be calculated as follows:

$$\text{Dimensions of } \rho = \frac{\text{Dimensions of } \gamma}{\text{Dimensions of } g} = \frac{\text{lb/ft}^3}{\text{ft/sec}^2} = \frac{\text{lb-sec}^2}{\text{ft}^4}$$

This algebraic use of the dimensions of quantities in the equation expressing physical relationship will be employed extensively and will prove to be an invaluable check on engineering calculations.[8]

The specific volume, defined as volume per unit of weight, will have dimensions of cubic feet per pound (ft^3/lb). This definition identifies specific volume as the reciprocal of specific weight.

[6] In American engineering practice, specific weight is frequently termed "density" and density, "mass density"; the author considers this unfortunate.

[7] For example, a variation in temperature from 32°F to 212°F will cause the specific weight of water to decrease 4 percent (Appendix II) and will cause the density of gases to decrease 37 percent (assuming no pressure variation).

[8] A summary of quantities and their dimensions is given in Appendix I.

Specific gravity, (s.g.), is the ratio of specific weight or density of a substance to the specific weight or density of water. Since all these items vary with temperature, temperatures must be quoted when specific gravity is used in precise calculations of specific weight or density. Specific gravities of a few common liquids are presented in

TABLE I

APPROXIMATE PROPERTIES OF SOME COMMON LIQUIDS AT 68°F AND STANDARD ATMOSPHERIC PRESSURE

	Specific Gravity, s.g.	Modulus of Elasticity, E, psi	Viscosity, $\mu \times 10^5$, lb-sec/ft^2	Surface Tension,[a] σ, lb/ft	Vapor Pressure, p_v, psia
Benzene	0.88	150,000	1.37	0.0020	1.45
Castor oil	0.96	210,000	2060	0.0027	—
Carbon tetrachloride	1.59	160,000	2.035	0.0018	1.74
Ethyl alcohol	0.79	175,000	2.51	0.0015	0.85
Glycerin	1.26	630,000	1800	0.0043	0.000002
Linseed oil	0.94	—	92	0.0023	—
Mercury	13.55	3,800,000	3.24	0.035	0.000025
Olive oil	0.91	230,000	176	0.0023	—
Turpentine	0.86	—	3.11	0.0018	0.0077
Water[b]	1.00	300,000	2.10	0.0050	0.34

[a] In contact with air.

[b] More complete information on properties of water will be found in Appendix II.

Table I,[9] from which the specific weights of liquids may be readily calculated by

$$\gamma = (\text{s.g.}) \times 62.4 \text{ lb/ft}^3 \tag{2}$$

The specific weights of perfect gases may be obtained by a combination of Boyle's and Charles' laws known as the *equation of state*; in terms of specific weight this is

$$\gamma = p/RT \tag{3}$$

in which p is the absolute pressure in pounds per square foot (psfa); T the temperature in degrees Rankine (°F + 459.6), and R the

[9] The reader should refer to the International or Smithsonian Physical Tables if precise specific gravities at other temperatures are required. The student may use the values of Table I in problem solutions even though the temperatures may not be exactly the same.

engineering gas constant in ft-lb/lb°R. The gas constant is constant only if the gas is a *perfect gas*. Common gases in the ordinary engineering range of pressure and temperature may be considered to be perfect for many engineering calculations, but departure from the simple equation of state is to be expected near the point of liquefaction, or at extremely high temperatures or low pressures.

Application of Avogadro's law, "all gases at the same pressure and temperature have the same number of molecules per unit of volume," allows the calculation of a *universal gas constant*. Consider two gases having constants R_1 and R_2, specific weights γ_1 and γ_2, and existing at the same pressure and temperature, p and T. Dividing their equations of state,

$$\frac{p/\gamma_1 T = R_1}{p/\gamma_2 T = R_2}$$

results in $\gamma_2/\gamma_1 = R_1/R_2$, but, according to Avogadro's principle, the specific weight of a gas must be proportional to its molecular weight; that is, $\gamma_2/\gamma_1 = m_2/m_1$ in which m_1 and m_2 are the respective molecular weights of the gases. Combining this equation with the preceding one gives $m_2/m_1 = R_1/R_2$, or

$$m_1 R_1 = m_2 R_2 \tag{4}$$

In other words, the product of molecular weight and gas constant is the same[10] for all gases. This product mR is called the *universal gas constant* and is preferred for general use by many engineers. Accepted values of these gas constants are given in Table II.

Illustrative Problem

Calculate the specific weight, specific volume, and density of chlorine gas at 80°F and pressure of 100 psia.

Solution. Atom. wt. chlorine = 35.45, mol. wt. = 2 × 35.45 = 70.9.

$$R = 1545/70.9 = 21.8 \text{ ft-lb/lb°R} \tag{4}$$

$$\gamma = 100 \times 144/21.8 \times 540 = 1.225 \text{ lb/cuft} \tag{3}$$

$$\text{Spec. vol.} = 1/1.225 = 0.815 \text{ cuft/lb}$$

$$\rho = 1.225/32.2 = 0.0381 \text{ slug/cuft} \tag{1}$$

[10] The constancy of mR applies particularly to the monatomic and diatomic gases. Gases having more than two atoms per molecule tend to deviate from the law mR = Constant. See Table II.

TABLE II

APPROXIMATE PROPERTIES OF SOME COMMON GASES

	Engineering Gas Constant, R, ft-lb/lb°R	Universal Gas Constant, mR	Adiabatic Exponent, k	Spec. Heat at Const. Pressure, c_p, ft-lb/lb°R	Viscosity at 68°F, $\mu \times 10^5$, lb-sec/ft^2
Carbon dioxide	34.9	1536	1.28	159.5	0.0307
Oxygen	48.3	1546	1.40	169.0	0.0419
Air	53.3	1545	1.40	186.5	0.0377
Nitrogen	55.1	1543	1.40	193.0	0.0368
Methane	96.3	1543	1.31	407	0.028
Helium	386	1544	1.66	945	0.0411
Hydrogen	767	1546	1.40	2685	0.0189

4. Compressibility, Elasticity. All fluids may be compressed by the application of pressure, elastic energy being stored in the process; assuming perfect energy conversions, such compressed volumes of fluids will expand to their original volumes when the applied pressure is released. Thus fluids are elastic media, and it is customary in engineering to summarize this property by defining a *modulus of elasticity* as is done for solid elastic materials such as steel. Since fluids do not possess rigidity of form, however, the modulus of elasticity must be defined on the basis of volume—such a modulus being termed a *bulk modulus.*

The mechanics of elastic compression of a fluid may be demonstrated by imagining the cylinder and piston of Fig. 3 to be perfectly rigid (inelastic) and to contain a volume of elastic fluid $\mathbf{V}_1$. Application of a force, F, to the piston will increase the pressure, p, in the fluid and cause the volume to decrease. Plotting p against $\mathbf{V}/\mathbf{V}_1$ produces the stress-strain diagram of Fig. 3 in which the modulus of

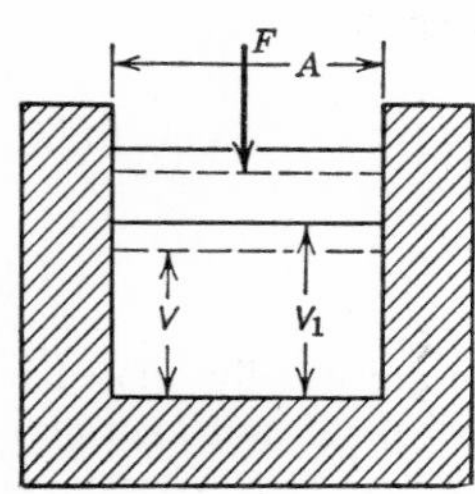

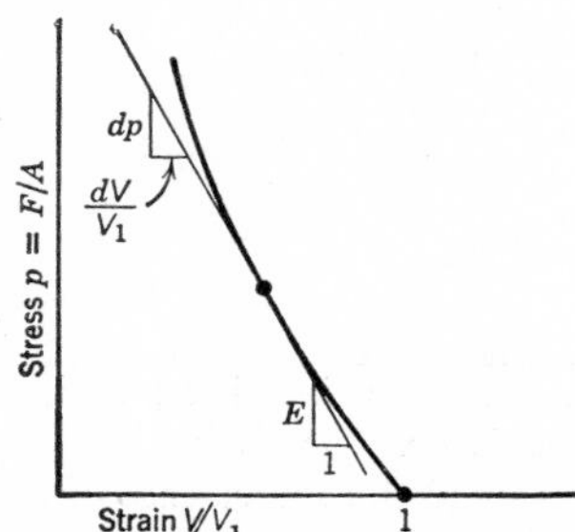

FIG. 3

elasticity of the fluid (at any point on the curve) is defined as the slope of the curve (at that point); thus

$$E = -\frac{dp}{d\mathbf{V}/\mathbf{V}_1} \tag{5}$$

The steepening of the curve with increasing pressure shows that as fluids are compressed they become increasingly difficult to compress further, a logical consequence of reducing the space between the molecules. Clearly the modulus of elasticity of a fluid cannot be constant but increases with increasing pressure.

Although the curve of Fig. 3 applies equally well to liquids and gases, the engineer is usually concerned only with the portion of the curve near $\mathbf{V}/\mathbf{V}_1 = 1$ for liquids. The slope of the curve in this region is taken as the modulus of elasticity for engineering use; such values of E for common liquids are given in Table I and may be used for most engineering problems involving pressures up to a few thousand pounds per square inch.

Compression and expansion of gases take place according to various laws of thermodynamics. A constant-temperature (*isothermal*) process is characterized by Boyle's law,

$$\frac{p}{\gamma} = \text{Constant} \tag{6}$$

whereas a frictionless process in which no heat is exchanged follows the *isentropic* relation

$$\frac{p}{\gamma^k} = \text{Constant} \tag{7}$$

in which k is the ratio of the two specific heats [11] of the gas, that at constant pressure, c_p, to that at constant volume, c_v. Values of k (frequently called the *adiabatic exponent*) for common gases are given in Table II.

Expressions for the modulus of elasticity of gases may be easily derived for isothermal and isentropic processes by writing the general form of equation 5 in terms of γ or ρ. Since the relative increases of γ or ρ are exactly equal to the relative decrease of volume,

$$E = \frac{dp}{d\gamma/\gamma} = \frac{dp}{d\rho/\rho} \tag{8}$$

[11] In American engineering practice, specific heats are usually quoted in Btu/lb°F. In this text these have been converted to ft-lb/lb°F and may be used as ft/°F because of the equivalence of lb (force) and lb (weight). Thermodynamics shows, for perfect gases, that $c_p - c_v = R$ which, in combination with $c_p/c_v = k$, yields $c_p = Rk/(k - 1)$.

When this equation is solved simultaneously with the differential forms of equations 6 and 7, the results are $E = p$ for the isothermal process and $E = kp$ for the isentropic one.

Pressure disturbances cannot be transmitted instantaneously between two points in a fluid unless the fluid is inelastic ($E = \infty$). Small pressure disturbances travel through elastic fluids at a finite velocity (or *celerity*) dependent upon the modulus of elasticity of the fluid. A small pressure disturbance will move at a celerity, a, given by [12]

$$a = \sqrt{\frac{dp}{d\rho}} = \sqrt{\frac{E}{\rho}} \tag{9}$$

in which a is frequently termed the *sonic* or *acoustic* velocity, since it is the velocity with which sound, a small pressure disturbance, travels.

The disturbance caused by a sound wave moving through a fluid is so small and rapid that heat exchange in the compression and expansion may be neglected and the process considered isentropic. Thus, for a perfect gas, the sonic velocity may be obtained by substituting kp for E in equation 9, giving

$$a = \sqrt{kp/\rho} \tag{10}$$

an equation which is accurately confirmed by experiment. This equation may be put into another useful form by substituting gRT for p/ρ (obtained from equations 1 and 3), which results in

$$a = \sqrt{kgRT} \tag{11}$$

and shows that the acoustic velocity in a perfect gas depends only upon the temperature of the gas.

In this era of high-speed flight, the reader is well aware of the *Mach Number*, $\mathbf{M}$, the ratio between flow velocity and sonic velocity, and that flow velocities are defined as *subsonic* for $\mathbf{M} < 1$ and *supersonic* for $\mathbf{M} > 1$. However, $\mathbf{M}$ is also a useful criterion of relative compressibility of the fluid, which permits decisions on whether or not fluids may be considered incompressible for engineering calculations. For the flow of an incompressible (inelastic) fluid, $\mathbf{M} = 0$ since $a = \infty$. Accordingly, as $\mathbf{M} \to 0$, compressibility becomes of decreasing importance; for most engineering calculations, experience has shown that the effects of compressibility may be safely neglected if $\mathbf{M} \gtrless 0.3$.

[12] For the derivation see Appendix III.

Illustrative Problem

Air at 60°F and 14.7 psia is compressed isentropically so that its volume is reduced 50 percent. Calculate the final pressure and temperature and the sonic velocities before and after compression.

Solution.
$$\gamma_1 = 14.7 \times 144/53.3 \times 520 = 0.0763 \text{ lb/cuft} \qquad (3)$$

$$\gamma_2 = 2 \times 0.0763 = 0.1526 \text{ lb/cuft}$$

$$\frac{p_2 \times 144}{(0.1526)^{1.4}} = \frac{14.7 \times 144}{(0.0763)^{1.4}}; \quad p_2 = 38.8 \text{ psia} \qquad (7)$$

$$\frac{38.8 \times 144}{53.3 \times T_2} = 0.1526; \quad T_2 = 687°\text{R } (227°\text{F}) \qquad (3)$$

$$a_1 = \sqrt{1.4 \times 32.2 \times 53.3 \times 520} = 49.0\sqrt{520}$$
$$= 1118 \text{ fps} \qquad (11)$$

$$a_2 = 49.0\sqrt{687} = 1285 \text{ fps}$$

5. Viscosity. All real fluids possess viscosity and therefore exhibit certain frictional phenomena when motion occurs. Viscosity is due fundamentally to cohesion and molecular momentum exchange between fluid layers, and, as flow occurs, these effects appear as tangential or shearing stresses between the moving layers. Consider the motion of a fluid of high viscosity along a solid boundary as in Fig. 4. Observations show that there is no velocity at the boundary and that velocity increases with increasing distance from the boundary. These facts are summarized on the *velocity profile*, which shows relative motion (and thus frictional action) between any two adjacent layers. Two such layers are shown having thickness dy, the lower layer moving with velocity v, the upper with velocity $v + dv$. It is evident that a frictional or shearing force must exist between these layers; it may be expressed as a *shearing* or *frictional stress* per unit of contact area. This stress, designated by τ, has been found for nonturbulent motion (in which viscosity plays a predominant role) to be proportional to the *velocity gradient*, dv/dy, with the *coefficient of viscosity*,[13] μ, defined as the constant of proportionality; thus

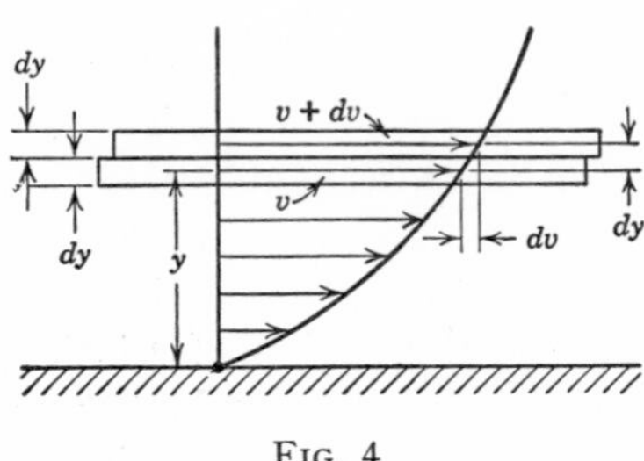

Fig. 4

$$\tau = \mu \frac{dv}{dy} \qquad (12)$$

[13] Also termed *absolute viscosity*, *coefficient of viscosity*, and *dynamic viscosity*.

Since this equation is basic to all problems of fluid resistance, its implications and restrictions are to be emphasized: (1) the nonappearance of pressure in the equation shows that both τ and μ are independent [14] of pressure, and that therefore fluid friction is drastically different from that between moving solids, where pressure plays a large part; (2) any shear stress τ, however small, will cause flow because applied tangential forces must produce velocity gradient, i.e., relative motion between adjacent fluid layers; (3) where $dv/dy = 0$, $\tau = 0$, irrespective of the magnitude of μ—the shearing stress in viscous fluids at rest will be zero, and thus its omission in the analysis of Fig. 2 is confirmed; (4) the velocity profile cannot be tangent to a solid boundary since this would imply infinite velocity gradient there and infinite shearing stress between fluid and solid; (5) the equation is limited to nonturbulent [15] (laminar) fluid motion, in which viscous action is strong. Also relevant to the use of equation 12 is the *observed* fact that the velocity at a solid boundary is zero, i.e., there is no "slip" between fluid and solid.

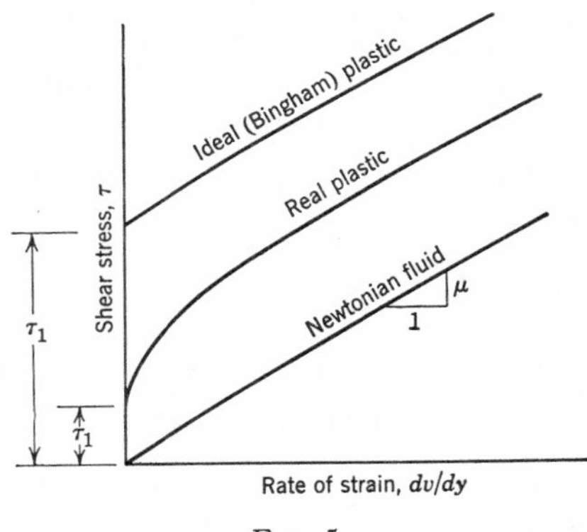

FIG. 5

Equation 12 may be usefully visualized on the plot of Fig. 5 on which μ is the slope of a straight line passing through the origin; here dv will be considered as displacement per unit time and the velocity gradient dv/dy as time rate of strain. Because of Newton's suggestion which led to equation 12, fluids which follow this law are commonly known as *true* or *Newtonian* fluids; it is these fluids with which fluid mechanics is concerned. The science of *rheology* covers the

[14] Viscosity usually increases very slightly with pressure, but the change is negligible in most engineering problems.

[15] The criterion defining laminar and turbulent flow is the Reynolds number, discussed in Chapter 7.

mechanics of the flow of *non-Newtonian* fluids, such as plastics, suspensions, paints, and foods, which *flow* but whose resistance is not characterized by equation 12. The relations between τ and dv/dy for two typical *plastics* are sketched [16] on Fig. 5, and the essential mechanical differences between fluid and plastic is seen to be the shear, τ_1, manifested by the latter, which must be overcome before flow can begin.

The viscosity μ having been defined by equation 12, its dimensions may be determined from dimensional homogeneity as follows:

$$\text{Dimensions of } \mu = \frac{\text{Dimensions of } \tau}{\text{Dimensions of } dv/dy} = \frac{\text{lb/ft}^2}{\text{ft/sec/ft}} = \text{lb-sec/ft}^2$$

However, from previous dimensional considerations (Art. 3), lb = slug-ft/sec^2, and if this is substituted [17] above, it becomes evident that viscosity, μ, may be quoted in slug/ft-sec and that this dimensional combination is equivalent to lb-sec/ft^2. The metric counterparts of these combinations are grams/cm-sec and dyne-sec/cm^2, which are also equivalent to one another and are given the special name *poises* (after Poiseuille, who did some of the first work on viscosity). Viscosities quoted in *poises* or *centipoises* may be readily converted into the English system from the definitions above and by use of basic conversion factors.

Viscosity varies widely with temperature, but temperature variation has an opposite effect upon the viscosities of liquids and gases because of their fundamentally different intermolecular characteristics. In *gases*, where intermolecular cohesion is usually negligible, the shear stress, τ, between moving layers of fluid results from an exchange of momentum between these layers brought about by molecular agitation normal to the general direction of motion. Since this molecular activity is known to increase with temperature, the shear stress, and thus viscosity of gases, will increase with temperature (Fig. 6). In liquids, momentum exchange due to molecular activity is small compared to the cohesive forces between the molecules, and thus shear stress, τ, and viscosity, μ, are primarily dependent on the magnitude of these cohesive forces. Since these forces decrease rapidly with increases of temperature, liquid viscosities decrease as temperature rises (Fig. 6).

[16] The reader should note the form of the curves but should attach no special significance to their relative positions.

[17] This substitution is somewhat artificial since viscous action involves no mass considerations.

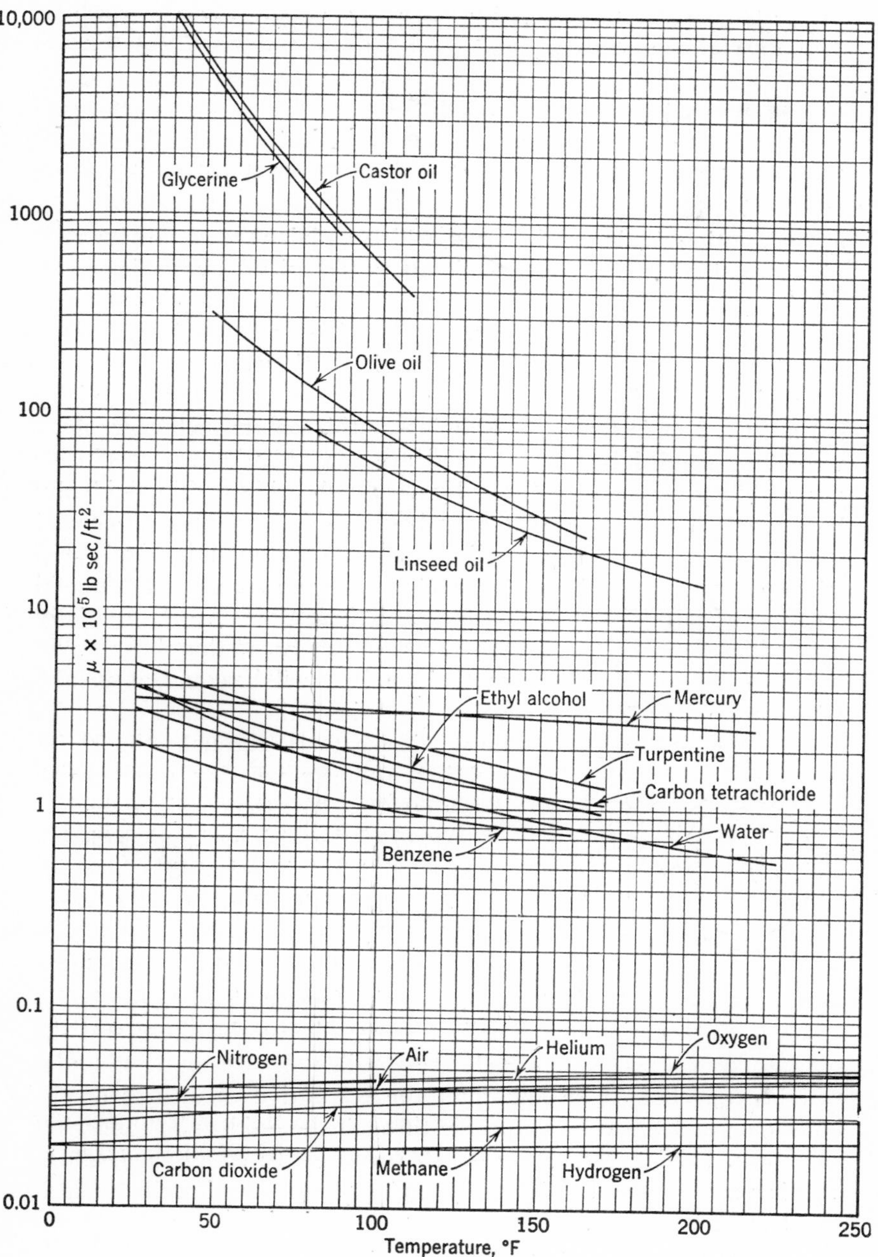

FIG. 6. Viscosities of some common fluids.

Owing to the appearance of the ratio μ/ρ in many of the equations of fluid flow, this term has been defined by

$$\nu = \frac{\mu}{\rho} \tag{13}$$

in which ν is called the *kinematic viscosity*.[18] Dimensional consideration of equation 13 shows the dimensions of ν to be square feet per second, a combination of kinematic terms, which explains the name *kinematic* viscosity. In the metric system, kinematic viscosity is quoted in square centimeters per second, and this dimensional combination is known as *stokes* (after Sir G. G. Stokes). It is fairly common in American practice to quote kinematic viscosities in *centistokes.*

Laminar flows in which the shear stress is constant (or essentially so) are the simplest ones to examine at this point; such *shear flows* were first studied by Couette and are generally known as *Couette flows.* They may be produced by slowly shearing a thin fluid film between two large flat plates or between the surfaces of coaxial cylinders. For constant V in Fig. 7*a*, the force F will be the same on both solid surfaces. If these plates are of essentially the same surface area, the shear stress on their surfaces will be equal, as will the velocity gradients there. These equalities must also extend through the fluid, yielding a linear velocity profile with $dv/dy = V/h$. For the coaxial cylinders of Fig. 7*b*, the driving and resisting torques T will be equal for constant velocity V. With cylinders of the same axial length, the surface area of the outer cylinder is larger than that of the inner one, and the former also has the larger radius. Accordingly the shear and the velocity gradient at the outer cylinder will be expected to be less than the respective quantities on the inner one, and the velocity profile through the fluid to be somewhat as shown. Evidently, $dv/dy \to V/h$ as $h \to 0$; equating dv/dy to V/h is a common and satisfactory engineering approximation in such problems if h is small.

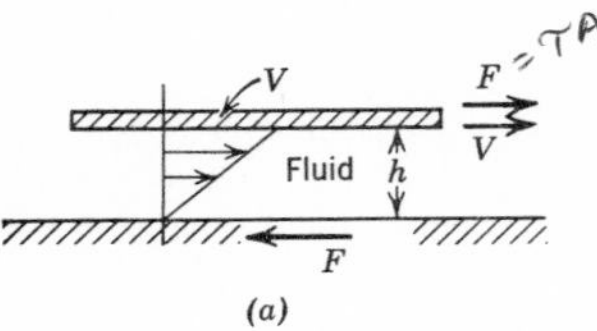

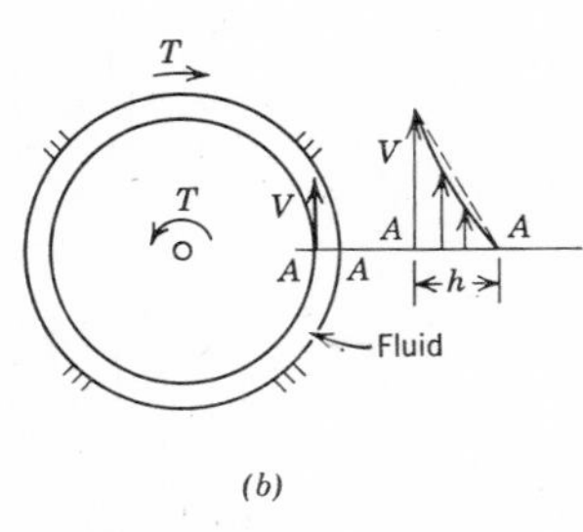

FIG. 7

[18] The wide use of the two viscositites, μ and ν, although of great convenience, may prove both surprising and troublesome to the beginner in the field. For example, water is more viscous than air in terms of μ, but air is more viscous than water in terms of ν.

Illustrative Problem

A cylinder 0.25 ft in radius and 2 ft in length rotates coaxially inside a fixed cylinder of the same length and 0.30-ft radius. Glycerin ($\mu = 0.018$ lb-sec/ft^2) fills the space between the cylinders. A torque of 0.25 ft-lb is applied to the inner cylinder. After constant velocity is attained, calculate the velocity gradients at the cylinder walls, the resulting rpm, and the power dissipated by fluid resistance. Ignore end effects.

Solution. Refer to Fig. 7*b* for a definition sketch. The torque of 0.25 ft-lb is transmitted from inner cylinder to outer cylinder through the fluid layers; therefore (r is the radial distance to any fluid layer)

$$0.25 = \tau(2\pi r \times 2)r; \quad \tau = 0.0199/r^2 = 0.018(dv/dy) \tag{12}$$

Consequently

$$dv/dy = 1.105/r^2, \quad (dv/dy)_i = 17.7 \text{ fps/ft}, \quad (dv/dy)_o = 12.3 \text{ fps/ft}$$

Substituting $-dr$ for dy in the equation for dv/dy and integrating between the appropriate limits as indicated,

$$\int_0^V dv = -1.105 \int_{0.30}^{0.25} r^{-2}\, dr; \quad V = 0.74 \text{ fps}, \quad \text{rpm} = 28.2$$

Assuming the velocity profile to be linear for an approximate calculation

$$V = 17.7/(0.30 - 0.25) = 0.885 \text{ fps}, \quad \text{rpm} = 33.7$$

Since this result differs from the former by nearly 20 percent, the approximation is not satisfactory in this case. The power dissipated in fluid friction will be

$$\text{hp} = 2\pi \times 0.25 \times 28.2/550 \times 60 = 0.00134$$

This power will appear as heat energy, tending to raise the fluid temperature and decrease its viscosity; evidently a suitable heat exchanger would be needed to preserve the steady-state conditions given.

6. Surface Tension, Capillarity. The apparent tension effects which occur on the surfaces of liquids, when the surfaces are in contact with another fluid or a solid, depend fundamentally upon the relative sizes of intermolecular cohesive and adhesive forces. Although such forces are negligible in many engineering problems, they may be predominant in others such as: the capillary rise of liquids in narrow spaces, the mechanics of bubble formation, the breakup of liquid jets, the formation of liquid drops, and the interpretation of results obtained on small models.

On a free liquid surface in contact with the atmosphere, surface tension manifests itself as an apparent "skin" over the surface which will support small loads.[19] The surface tension, σ, is the force in the liquid surface normal to a *line* of unit length drawn in the surface; thus it will have dimensions of pounds per foot. Since surface tension is directly dependent upon intermolecular cohesive forces, its magnitude will decrease as temperature increases.[20] Surface tension is also dependent upon the fluid in contact with the liquid surface; thus surface tensions are usually quoted *in contact with air* as indicated in the footnote to Table I.

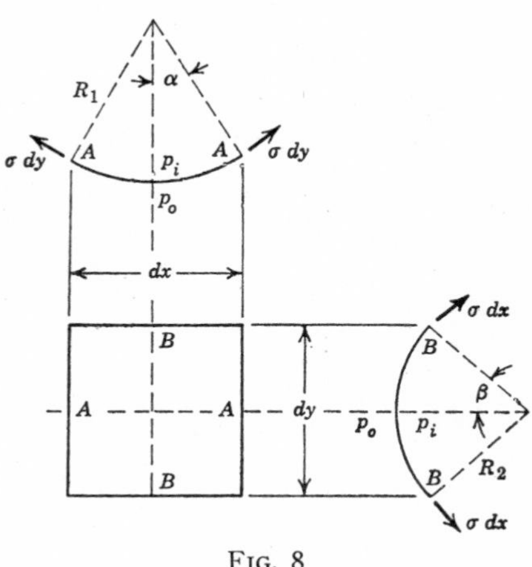

FIG. 8

Consider now (Fig. 8) the general case of a small element $dx\,dy$ of a surface of double curvature with radii R_1 and R_2. Evidently a pressure difference $(p_i - p_o)$ must accompany the surface tension for static equilibrium of the element. A relation between the pressure difference and the surface tension may be derived from this equilibrium by taking $\Sigma F = 0$ for the force components normal to the element:

$$(p_i - p_o)\,dx\,dy = 2\,\sigma\,dy \sin\alpha + 2\sigma\,dx \sin\beta$$

in which α and β are small angles. However, from the geometry of the element, $\sin\alpha = dx/2R_1$ and $\sin\beta = dy/2R_2$. When these values are substituted above, the basic relation between surface tension and pressure difference is obtained; it is

$$p_i - p_o = \sigma\left(\frac{1}{R_1} + \frac{1}{R_2}\right) \tag{14}$$

[19] A small needle placed gently upon a water surface will not sink but will be supported by the tension in the liquid surface. Note, however, that the surface is depressed slightly in the process; this causes localized surface curvature.

[20] See Appendix II.

From this equation the pressure (caused by surface tension) in droplets and tiny jets may be calculated and the rise of liquids in capillary spaces estimated; for a spherical droplet, $R_1 = R_2$; for a cylindrical jet, one R is infinite and the other is the radius of the jet. For the cylindrical capillary tube of Fig. 9 (assuming the liquid surface to be a section of a sphere), $p_o = -\gamma h$, $p_i = 0$, and $p_i - p_o = \gamma h$; also $R_1 = R_2 = R$ and $r/R = \cos\theta$. Substituting these in equation 14 yields

$$h = \frac{2\sigma\cos\theta}{\gamma r} \tag{15}$$

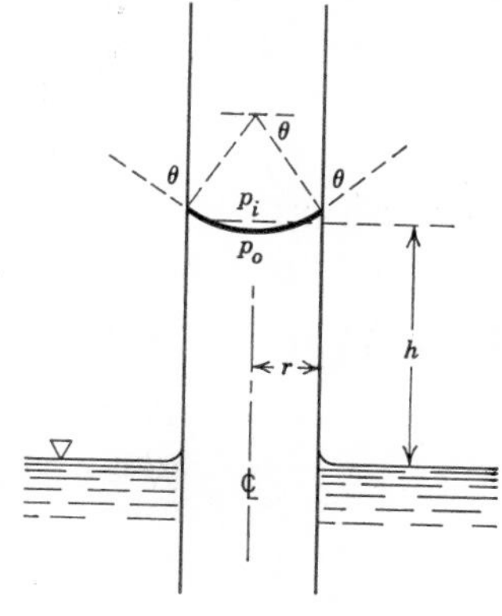

Fig. 9

This result immediately raises several questions: the meaning of the angle θ, the limitations of the equation, and its confirmation by experiment. From the assumption of spherical liquid surface it is clear that the equation is limited to very small tubes; in large tubes the liquid surface will be far from the spherical form. The angle θ is known as the *angle of contact*, and it results from surface tension phenomena of complex nature. Figure 10 describes the situation when

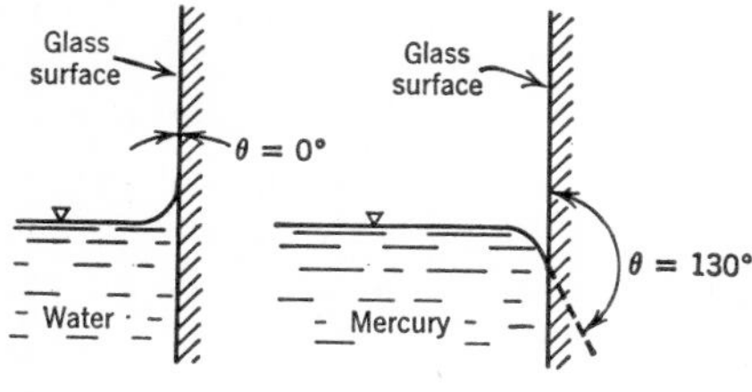

Fig. 10

mercury and water surfaces contact a vertical glass surface. Evidently the mercury molecules possess a greater affinity for each other (cohesion) than for the glass (adhesion), whereas the opposite condition obtains for the water and glass. Although the detailed character of these molecular interactions is not completely understood, the contact angles have been measured and found (for pure substances) to be as indicated. Good experimental confirmation of equation 15 is obtained for small tubes ($r < 0.1$ in.) providing liquids and tube surfaces are extremely clean; in engineering, however, such cleanliness is virtually never encountered and h will be found to be considerably *smaller* than given by equation 15. Thus the equation is useful for making conservative estimates of capillary errors. In the sizing of tubes for pressure measurement, the capillarity problem may be avoided entirely by providing tubes large enough to render the capillarity correction negligible for the desired accuracy.

Illustrative Problem

Of what diameter must a droplet of water (70°F) be to have the pressure within it 0.1 psi greater than that outside?

Solution. From Appendix II, $\sigma = 0.00498$ lb/ft.

$$p_i - p_o = 0.1 \times 144 = 0.00498(2/R) \qquad (15)$$

$$R = 0.00069 \text{ ft}, \quad d = 0.0166 \text{ in.}$$

7. Vapor Pressure. In the mechanics of *liquids*, the physical property of vapor pressure is frequently important in the analysis of problems. All liquids possess a tendency to vaporize, i.e., to change from the liquid to the gaseous phase. Such vaporization occurs because molecules are continually projected through the free liquid surface and lost from the body of the liquid. The ejected molecules, being gaseous, then exert their own partial pressure, which is known as the *vapor pressure* (p_v) of the liquid. Because of the increase of molecular activity with temperature, vapor pressure will be expected to increase with temperature; for water this variation is given in Appendix II. *Boiling* will occur (whatever the temperature) when the external absolute pressure imposed on the liquid is equal to or less than the vapor pressure of the liquid. This means that the boiling point of a liquid is dependent upon the imposed pressure as well as upon temperature.[21]

Table I offers a comparison of the vapor pressures of a few common liquids at the same temperature. The low vapor pressure of mercury along with its high density makes this liquid well suited for use in barometers and other pressure-measuring devices. The more *volatile* liquids, which vaporize more easily, possess the higher vapor pressures.

Illustrative Problem

A vertical cylinder 12-in. in diameter is fitted (at the top) with a tight but frictionless piston and is completely filled with water at 150°F. The outside

[21] For example, water boils at 212°F when exposed to an atmospheric pressure of 14.7 psia but will boil at 200°F if the imposed pressure is reduced to 11.52 psia. See Appendix II.

of the piston is exposed to an atmospheric pressure of 14.52 psia. Calculate the minimum force applied to the piston which will cause the water to boil.

Solution. The force must be applied slowly (to avoid acceleration) in a direction to withdraw the piston from the cylinder. Since the water cannot expand, a space filled with water vapor will be created beneath the piston, whereupon the water will boil. The pressure on the inside of the piston will then be (Appendix II) 3.72 psia and the force on the piston $(14.52 - 3.72)\pi(12)^2/4 = 1225$ lb.

REFERENCES

General and Historical

J. R. Freeman, "Historical Development of Hydraulic Researches and Some Noteworthy Hydraulic Laboratories in America," *Hydraulic Laboratory Practice*, Chapter I, A.S.M.E., 1929.

W. F. Durand, "Historical Sketch of the Development of Aerodynamic Theory," *Trans. A.S.M.E.*, AER-51-3, 1929.

L. F. Moody and B. R. Van Leer, "Fifty Years' Progress in Hydraulics," *Mech. Eng.*, Vol. 52, No. 4, April, 1930.

W. F. Durand, "The Development of Our Knowledge of the Laws of Fluid Mechanics," *Science*, Vol. 78, No. 2025, p. 343, October 20, 1933.

R. Giacomelli and E. Pistolesi, *Aerodynamic Theory* (W. F. Durand, Ed.), Vol. I, Julius Springer, 1934.

C. E. Bardsley, "Historical Résumê of the Development of the Science of Hydraulics," *Pub. 39, Engineering Experiment Station of Oklahoma Agricultural and Mechanical College*, April, 1939.

W. F. Durand, "The Outlook in Fluid Mechanics," *Jour. Franklin Institute*, Vol. 228, No. 2, August, 1939.

B. A. Bakhmeteff, "The Significance of Fluid Mechanics to the Hydraulic Engineer," Proc. Second Hydraulics Conf., *Bul. 27 of the State University of Iowa*, 1942.

Th. von Kármán, "The Role of Fluid Mechanics in Modern Warfare," Proc. Second Hydraulics Conf., *Bul. 27 of the State University of Iowa*, 1942.

J. C. Hunsaker, "Newton and Fluid Mechanics," *Jour. Aero. Sci.*, Vol. 13, No. 12, December, 1946.

H. Rouse, "Present Day Trends in Hydraulics," *Appl. Mech. Revs.*, Vol. 5, No. 2, February, 1952.

H. Rouse and S. Ince, *History of Hydraulics*, Iowa Institute of Hydraulic Research, 1957.

Physical Properties

E. C. Bingham, *Fluidity and Plasticity*, McGraw-Hill Book Co., 1922.

E. Hatschek, *The Viscosity of Liquids*, D. Van Nostrand, 1928.

N. E. Dorsey, *Properties of Ordinary Water-Substance*, Reinhold Publishing Corp., 1940.

R. Weller, D. J. Middlehurst, and R. Steiner, "The Photoviscous Properties of Fluids," *N.A.C.A., Tech. Note 841*, February, 1942.

P. W. Bridgman, "Recent Work in the Field of High Pressures," *Amer. Scientist*, Vol. 31, No. 1, January, 1943.

M. D. Hersey and R. F. Hopkins, "Viscosity of Lubricants under High Pressure," *Mech. Eng.*, December, 1945.

R. S. Burdon, *Surface Tension and the Spreading of Liquids*, 2nd Edition, Cambridge University Press, 1949.

S. L. Kerr, L. H. Kessler, and M. B. Gamet, "New Method for Bulk Modulus Determination," *Trans. A.S.M.E.*, November, 1950.

J. J. Donoghue, R. E. Vollrath, and E. Gerjuoy, "The Tensile Strength of Benzene," *Jour. Chem. Phys.*, Vol. 19, No. 1, January, 1951.

PROBLEMS

1. If 186 cuft of a certain oil weigh 9860 lb, calculate the specific weight, density, and specific gravity of this oil.

2. Calculate the specific weight and density of mercury at 68°F.

3. The density of alcohol is 1.53 slugs/cuft. Calculate its specific weight, specific gravity, and specific volume.

4. The specific gravity of a liquid is 3.0. What is its specific volume?

5. A cubic foot of air at 14.7 psia and 59°F weighs 0.0765 lb/cuft. What is its specific volume?

6. Calculate the specific weight, specific volume, and density of air at 40°F and 50 psia.

7. Calculate the density, specific weight, and specific volume of carbon dioxide at 100 psia and 200°F.

8. Calculate the density of carbon monoxide at 20 psia and 50°F.

9. The specific volume of a certain perfect gas at 30 psia and 100°F is 10 cuft/lb. Calculate its gas constant and molecular weight.

10. If $h = p/\gamma$, what are the dimensions of h?

11. If $V = \sqrt{2gh}$, calculate the dimensions of V.

12. If $F = Q\gamma V/g$, what are the dimensions of F?

13. Twelve cubic feet of water are placed under a pressure of 1000 psi. Calculate the volume at this pressure.

14. If the volume of a liquid is reduced 0.035 percent by application of a pressure of 100 psi, what is its modulus of elasticity?

15. What pressure must be applied to water to reduce its volume 1 percent?

16. Calculate the specific gravity of carbon tetrachloride at 68°F and 3000 psi.

17. When a volume of 1.0212 cuft of alcohol is subjected to a pressure of 7350 psi, it will contract to 0.9784 cuft. Calculate the modulus of elasticity.

18. Ten cubic feet of nitrogen at 100°F and 50 psia are compressed isothermally to 2 cuft. What is the pressure when the nitrogen is reduced to this volume? What is the modulus of elasticity at the beginning and end of the compression?

19. If the nitrogen in the preceding problem is compressed isentropically to 2 cuft, calculate the final pressure and temperature and the modulus of elasticity at the beginning and end of the compression.

20. Calculate the velocity of sound in air at 32°F and 14.7 psia.

21. Calculate the velocity of sound in water at 68°F.

22. Calculate the Mach numbers for an airplane flying at 700 mph through still air at altitudes 5000 and 50,000 ft. (See Appendix IV.)

23. Calculate the kinematic viscosity of turpentine at 68°F.

24. Calculate the kinematic viscosity of nitrogen at 100°F and 80 psia.

25. What is the ratio between the viscosities of air and water at 50°F? What is the ratio between their kinematic viscosities at this temperature and standard barometric pressure?

26. Using data from Appendix II, calculate the kinematic viscosity of water at 50°F and 5000 psi.

27. If a certain liquid has viscosity 0.0001 lb-sec/ft^2 and kinematic viscosity 0.0000375 ft^2/sec, what is its specific gravity?

28. A certain diatomic gas has a kinematic viscosity of 0.00018 ft^2/sec and viscosity of 0.00000060 lb-sec/ft^2 at 14.7 psia and 100°F. Calculate its molecular weight.

29. The kinematic viscosity and specific gravity of a certain liquid are 0.006 ft^2/sec and 2.00, respectively. Calculate the viscosity of this liquid.

30. Nitrogen at 15 psia and 100°F is compressed adiabatically to 45 psia. Calculate kinematic viscosities and acoustic velocities before and after compression.

31. Calculate velocity gradients for $y = 0$, 2, 4, and 6 in. if the velocity profile is a quarter-circle having its center 6 in. from the boundary.

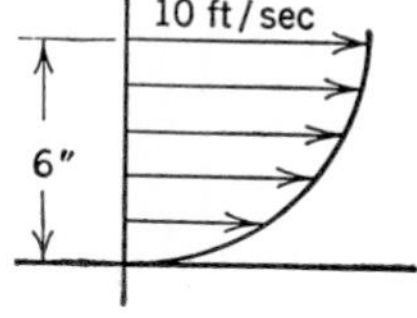

PROBLEM 31

32. Calculate the velocity gradients of the preceding problem, assuming that the velocity profile is a parabola with the vertex 6 in. from the boundary. Also calculate the shear stresses at these points if the fluid viscosity is 0.02 lb-sec/ft^2.

33. If the equation of a velocity profile is $v = 4y^{\frac{2}{3}}$ (v fps, y ft), what is the velocity gradient at the boundary and at 3 in. and 6 in. from it?

34. If the viscosity of a liquid is 0.002 lb-sec/ft^2, what is its viscosity in poises? in centipoises?

35. If the viscosity of a liquid is 3.00 centipoises, what is its viscosity in lb-sec/ft^2?

36. If the kinematic viscosity of an oil is 1000 centistokes, what is its kinematic viscosity in ft^2/sec? If its specific gravity is 0.92, what is its viscosity in lb-sec/ft^2?

37. If the kinematic viscosity of an oil is 0.004 ft^2/sec, what is its kinematic viscosity in stokes? In centistokes?

38. At a point in a viscous flow the shearing stress is 5.0 psi and the velocity gradient 6000 fps/ft. If the specific gravity of the liquid is 0.93, what is its kinematic viscosity?

39. A space 1 in. wide between two large plane surfaces is filled with glycerin at 68°F. What force is required to drag a very thin plate 5 sqft in area between the surfaces at a speed of 0.5 fps if this plate remains equidistant from the two surfaces? if it is at a distance of 0.25 in. from one of the surfaces?

40. A very large thin plate is centered in a gap of width 0.2 ft with different oils of unknown viscosities above and below; one viscosity is twice the other. When the plate is pulled at a velocity of 1 fps, the resulting force on one square foot of plate due to the viscous shear on both sides is 0.6 lb. Assuming viscous flow and neglecting all end effects, calculate the viscosities of the oils.

41. A thin plate of large area is placed midway in a gap of height h filled with oil of viscosity μ_o and the plate is pulled at a constant velocity V. If a lighter oil of viscosity μ_1 is then substituted in the gap, it is found that for the same velocity V the drag force will be the same as before if the plate is located unsymmetrically in the gap but parallel to the walls. Find μ_1 in terms of μ_o and the distance from the nearer wall to the plate.

42. Through a very narrow gap of height h a thin plate of very large extent is being pulled at constant velocity V. On one side of the plate is oil of viscosity μ

and on the other side oil of viscosity $k\mu$. Calculate the position of the plate so that the drag force on it will be a minimum.

43. A vertical gap 1 in. wide of infinite extent contains oil of specific gravity 0.95 and viscosity 0.05 lb-sec/ft^2. A metal plate 5 ft x 5 ft x $\frac{1}{16}$ in. weighing 10 lb is to be lifted through the gap at a constant speed of 0.2 fps. Estimate the force required.

44. A cylinder 8 in. in diameter and 3 ft long is concentric with a pipe of 8.25 in. i.d. Between cylinder and pipe there is an oil film. What force is required to move the cylinder along the pipe at a constant velocity of 3 fps? The kinematic viscosity of the oil is 0.006 ft^2/sec; the specific gravity is 0.92.

45. Castor oil at 68°F fills the space between two concentric cylinders 10 in. high and with diameters of 6 in. and 6.25 in. What torque is required to rotate the inner cylinder at 12 rpm, the outer cylinder remaining stationary?

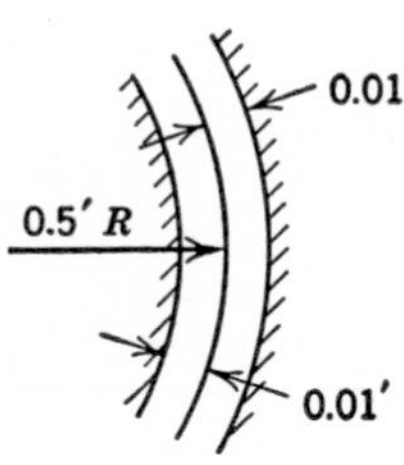

PROBLEM 46

46. A torque of 3 ft-lb is required to rotate the intermediate cylinder at 30 rpm. Calculate the viscosity of the oil. All cylinders are 18 in. long. Neglect end effects.

47. The viscosity of the oil in the preceding problem is 0.005 lb-sec/ft^2. What torque is required to rotate the intermediate cylinder at a constant speed of 40 rpm?

48. A tube of 6-in. i.d. and 6.5-in. o.d. has flat ends and rotates about its axis. Stationary plane surfaces are located $\frac{1}{16}$ in. from each end of this tube, and between surfaces and tube there are oil films. If the oil has viscosity 0.02 lb-sec/ft^2, what torque is required to rotate the tube at a constant speed of 50 rpm?

49. A circular disk of diameter d is slowly rotated in a liquid of large viscosity μ at a small distance h from a fixed surface. Derive an expression for the torque T necessary to maintain an angular velocity ω. Neglect centrifugal effects.

50. The fluid drive shown transmits a torque T for steady-state conditions (ω_1, and ω_2 constant). Derive an expression for the slip ($\omega_1 - \omega_2$) in terms of T, μ, d, and h.

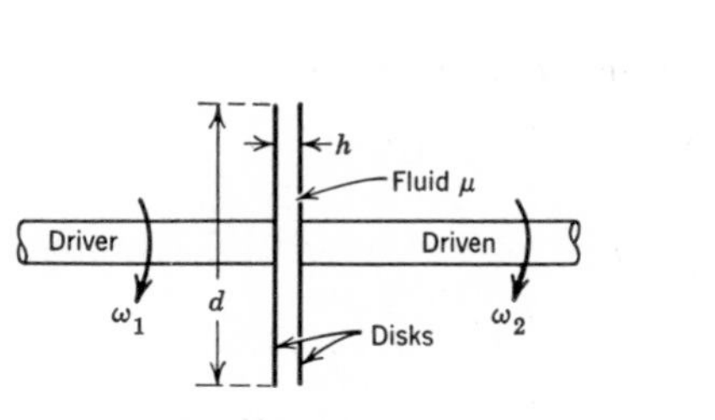

PROBLEM 50

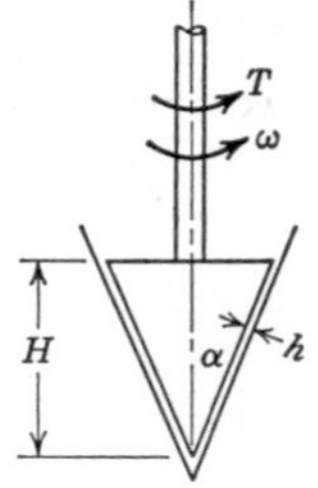

PROBLEM 51

51. Oil of viscosity μ fills the gap h, which is very small. Calculate the torque T required to rotate the cone at constant speed ω.

52. A piece of pipe 12 in. long weighing 3 lb and having i.d. of 2.05 in. is slipped over a vertical shaft 2.00 in. in diameter and allowed to fall. Calculate the approximate velocity attained by the pipe if a film of oil of viscosity 0.5 lb-sec/ft^2 is maintained between pipe and shaft.

53. A piece of pipe of 2.04 in. i.d. and 10 in. long slides down a vertical shaft of 2.00-in. diameter at a constant speed of 0.1 fps. A vertical force of 1 lb is required to pull the pipe back up the shaft at the same constant speed. Calculate the approximate viscosity of oil which fills the small gap between pipe and shaft.

54. The lubricant has a kinematic viscosity of 0.0003 ft²/sec and (s.g.) 0.92. If the mean velocity of the piston is 20 fps, approximately what is the horsepower dissipated in friction?

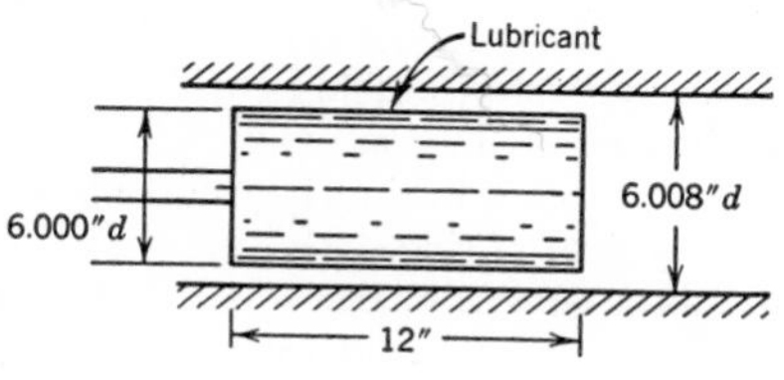

PROBLEM 54

55. Calculate the approximate viscosity of the oil.

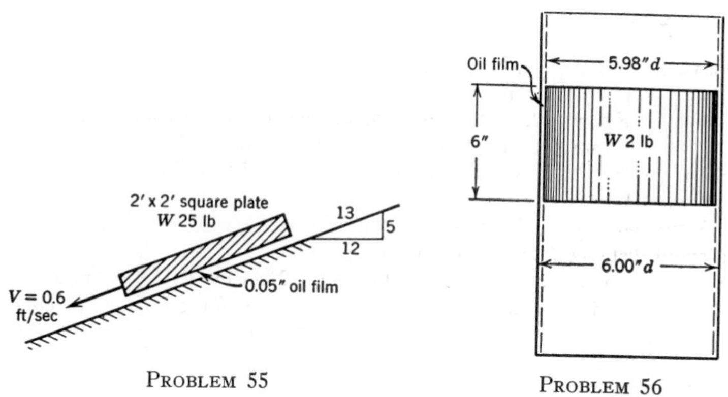

PROBLEM 55 PROBLEM 56

56. The weight falls at a constant velocity of 0.15 fps. Calculate the approximate viscosity of the oil.

57. Calculate the approximate horsepower lost in friction in this bearing.

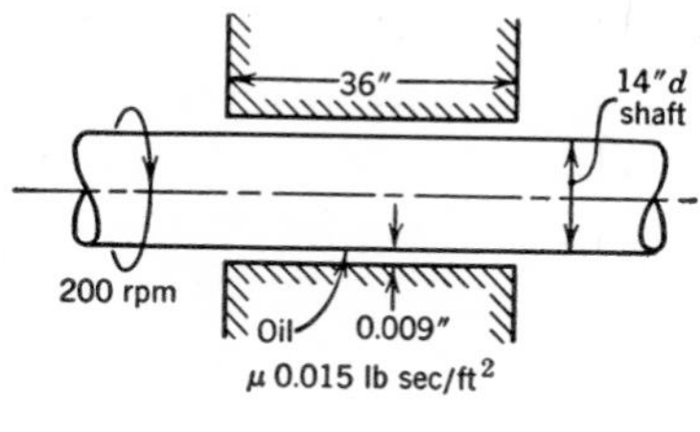

PROBLEM 57

58. What excess pressure may be caused within a cylindrical jet of water 0.20 in. in diameter by surface tension?

59. Calculate the maximum capillary rise of water (68°F) to be expected in a vertical glass tube 1 mm in diameter.

60. Calculate the maximum capillary rise of water (68°F) to be expected between two vertical, clean glass plates spaced 1 mm apart.

61. Derive an equation for theoretical capillary rise between vertical parallel plates.

62. Calculate the maximum capillary depression of mercury to be expected in a vertical glass tube 1 mm in diameter at 68°F.

63. A soap bubble 2 in. in diameter contains a pressure (in excess of atmospheric) of 0.003 psi. Calculate the tension in the soap film.

64. What force is necessary to lift a thin wire ring 1 in. in diameter from a water surface at 68°F? Neglect weight of ring.

65. Using the assumptions of Art. 6, derive an expression for capillarity correction h for an interface between liquids in a vertical tube.

66. What is the minimum absolute pressure which may be maintained in the space above the liquid in a can of ethyl alcohol at 68°F?

67. To what value must the absolute pressure over carbon tetrachloride be reduced to make it boil at 68°F?

68. What reduction below standard atmospheric pressure must occur to cause water to boil at 150°F?

69. At what temperature will water boil at an altitude of 20,000 ft? See Appendices II and IV.

chapter 2

Fluid Statics

Fluid statics is the study of fluid problems in which there is no relative motion between fluid elements. With no relative motions between such elements (and thus no velocity gradients) no shear stress can exist, whatever the viscosity of the fluid.[1] Thus viscosity can have no effect and real fluids at rest may be treated as ideal fluids, allowing exact analytical solutions to such problems without the necessity of experimental confirmation.

8. Pressure-Density-Height Relationships. The fundamental equation of fluid statics is that relating pressure, density, and vertical distance in a fluid. This equation may be derived readily by considering the static equilibrium of a typical differential element[2] of fluid (Fig. 11). Applying $\Sigma F_x = 0$ and $\Sigma F_z = 0$ to the element,

$$\Sigma F_x = \left(\frac{p_A + p_D}{2}\right) dz - \left(\frac{p_B + p_C}{2}\right) dz = 0 \tag{16}$$

$$\Sigma F_z = \left(\frac{p_A + p_B}{2}\right) dx - \left(\frac{p_C + p_D}{2}\right) dx - dW = 0 \tag{17}$$

in which p and γ must be presumed to be functions of x and z. Using the partial derivative notation[3] to account for this, the pressures at the corners of the element are

$$p_A = p, \quad p_B = p + \frac{\partial p}{\partial x} dx$$

[1] See equation 12.

[2] A two-dimensional element is chosen for simplicity and convenience; a three-dimensional one will yield the same result.

[3] See Appendix V.

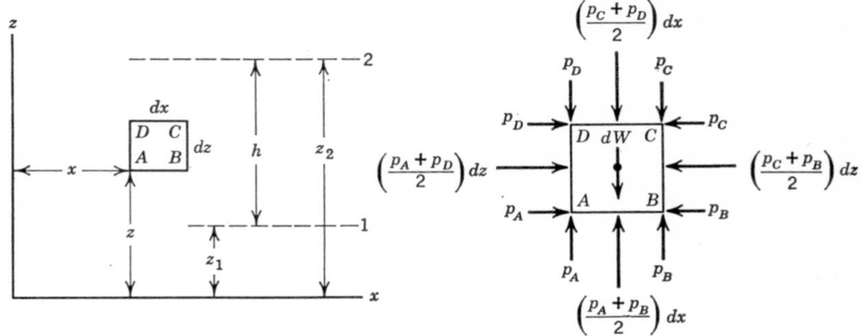

FIG. 11

$$p_C = p + \frac{\partial p}{\partial x} dx + \frac{\partial p}{\partial z} dz, \quad p_D = p + \frac{\partial p}{\partial z} dz$$

and the specific weights γ_A, γ_B, γ_C, and γ_D may be similarly stated by replacing p by γ in the expressions above. The mean specific weight of the element may be calculated from

$$\gamma_{\text{mean}} = \tfrac{1}{4}\,(\gamma_A + \gamma_B + \gamma_C + \gamma_D)$$

which yields

$$\gamma_{\text{mean}} = \gamma + \frac{1}{2}\frac{\partial \gamma}{\partial x} dx + \frac{1}{2}\frac{\partial \gamma}{\partial z} dz$$

showing that γ approaches γ_{mean} as dx and dz both approach zero. Accordingly $\gamma = \gamma_{\text{mean}}$ and the weight dW of the element is $\gamma\, dx\, dz$. Substituting this and the expressions for pressure into equations 16 and 17, they reduce to

$$\frac{\partial p}{\partial x} = 0 \quad \text{and} \quad \frac{\partial p}{\partial z} = \frac{dp}{dz} = -\gamma \tag{18}$$

The first of these equations shows there is no variation of pressure with horizontal distance, i.e., pressure is constant over horizontal surfaces in a static fluid; therefore pressure is a function of z only and the total derivative may replace the partial derivative in the second equation. The latter is the basic equation of fluid statics, which must be integrated for the solution of engineering problems.

When the integration has been performed, there results

$$z_2 - z_1 = \int_{p_2}^{p_1} \frac{dp}{\gamma} \tag{19}$$

For a *fluid of constant density* (this may be safely assumed for liquids over large vertical distances and for gases over small ones) the integration yields

$$z_2 - z_1 = h = \frac{p_1 - p_2}{\gamma} \quad \text{or} \quad p_1 - p_2 = \gamma(z_2 - z_1) = \gamma h \quad (20)$$

permitting ready calculation of the increase of pressure with depth in a fluid of constant density. Equation 20 also shows that pressure differences $(p_1 - p_2)$ may be readily converted to *head* h of fluid of specific weight γ. Thus pressures may be quoted in *inches of mercury*, *feet of water*, etc. The relation of pressure to head[4] is illustrated by the open *manometer* and *piezometer columns* of Fig. 12.

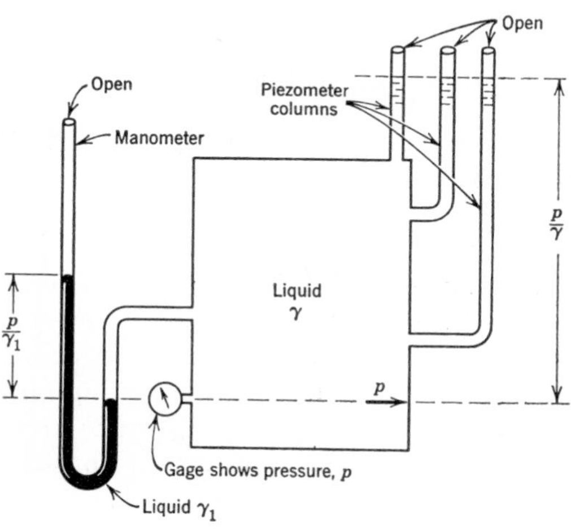

FIG. 12

Equation 20 may be rearranged fruitfully to

$$\frac{p_1}{\gamma} + z_1 = \frac{p_2}{\gamma} + z_2 = \text{Constant} \quad (21)$$

for later comparison with equations of fluid flow. Taking points 1 and 2 as typical, it is evident from equation 21 that the quantity $(z + p/\gamma)$ will be the same for all points in a static liquid. This may

[4] For use in problem solutions it is advisable to keep in mind certain pressure and head equivalents for common liquids. The use of "conversion factors," whose physical significance is rapidly lost, may be avoided by remembering that standard atmospheric pressure at sea level is 14.70 psia, 29.92 in. of mercury (32°F), or 33.9 ft of water (60°F). The establishment of this standard atmospheric pressure also allows pressures to be quoted in *atmospheres*; e.g., a pressure of 147 psia could be stated as 10 atmospheres.

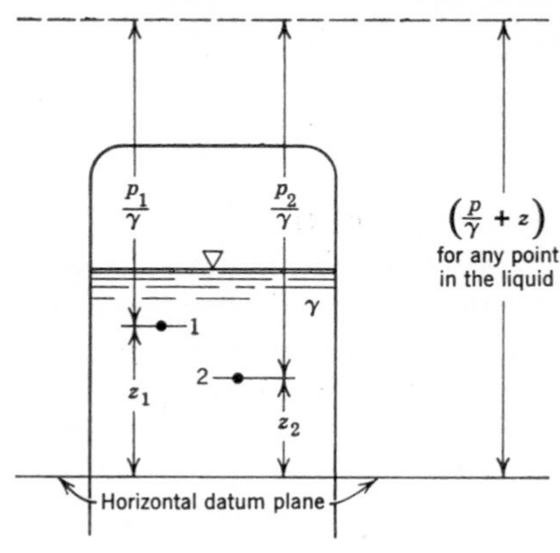

FIG. 13

be visualized geometrically as shown on Fig. 13. Frequently in engineering problems the liquid surface is exposed to atmospheric pressure; if the latter is taken to be zero (see Art. 9) the dashed line of Fig. 13 will necessarily coincide with the liquid surface.

For a fluid of variable density, integration of equation 19 cannot be accomplished until a relationship between p and γ is known. This problem is encountered in the fields of oceanography and meteorology; in the former, a suitable relationship may be obtained from elasticity considerations (equation 8) or empirical relations between pressure and density, and, in the latter, from certain gas laws. For gases (as in the atmosphere) the polytropic equation [5]

$$\frac{p}{\gamma^n} = \text{Constant} \tag{22}$$

may be employed to develop relations between pressure, density, temperature, and altitude. One of the most important of these is $-dT/dz$, the rate of temperature change with altitude, termed the *temperature lapse rate*, which may be derived as follows: substituting γRT for p in equation 22 and differentiating yields

$$\frac{T\,d\gamma}{\gamma\,dT} = \frac{1}{n-1}$$

Substituting $d(\gamma RT)$ for dp in equation 18 gives

$$\frac{dz}{dT} = R\left(\frac{T\,d\gamma}{\gamma\,dT} + 1\right)$$

[5] For the adiabatic (equation 7), $n = k$, and, for the isothermal (equation 6), $n = 1$.

Substituting the first of these equations into the second results in

$$-\frac{dT}{dz}=\frac{n-1}{nR} \tag{23}$$

For $n > 1$, $-dT/dz > 0$, which is the familiar situation in the lower portion of the earth's atmosphere (the *troposphere*) where temperature declines with increasing altitude. Between altitudes 36,000 ft and 80,000 ft (in the *stratosphere*), however, the temperature has been observed to be essentially constant at $-69.7°F$; here the atmosphere is *isothermal*, $n = 1$, and $-dT/dz = 0$. Through the troposphere the mean lapse rate has been found to be practically constant, and this has led to the definition of a *standard atmosphere* which closely approximates the yearly mean at latitude 40°. The ICAO Standard Atmosphere [6] assumes a sea level pressure of 14.70 psia and a constant lapse rate through the troposphere of 0.00356°F/ft from 59°F at sea level to $-69.7°F$ at altitude 36,150 ft, at which the stratosphere begins. From the lapse rate above, and taking R to be 53.3 ft-lb/lb°F, n is found (from equation 23) to be 1.235, from which pressure and air density may be calculated throughout the (standard) troposphere. The engineer assumes this standard (and static) atmosphere for design calculations and performance predictions on high altitude aircraft and their appurtenances; however, the earth's atmosphere with its winds and air currents is not, of course, precisely static but this is usually a satisfactory approximation for the prediction of pressures and densities; for violent disturbances (e.g., tornadoes) the assumption of a static atmosphere is clearly untenable.

The lapse rate for an *adiabatic atmosphere* may be calculated from equation 23 using $n = k = 1.40$ and $R = 53.3$ ft-lb/lb°F providing the air is dry;[7] the result is $-dT/dz = 0.00535°F/ft$, which is known as the *adiabatic lapse rate* and will be shown to be a criterion of atmospheric stability. Suppose that in an adiabatic atmosphere a mass of fluid is moved from one altitude to another. If it is moved upward, it will expand almost without acceptance or rejection of heat (i.e., adiabatically) because of its poor conduction; accordingly at its new altitude it will have the same density as the surrounding air and thus possess no tendency to move from its new position. The adiabatic atmosphere is thus in a state of *neutral equilibrium*

[6] See Appendix IV.

[7] Although atmospheric moisture is a critical factor in many meteorology problems, it will be disregarded here since it does not change the sense of the development; it will, of course, change the numerical values.

and is inherently *stable*. When this process is imagined for a lapse rate $(-dT/dz)$ larger than the adiabatic, the expansion will tend to be adiabatic as before, but in its new position the density of the fluid mass will be smaller than that of its surroundings and its greater buoyancy will cause it to rise further; such an atmosphere is inherently *unstable*—and this fact leads to the expectation that a stable atmosphere can occur only for lapse rates less than the adiabatic. This expectation is confirmed when the foregoing reasoning is applied to an atmosphere with lapse rate less than adiabatic; here displacements of fluid masses produce density changes which tend to restore the air masses to their original position. Thus for diminishing lapse rates atmospheric stability steadily increases, becoming greatest in the case of an *inversion*, when the lapse rate is negative.

Illustrative Problems

A closed tank is partially filled with carbon tetrachloride at 68°F. The pressure on the surface of the liquid is 10 psi. Calculate the pressure 15 ft below the surface.

Solution. $\gamma = 1.59 \times 62.4 = 99.2 \text{ lb/cuft}$ (2)

$$p_1 = 99.2 \times 15 + 10 \times 144 = 2930 \text{ psf} = 20.3 \text{ psi} \qquad (20)$$

Calculate pressure and specific weight of air in the ICAO Standard Atmosphere at altitude 35,000 ft.

Solution. At sea level (see Appendix IV),

$$p_1 = 14.7 \text{ psia}, \quad T_1 = 519°\text{R}, \quad \gamma_1 = 0.0765 \text{ lb/cuft}$$

$$\frac{p}{\gamma^{1.235}} = \frac{14.7 \times 144}{(0.0765)^{1.235}}, \quad \frac{1}{\gamma} = \frac{6450}{p^{0.81}} \qquad (22)$$

$$35{,}000 = 6450 \int_{p_2}^{14.7 \times 144} p^{-0.81}\, dp, \quad p_2 = 504 \text{ psfa} = 3.50 \text{ psia} \qquad (19)$$

$$T_2 = 519 - 35{,}000 \times 0.00356 = 394°\text{R}$$

$$\gamma_2 = 504/53.3 \times 394 = 0.024 \text{ lb/cuft} \qquad (3)$$

Compare results with the values of Appendix IV.

9. Absolute and Gage Pressures. Pressures, like temperatures, are measured and quoted in two different systems, one relative (gage), and the other absolute; no confusion results if the relation between the systems and the common methods of measurement are completely understood.

The Bourdon pressure gage and the aneroid barometer (shown

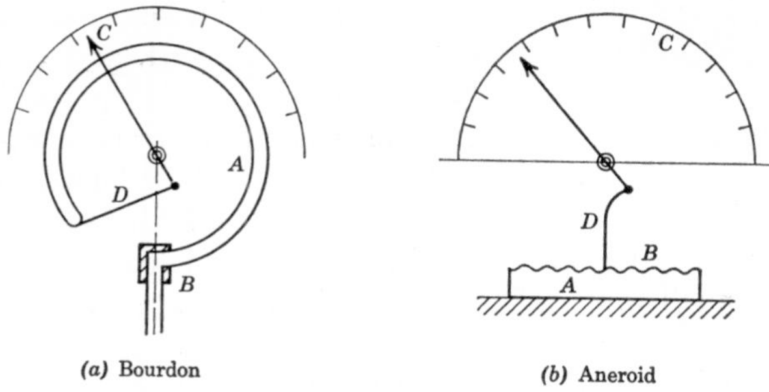

FIG. 14. Mechanical pressure gages.

schematically in Fig. 14) are the standard mechanical devices for measuring gage and absolute pressures, respectively. In the pressure gage a bent tube (A) of elliptical cross section is held rigidly at B and its free end is connected to a pointer (C) by a link (D). When pressure is admitted to the tube, its cross section tends to become circular, causing the tube to straighten and move the pointer over the graduated scale. If the gage is in proper adjustment, the pointer rests at zero on the scale *when the gage is disconnected*; in this condition the pressure inside and outside of the tube will be the same, and thus there is no tendency for the tube to deform. Since atmospheric pressure *usually* exists outside the tube, it is apparent that such pressure gages are actuated by the *difference* between the pressure inside and that outside the tube. Thus, in the gage, or relative, system of pressure measurement, the local (not standard) atmospheric pressure becomes the zero of pressure. For pressure greater than local atmospheric the pointer will move to the right; for pressure less than local atmospheric the tube will tend to contract, moving the pointer to the left. The reading for pressure greater than local atmospheric is positive and is called *gage pressure*, or simply *pressure*,[8] and it is usually measured in pounds per square inch (psi); the reading for pressure below local atmospheric is negative, designated as *vacuum*, and is usually measured in inches of mercury.

The aneroid gage is a device for measuring absolute pressure. The essential element is a short cylinder (A) with one end an elastic diaphragm (B). The cylinder is evacuated[9] so that the pressure

[8] Throughout the remainder of the book *pressure* should be understood to mean gage pressure (psi); when *absolute pressure* (psia) is meant it will be designated as such.

[9] If the cylinder could be completely evacuated, the pressure therein would be the lowest possible (absolute zero) since there would be no fluid molecules to exert pressure.

therein is close to absolute zero; pressures imposed on the outside of the diaphragm will cause it to deflect inward, and these deflections then become a direct measure of the applied pressures, which may be transferred to a suitable scale (C) through appropriate linkages (D); here the pressures recorded will be relative to absolute zero, and they are called *absolute pressures.* Although the aneroid cylinder as conventionally used in barometers is capable of measuring only a small range of pressures, the basic idea may be utilized for absolute pressure gages for more general use.

Liquid devices which measure gage and absolute pressures are shown on Fig. 15; these are the open U-tube and the conventional

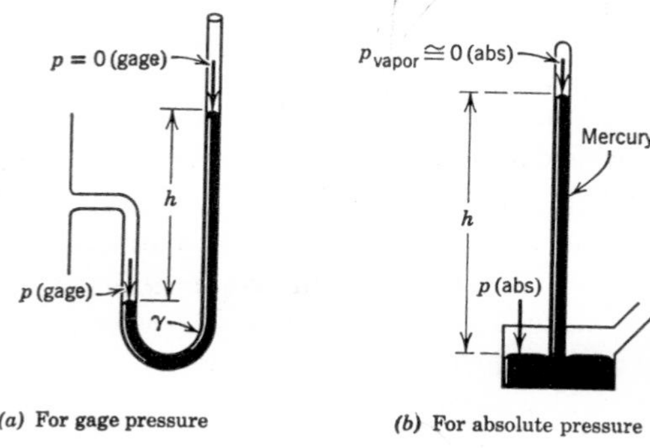

FIG. 15. Liquid pressure gages.

mercury barometer. With the U-tube open, atmospheric pressure will act on the upper liquid surface; if this pressure is taken to be zero, the applied gage pressure p will equal γh and h will thus be a direct measure of gage pressure.

The mercury barometer (invented by Torricelli, 1643) is constructed by filling the tube with air-free mercury and inverting it with its open end beneath the mercury surface in the receptacle. Ignoring the small pressure of the mercury vapor,[10] the pressure in the space above the mercury will be at absolute zero and again $p = \gamma h$; here the height h is a direct measure of the absolute pressure, p. Although conventional use of the barometer is for the measurement of local atmospheric pressure, the basic scheme is frequently used in industry for the direct measurement of any absolute pressure.

From the foregoing descriptions an equation relating gage and absolute pressures may now be written,

$$\text{Absolute pressure} = \text{Atmospheric pressure} \begin{array}{l} - \text{Vacuum} \\ + \text{Gage pressure} \end{array} \tag{24}$$

[10] See Table I.

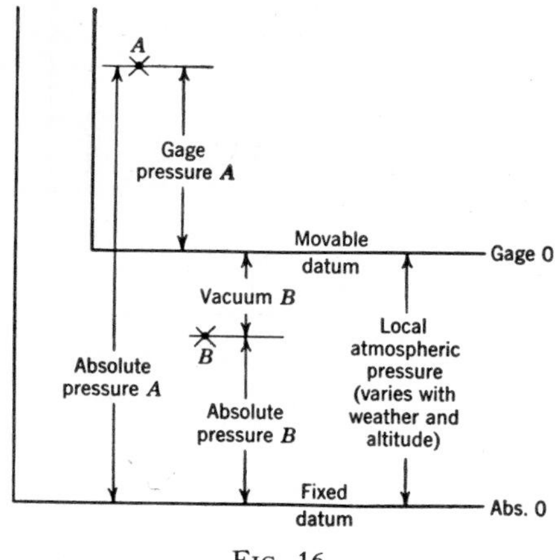

FIG. 16

which allows easy conversion from one system to the other. Possibly a better picture of these relationships can be gained from a diagram such as that of Fig. 16 in which are shown two typical pressures, A and B, one above, the other below, atmospheric pressure, with all the relationships indicated graphically.

ILLUSTRATIVE PROBLEM

A Bourdon gage registers a vacuum of 12.5 in. of mercury when the barometric pressure is 14.50 psia. Calculate the corresponding absolute pressure.

Solution. Absolute pressure = 14.50 − 12.5(14.70/29.92) = 8.35 psia (24)

10. Manometry. Bourdon and aneroid pressure gages, owing to their inevitable mechanical limitations, are not usually adequate for precise measurements of pressure; when greater precision is required, *manometers* like those of Fig. 17 may be effectively employed.

Consider the U-tube manometer of Fig. 17*a* in which all distances and densities are known, and pressure p_x is to be found. Since, over horizontal planes *within continuous columns of the same fluid,* pressures are equal, it is evident at once that $p_1 = p_2$, and, from equation 20,

$$p_1 = p_x + \gamma l \quad \text{and} \quad p_2 = 0 + \gamma_1 h$$

Equating these expressions for p_1 and p_2 results in

$$p_x = \gamma_1 h - \gamma l$$

allowing pressure, p_x, to be calculated.[11]

[11] The use of derived formulas for manometer solutions is not recommended until experience has been gained in their limitations.

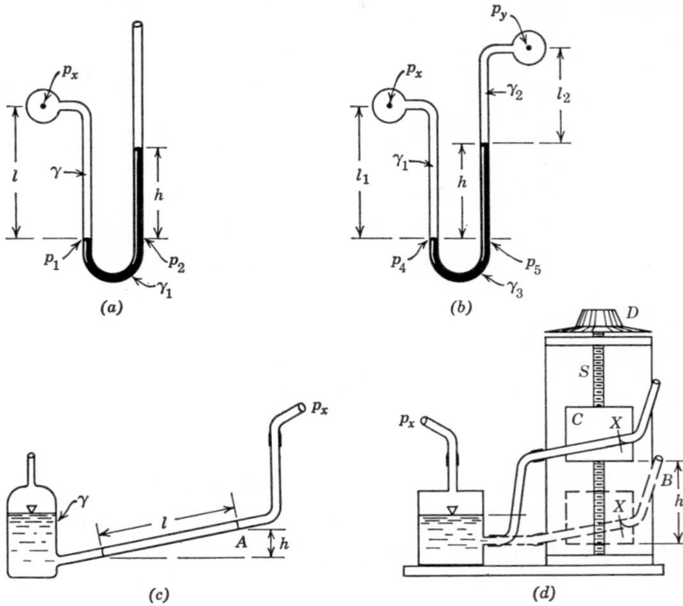

FIG. 17

U-tube manometers are frequently used to measure the difference between two unknown pressures, p_x and p_y, as in Fig. 17*b*. Here

$$p_x + \gamma_1 l_1 = p_4 = p_5 = p_y + \gamma_2 l_2 + \gamma_3 h$$

whence

$$p_x - p_y = \gamma_2 l_2 + \gamma_3 h - \gamma_1 l_1$$

thus allowing direct calculation of the pressure difference, $p_x - p_y$. Differential manometers of the type above are sometimes made with the U-tube inverted, a liquid of small density existing in the top of the inverted U; the pressure difference measured by manometers of this type may be readily calculated by application of the foregoing principles. When large pressures or pressure differences are to be measured, a number of U-tube manometers may be connected in series. Several applications of these principles will allow solution for the unknown pressure or pressure difference.

There are many forms of precise manometers; two of the most common are shown in Fig. 17. Figure 17*c* represents the ordinary *inclined gage* used in measuring the comparatively small pressures in low-velocity gas flows. Its equilibrium position is shown at A, and when it is submitted to a pressure, p_x, a vertical deflection, h, is obtained in which $p_x = \gamma h$. In this case, however, the liquid is forced down a gently inclined tube so that the deflection, l, is much greater than h and, therefore, more accurately read. This type of manometer, when calibrated to read directly in inches of water, is frequently called a *draft gage.*

The principle of the sloping tube is also employed in the alcohol micromanometer of Fig. 17*d*, used in aeronautical research work. Here the gently sloping glass tube is mounted on a carriage, C, which is moved vertically by turning the dial, D, which actuates the screw, S. When p_x is zero, the carriage is adjusted so that the liquid in the tube is brought to the hair line, X, and the reading on the dial is recorded. When the unknown pressure, p_x, is admitted to the reservoir, the alcohol runs upward in the tube toward B and the carriage is then raised until the liquid surface in the tube rests again at the hair line, X. The difference between the dial reading at this point and the original reading gives the vertical travel of the carriage, h, which is the head of alcohol equivalent to the pressure p_x.

Along with these principles of manometry the following practical considerations should be appreciated: (1) manometer liquids, in changing their specific gravities with temperature, will induce errors in pressure measurements if this factor is overlooked; (2) errors due to capillarity may frequently be cancelled by selecting manometer tubes of uniform size; (3) although some liquids appear excellent (from density considerations) for use in manometers, their surface-tension effects may give poor menisci and thus inaccurate readings; (4) fluctuations of the manometer liquids will reduce accuracy of pressure measurement, but these fluctuations may be reduced by a throttling device in the manometer line (a short length of small tube is excellent for this purpose); and (5) when fluctuations are negligible, refined optical devices and verniers may be used for extremely precise readings of the liquid surfaces.

Illustrative Problem

This vertical pipe line with attached gage and manometer contains oil and mercury as shown. The manometer is open to the atmosphere. There is no flow in the pipe. What will be the gage reading, p_x?

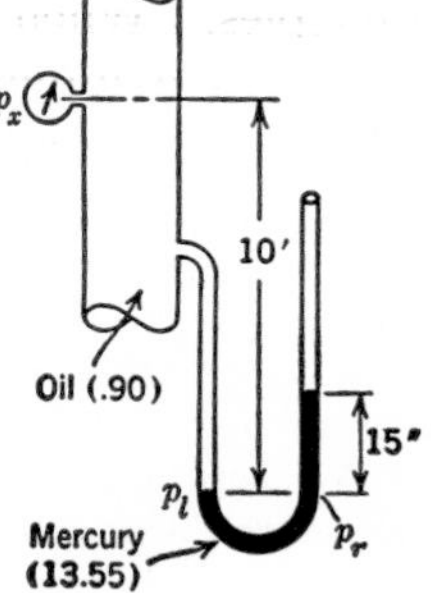

Solution. $$p_l = p_x + (0.90 \times 62.4)10 \qquad (20)$$

$$p_r = (13.55 \times 62.4)\tfrac{15}{12}. \qquad (20)$$

Since $$p_l = p_r$$

$$p_x = 495 \text{ psf} = 3.44 \text{ psi}$$

11. Forces on Submerged Plane Surfaces. The calculation of the magnitude, direction, and location of the total forces on surfaces submerged in a liquid is essential in the design of dams, bulkheads, gates, tanks, ships, etc.

For a submerged, plane, *horizontal* area the calculation of these force properties is simple because the pressure does not vary over the area; for nonhorizontal planes the problem is complicated by pressure variation. Pressure in liquids,[12] however, has been shown to vary *linearly* with depth (equation 20), producing the typical

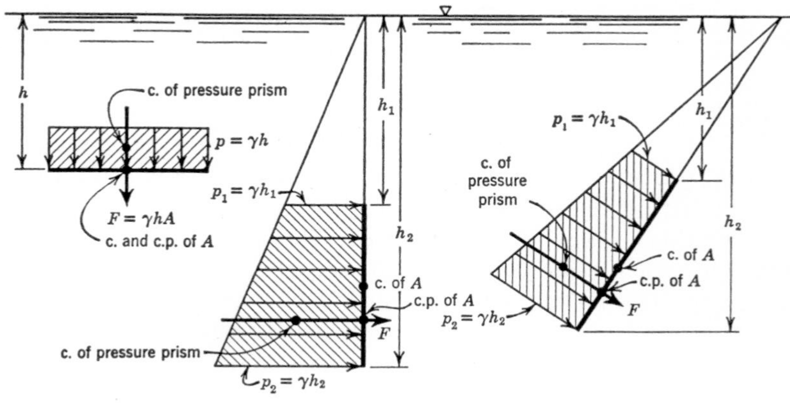

FIG. 18

pressure diagrams and resultant forces of Fig. 18. The shaded areas, appearing as trapezoids, are really volumes, known as *pressure prisms*. In mechanics it has been shown that the resultant force, F, is equal to the volume of the pressure prism and passes through its centroid.[13]

Now consider the general case[14] of a plane submerged area, A, such as that of Fig. 19, located in any inclined plane, L–L. Let the centroid of this area be located, as shown, at a depth h_c and at a distance l_c from the line of intersection, O–O, of plane L–L and the

[12] In gases, pressure variation with depth is so small that it is usually ignored in the calculation of resultant forces in engineering problems.

[13] Note that the resultant force also passes through a point on the plane defined as the *center of pressure* (c.p.). For nonhorizontal areas (with liquid on one side), the center of pressure of an area is always below its centroid. When the same liquid *covers both sides* of the area, examination of the pressure prisms shows that the resultant force of liquid on the area passes through the centroid of the area.

[14] A general solution for the magnitude, direction, and location of the resultant force on this area will allow easy calculation of the forces on areas of more regular shape.

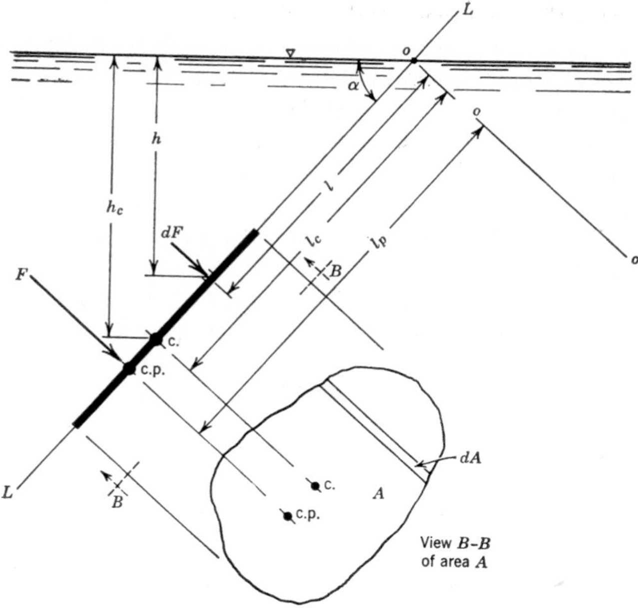

FIG. 19

liquid surface. The force, dF, on the area, dA, will be given by $p\,dA = \gamma h\,dA$; but, since $h = l \sin \alpha$, dF may be expressed as

$$dF = \gamma l\,dA \sin \alpha \tag{25}$$

and the total force on the area A will result from the integration of this expression over the area, which gives

$$F = \gamma \sin \alpha \int^{A} l\,dA \tag{26}$$

in which $\int^{A} l\,dA$ is recognized as the moment of the area A, about the line O–O, which is also given by the product of the area, A, and the perpendicular distance, l_c, from O–O to the centroid (c.) of the area. Thus

$$\int^{A} l\,dA = l_c A$$

and, when this is substituted in equation 26,

$$F = \gamma A l_c \sin \alpha$$

However, $h_c = l_c \sin \alpha$, which reduces the foregoing equation to

$$F = \gamma h_c A \tag{27}$$

and indicates that the magnitude of the resultant force on (one side of) any submerged plane area may be calculated by multiplying the area, A, by the pressure at its centroid, γh_c.

The magnitude of the resultant force having been calculated, its direction and location must be considered. Its direction, because of the nonexistence of shear stress, is necessarily normal to the plane, and the distance, l_p, between its line of action and axis O–O may be found by dividing the moment of the force (about this axis) by the magnitude of the force.

Referring again to Fig. 19, the moment, dM, of the force, dF, about line O–O is equal to $l\,dF$. Substituting for dF its value from equation 25,

$$dM = \gamma l^2\,dA \sin \alpha$$

and integrating to obtain the total moment, M,

$$M = \gamma \sin \alpha \int^A l^2\,dA$$

in which $\int^A l^2\,dA$ is the moment of inertia, $I_{O\text{-}O}$, of the area, A, about the line $_{O\text{-}O}$; thus

$$M = \gamma I_{O\text{-}O} \sin \alpha$$

From "moment arm = moment divided by force" it is apparent that $l_p = M/F$, and, when the expressions above for M and F are substituted,

$$l_p = I_{O\text{-}O}/l_c A \qquad (28)$$

Thus the resultant force is located with respect to line O–O and the general solution of the problem is completed.

Equation 28 may be made more usable by putting it in terms of moment of inertia,[15] I_c, about an axis lying in the area, parallel to O–O, and through the centroid of the area. Using the transfer equation for moment of inertia,

$$I_{O\text{-}O} = I_c + l_c^2 A$$

which, when substituted in equation 28, gives

$$l_p - l_c = I_c/l_c A \qquad (29)$$

allowing direct calculation of the distance (down the L–L plane) between centroid and center of pressure. This equation also indicates

[15] A summary of I_c's for common areas will be found in Appendix VI

that the center of pressure is always *below* the centroid except for a horizontal area, but that the distance between center of pressure and centroid diminishes as the depth of submergence of the area is increased.[16]

The lateral location of the center of pressure for regular plane areas, such as that of Fig. 20, is readily calculated by considering the area to be composed of a large number of rectangles of differential height, *dl*. The centroid and center of pressure of each of these small

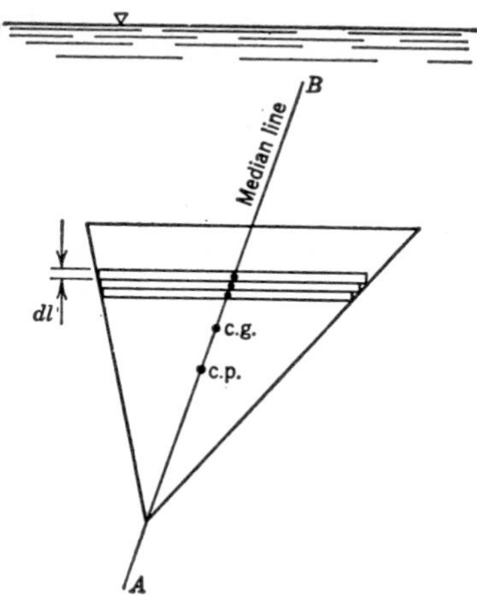

FIG. 20

rectangles will be coincident at the center of the rectangle, and, therefore, all the forces on the rectangles will act on the median line, AB; thus (from the statics of parallel forces) the line of action of the resultant of these forces must also intersect the median line; this intersection is the center of pressure of the area.

Although a general theorem (in terms of *product of inertia*) may be developed to find the centers of pressure of areas of more irregular forms, this is not essential to the solution of the problem. Frequently such areas may be divided into simple areas, the forces being located on them, and the location of their resultant being found by the methods of statics. The point where the line of action of the resultant force intersects the area is the center of pressure for the composite area. The following illustrative problem demonstrates this method.

[16] This fact allows the approximation made for small areas under great submergence, or pressure, that the resultant force acts at their centroids.

Illustrative Problem

Calculate magnitude, direction, and location of the total force exerted by the water on one side of this composite area which lies in a vertical plane.

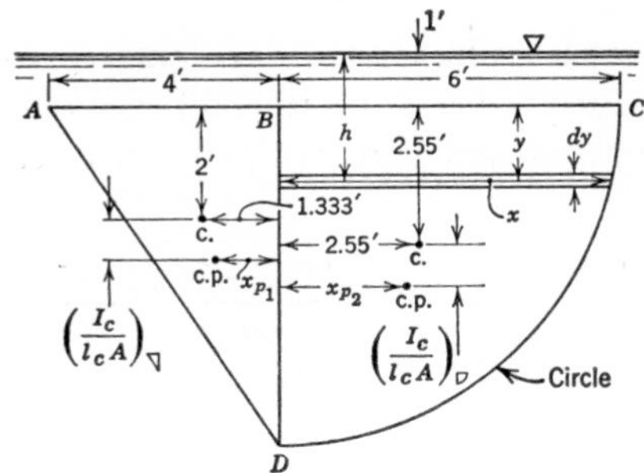

Solution. By inspection the direction of the force is normal to the area.

Magnitude (see Appendix VI):

Force on triangle	$= 12 \times 62.4 \times 3$	$= 2245$ lb	(27)
Force on quadrant	$= (36\pi/4)62.4 \times 3.55$	$= 6275$ lb	(27)
Total force on composite area		8520 lb	

Vertical location of resultant force (see Appendix VI):

$$I_c/l_cA \text{ for triangle } = \frac{4 \times 6^3}{36 \times 3 \times 12} = 0.667 \text{ ft} \qquad (29)$$

$$I_c/l_cA \text{ for quadrant } = \frac{70}{3.55 \times 28.3} = 0.695 \text{ ft} \qquad (29)$$

Taking moments about line AC,

$$2245 \times 2.667 + 6275 \times 3.245 = 8520(l_p - 1); \quad l_p = 4.10 \text{ ft}$$

Lateral location of resultant force: Since the center of pressure of the triangle is on the median line, x_{p_1} is given (from similar triangles) by

$$(1.333 - x_{p_1})/2 = 0.667/6; \quad x_{p_1} = 1.111 \text{ ft}$$

Dividing the quadrant into horizontal strips of differential height, dy, the moment about BD of the force on any one of them is

$$dM = (x\, dy)\gamma h(x/2)$$

in which $h = y + 1$ and $x^2 + y^2 = 36$. Substituting and integrating gives the moment about BD of the total force on the quadrant,

$$M = \frac{62.4}{2}\int_0^6 (36 - y^2)(y + 1)\, dy = 14{,}600 \text{ ft-lb}$$

and thus

$$x_{p_2} = 14{,}600/6275 = 2.33 \text{ ft}$$

Finally, taking moments about line BD,

$$2.33 \times 6275 - 2245 \times 1.111 = 8520\, x_p; \quad x_p = 1.425 \text{ ft}$$

Thus the center of pressure of the composite figure is 1.425 ft to the right of BD and 4.10 ft below the water surface.

12. Forces on Submerged Curved Surfaces. Resultant forces on submerged curved surfaces cannot be calculated by the foregoing methods but may be readily determined through computation of

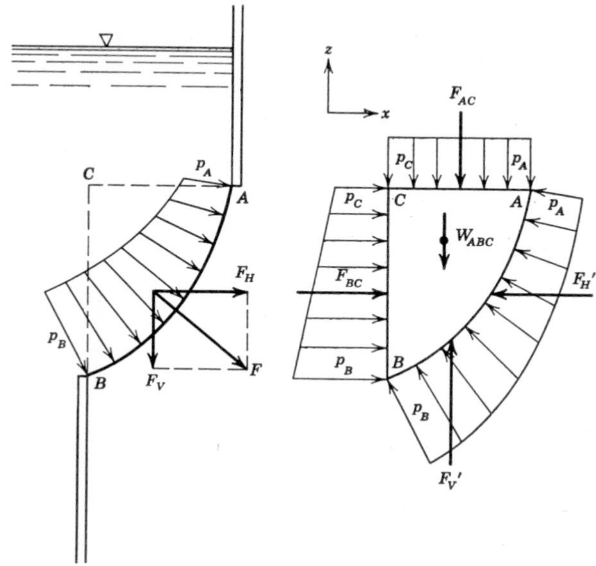

FIG. 21

their horizontal and vertical components. Consider the curved surface AB of Fig. 21. On each element of surface, the magnitude, direction, and location of the pressure force may be determined by foregoing principles, and they lead to the indicated pressure distribution which may be reduced to a single resultant force F with components F_H and F_V as shown.

Analysis of the free body of fluid ABC will allow prediction of F_H' and F_V', the components of the resultant force exerted *by the surface AB on the fluid*, and the equal-and-opposite of F_H and F_V, respectively. From the static equilibrium of the free body,

$$\Sigma F_x = F_{BC} - F_H' = 0$$

$$\Sigma F_z = F_V' - W_{ABC} - F_{AC} = 0$$

whence $F_H' = F_{BC}$ and $F_V' = W_{ABC} + F_{AC}$. From the inability of the free body of fluid to support shear stress, it follows that F_H' must be collinear with F_{BC} and F_V' collinear with the resultant of W_{ABC} and F_{AC}. The foregoing analysis reduces the problem to one of computation of magnitude and location of F_{BC}, F_{AC}, and W_{ABC}; for F_{AC} and F_{BC} the methods of Art. 11 may be used, whereas W_{ABC} is merely the weight of the free body of fluid and necessarily acts through its center of gravity.

When the same liquid covers both sides of a curved area but the liquid surfaces are at different levels for the two sides, the net effective pressure distribution is uniform since the effective pressure at any point on the area is dependent only upon the difference in surface levels. The resultant force on such an area may be obtained by application of the methods above. The horizontal component will be found to pass through the centroid of the vertical projection of the area, and the vertical component will pass through the centroid of the horizontal projection.

Engineers frequently employ graphical techniques to obtain the components of resultant forces produced by pressure distributions on curved areas. The method may be learned by the following development applied to the area AB of Fig. 22; the pressure may vary in any manner from p_A at A to p_B at B, but the pressure on any element of area dA is normal to dA. The differential force on dA is $p\,dA$, and the angle θ defines the slope of dA relative to the system of axes x and y. The components of this force parallel to x and y will be $p\,dA \sin\theta$ and $p\,dA \cos\theta$, respectively, and their sums (integrals) will be the components F_x and F_y of the resultant force on area AB; thus

$$F_x = \int^{AB} p\,dA \sin\theta \quad \text{and} \quad F_y = \int^{AB} p\,dA \cos\theta$$

However, $dA \sin\theta$ and $dA \cos\theta$ are readily identified on the figure, and it becomes evident that the hatched areas are direct measures of the force components F_x and F_y; hence these areas may be planimetered and the magnitude of the force components obtained. By the usual methods of statics the locations of these force components may be found; they will pass through the centroids of the hatched areas. These methods may be used effectively in problems of fluid statics by those who favor graphical solutions; however, as such methods are not restricted to linear pressure distributions they may be used in problems of fluid dynamics as well. For example, the determination of the lift of an airfoil from its measured pressure distribution is usually accomplished in this manner.

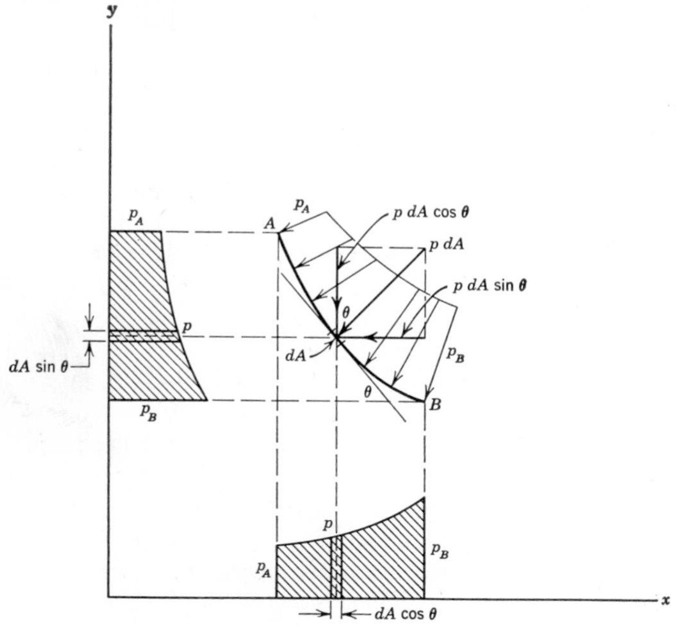

FIG. 22

ILLUSTRATIVE PROBLEM

Calculate magnitude, direction, and location of the resultant force exerted by the water on the surface AB, which is a quarter cylinder 8 ft in length (normal to plane of paper).

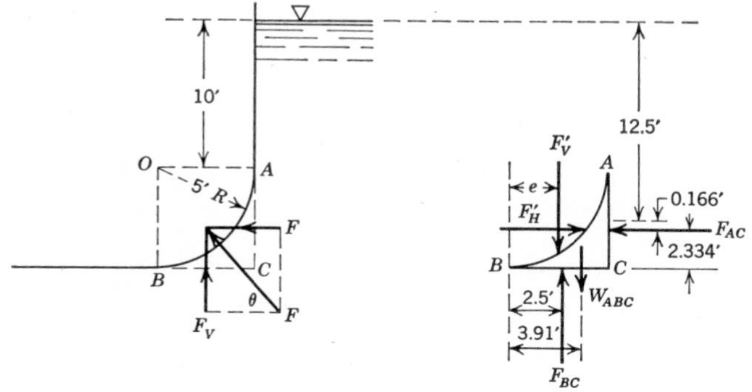

Solution. Isolate the free body of water ABC and show forces thereon.

Horizontal component: By inspection, this force (on the free body) has direction to the right:

$$F_H' = F_{AC} = 8 \times 5 \times 62.4 \times 12.5 = 31{,}200 \text{ lb} \tag{27}$$

$$(l_p - l_c)_{AC} = 83.2/12.5 \times 40 = 0.166 \text{ ft} \tag{29}$$

Vertical component: By inspection, this force (on the free body) has a downward direction:

$$\Sigma F_z = 15 \times 62.4 \times 40 - F_V' - 8 \times 62.4(25 - 25\pi/4) = 0;$$

$$F_V' = 34{,}760 \text{ lb}$$

From statics the center of gravity of ABC is found to be 3.91 ft to the right of B, and, taking moments of the forces (on the free body) about O,

$$34{,}760 \times e - 2680 \times 3.91 - 37{,}440 \times 2.5 = 0; \quad e = 2.40 \text{ ft}$$

Resultant force of water on AB:

Direction: upward to the left, $\theta = \arctan 34{,}760/31{,}200 = 48°$
Magnitude:

$$F = 1000\sqrt{(31.2)^2 + (34.76)^2} = 46{,}600 \text{ lb}$$

Location: through a point 2.334 ft above and 2.40 ft to the right of B. Since the pressure forces on the elements of the cylinder, although different in magnitude, all pass through O, and thus form a concurrent force system, their resultant will also be expected to pass through O. Since $(5.00 - 2.334)/2.40 = 34{,}760/31{,}200$, this expectation is confirmed.

13. Buoyancy and Flotation. The familiar laws of buoyancy (Archimedes' principle) and flotation are usually stated: (1) *a body immersed in a fluid is buoyed up by a force equal to the weight of fluid displaced;* and (2) *a floating body displaces its own weight of the liquid in which it floats.* These laws are corollaries of the general principles of Art. 12 and may be readily proved by application of those principles.

A body $ABCD$ suspended in a fluid of specific weight γ is illustrated in Fig. 23. Isolating a free body of fluid with vertical sides tangent to the body allows identification of the vertical forces exerted by the lower (ADC) and upper (ABC) surfaces of the body on the surrounding fluid. These are F_1' and F_2' with $(F_1' - F_2')$ the buoyant force on the body. For the upper portion of the free body

$$\Sigma F_z = F_2' - W_2 - p_2 A = 0$$

and for the lower portion

$$\Sigma F_z = F_1' + W_1 - p_1 A = 0$$

whence (by subtraction of the equations)

$$F_B = F_1' - F_2' = (p_1 - p_2)A - (W_1 + W_2)$$

However, $p_1 - p_2 = \gamma h$ and $\gamma h A$ is the weight of a cylinder of fluid extending between horizontal planes 1 and 2, and the right side of

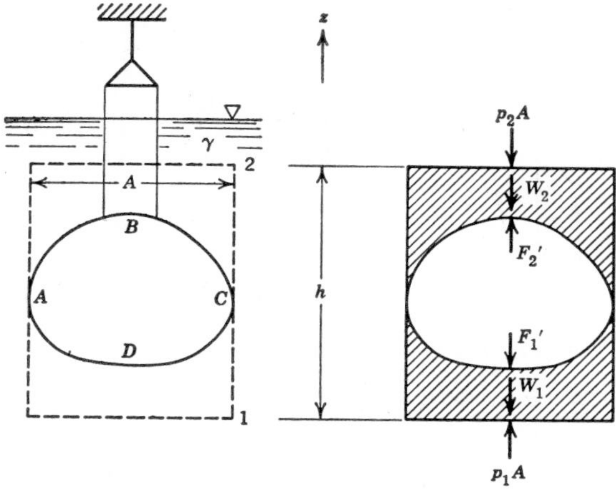

FIG. 23

the equation for F_B is identified as the weight of a volume of fluid exactly equal to that of the body. Accordingly[17]

$$F_B = \gamma(\text{Volume of object}) \tag{30}$$

and the law of buoyancy is proved.

For the floating object of Fig. 24 a similar analysis will show that

$$F_B = \gamma(\text{Volume displaced}) \tag{31}$$

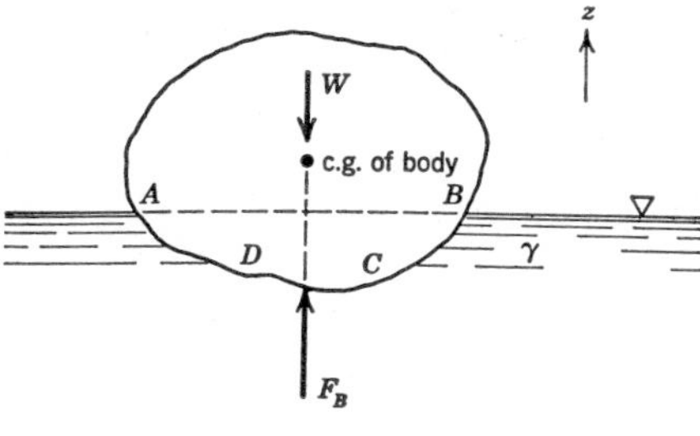

FIG. 24

and, from static equilibrium of the object, its weight must be equal to this buoyant force; thus the object displaces its own weight of the liquid in which it floats.

[17] Note carefully that this calculation for buoyant force presumes that the fluid completely surrounds the body; if it does not, the concept of the buoyant force must be adjusted accordingly by reapplication of the same principles. A sunken ship embedded in the ocean floor is a classic example of this; here the water does not completely surround the hull.

The principles above find many applications in engineering, for example, in calculations of the draft of surface vessels, the increment in weight of a ship's cargo from the increment in depth of flotation, and the lift of airships and balloons.

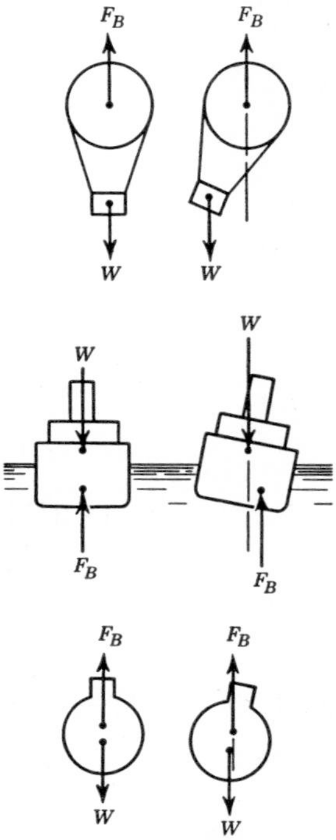

The stability of submerged or floating bodies is dependent upon the relative location of the buoyant force and the weight of the body. The buoyant force acts upward through the center of gravity of the displaced volume;[18] the weight acts downward at the center of gravity of the body. Stability or instability will be determined by whether a righting or overturning moment is developed when the center of gravity and center of buoyancy move out of vertical alignment. Obviously, for the submerged bodies, such as the balloon and submarine of Fig. 25, stability requires the center of buoyancy to be above the center of gravity. In surface vessels, however, the center of gravity is usually above the center of buoyancy, and stability exists because of movement of the center of buoyancy to a position outboard of the center of gravity as the ship "heels over," producing a righting moment. An overturning moment, resulting in capsizing, occurs if the center of gravity moves outboard of the center of buoyancy.

FIG. 25

ILLUSTRATIVE PROBLEMS

A ship has a cross-sectional area of 4000 sqft at the water line when the draft is 10 ft. How many pounds of cargo will increase the draft 2 in.? Assume salt water of specific weight 64.0 lb/cuft.

Solution. Since the ship floats, the weight of water displaced by the cargo equals the weight of the cargo. Therefore

$$\text{Weight of cargo} = 4000 \times \tfrac{2}{12} \times 64.0 = 42{,}700 \text{ lb}$$

The solid wooden sphere ($\gamma = 53.0$ lb/cuft) of diameter 17 in. is held in the orifice (12-in. diameter) by the water. Calculate the force exerted be-

[18] This is evident from the fact that the buoyant force is the *weight* of the displaced volume; it needs no formal proof.

tween sphere and orifice plate when the depth is 2 ft. The sphere will float away if this force becomes zero; can this ever happen?

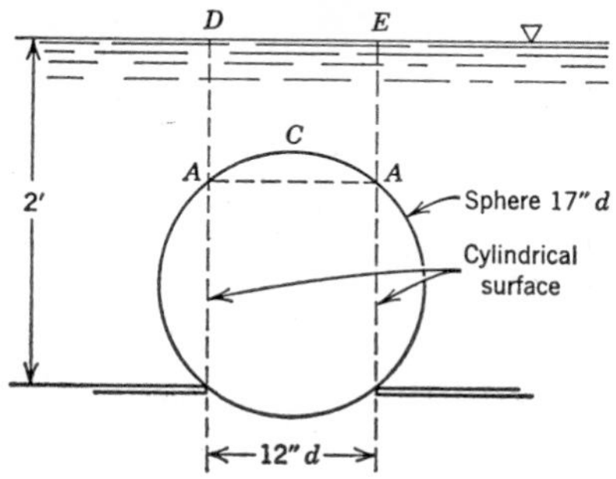

Solution. Volume of sphere $= (\pi/6)(\frac{17}{12})^3 = 1.435$ cuft

Weight of sphere $= 1.435 \times 53.0 = 76.0$ lb downward

Volume of section of sphere outside of cylindrical surface (from solid geometry) is 0.584 cuft. From this volume the buoyant force F_B may be calculated:

$$F_B = 0.584 \times 62.4 = 36.4 \text{ lb upward}$$

The force downward on surface ACA is computed from the volume of $ACADE$, which (from solid geometry) is found to be 0.711 cuft:

$$F_{ACA} = 0.711 \times 62.4 = 44.4 \text{ lb downward}$$

For static equilibrium of the sphere,

$$\text{Force between sphere and plate} = 76.0 + 44.0 - 36.4 = 84.0 \text{ lb}$$

When the water surface coincides with the horizontal plane A–A, the force on surface ACA will be zero. Here the force between sphere and plate will be $76.0 - 36.4 = 39.6$ lb. Below this point the buoyant force will be less than 36.4 lb; accordingly the force between sphere and plate can never be zero.

14. Fluid Masses Subjected to Acceleration. Fluid masses may be subjected to various types of acceleration without occurrence of relative motion between fluid particles or between fluid particles and boundaries. Such fluid masses will be found to conform to the laws of fluid statics, modified to allow for the effects of acceleration, and they may often be treated by assuming a change in the magnitude and direction of g.

A generalized approach to this problem may be obtained by applying Newton's second law to the fluid element if Fig. 26 which is being accelerated in such a way that its components of acceleration are

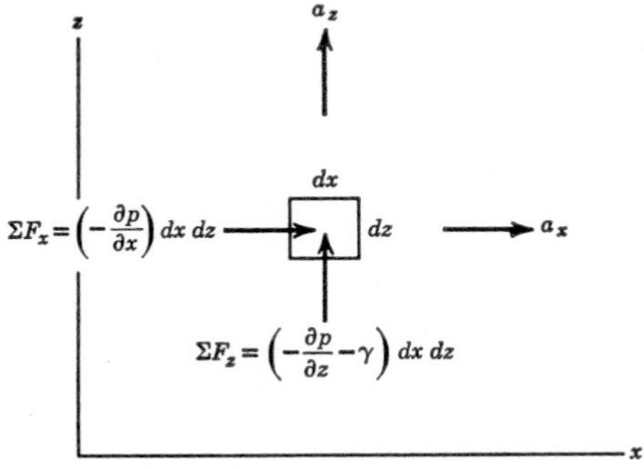

FIG. 26

a_x and a_z. The summation of force components on such an element has been indicated in Art. 8, and will be seen to be

$$\Sigma F_x = \left(-\frac{\partial p}{\partial x}\right) dx\,dz \tag{32}$$

$$\Sigma F_z = \left(-\frac{\partial p}{\partial z} - \gamma\right) dx\,dz \tag{33}$$

With the mass of the element equal to $(\gamma/g)\,dx\,dz$, Newton's second law may be written

$$\left(-\frac{\partial p}{\partial x}\right) dx\,dz = \frac{\gamma}{g}\,a_x\,dx\,dz$$

$$\left(-\frac{\partial p}{\partial z} - \gamma\right) dx\,dz = \frac{\gamma}{g}\,a_z\,dx\,dz$$

which easily reduce to

$$-\frac{\partial p}{\partial x} = \frac{\gamma}{g}\,(a_x) \tag{34}$$

$$-\frac{\partial p}{\partial z} = \frac{\gamma}{g}\,(a_z + g) \tag{35}$$

These equations characterize the pressure variation through an accelerated mass of fluid, and with them specific applications may be studied.

One other useful generalization may be derived from the foregoing equations: this is a property of a line of constant pressure. Using the mathematical relation [19] for dp in terms of its partial derivatives,

[19] See Appendix V.

$$dp = \frac{\partial p}{\partial x} dx + \frac{\partial p}{dz} dz$$

and substituting the above expressions for $\partial p/\partial x$ and $\partial p/\partial z$,

$$dp = -\frac{\gamma}{g}(a_x)\, dx - \frac{\gamma}{g}(a_z + g)\, dz \tag{36}$$

However, along a line of constant pressure $dp = 0$ and hence, for such a line,

$$\frac{dz}{dx} = -\left(\frac{a_x}{g + a_z}\right) \tag{37}$$

Thus the *slope* (dz/dx) of a line of constant pressure is defined; its *position* must be determined from external (boundary) conditions in specific problems.

From the foregoing generalizations some situations of engineering significance may now be examined.

(*a*) *Constant Linear Acceleration with* $a_x = 0$. Here a container of liquid is accelerated vertically upward, $\partial p/\partial x = 0$, and with no change of pressure with x, equation 35 becomes

$$\frac{dp}{dz} = -\gamma\left(\frac{g + a_z}{g}\right)$$

For a_z constant, this equation shows that the characteristic linear pressure variation of fluid statics is preserved but that magnitudes of pressure will now depend upon a_z. The quantitative aspects of this are shown in Fig. 27 for $a_z > 0$ and $a_z < 0$. The latter case is of particular interest when $a_z = -g$, yielding $dp/dz = 0$, and showing that the pressure will be constant throughout a freely falling mass of fluid; for an *unconfined* mass of freely falling fluid the pressure will therefore be equal to that surrounding it—if the surrounding pressure is zero, all pressures within the fluid mass will be zero, a fact which will have many applications in subsequent problems.

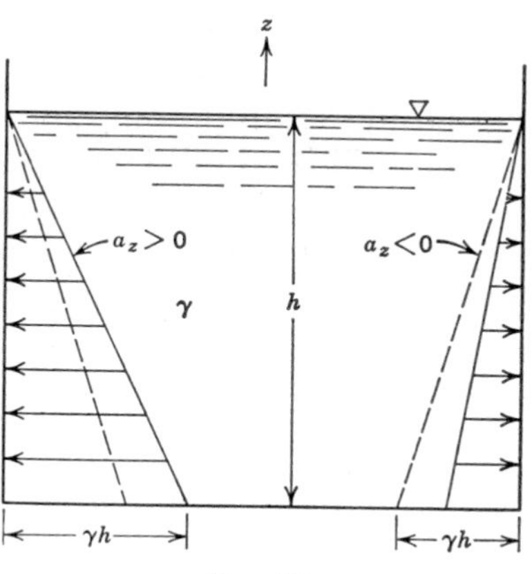

FIG. 27

ILLUSTRATIVE PROBLEM

An open tank of water is accelerated vertically upward at 15 ft/sec². Calculate the pressure at a depth of 5 ft.

Solution.

$$\frac{dp}{dz} = -62.4\left(\frac{15 + 32.2}{32.2}\right) = -91.5 \text{ lb/cuft} \tag{35}$$

$$p = \int_0^p dp = -\int_0^{-5} 91.5\, dz = 457 \text{ psf}$$

(*b*) *Constant Linear Acceleration.* Here the slope of a liquid surface (which is a line of constant pressure) is given (from equation 37) by $-a_x/(a_z + g)$ as shown in Fig. 28, and other lines of constant

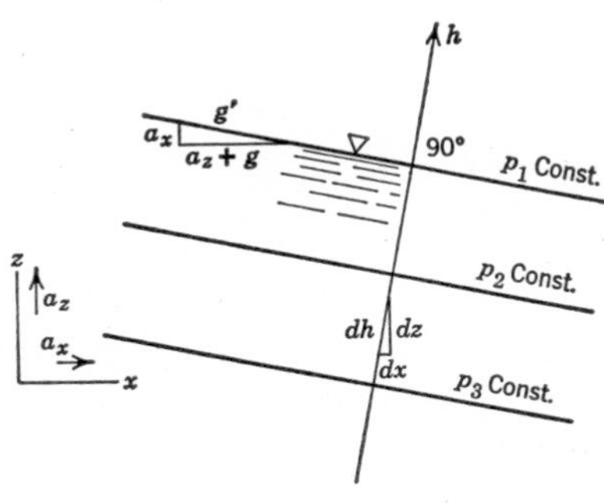

FIG. 28

pressure will be parallel to the surface. Along x and z the (linear) pressure variations may be computed from equations 34 and 35, respectively. In the direction h, normal to the lines of constant pressure, equation 36 may be employed. Dividing this equation by dh,

$$\frac{dp}{dh} = -\gamma\left[\frac{a_x}{g}\frac{dx}{dh} - \frac{g + a_z}{g}\frac{dz}{dh}\right]$$

but, from the similar triangles of Fig. 28,

$$dx/dh = a_x/g' \quad \text{and} \quad dz/dh = (a_z + g)/g'$$

Substituting these expressions above gives

$$dp/dh = -\gamma(g'/g)$$

which shows that the pressure variation along h is linear and allows computation of pressures as in statics (equation 18), $\gamma g'/g$ being used for the specific weight of the fluid.

Illustrative Problem

This open tank moves up the plane with constant acceleration. Calculate the acceleration required for the water surface to move to the position indicated. Calculate the pressure in the corner of the tank at A before and after acceleration.

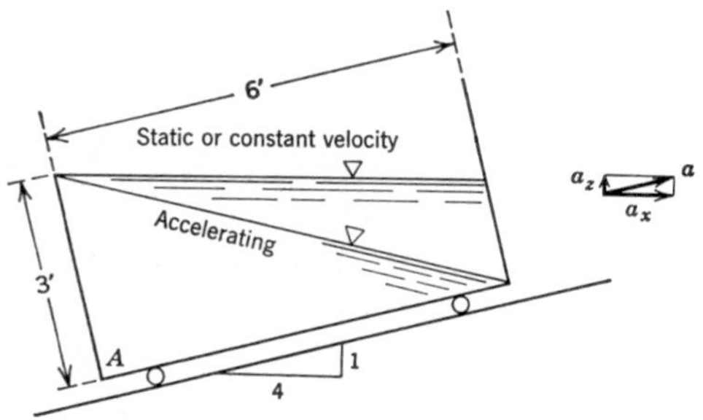

Solution. From geometry, the slope of the water surface during acceleration is -0.229. From the slope of the plane, $a_x = 4a_z$. Using equation 37,

$$-0.229 = -4a_z/(a_z + 32.2)$$

and from the foregoing

$$a_z = 1.96 \text{ ft/sec}^2, \quad a_x = 7.84 \text{ ft/sec}^2, \quad a = 8.08 \text{ ft/sec}^2$$

From geometry, the depth of water vertically above corner A before acceleration is 2.91 ft; hence the pressure there is $2.91 \times 62.4 = 181.5$ psf. After acceleration this depth is 2.75 ft; from equation 35,

$$\partial p/\partial z = -(32.2 + 1.96)(62.4/32.2) = -66.0 \text{ lb/cuft}$$

Therefore $p = 2.75 \times 66.0 = 181.5$ psf.

The fact that the pressures at A are the same before and after acceleration is no coincidence; general proof may be offered that p_A does not change whatever the acceleration. This means that the force exerted by the end of the tank on the water is constant for all accelerations. However, this is no violation of Newton's second law, since the mass of liquid diminishes with increased acceleration so that the product of mass and acceleration remains constant and equal to the applied force.

(c) *Centripetal Acceleration with Constant Angular Velocity about a Vertical Axis, and* $a_z = 0$. Equations 34, 35, and 37 may be written[20] with radial distance, r, substituted for x and a_r for a_x to give (for $a_z = 0$): $-\partial p/\partial r = \gamma a_r/g$, $-\partial p/\partial z = \gamma$, and, for surfaces of constant pressure, $dz/dr = -a_r/g$. From kinematics, $a_r = -\omega^2 r$, in which ω is the angular velocity. Substituting this in the foregoing

[20] For rigorous proof of the validity of this, a complete analysis should be made using the conventional element of polar coordinates; however (after discarding the negligible terms), the same result is obtained.

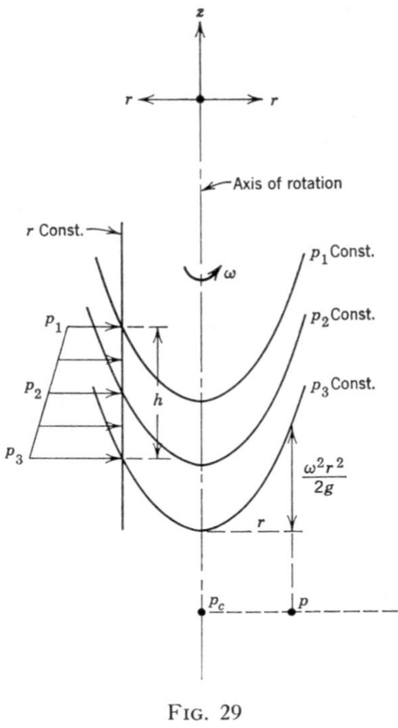

FIG. 29

equations,

$$\frac{\partial p}{\partial r} = \frac{\gamma \omega^2 r}{g} \tag{38}$$

$$\frac{\partial p}{\partial z} = -\gamma \tag{39}$$

For surfaces of constant pressure,

$$\frac{dz}{dr} = \frac{\omega^2 r}{g} \tag{40}$$

The pressure gradient along r and z may be computed from the first two equations; the third may be easily integrated to

$$z = \frac{\omega^2 r^2}{2g} + \text{Constant} \tag{41}$$

showing that lines of constant pressure are parabolas (Fig. 29) symmetrical about the axis of rotation. The second equation shows that pressure variation in the vertical is that of fluid statics (equation 20) so that $p_3 - p_1 = \gamma h$. The second equation may be integrated (for z = constant) from the axis of rotation (where the pressure is p_c and r is zero) to any radius r where the pressure is p. The result is

$$\frac{p - p_c}{\gamma} = \frac{\omega^2 r^2}{2g} \tag{42}$$

which may also be deduced directly from Fig. 29.

The foregoing analysis shows the possibility of pressure being created by the rotation of a fluid mass. This principle is utilized in centrifugal pumps and blowers to produce a pressure difference in order to cause fluids to flow.

ILLUSTRATIVE PROBLEMS

This tank is fitted with three piezometer columns of the same diameter. Calculate the pressure heads at points A and B when the tank is rotating at a constant speed of 100 rpm.

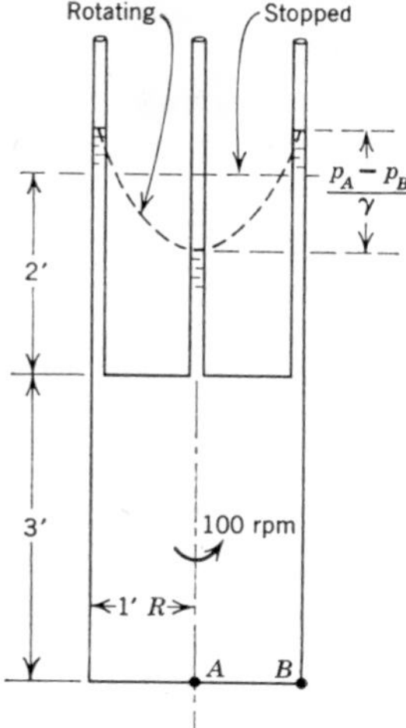

Solution. During rotation the surface in the piezometer columns must be on a parabola of constant pressure, and liquid lost from the central column must appear in the other two. Accordingly

$$(p_B - p_A)/\gamma = (2\pi \times 100/60)^2(1)^2/2g = 1.71 \text{ ft} \tag{42}$$

$$p_A/\gamma = 5 - \tfrac{2}{3}(1.71) = 3.86 \text{ ft}$$

$$p_B/\gamma = 5 + \tfrac{1}{3}(1.71) = 5.57 \text{ ft}$$

A cylinder of radius r_o contains a gas at constant temperature and rotates about its (vertical) axis at constant angular speed. Derive a relation between the pressures on the cylindrical surface and on the axis of rotation.

Solution. Ignoring pressure variation in the z-direction and writing $\partial p/\partial r = dp/dr$,

$$dp = \gamma\omega^2 r\, dr/g \tag{38}$$

Substituting p/RT for γ (in which T is constant), separating variables, and noting that, for $p = p_c$, $r = 0$ and, for $p = p_o$, $r = r_o$, the equation may be integrated between appropriate limits to yield

$$\ln (p_o/p_c) = \omega^2 r_o^2/2gRT$$

PROBLEMS

70. Calculate the pressure in an open tank of benzene at a point 8 ft below the liquid surface.

71. If the pressure 10 ft below the free surface of a liquid is 20 psi, calculate its specific weight and specific gravity.

72. If the pressure at a point in the ocean is 200 psi, what is the pressure 100 ft below this point? Specific weight of salt water is 64.0 lb/cuft.

73. An open vessel contains carbon tetrachloride to a depth of 6 ft and water on the carbon tetrachloride to a depth of 5 ft. What is the pressure at the bottom of the vessel?

74. How many inches of mercury are equivalent to a pressure of 20 psi? how many feet of water?

75. One vertical foot of air at 59°F and 14.7 psia is equivalent to how many pounds per square inch? inches of mercury? feet of water?

76. The barometric pressure at sea level is 30.00 in. of mercury when that on a mountain top is 29.00 in. If the specific weight of air is assumed constant at 0.075 lb/cuft, calculate the elevation of the mountain top.

77. If at the surface of a liquid the specific weight is γ_o, with z and p both zero, show that, if E = constant, the specific weight and pressure are given by

$$\gamma = E/(z + E/\gamma_o) \quad \text{and} \quad p = -E \ln (1 + \gamma_o z/E)$$

Calculate specific weight and pressure at a depth of 2 miles assuming $\gamma_o = 64.0$ lb/cuft and $E = 300{,}000$ psi.

78. The specific weight of water in the ocean may be calculated from the empirical relation $\gamma = \gamma_o + K\sqrt{h}$ (in which h is the depth below the ocean surface). Derive an expression for the pressure at any point h and calculate specific weight and pressure at a depth of 2 miles assuming $\gamma_o = 64.0$ lb/cuft, h in feet, and $K = 0.025$.

79. If the specific weight of a liquid varies linearly with depth below the liquid surface ($\gamma = \gamma_o + Kh$), derive an expression for pressure as a function of depth.

80. If atmospheric pressure at the ground is 14.7 psia and temperature is 59°F, calculate the pressure 25,000 ft above the ground, assuming (*a*) no density variation, (*b*) isothermal variation of density with pressure, and (*c*) adiabatic variation of density with pressure.

81. Calculate pressures and densities of air in the ICAO Standard Atmosphere at 25,000 ft and 50,000 ft. Check results with the values in Appendix IV.

82. Calculate the depth of an adiabatic atmosphere if temperature and pressure at the ground are, respectively, 59°F and 14.7 psia.

83. Derive a relation between pressure and altitude: (*a*) for an isothermal atmosphere and (*b*) for the ICAO Standard Atmosphere to altitude 35,000 ft.

84. Show that the temperature lapse rate in an adiabatic atmosphere is the reciprocal of the specific heat at constant pressure.

85. Assuming a linear rise of temperature in the ICAO Standard Atmosphere between altitudes of 85,000 and 100,000 ft, calculate the value of the polytropic exponent n over this range of altitude.

86. With atmospheric pressure at 14.5 psia, what absolute pressure corresponds to a gage pressure of 20 psi?

87. When the barometer reads 30 in. of mercury, what absolute pressure corresponds to a vacuum of 12 in. of mercury?

88. If a certain absolute pressure is 12.35 psia, what is the corresponding vacuum if atmospheric pressure is 29.92 in. of mercury?

89. A Bourdon pressure gage attached to a closed tank of air reads 20.47 psi with the barometer at 30.50 in. of mercury. If barometric pressure drops to 29.18 in. of mercury, what will the gage read?

90. A Bourdon gage is connected to a tank in which the pressure is 40.0 psi above atmospheric at the gage connection. If the pressure in the tank remains unchanged but the gage is placed in a chamber where the air pressure is reduced to a vacuum of 25 in. of mercury, what gage reading will be expected?

91. The compartments of these tanks are closed and filled with air. Gage A reads 30 psi. Gage B registers a vacuum of 10 in. of mercury. What will gage C read if it is connected to compartment 1 but inside compartment 2? Barometric pressure is 14.6 psia.

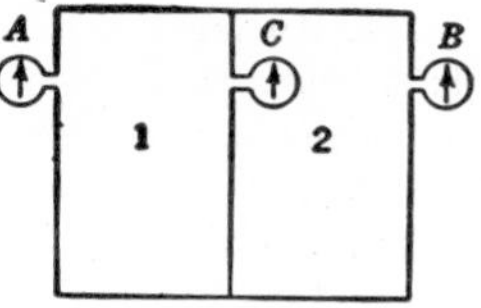

PROBLEM 91

92. If the barometric pressure is 28 in. of mercury, what will a carbon tetrachloride barometer read at 68°F?

93. If the barometer of Fig. 15 is filled with benzene at 68°F, calculate h if the barometric pressure is 14.7 psia.

94. Calculate the height of the column of a water barometer for an atmospheric pressure of 14.70 psia when the water is at 50°F, 150°F, and 212°F.

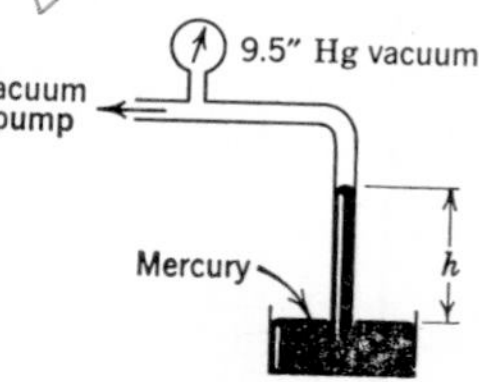

PROBLEM 95

95. Barometric pressure is 29.43 in. of mercury. Calculate h.

96. Calculate the pressure p_x in Fig. 17a if l = 30 in., h = 20 in.; liquid γ is water, and γ_1 mercury.

97. With the manometer reading as shown, calculate p_x.

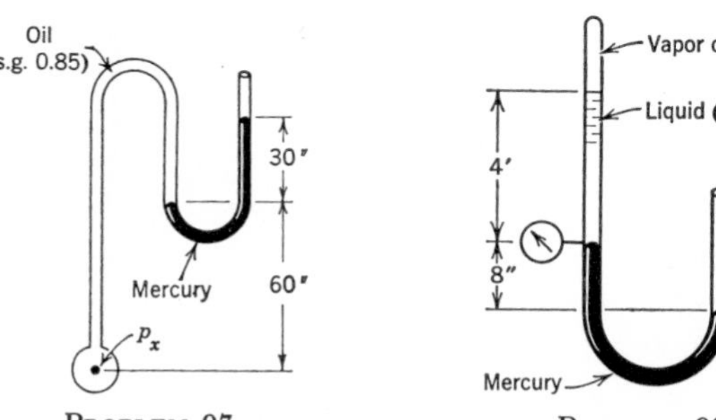

PROBLEM 97 PROBLEM 98

98. Barometric pressure is 13.2 psia. Calculate the vapor pressure of the liquid and the gage reading.

99. In Fig. 17b, l_1 = 50 in., h = 20 in., l_2 = 30 in., liquid γ_1 is water, γ_2 benzene, and γ_3 mercury. Calculate $p_x - p_y$.

100. Calculate $p_x - p_y$ for this inverted U-tube manometer.

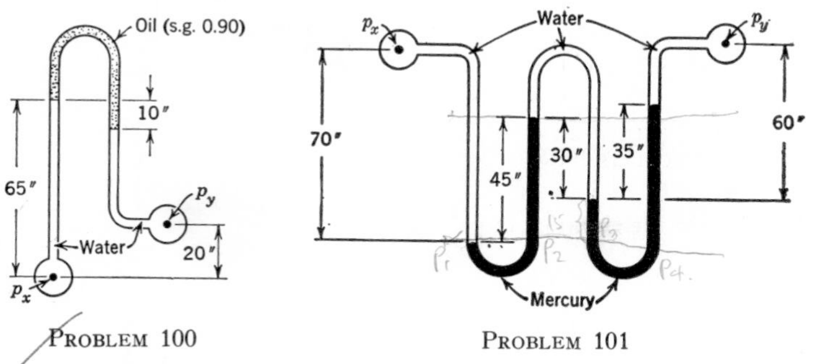

PROBLEM 100 PROBLEM 101

101. Two manometers as shown are connected in series. Calculate $p_x - p_y$.

102. An inclined gage (Fig. 17*c*) having a tube of $\frac{1}{8}$-in. bore, laid on a slope of 1:20, and a reservoir of 1 in. diameter contains linseed oil. What distance will the oil move along the tube when a pressure of 1 in. of water is connected to the gage?

103. The meniscus between the oil and water is in the position shown when $p_1 = p_2$. Calculate the pressure difference $(p_1 - p_2)$ which will cause the meniscus to rise 2 in.

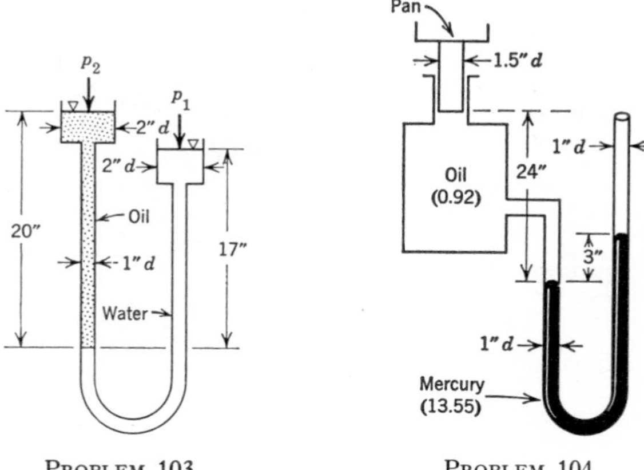

PROBLEM 103 PROBLEM 104

104. Predict the manometer reading after a 1-lb weight is placed on the pan. Assume no leakage or friction between piston and cylinder.

105. Calculate the gage reading.

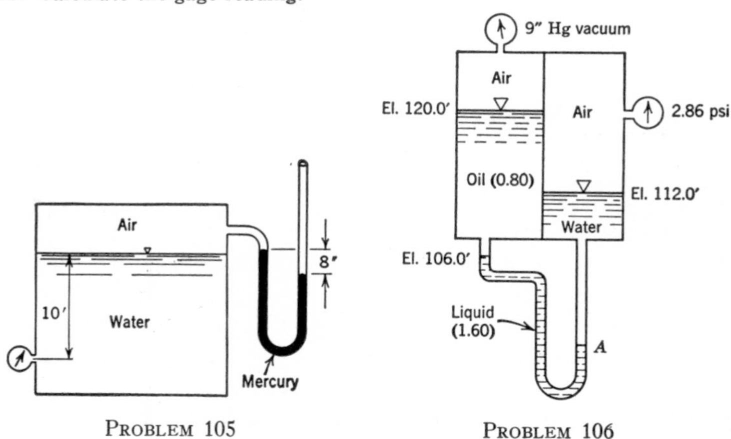

PROBLEM 105 PROBLEM 106

106. Calculate the elevation of point *A*.

107. The sketch shows a sectional view through a submarine. Calculate the depth of submergence, y. Assume specific weight of sea water 64.0 lb/cu ft.

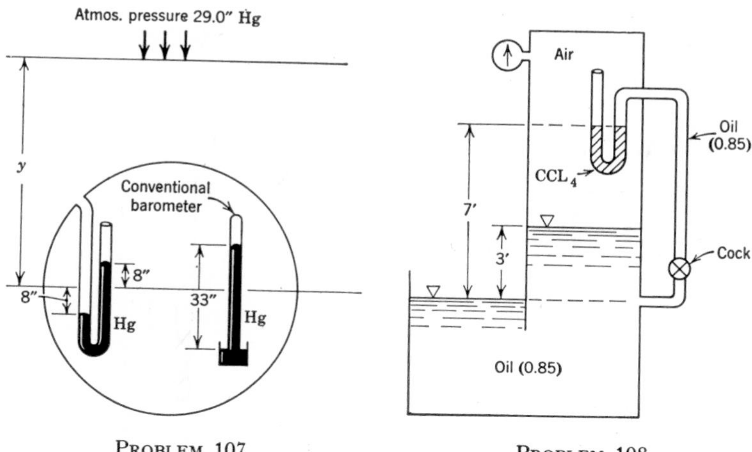

PROBLEM 107 PROBLEM 108

108. Calculate the gage reading and the manometer reading after the cock is opened.

109. Calculate the gage reading. Specific gravity of the oil is 0.85. Barometric pressure is 29.75 in. of mercury.

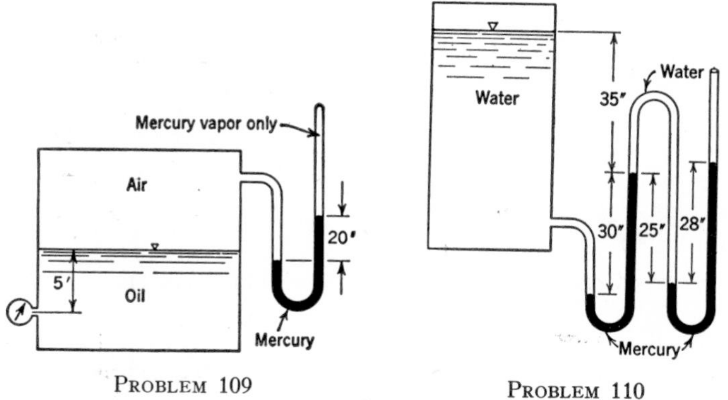

PROBLEM 109 PROBLEM 110

110. Calculate absolute and gage pressures in the tank above the water surface. Barometric pressure is 28 in. of mercury.

111. The manometer stands as shown when the air space is open to the atmosphere. Calculate the manometer reading when the tank is closed and the pressure in the air space is raised to 10 psi.

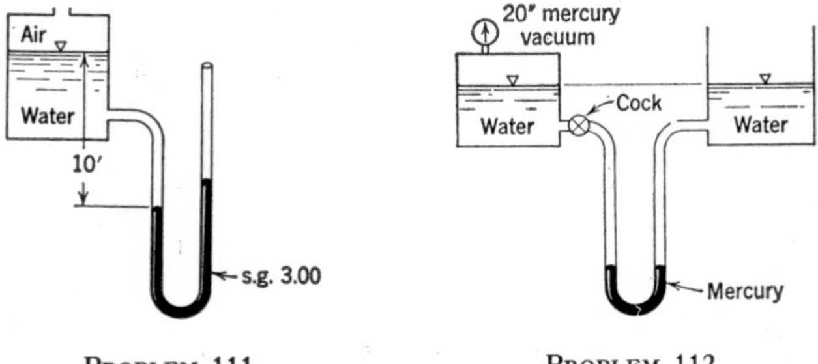

PROBLEM 111 PROBLEM 112

112. Calculate magnitude and direction of manometer reading when the cock is opened. The tanks are very large compared to the manometer tubes.

113. The manometer reading is 6 in. when the tank is empty (water surface at *A*). Calculate the manometer reading when the tank is filled with water.

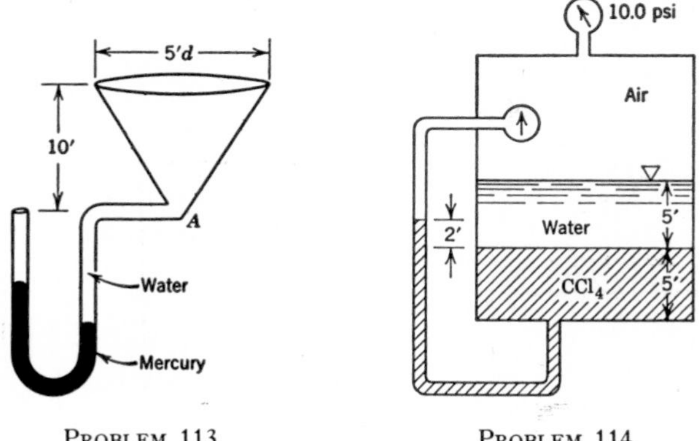

PROBLEM 113 PROBLEM 114

114. The Bourdon pressure gage connected to the manometer is inside the sealed tank. Atmospheric pressure is 14.25 psia. Calculate the gage reading.

115. Barometric pressure is 28 in. of mercury. The cock is opened and the air space pumped out so that the gage reads 20 in. of mercury vacuum. Calculate the absolute pressure in the tank and the manometer reading. Neglect change of water surface in the tank.

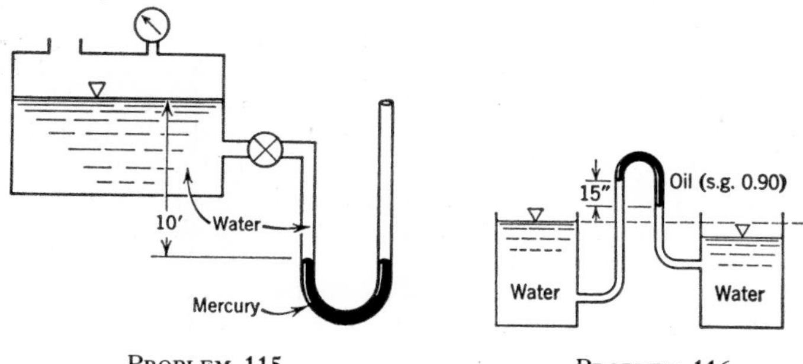

PROBLEM 115 PROBLEM 116

116. This manometer is used to measure the difference in water level between the two tanks. Calculate this difference.

117. Two reservoirs with different surface levels contain the same liquid (of specific weight γ) and are connected by two manometers as shown. Derive an expression for γ in terms of the manometer readings and γ_A and γ_B.

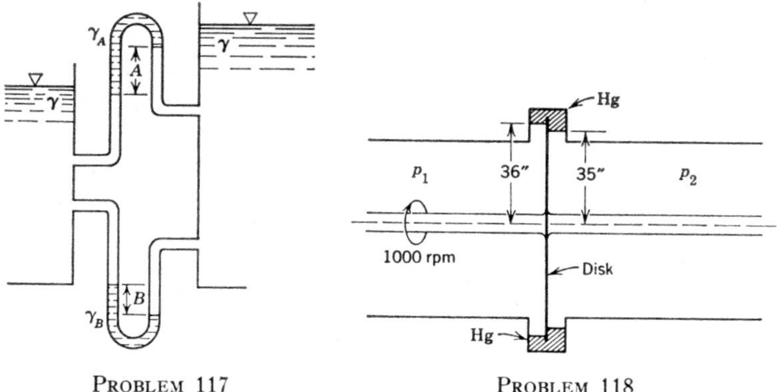

PROBLEM 117 PROBLEM 118

118. The mercury seal shown is used to support a pressure difference $(p_1 - p_2)$ across the rotating disk. Calculate this pressure difference for a speed of 1000 rpm, assuming that the mercury rotates with the disk.

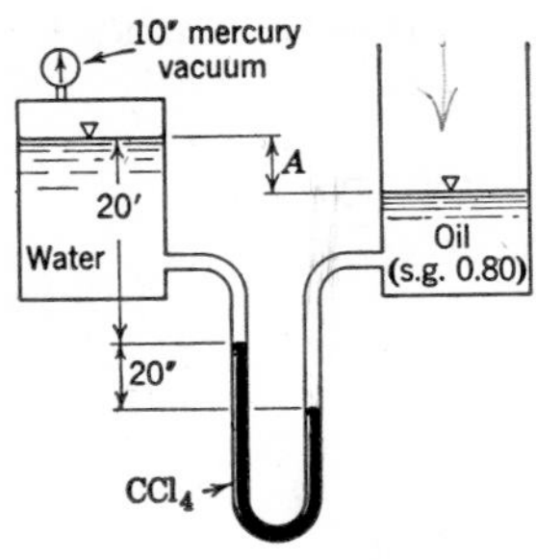

PROBLEM 119

119. Calculate A.

120. A rectangular gate 6 ft long and 4 ft high lies in a vertical plane with its center 7 ft below a water surface. Calculate magnitude, direction, and location of the total force on the gate.

121. A circular gate 10 ft in diameter has its center 8 ft below a water surface and lies in a plane sloping at 60°. Calculate magnitude, direction, and location of total force on this gate.

122. An isosceles triangle of 12-ft base and 15-ft altitude is located in a vertical plane. Its base is vertical, and its apex is 8 ft below the water surface. Calculate magnitude and location of the force of the water on the triangle.

123. A triangular area of 6-ft base and 5-ft altitude has its base horizontal and lies in a 45° plane with its apex below the base and 9 ft below a water surface. Calculate magnitude, direction, and location of the resultant force on this area.

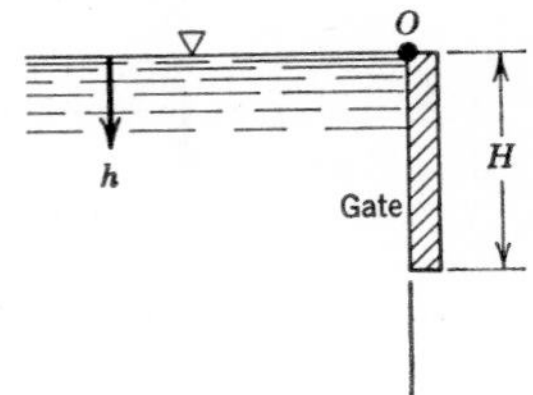

PROBLEM 124

124. If the specific weight of a liquid varies linearly with depth h according to the equation $\gamma = \gamma_o + Kh$, derive expressions for resultant force on the rectangular gate (1 ft wide) and the moment of this force about O.

125. A square 9 ft by 9 ft lies in a vertical plane. Calculate the distance between the center of pressure and centroid, and the total force on the square, when its upper edge is (a) in the water surface, and (b) 50 ft below the water surface.

126. Calculate the vertical and lateral location of the center of pressure of this triangle, which is located in a vertical plane.

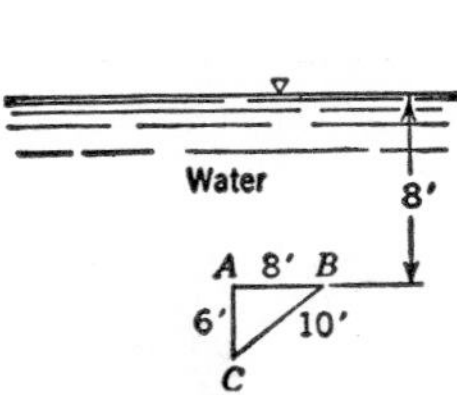

PROBLEM 126

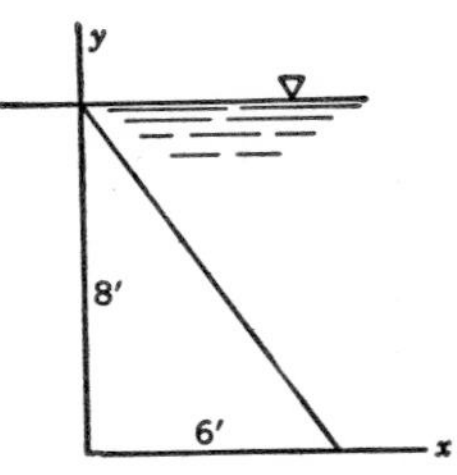

PROBLEM 127

127. Calculate the x- and y-coordinates of the center of pressure of this vertical right triangle.

128. Calculate $(W \times l)$ and the exact position of the pointer to hold this triangular gate in equilibrium.

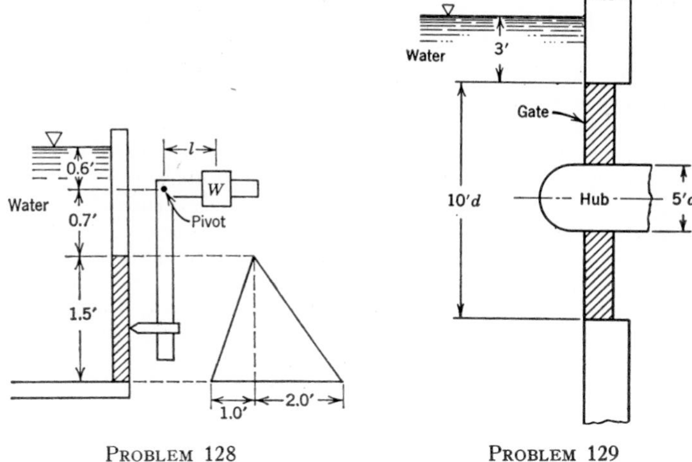

PROBLEM 128 PROBLEM 129

129. Calculate magnitude and location of the resultant force of water on this annular gate.

130. Calculate magnitude and location of the resultant force of the liquid on this tunnel plug.

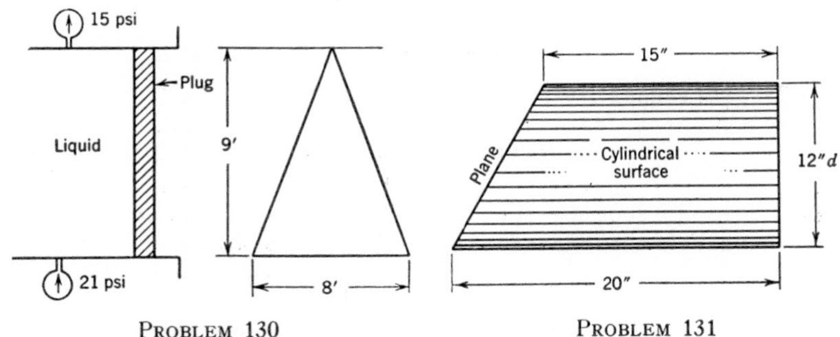

PROBLEM 130 PROBLEM 131

131. Using hydrostatics principles, find the vertical distance between centerline and centroid of this figure.

132. The center of pressure of an isosceles triangle of 9-ft altitude and 6-ft base, lying in a vertical plane, is at a depth of 12 ft. Calculate the depth of water over the apex of the triangle. The base of the triangle is horizontal, and the apex is above the base.

133. Calculate the force exerted by water on the largest completely submerged circle (located in a vertical plane) having a distance of 1 ft between its centroid and center of pressure.

134. A submerged rectangle located in a vertical plane has its base 20 ft below, and its center of pressure 15 ft below, a water surface. Calculate the height of the rectangle.

135. A vertical rectangular gate 10 ft high and 6 ft wide has a depth of water on its upper edge of 15 ft. What is the location of a horizontal line which divides this area (*a*) so that the forces on the upper and lower portions are the same, (*b*) so that the moments of the forces about the line are the same?

136. Calculate magnitude and location of the total force on one side of a semicircular area, located in a vertical plane, with straight side vertical. Radius of semicircle is 6 ft, and centroid is 6 ft below the water surface.

137. A horizontal tunnel of 10-ft diameter is closed by a vertical gate. Calculate magnitude, direction, and location of the total force of water on the gate when the tunnel is (*a*) one-half full, (*b*) one-fourth full, and (*c*) three-fourths full.

138. Calculate magnitude, direction, and location of the total force on an isosceles triangle of 5-ft base and 6-ft altitude, located in a 60° plane with base horizontal, (*a*) when the apex is in the water surface, (*b*) when the base is in the water surface.

139. Calculate the magnitude, direction, and location of the force of the water on one side of this area located in a vertical plane.

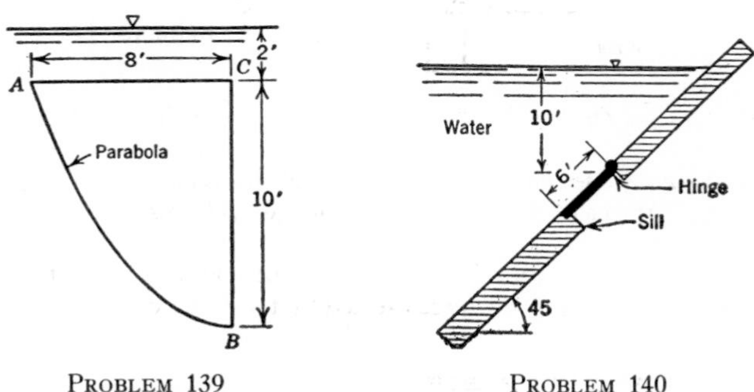

Problem 139 Problem 140

140. This rectangular gate is hinged at the upper edge and is 4 ft wide. Calculate the total force on the sill, neglecting the weight of the gate.

141. A vertical rectangular gate 8 ft wide and 9 ft high is subjected to water pressure on one side, the water surface being at the top of the gate. The gate is hinged at the bottom and is held by a horizontal chain at the top. What is the tension in the chain?

142. A sliding gate 10 ft wide and 5 ft high situated in a vertical plane has a coefficient of friction, between itself and guides, of 0.20. If the gate weighs 2 tons and if its upper edge is at a depth of 30 ft, what vertical force is required to raise it? Neglect buoyancy force on the gate.

143. A butterfly valve, consisting essentially of a circular area pivoted on a horizontal axis through its center (and in the plane of the valve), is 7 ft in diameter and lies in a 60° plane with its center 10 ft below a water surface. What torque must be exerted on the valve's axis to just open it?

144. Calculate the h at which this gate will open.

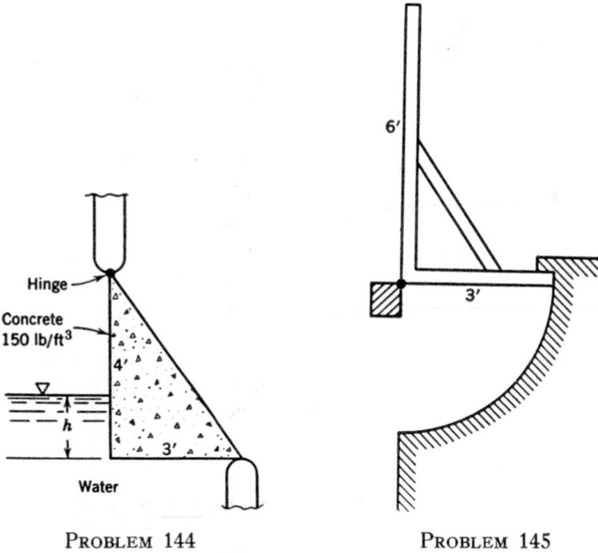

Problem 144 Problem 145

145. As the water rises on the left-hand side of this rectangular gate it will open automatically. At what depth above the hinge will this occur? Neglect the weight of the gate.

146. ABC is a gate in a vertical wall with water behind it. Calculate the moment (about the hinge line AC) of the force exerted by the water on the gate.

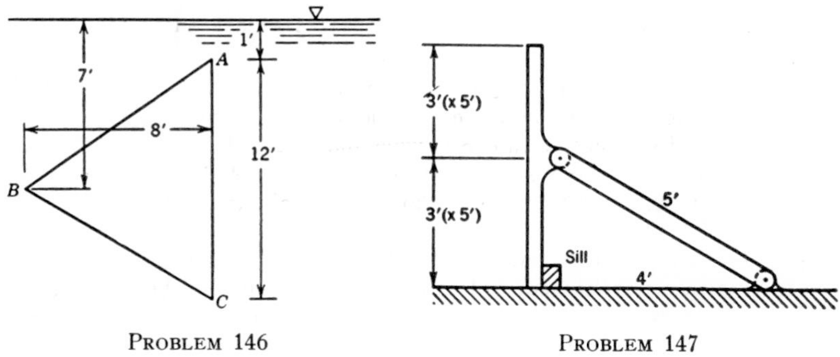

Problem 146 Problem 147

147. This wicket dam will collapse as the water behind it rises. If the coefficient of static friction between the dam and sill is 0.4, at what water depth will the dam become unstable? Neglect the weight of the dam.

148. Calculate the height h to which the water must rise to tip this (weightless) gate.

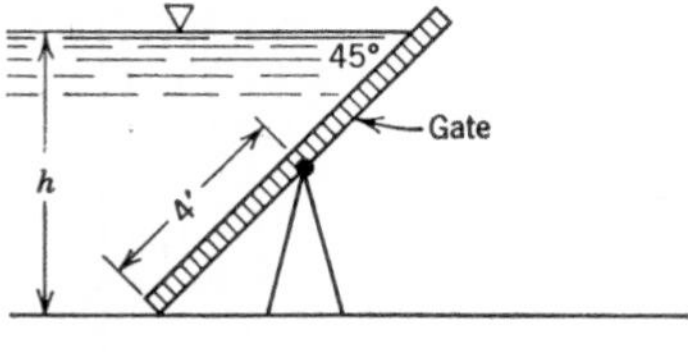

PROBLEM 148

149. These (rectangular) miter gates at the entrance to a canal lock are 15 ft high. Calculate the reactions at the pivots when the water surface is 3 ft below the top of the gates.

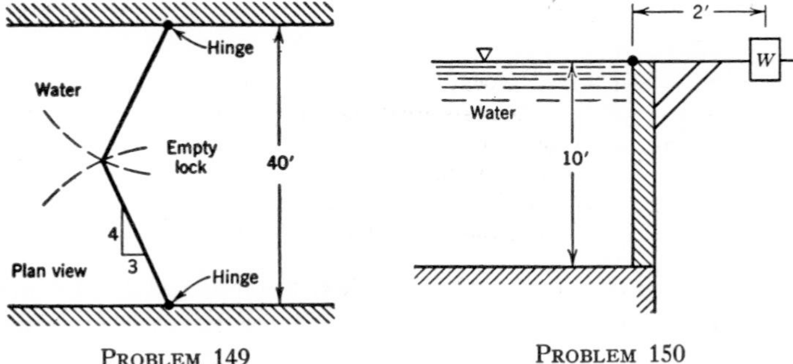

PROBLEM 149 PROBLEM 150

150. This vertical gate closes the end of a triangular flume (top width 8 ft) and is pivoted about its upper edge. When the flume is full, how large must the weight be to keep the gate closed?

151. The wire alone keeps the gate closed. Calculate the tension force in the wire.

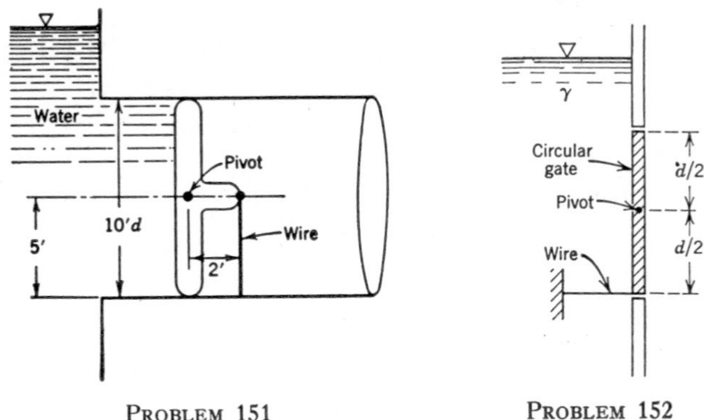

PROBLEM 151 PROBLEM 152

152. Derive an algebraic expression for the force in the wire.

153. This wicket dam is 15 ft high and 4 ft wide and is pivoted at its center. Assume a hydrostatic pressure distribution, and calculate the vertical and horizontal reactions at the two sills. Neglect the weight of the dam, and consider all joints to be pin connected.

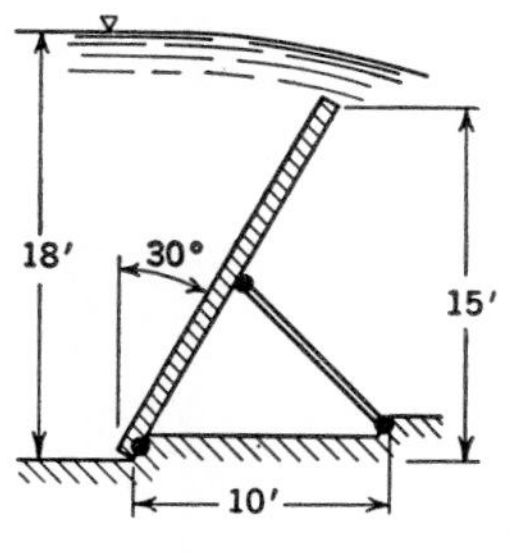

PROBLEM 153

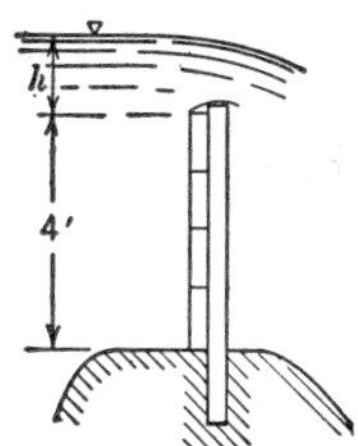

PROBLEM 154

154. The flashboards on a spillway crest are 4 ft high and supported on steel posts spaced 2 ft on centers. The posts are designed to fail under a bending moment of 5000 ft-lb. What depth over the flashboards will cause the posts to fail? Assume hydrostatic pressure distribution.

155. This rectangular gate will open automatically when the depth of water, d, becomes large enough. What is the minimum depth which will cause the gate to open?

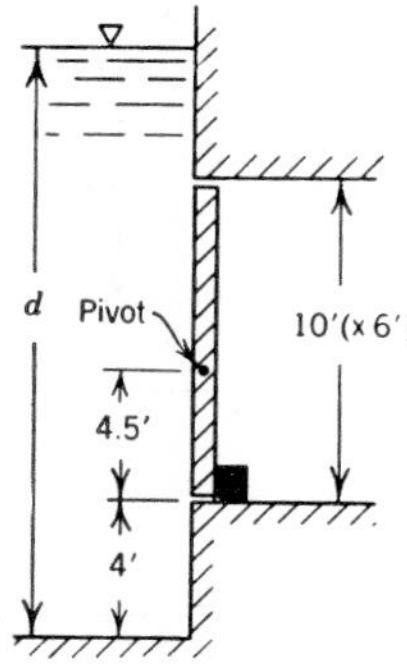

PROBLEM 155

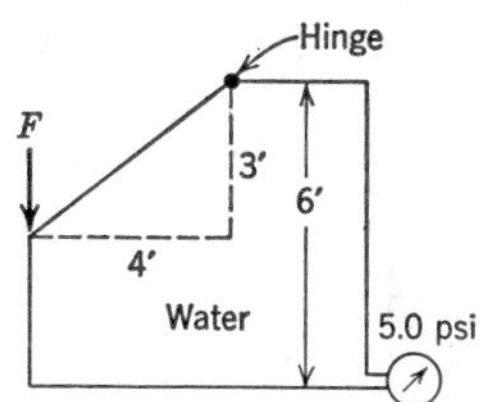

PROBLEM 156

156. Calculate the minimum vertical force F required to keep the cover of this box closed. The cover is 10 ft wide perpendicular to the plane of the paper.

157. Calculate the minimum volume of concrete block (γ = 150 lb/cuft) which will hold this circular gate in place. The block is submerged in the water.

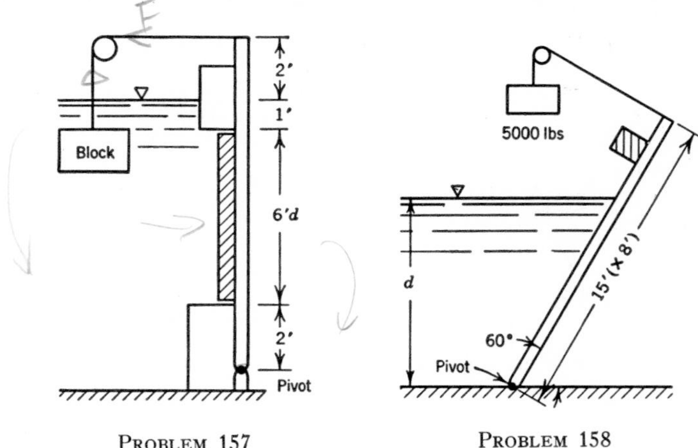

PROBLEM 157 PROBLEM 158

158. What depth of water will cause this rectangular gate to fall? Neglect the weight of the gate.

159. Calculate magnitude and location of the total force on one side of this vertical plane area.

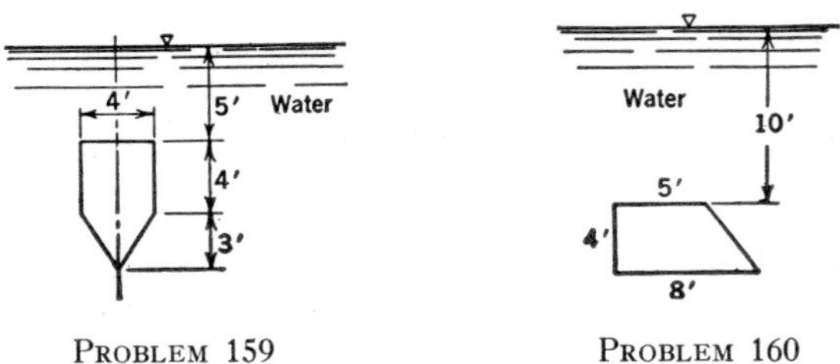

PROBLEM 159 PROBLEM 160

160. Calculate magnitude and location of the total force on one side of this vertical plane area.

161. A rectangle 3 ft by 4 ft lies in a vertical plane with one diagonal horizontal and 7 ft below the water surface. Calculate magnitude and location of total force on the rectangle.

162. This area lies in a vertical plane beneath a water surface. Calculate magnitude and location of the force of the water on the area. Compare results with those of the preceding problem.

163. A horizontal open channel of U-shaped cross section is closed by a vertical gate. If the channel is 8 ft wide, calculate magnitude and location of the total force on the gate when the water is 8 ft deep. Assume the lower portion of the channel to be a semicircle.

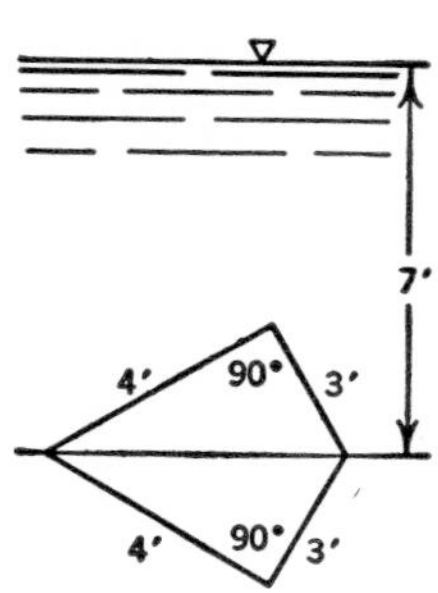

PROBLEM 162

164. This 6-ft by 6-ft square gate is hinged at the upper edge. Calculate the total force on the sill.

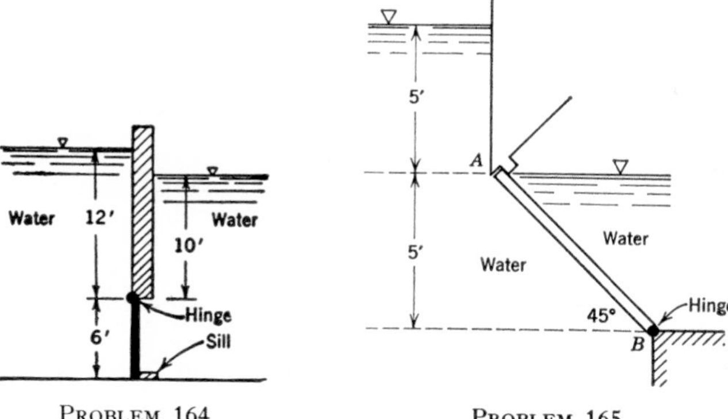

PROBLEM 164 PROBLEM 165

165. Calculate the force exerted against the stop at A by the weightless rectangular gate AB.

166. Calculate the magnitude, direction, and location of the total force on the gate of problem 137 when on one side the water surface is 18 ft above the tunnel invert and on the other side 5 ft above the tunnel invert. (The "invert" is the low point of the tunnel cross section.)

167. A vertical gate in a tunnel 20 ft in diameter has water on one side and air on the other. The water surface is 35 ft above the invert, and the air pressure is 16 psi. Where could a single support be located to hold this gate in position?

168. Calculate the total force on the gate of problem 137*a* when the air pressure on the water surface is 5 psi.

169. A rectangular tank 5 ft wide, 6 ft high, and 10 ft long contains water to a depth of 3 ft and oil (s.g. = 0.85) on the water to a depth of 2 ft. Calculate magnitude and location of the force on one end of the tank.

170. A rectangular tunnel 6 ft wide and 10 ft high is closed by a vertical gate. The tunnel is half-full of water, and the air above the water is at a pressure of 2 psi. Calculate the resultant force of the fluids on one side of the gate.

171. Calculate the force exerted by the gate AB on the stop at B. The gate is 10 ft wide normal to the plane of the paper.

172. A circular tunnel of 10-ft diameter is closed by a bulkhead, having water on one side and air on the other. If the gage pressures at the top of the tunnel on each side of the bulkhead are 2 psi, calculate the resultant force of the fluids on the bulkhead.

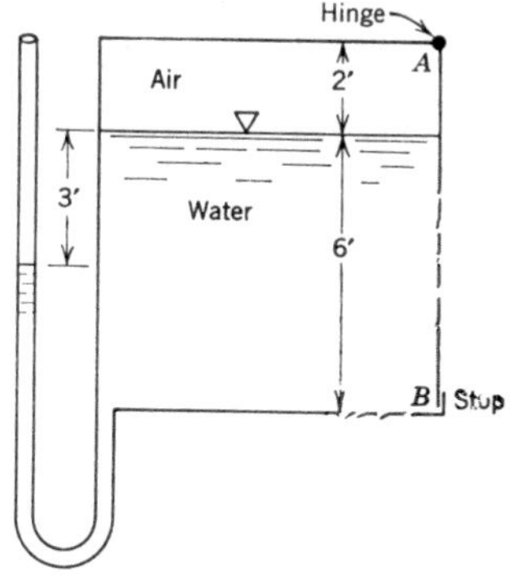

PROBLEM 171

173. At what gage reading will this weightless gate be in equilibrium in the position shown?

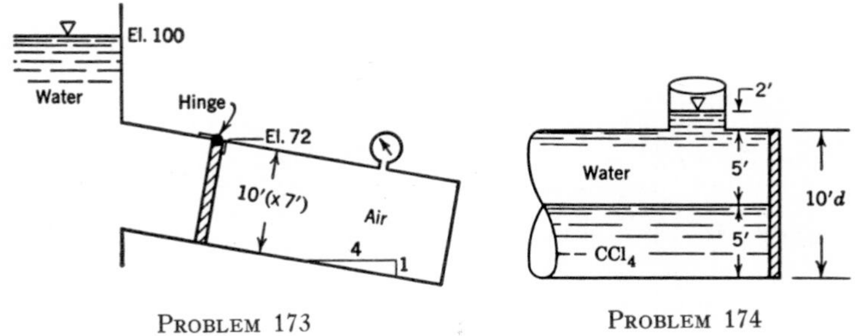

PROBLEM 173 PROBLEM 174

174. Calculate magnitude and location of the resultant force of the liquids on the end of this cylindrical tank.

175. Calculate magnitude, direction, and location of the resultant force exerted by the fluids on the end of this cylindrical tank.

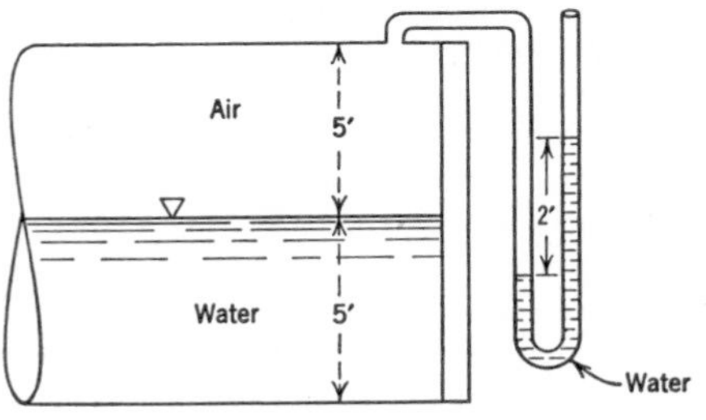

PROBLEM 175

176. Using force components, calculate the load in the strut AB if these struts have 5-ft spacing along the small dam AC. Consider all joints to be pin connected.

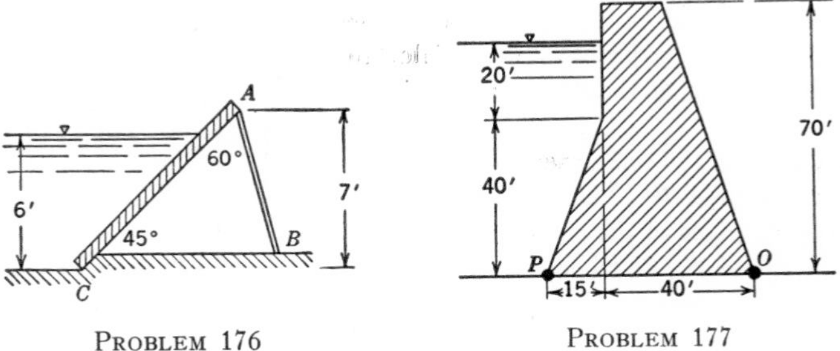

PROBLEM 176 PROBLEM 177

177. Using the method of components, calculate the magnitude, direction, and location of the total force on the upstream face of a section of this dam 1 ft wide. What is the moment of this force about O?

178. Using hydrostatics principles (not geometry or calculus), calculate the volume $BCDE$ (cuft).

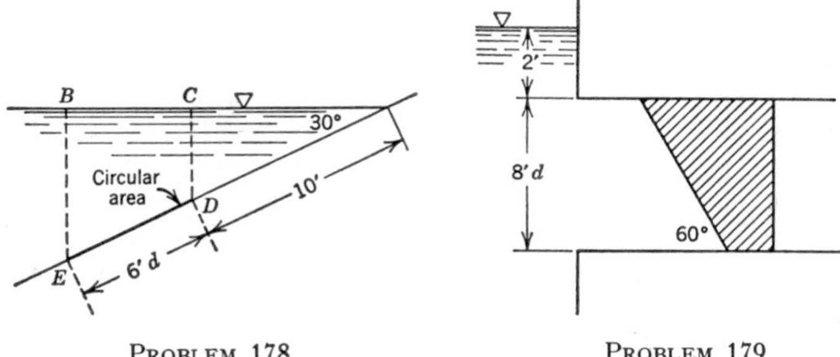

PROBLEM 178 PROBLEM 179

179. Calculate magnitude, direction, and location of the resultant force exerted by the water on this tunnel plug.

180. Calculate the magnitude of the total force in problem 123 by the method of components.

181. A concrete pedestal, having the shape of the frustum of a right pyramid of lower base 4 ft square, upper base 2 ft square, and height 3 ft, is to be poured. Taking γ for concrete to be 150 lb/cuft, calculate the vertical force of uplift on the forms.

182. This tainter gate is pivoted at O and is 30 ft long. Calculate the magnitudes of horizontal and vertical components of force on the gate. The pivot is at the same level as the water surface.

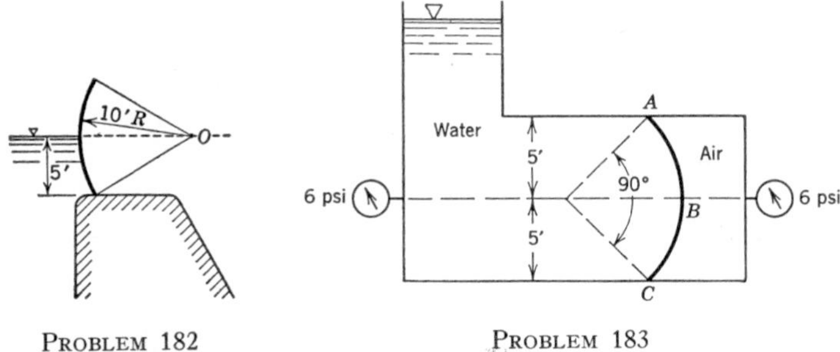

PROBLEM 182 PROBLEM 183

183. Calculate magnitude, direction, and location of the resultant force of the air and water on the quarter-cylindrical surface ABC. Assume 10 ft normal to plane of paper.

184. Calculate magnitude, direction, and location of horizontal and vertical components of force exerted by the water on this gate.

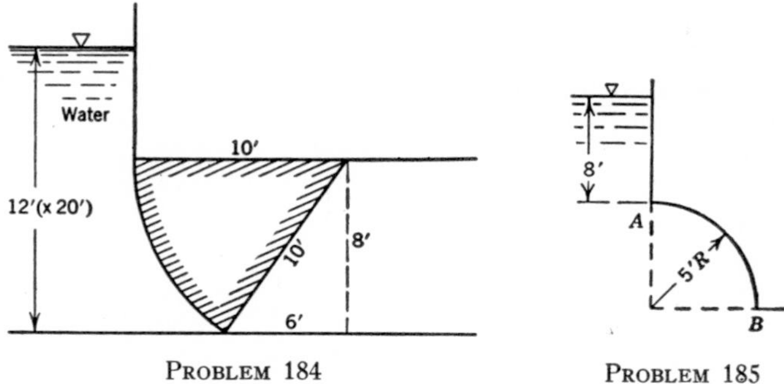

PROBLEM 184 PROBLEM 185

185. The quarter cylinder AB is 10 ft long. Calculate magnitude, direction, and location of the resultant force of the water on AB.

186. Calculate the magnitude, direction, and location of the resultant force exerted by the water on the three-quarter cylinder, which is 10 ft long perpendicular to the paper.

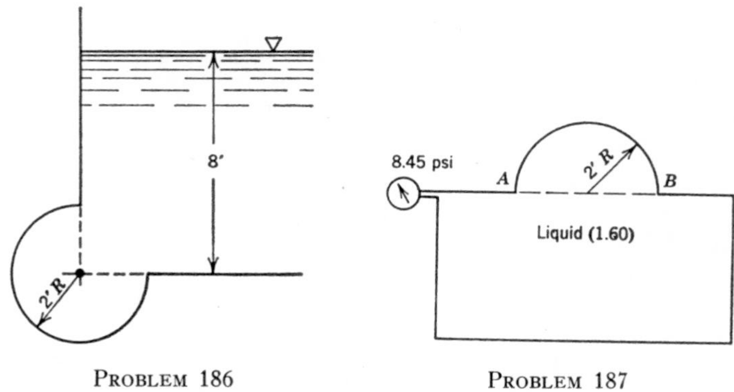

PROBLEM 186 PROBLEM 187

187. Calculate the vertical force exerted by the liquid on this semicylindrical dome AB, which is 5 ft long.

188. Calculate magnitude, direction, and location of the resultant forces on the surfaces AB, AC, and BC of this quarter cylinder, which is 6 ft long.

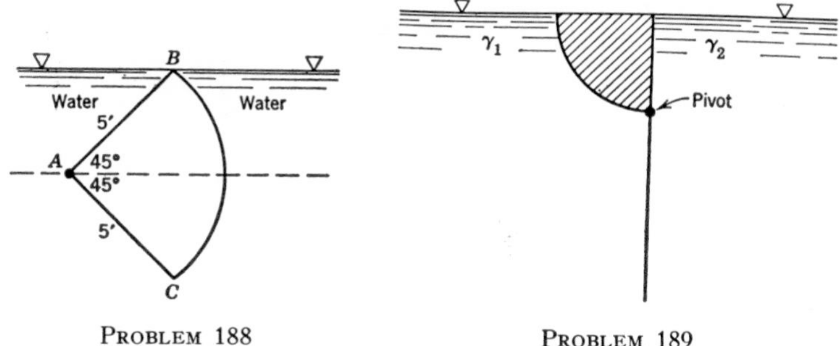

PROBLEM 188 PROBLEM 189

189. If this weightless quarter-cylindrical gate is in static equilibrium, what is the ratio between γ_1 and γ_2?

190. Calculate the moment about O of the resultant force exerted by the water on this half cylinder, which is 10 ft long.

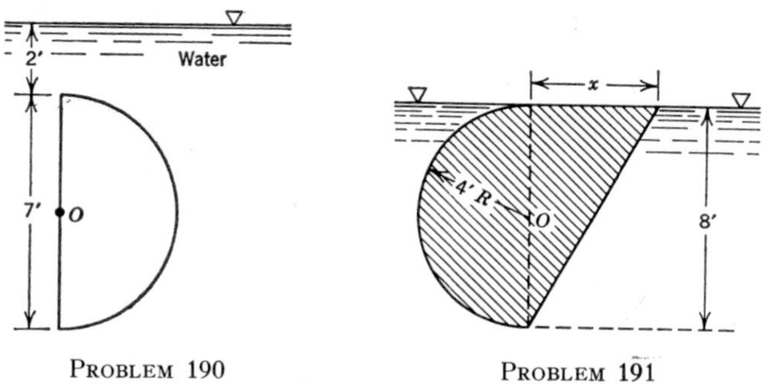

PROBLEM 190 PROBLEM 191

191. This weightless two-dimensional solid body is suspended on a pivot at point O. What should be the value of x so that the body will have no tendency to rotate?

192. This cylindrical gate is 10 ft long. Calculate the magnitude, location, and direction of the resultant water force exerted on the gate. Also calculate the least weight of the gate to ensure that it will not float away from the floor.

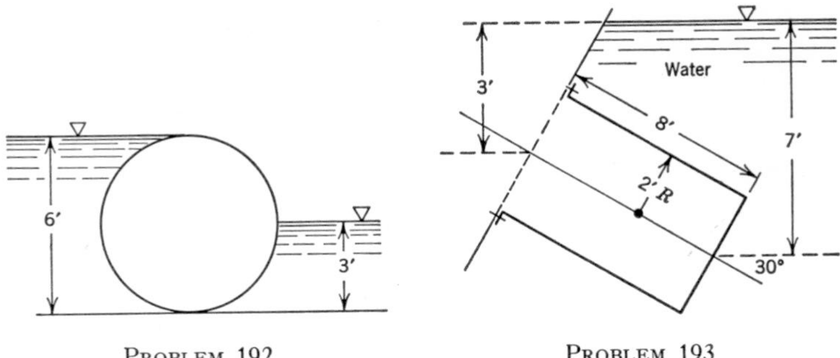

PROBLEM 192 PROBLEM 193

193. Calculate magnitude and direction of the resultant forces exerted by the water (*a*) on the end of the cylinder and (*b*) on the curved surface of the cylinder.

194. Solve problem 185, assuming that there is also water to the right of *AB* with a surface level 3 ft above *A*.

195. The cylinder is 8 ft long and is pivoted at *O*. Calculate the moment (about *O*) required to hold it in position.

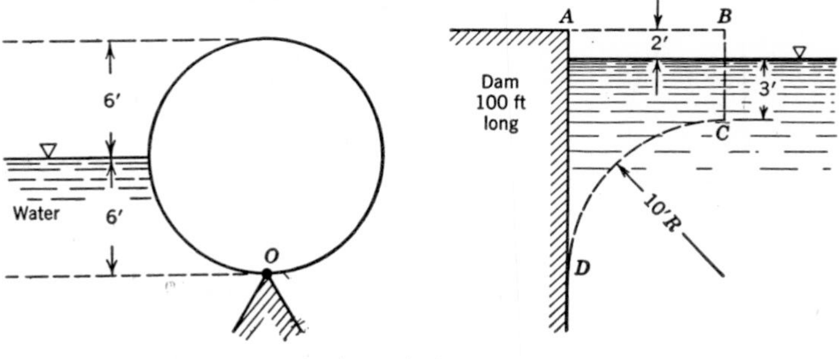

PROBLEM 195 PROBLEM 196

196. If this solid concrete (150 lb/cuft) overhang *ABCD* is added to the dam, what *additional* force (magnitude and direction) will be exerted on the dam?

197. If the liquid shown above *B* in problem 185 consists of a 7-ft layer of water on top of a 6-ft layer of carbon tetrachloride, calculate magnitude, direction, and location of horizontal and vertical components of force on *AB*.

198. If the line *AB* in problem 139 represents the edge view of a surface 6 ft wide (normal to the plane of the paper), calculate magnitude, direction, and location of the vertical and horizontal force components on the left-hand side of this surface.

199. A hemispherical shell 4 ft in diameter is connected to the vertical wall of a tank containing water. If the center of the shell is 6 ft below the water surface, what are the vertical and horizontal force components on the shell? on the top half of the shell?

200. This half-conical buttress is used to support a half-cylindrical tower on the upstream face of a dam. Calculate the magnitude, direction, and location of the vertical and horizontal components of force exerted by the water on the buttress (*a*) when the water surface is at the base of the half cylinder, (*b*) when it is 4 ft above this point.

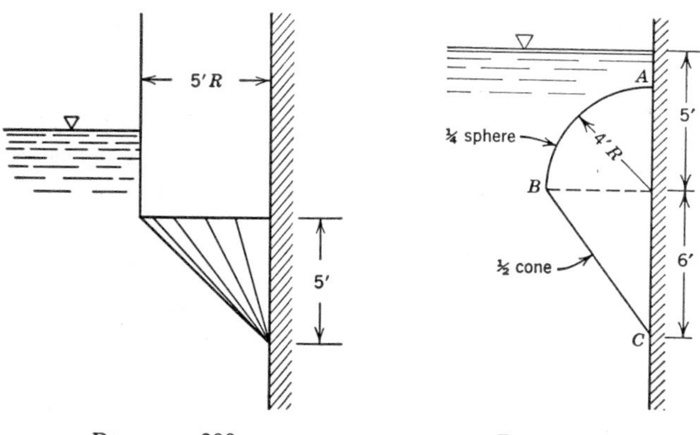

PROBLEM 200 PROBLEM 201

201. Calculate the magnitude and location of the vertical and horizontal components of resultant force exerted by the water on the curved surface *ABC*.

202. A hole 12 in. in diameter in a vertical wall between two water tanks is closed by a sphere 18 in. in diameter in the tank of higher water-surface elevation. The difference in the water-surface elevations in the two tanks is 5 ft. Calculate the horizontal component of force exerted by the water on the sphere.

203. Calculate the magnitude, direction, and location of the force of water on *AB* in problem 185 if *AB* is a quarter sphere.

204. If the half cone of problem 200 is replaced by a quarter sphere, what are the magnitudes of the vertical and horizontal force components on the buttress?

205. Calculate *H* at which this cone valve will start to leak. The valve weighs 500 lb.

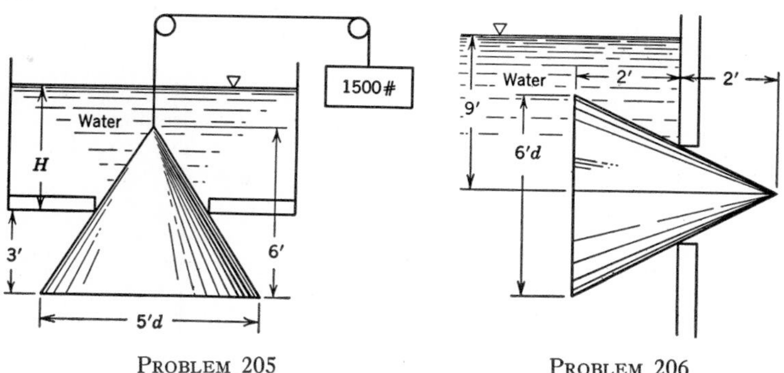

PROBLEM 205 PROBLEM 206

206. Calculate magnitude and direction of the resultant force of the water on this solid conical plug

207. Calculate the horizontal component of resultant force of water on a 90° pipe bend in a horizontal 24-in. pipe when the water is at rest and the pressure on the centerline of the pipe is 40 psi.

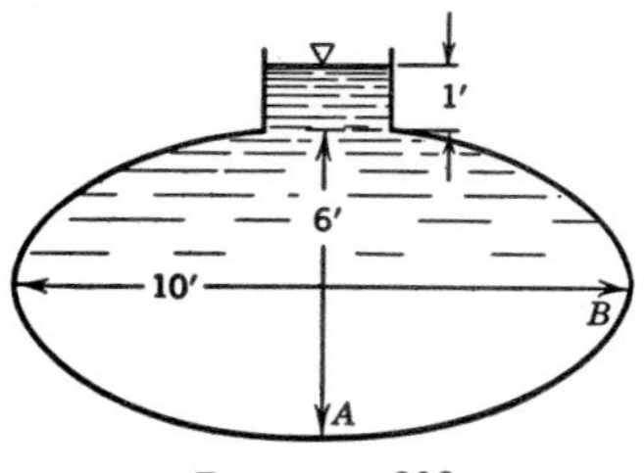

PROBLEM 208

208. The tank is an elliptic cylinder of 8-ft length. Calculate magnitude and direction of the vertical and horizontal components of force on AB.

209. Solve problem 174 assuming that the tank has a hemispherical end.

210. A stone weighs 60 lb in air and 40 lb in water. Calculate its volume and specific gravity.

211. Six cubic inches of lead (s.g. 11.4) are suspended from the apex of a conical can having a height of 12 in. and a base of 6-in. diameter, and weighing 1 lb. When placed in water, to what depth will the can be immersed? The apex is below the base.

212. A cylindrical can 3 in. in diameter and 6 in. high, weighing 4 oz, contains water to a depth of 3 in. When this can is placed in water, how deep will it sink?

213. The timber weighs 40 lb/cuft and is held in a horizontal position by the concrete (150 lb/cuft) anchor. Calculate the minimum total weight which the anchor may have.

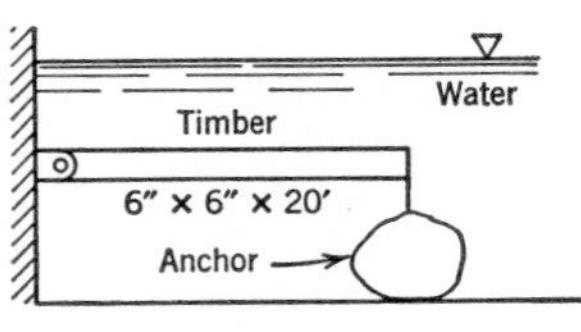

PROBLEM 213

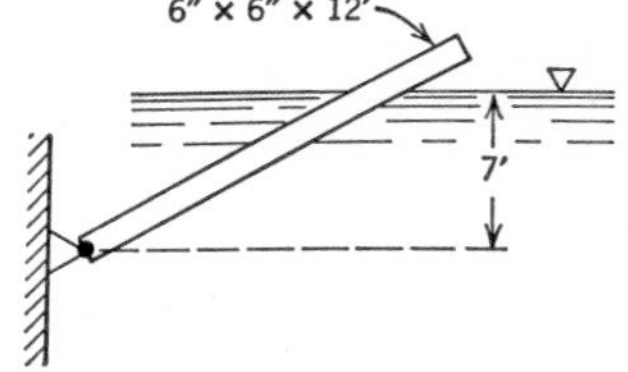

PROBLEM 214

214. If the timber weighs 150 lb, calculate its angle of inclination when the water surface is 7 ft above the pivot. Above what depth will the timber stand vertically?

215. The barge shown weighs 40 tons and carries a cargo of 40 tons. Calculate its draft in fresh water.

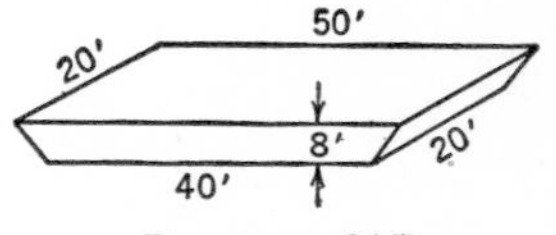

PROBLEM 215

216. A tank nearly full of water is placed on a platform weighing scales and a reading of 1000 lb noted. If a steel (s.g. 7.80) bar of 6-in. by 6-in. cross section is then suspended from the ceiling so that its lower end extends 2 ft into the water, calculate the new reading.

217. The plug weighs 0.28 lb/in.^{3}. Calculate D for the valve to open when the water is 5 ft deep. Consider the cable and sphere to be weightless.

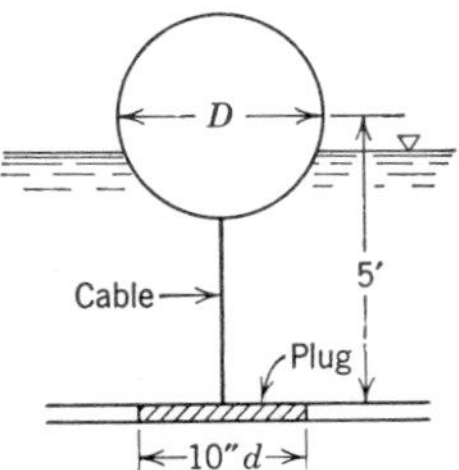

PROBLEM 217

218. What is the minimum weight required for this solid cone if it is to remain in the position shown?

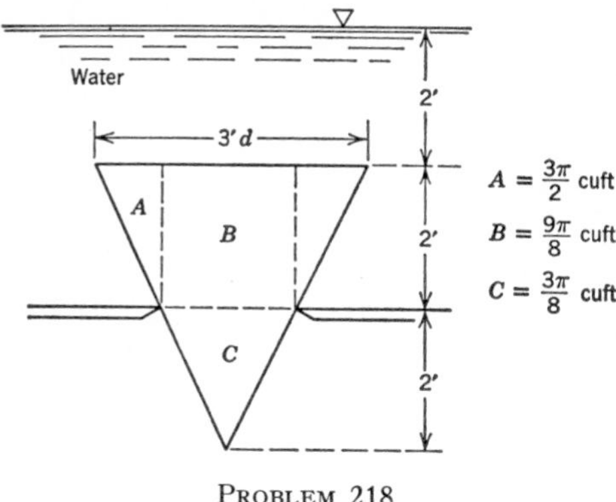

PROBLEM 218

219. Calculate the required weight of this object for it to "float" at the water-carbon tetrachloride interface as shown.

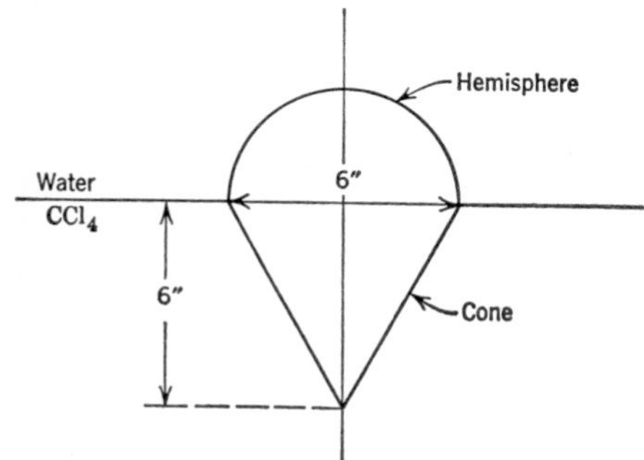

PROBLEM 219

220. To what depth will a rigid homogeneous object (γ = 66.0 lb/cuft) sink in the ocean if the specific weight therein varies with depth according to the empirical relation $\gamma = 64.0 + 0.025\sqrt{\text{depth (ft)}}$.

221. A balloon having a total solid weight of 800 lb contains 15,000 cuft of hydrogen. How many pounds of ballast are necessary to hold the balloon on the ground? Barometric pressure is 14.7 psia; temperature of air and hydrogen, 60°F. Assume hydrogen to be at barometric pressure.

222. A balloon has a weight (including crew but not gas) of 500 lb and a gas-bag capacity of 20,000 cuft. At the ground it is (partially) inflated with 100 lb of helium. How high can this balloon rise in the ICAO Standard Atmosphere (Appendix IV) if the helium always assumes the pressure and temperature of the atmosphere?

223. An open cylindrical container holding 1.0 cuft of water at a depth of 2 ft is accelerated vertically upward at 20 ft/sec^2. Calculate the pressure and total force on the bottom of the container. Also calculate this total force by application of Newton's second law.

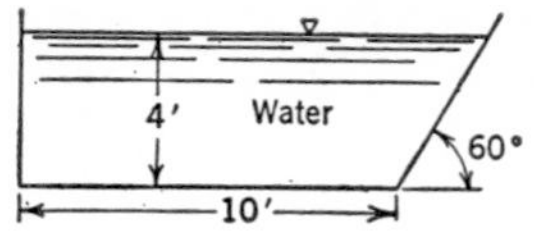

PROBLEM 224

224. Calculate the total forces on the ends and bottom of this container while at rest and when being accelerated vertically upward at 10 ft/sec^2. The container is 5 ft wide.

225. An open conical container 6 ft high is filled with water and moves vertically downward with a deceleration of 10 ft/sec^2. Calculate the pressure at the bottom of the container.

226. A rectangular tank 5 ft wide, 10 ft long, and 6 ft deep contains water to a depth of 4 ft. When it is accelerated horizontally at 10 ft/sec^2 in the direction of its length, calculate the depth of water at each end of the tank and the total force on each end of the tank. Check the difference between these forces by calculating the inertia force of the accelerated mass. Repeat these calculations for an acceleration of 20 ft/sec^2.

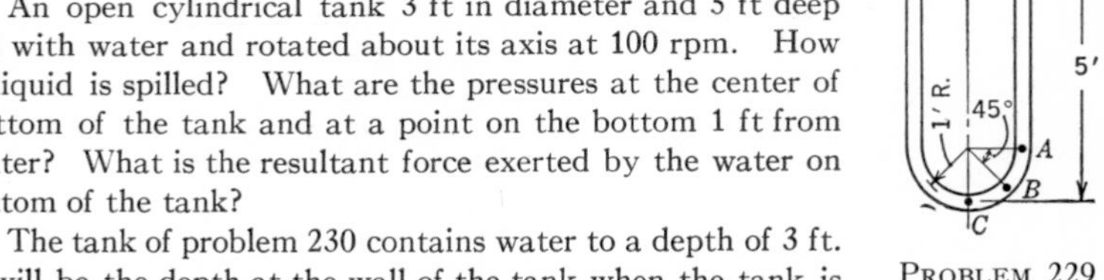

227. A closed rectangular tank 4 ft high, 8 ft long, and 5 ft wide is three-fourths full of water and the pressure in the air space above the water is 20 psi. Calculate the pressures in the corners of this tank when it is accelerated horizontally along the direction of its length at 15 ft/sec^2. Using Newton's second law, calculate the forces on the ends of the tank and check their difference.

228. An open container of liquid accelerates down a 30° inclined plane at 15 ft/sec^2. What is the slope of its free surface? What is the slope for the same acceleration up the plane?

229. This U-tube containing water is accelerated horizontally to the right at 10 ft/sec^2; what are the pressures at A, B, and C? Assume that the tubes are long enough that no liquid is spilled.

PROBLEM 229

230. An open cylindrical tank 3 ft in diameter and 5 ft deep is filled with water and rotated about its axis at 100 rpm. How much liquid is spilled? What are the pressures at the center of the bottom of the tank and at a point on the bottom 1 ft from the center? What is the resultant force exerted by the water on the bottom of the tank?

231. The tank of problem 230 contains water to a depth of 3 ft. What will be the depth at the wall of the tank when the tank is rotated at 60 rpm?

232. The tank of problem 230 contains water to a depth of 1 ft. At what speed must it be rotated to uncover a bottom area 1 ft in diameter?

233. A vertical cylindrical tank 5 ft high and 3 ft in diameter is filled with water to a depth of 4 ft. The tank is then closed and the pressure in the space above the water raised to 10 psi. Calculate the pressure at the intersection of wall and tank bottom when the tank is rotated about a central vertical axis at 150 rpm. Also calculate the resultant force exerted by the water on the bottom of the tank.

234. A tube 2 in. in diameter and 4 ft long is filled with water and closed. It is then rotated at 150 rpm in a horizontal plane about one end as a pivot. Calculate the pressure on the outer end of the tube, using the equation in the text, and check by calculating the centrifugal force of the rotating mass.

235. The impeller of a closed filled centrifugal water pump is rotated at 1750 rpm. If the impeller is 2 ft in diameter, what pressure is developed by rotation?

236. When the U-tube of problem 229 is rotated at 200 rpm about its central axis, what are the pressures at points A, B, and C?

chapter 3

Kinematics of Fluid Motion

The objective of this chapter is to treat the kinematics of somewhat idealized fluid motion along streamlines and in flowfields. As in particle mechanics, kinematics describes motion in terms of displacements, velocities, and accelerations without regard to the forces which are associated with these variables. However, no attempt will be made here to describe the kinematics of turbulence or of the motion of large-scale eddies which the reader has no doubt observed in real fluid flows; these topics will be discussed later in the text.

15. Steady and Unsteady Flow. Streamlines and Streamtubes. Fluid flow may be steady or unsteady. Steady flow exists when none of the variables in a flow problem changes with time; if any variable does change with time, the condition of unsteady flow obtains. For example, in the pipe of Fig. 30 leading from an infinite [1] reservoir of fixed surface elevation, unsteady flow exists while the valve A is being opened or closed; with the valve opening fixed, steady flow occurs—under the former condition, pressures, velocities, etc., vary with time; under the latter they do not. Problems of steady flow are more elementary than those of unsteady flow and have wider engi-

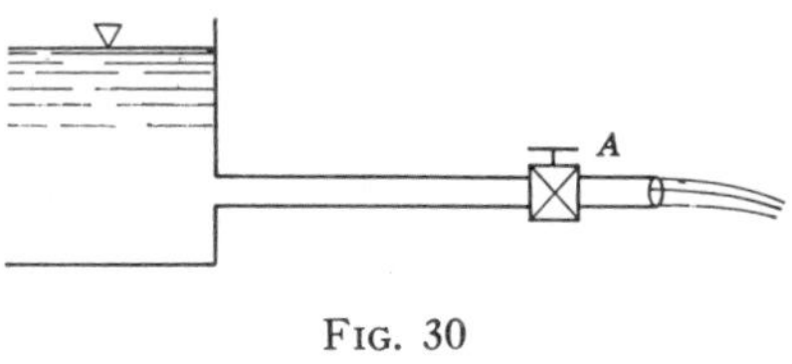

Fig. 30

[1] To preserve steady flow with a finite reservoir the same flowrate must be supplied to the reservoir as flows out of the pipe.

neering application; therefore unsteady-flow situations will not be included in this elementary treatise.

If curves are drawn in a steady flow in such a way that the tangent at any point is in the direction of the velocity at that point, such curves are called *streamlines*, and the sketching of such streamlines results in a *streamline picture* or *flowfield* (Fig. 31). Such streamline pictures are of both qualitative and quantitative value to the engineer; they allow him to visualize fluid flow and to locate regions of high and low velocity and, from these, zones of high and low pressure.

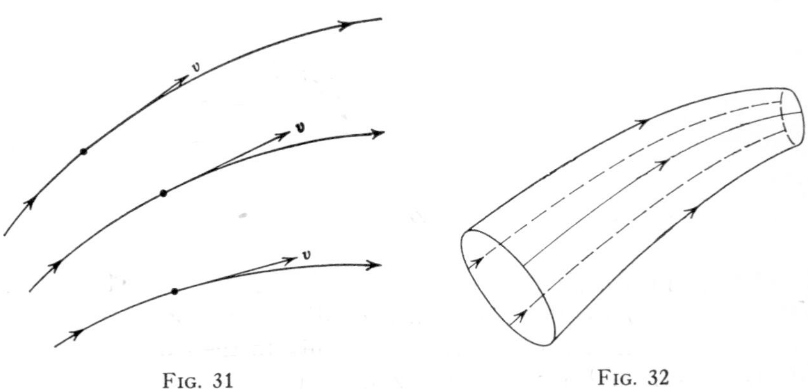

FIG. 31 FIG. 32

When streamlines are drawn through a closed curve (Fig. 32) in a steady flow, they form a boundary across which fluid particles cannot pass. Thus the space between the streamlines becomes a tube or passage called a *streamtube*, and such a tube may be treated as if isolated from the adjacent fluid. The use of the streamtube concept will be found to broaden the application of fluid-flow principles; for example, it allows treating two apparently different problems such as flow in a passage and flow about an immersed object with the same laws. Also, since a streamtube of differential size essentially coincides with its axis (which is a streamline), it is to be expected that many of the equations developed for a small streamtube will apply equally well to a streamline.

16. One-, Two-, and Three-Dimensional Flows. A streamline, being a mathematical line, obviously possesses only one dimension. Thus the flow along individual streamlines (however curved) is termed *one-dimensional*. In such flows there can be no variation of pressure, velocity, etc., except *along* the streamline, since dimensions transverse to the streamline are not definable. Later it will be shown that some two- and three-dimensional flows may be treated very effectively for engineering purposes as one-dimensional in zones where the streamlines of the flow picture are all essentially straight and parallel.

Two- and three-dimensional flow pictures describe *flowfields*, the former when the flow is completely defined by the streamlines in a single plane, the latter in (three-dimensional) space. Examples of two-dimensional flows are shown over the weir and about the wing of Fig. 33. Here the velocities, pressures, etc., vary throughout the

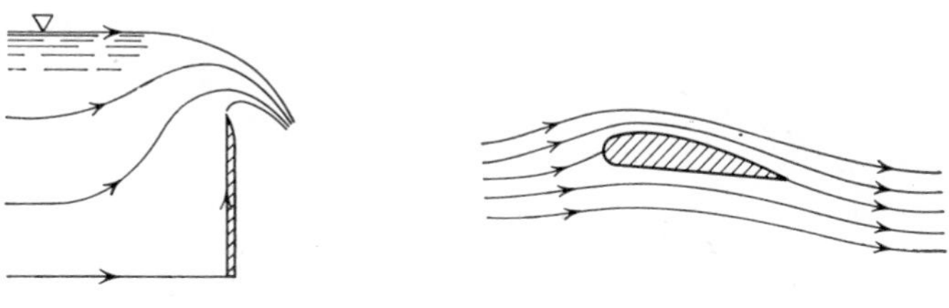

FIG. 33

flowfield and thus are functions of position in the field. The reader will also note that such two-dimensional flows are approximations to reality in that they are strictly correct only to the extent that end effects on weir and wing are negligible; this may also be visualized by assuming weir and wing to be infinitely long perpendicular to the plane of the paper. To the extent that such approximations are valid, the flow is completely described by a streamline picture drawn in a single plane.

Two axisymmetric three-dimensional flows are depicted in Fig. 34. Here the streamlines are really stream surfaces and the streamtubes are of annular cross section. On planes passed through the axis of

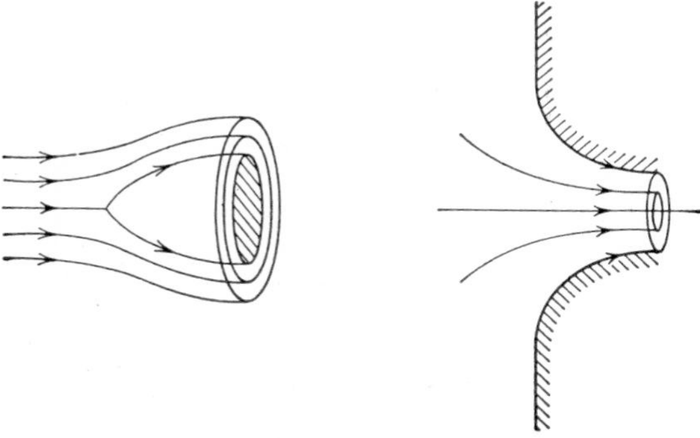

FIG. 34

such flows, streamline pictures may be drawn which superficially resemble two-dimensional flows; such pictures may be used fruitfully for streamline visualization, but they do not represent a reduction of a three-dimensional problem to a two-dimensional one. Non-axisymmetric flows such as that into the air inlets of a single-jet airplane are *three-dimensional flowfields* of the most general character. Obviously these flows are more difficult to visualize and most difficult to predict. To generalize the kinematics and dynamics of the flowfield so that derived equations will allow all possible flow configurations would lead to mathematical complexities far beyond the scope of an elementary treatment of fluid mechanics and would obscure the physical picture so essential to a real understanding of the problem. Accordingly the treatment hereinafter is restricted primarily to one, two-, and axisymmetric three-dimensional problems.

17. Velocity and Acceleration. For one-dimensional flow along a streamline (Fig. 35), velocity and acceleration may be readily defined from past experience in engineering mechanics with the motion of single particles. Select a fixed point O as a reference point and define the displacement s of a fluid particle along the streamline in the direction of motion. In time dt the particle will cover a differential distance ds along the streamline. The velocity, v, of this particle over the distance ds is given by $v = ds/dt$, which also shows v to be along the direction of ds (confirming the definition of a streamline), since v and ds are the only vector quantities involved. Components of acceleration along (tangent to) and across (normal to) the element ds may also be written:

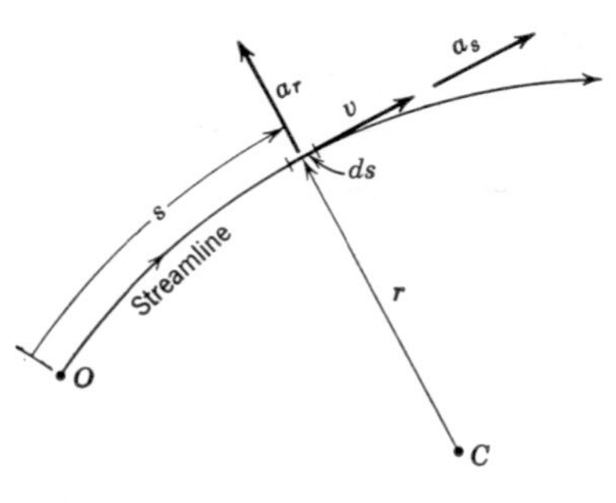

FIG. 35

$$a_s = \frac{d^2s}{dt^2} = \frac{d}{dt}\left(\frac{ds}{dt}\right) = \frac{dv}{dt} = \frac{ds}{dt}\frac{dv}{ds} = v\frac{dv}{ds} \tag{43}$$

and, from particle mechanics,

$$a_r = -\frac{v^2}{r} \tag{44}$$

in which r is the radius of curvature of the element ds.

ILLUSTRATIVE PROBLEMS

Along the straight streamline shown, the velocity is given by $v = 3\sqrt{x^2 + y^2}$. Calculate the velocity and acceleration at the point (8, 6).

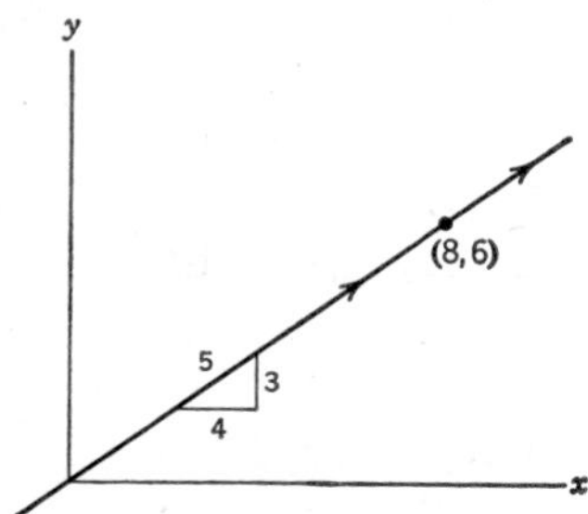

Solution. Since $s = \sqrt{x^2 + y^2}$, $v = 3s$.
At (8, 6), $s = 10$; therefore $v = 3 \times 10 = 30$ fps.

$$a_t = (3s)3 = 9s = 90 \text{ ft/sec}^2 \tag{43}$$

a_r is obviously zero since the radius of curvature of the streamline is infinite.

The velocity along the circular streamline shown is constant at 5 fps. Calculate the tangential and radial components of acceleration at any point on the streamline.

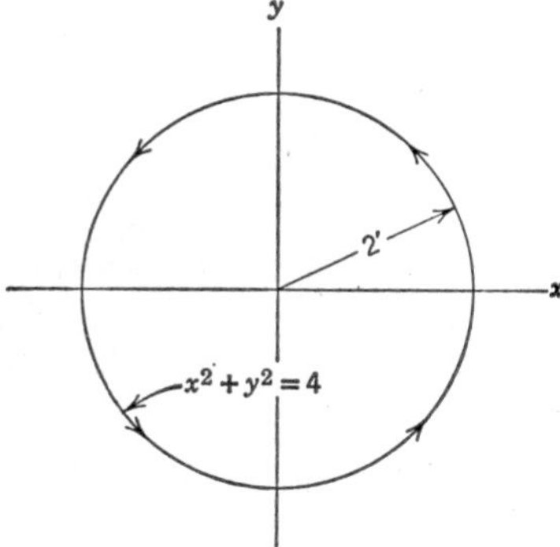

Solution. Since the velocity is constant, $a_t = 0$. Since the radius of the streamline is 2 ft and the velocity along it is 5 fps,

$$a_r = (5)^2/2 = 12.5 \text{ ft/sec}^2 \tag{44}$$

directed toward the center of the circle.

In flowfields velocity and acceleration are somewhat more difficult to define since a generalization is required which is applicable to the whole flowfield. Consider a flow in the x–y plane of Fig. 36. In

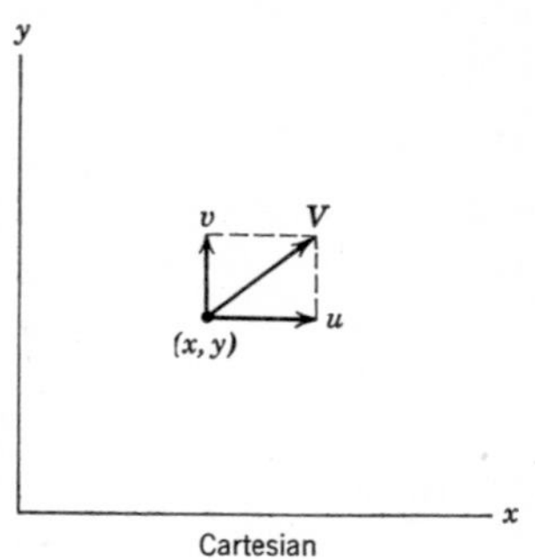

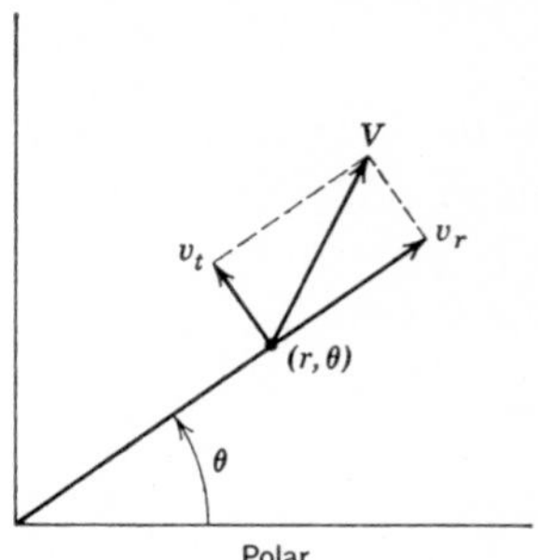

FIG. 36

general, the velocities are everywhere different in magnitude and direction at different points in the flowfield. At each point, however, each velocity will have components u and v parallel to the x- and y-axes, respectively. If the velocities depend on x and y, their components are also functions of x and y. Written mathematically,

$$u = u(x, y) \quad \text{and} \quad v = v(x, y)$$

In terms of displacement and time, however,

$$u = \frac{dx}{dt} \quad \text{and} \quad v = \frac{dy}{dt}$$

Accelerations a_x and a_y are

$$a_x = \frac{du}{dt} \quad \text{and} \quad a_y = \frac{dv}{dt} \tag{45}$$

Writing the differentials du and dv in terms of partial derivatives,[2]

$$du = \frac{\partial u}{\partial x} dx + \frac{\partial u}{\partial y} dy \quad \text{and} \quad dv = \frac{\partial v}{\partial x} dx + \frac{\partial v}{\partial y} dy$$

Substituting these relations in equations 45 and recognizing the velocities u and v as dx/dt and dy/dt, respectively,

$$a_x = u \frac{\partial u}{\partial x} + v \frac{\partial u}{\partial y} \quad \text{and} \quad a_y = u \frac{\partial v}{\partial x} + v \frac{\partial v}{\partial y} \tag{46}$$

A similar analysis for polar coordinates, in which v_r and v_t are both functions of r and θ (Fig. 36) leads to

$$v_r = \frac{dr}{dt} \quad \text{and} \quad v_t = r \frac{d\theta}{dt} \tag{47}$$

[2] See Appendix V.

and, for the components of acceleration,

$$a_r = v_r \frac{\partial v_r}{\partial r} + v_t \frac{\partial v_r}{r\,\partial \theta} - \frac{v_t^2}{r} \tag{48}$$

$$a_t = v_r \frac{\partial v_t}{\partial r} + v_t \frac{\partial v_t}{r\,\partial \theta} + \frac{v_r v_t}{r} \tag{49}$$

Illustrative Problem

For the circular streamline along which the velocity is 5 fps, calculate the horizontal, vertical, tangential, and normal components of the velocity and acceleration at the point P (2, 60°).

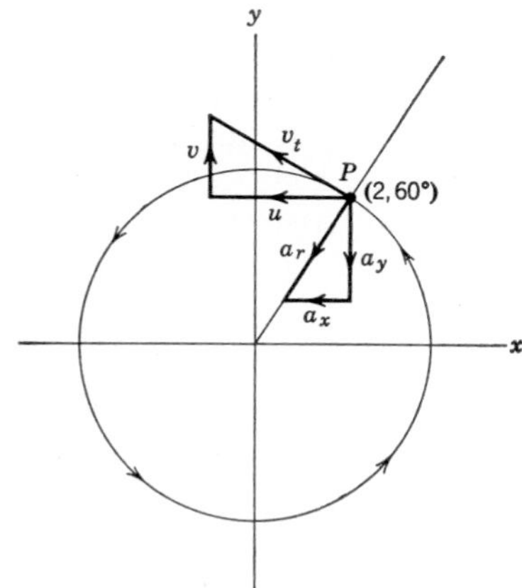

Solution. From similar triangles,

$$u = \frac{-5y}{\sqrt{x^2 + y^2}} \quad \text{and} \quad v = \frac{5x}{\sqrt{x^2 + y^2}}$$

At P, $x = 1$, $y = \sqrt{3}$, so

$$u = -4.33 \text{ fps}, \quad v = 2.50 \text{ fps}$$

Using equations 46,

$$a_x = -\frac{5y}{\sqrt{x^2 + y^2}} \frac{\partial}{\partial x}\left(\frac{-5y}{\sqrt{x^2 + y^2}}\right) + \frac{5x}{\sqrt{x^2 + y^2}} \frac{\partial}{\partial y}\left(\frac{-5y}{\sqrt{x^2 + y^2}}\right) = -\frac{50x}{8} \tag{46}$$

$$a_y = -\frac{5y}{\sqrt{x^2 + y^2}} \frac{\partial}{\partial x}\left(\frac{5x}{\sqrt{x^2 + y^2}}\right) + \frac{5x}{\sqrt{x^2 + y^2}} \frac{\partial}{\partial y}\left(\frac{5x}{\sqrt{x^2 + y^2}}\right) = -\frac{50y}{8} \tag{46}$$

Substituting $x = 1$, $y = \sqrt{3}$,

$$a_x = -6.25 \text{ ft/sec}^2, \quad a_y = -10.83 \text{ ft/sec}^2$$

By inspection,

$$v_t = 5 \text{ fps}, \quad v_r = 0 \text{ fps}$$

Using equations 48 and 49,

$$a_r = 0 \frac{\partial}{\partial r}(0) + 5 \frac{\partial}{r\,\partial\theta}(0) - \frac{5 \times 5}{2} = -12.5 \text{ ft/sec}^2 \tag{48}$$

$$a_t = 0 \frac{\partial}{\partial r}(5) + 5 \frac{\partial}{r\,\partial\theta}(5) + \frac{0 \times 5}{2} = 0 \text{ ft/sec}^2 \tag{49}$$

Note that a_x and a_y might have been obtained more easily (in this problem) by calculating them as the horizontal and vertical components of a_r.

18. Equation of Continuity—One-Dimensional Flow. The application of the principle of conservation of mass ("matter can neither be created nor destroyed") to a steady flow in a streamtube results in the *equation of continuity*, which expresses the continuity of flow from section to section of the streamtube.

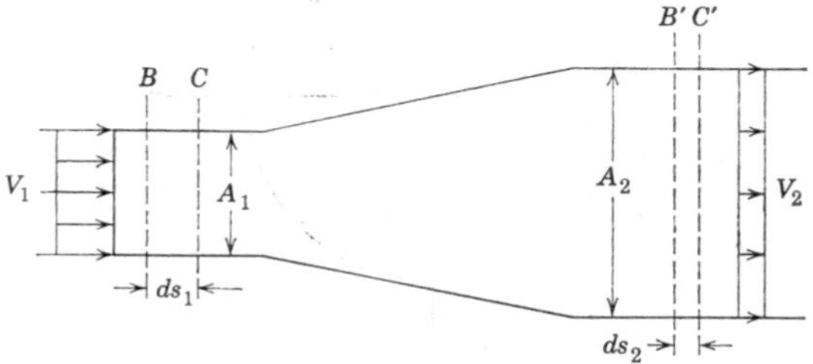

Fig. 37

Consider the element of streamtube in Fig. 37 through which passes a steady flow of compressible fluid. At section 1 the cross-sectional area and mean fluid density are A_1 and ρ_1, respectively, and at section 2, A_2 and ρ_2. If the fluid mass occupying position BB' moves to position CC' in time dt, conservation of mass considerations yield

$$\rho_1 A_1 \, ds_1 = \rho_2 A_2 \, ds_2$$

Dividing by dt,

$$\rho_1 A_1 \frac{ds_1}{dt} = \rho_2 A_2 \frac{ds_2}{dt}$$

However, ds_1/dt and ds_2/dt are recognized as the mean velocities past sections 1 and 2, respectively; therefore

$$\rho_1 A_1 V_1 = \rho_2 A_2 V_2 \tag{50}$$

which is the equation of continuity. In words, it expresses the fact that in steady flow the *mass flowrate*, $A\rho V$, passing all sections of a streamtube is constant. This equation may also be written

$$A\rho V = \text{Constant}, \quad d(A\rho V) = 0, \quad \text{or} \quad \frac{dA}{A} + \frac{d\rho}{\rho} + \frac{dV}{V} = 0 \qquad (51)$$

Multiplication of equation 50 by g gives

$$G = A_1\gamma_1 V_1 = A_2\gamma_2 V_2 \qquad (52)$$

in which the product, G, will be found to have dimensions of pounds (weight) per second, is termed the *weight flowrate*, and is a convenient and concise means of expressing the flowrate of a gas in which the specific weight γ may vary along the streamtube with changes of pressure and temperature.

For liquids and for gas flows where variation of specific weight is negligible the equation of continuity reduces to

$$Q = A_1 V_1 = A_2 V_2 \qquad (53)$$

Thus for fluids of constant density it is seen that along a streamtube the product of velocity and cross-sectional area will be constant. This product, Q, is designated as (volume) *flowrate* and will have dimensions of cubic feet per second.[3] This quantity may, of course, be computed in compressible flow problems, but this is not generally useful because of its variation from section to section along the streamtube.

For two-dimensional flows the flowrate is usually quoted *per unit distance normal to the plane of the flow.* Taking b to be the distance between any two parallel flow planes and h the distance between streamlines, equation 52 becomes

$$G/b = \gamma_1 h_1 V_1 = \gamma_2 h_2 V_2 \qquad (54)$$

in which G/b is known as the two-dimensional (weight) flowrate. The counterpart of this for the incompressible fluid (from equation 53) is

$$Q/b = q = h_1 V_1 = h_2 V_2 \qquad (55)$$

[3] British practice abbreviates cubic feet per second with the term "cusec." In American practice cfs and second-feet are widely used (the latter is unfortunate as it is dimensionally meaningless); other common means of expressing volume flowrates are cubic feet per minute (cfm), gallons per minute (gpm), and millions of gallons per day (mgd).

in which Q/b (hereinafter termed q) is the two-dimensional (volume) flowrate, the dimensions of which are evidently cfs/ft (or ft^2/sec).

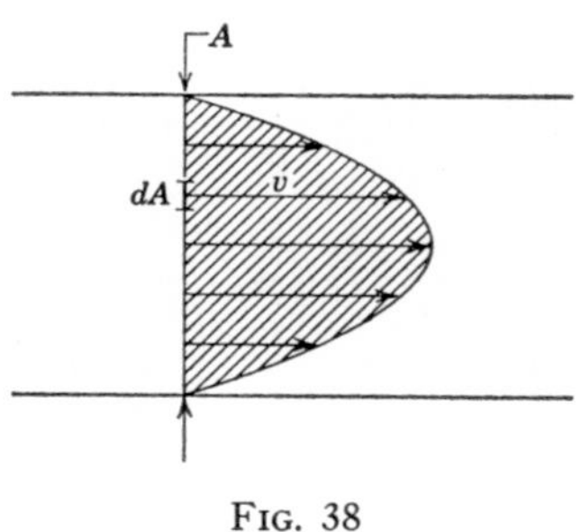

FIG. 38

Frequently in fluid flows the velocity distribution through a flow cross section may be *nonuniform*, as shown in Fig. 38. From considerations of conservation of mass, it is evident at once that nonuniformity of velocity distribution does not invalidate the continuity principle as presented above. Thus, for steady flow of the incompressible fluid, equation 53 applies as before. Here, however, the velocity V is the *mean velocity* defined by $V = Q/A$ in which the flowrate Q is obtained from the summation of the differential flowrates, dQ, passing through the differential areas, dA. Thus

$$V = \frac{1}{A}\int^{A} v\,dA \tag{56}$$

from which the mean velocity may be obtained by performing the indicated integration. With the velocity profile mathematically defined, formal integration may be employed; when the velocity profile is known but not mathematically defined,[4] graphical methods may be used to evaluate the integral. If the flow of Fig. 38 is two-dimensional, this integration is especially easy, since the integral is seen to be the hatched area—which may be planimetered to yield q directly.

The fact that the product AV remains constant along a streamtube (in a fluid of constant density) allows partial interpretation of streamline pictures. As the cross-sectional area of a streamtube increases, the velocity must decrease; hence the conclusion: streamlines widely spaced indicate regions of low velocity, streamlines closely spaced indicate regions of high velocity.

ILLUSTRATIVE PROBLEMS

Six hundred pounds of water per second flow through this pipe-line reducer. Calculate the flowrate in cubic feet per second and the mean velocities in the 12-in. and 8-in. pipes.

[4] For example, when obtained from experimental measurements; see Art. 102.

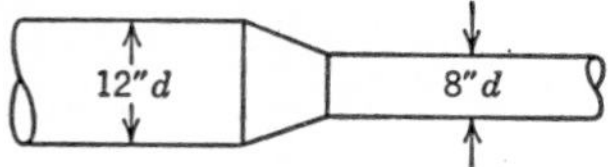

Solution.

$$Q = \frac{600}{62.4} = 9.61 \text{ cfs} \qquad (52 \text{ and } 53)$$

$$V_{12} = \frac{9.61}{\frac{\pi}{4}\left(\frac{12}{12}\right)^2} = 12.23 \text{ fps} \qquad (53)$$

$$V_8 = \frac{9.61}{\frac{\pi}{4}\left(\frac{8}{12}\right)^2} = 27.55 \text{ fps} \quad \text{or} \quad V_8 = 12.23\left(\frac{12}{8}\right)^2 = 27.55 \text{ fps} \qquad (53)$$

Six pounds of air per second flow through the reducer of the preceding problem, the air in the 12-in. pipe having a specific weight of 0.0624 lb/cuft. In flowing through the reducer, the pressure and temperature will fall, causing the air to expand and producing a reduction of density; assuming that the specific weight of the air in the 8-in. pipe is 0.050 lb/cuft, calculate the (volume) flowrates and velocities in the two pipes.

Solution.

$$Q_{12} = \frac{6.0}{0.0624} = 96.1 \text{ cfs}, \quad Q_8 = \frac{6.0}{0.050} = 120 \text{ cfs} \qquad (52 \text{ and } 53)$$

$$V_{12} = \frac{96.1}{\frac{\pi}{4}\left(\frac{12}{12}\right)^2} = 122.3\text{fps}, \quad V_8 = \frac{120}{\frac{\pi}{4}\left(\frac{8}{12}\right)^2} = 344 \text{ fps} \qquad (53)$$

Compare this problem with the preceding one, noting similarities and differences.

Taking Fig. 38 to represent an axisymmetric parabolic velocity distribution in a cylindrical pipe of radius R, calculate the mean velocity in terms of the maximum velocity, v_c.

Solution. Taking r as radial distance to any local velocity, v, and element of area dA, $dA = 2\pi r\, dr$, the equation of the parabola is $v = v_c(1 - r^2/R^2)$.

$$V = \frac{1}{\pi R^2}\int_0^R v_c\left(1 - \frac{r^2}{R^2}\right)2\pi r\, dr \qquad (56)$$

Performing the integration,

$$V = v_c/2$$

19. Equation of Continuity—Two-Dimensional Flow. For two-dimensional flow the equation of continuity may be derived by considering the flowrate into and out of a box of differential size drawn

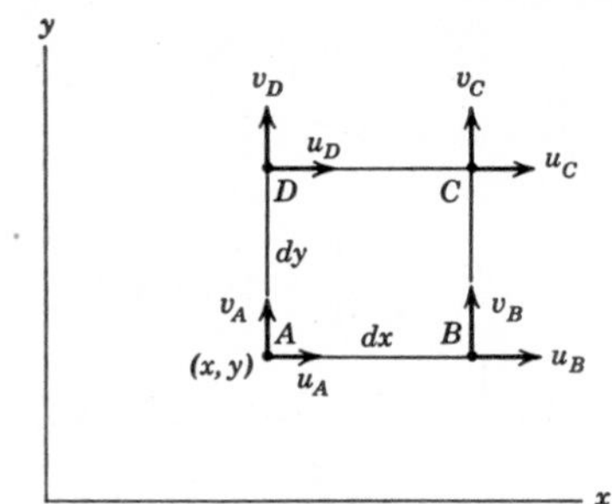

	x-component of velocity	y-component of velocity	Density	Mean velocity across	Mean density along	Mean velocity along
A	$u_A = u$	$v_A = v$	$\rho_A = \rho$			
AB				$\frac{v_A+v_B}{2}$	$\frac{\rho_A+\rho_B}{2}$	$\frac{u_A+u_B}{2}$
B	$u_B = u+\frac{\partial u}{\partial x}dx$	$v_B = v+\frac{\partial v}{\partial x}dx$	$\rho_B = \rho+\frac{\partial \rho}{\partial x}dx$			
BC				$\frac{u_B+u_C}{2}$	$\frac{\rho_B+\rho_C}{2}$	$\frac{v_B+v_C}{2}$
C	$u_C = u+\frac{\partial u}{\partial x}dx+\frac{\partial u}{\partial y}dy$	$v_C = v+\frac{\partial v}{\partial x}dx+\frac{\partial v}{\partial y}dy$	$\rho_C = \rho+\frac{\partial \rho}{\partial x}dx+\frac{\partial \rho}{\partial y}dy$			
CD				$\frac{v_C+v_D}{2}$	$\frac{\rho_C+\rho_D}{2}$	$\frac{u_C+u_D}{2}$
D	$u_D = u+\frac{\partial u}{\partial y}dy$	$v_D = v+\frac{\partial v}{\partial y}dy$	$\rho_D = \rho+\frac{\partial \rho}{\partial y}dy$			
DA				$\frac{u_A+u_D}{2}$	$\frac{\rho_A+\rho_D}{2}$	$\frac{v_A+v_D}{2}$

FIG. 39

(Fig. 39) anywhere in the flowfield. Here the horizontal and vertical components of velocity and the density are set down in the accompanying table, these quantities at point A being taken as u, v, and ρ, respectively. At the other corners of the box they are computed from

$$\begin{bmatrix}\text{Original}\\ \text{quantity}\\ u, v, \text{or } \rho\end{bmatrix} + \begin{bmatrix}\text{Rate of change}\\ \text{of quantity}\\ \text{along } x\end{bmatrix} dx + \begin{bmatrix}\text{Rate of change}\\ \text{of quantity}\\ \text{along } y\end{bmatrix} dy$$

in which the *rate of change* is by definition the appropriate partial derivative. The continuity principle states that for steady flow the mass flowrate into the differential box must be exactly equal to the mass flowrate out. Assuming that the flow picture is of unit thickness normal to the paper, this relation will be expressed by

$$\begin{bmatrix}\text{Mean velocity}\\ \text{across } AB\end{bmatrix} \times \begin{bmatrix}\text{Mean density}\\ \text{along } AB\end{bmatrix} dx + \begin{bmatrix}\text{Mean velocity}\\ \text{across } AD\end{bmatrix} \times \begin{bmatrix}\text{Mean density}\\ \text{along } AD\end{bmatrix} dy = \begin{bmatrix}\text{Mean velocity}\\ \text{across } BC\end{bmatrix} \times \begin{bmatrix}\text{Mean density}\\ \text{along } BC\end{bmatrix} dy + \begin{bmatrix}\text{Mean velocity}\\ \text{across } CD\end{bmatrix} \times \begin{bmatrix}\text{Mean density}\\ \text{along } CD\end{bmatrix} dx$$

Substituting values from the table of Fig. 39, expanding the products, and retaining only the terms of highest order of magnitude, the continuity equation for two-dimensional steady compressible flow is obtained; it is

$$\frac{\partial}{\partial x}(\rho u) + \frac{\partial}{\partial y}(\rho v) = 0 \qquad (57)$$

For steady incompressible flow, ρ is constant and the equation reduces to

$$\frac{\partial u}{\partial x} + \frac{\partial v}{\partial y} = 0 \qquad (58)$$

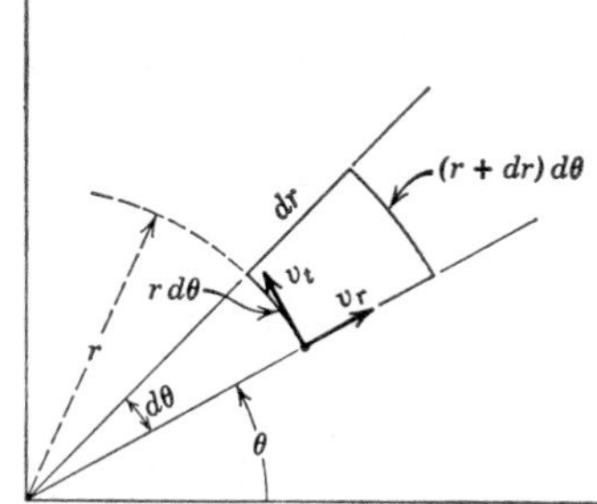

FIG. 40

Application of the same principles to the polar element of Fig. 40 yields the two-dimensional continuity equation in polar coordinates. For the compressible fluid this is

$$\frac{1}{r}(\rho v_r) + \frac{\partial}{\partial r}(\rho v_r) + \frac{\partial}{r\,\partial\theta}(\rho v_t) = 0 \qquad (59)$$

which, for the incompressible fluid, reduces to

$$\frac{v_r}{r} + \frac{\partial v_r}{\partial r} + \frac{\partial v_t}{r\,\partial\theta} = 0 \qquad (60)$$

ILLUSTRATIVE PROBLEMS

A two-dimensional incompressible flowfield is described by the equations: $u = 4x$, $v = -4y$. Sketch this flow and show that it satisfies the equation of continuity.

Solution. From the equations it is seen that the x-component of velocity increases with x and the y-component increases with negative y. Accordingly the flow may be sketched as shown in all four quadrants. It may be noted that the flow in any one quadrant (with the axes taken as solid boundaries)

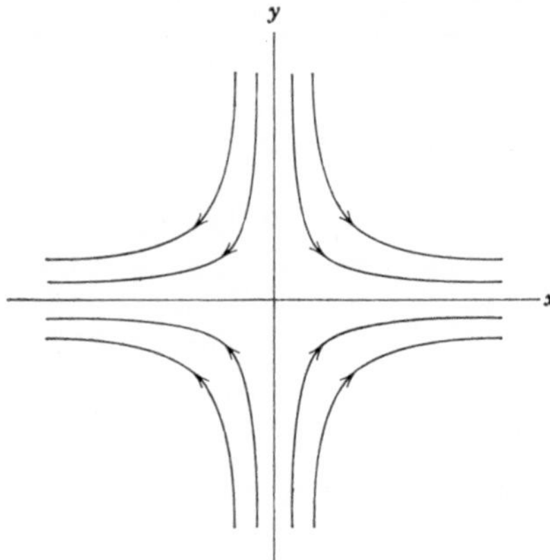

describes a possible flow in a square corner, providing that it satisfies the continuity equation. Substituting $4x$ and $-4y$ for u and v in equation 58,

$$\frac{\partial}{\partial x}(4x) + \frac{\partial}{\partial y}(-4y) = 4 - 4 = 0$$

it is noted that continuity is satisfied and the flow is thus physically possible.

A two-dimensional incompressible flowfield is described by the equations $v_t = \omega r$ and $v_r = 0$, in which ω is a constant. Sketch this flow and show that it satisfies the continuity equation.

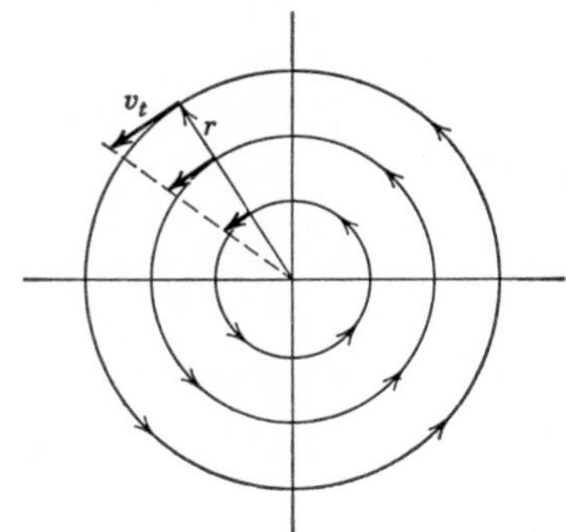

Solution. Evidently the streamlines of this flow must be concentric circles since the radial component of velocity is everywhere zero. This flow is known as a *forced vortex;* some of its features have been examined in Art. 14 and others are discussed in Art. 109. Substituting $v_r = 0$ and $v_t = \omega r$ in equation 60,

$$\frac{0}{r} + \frac{\partial}{\partial r}(0) + \frac{\partial}{r\,\partial\theta}(\omega r) = 0 + 0 + 0 = 0 \tag{60}$$

it is noted that the equation of continuity is satisfied and that this flow is physically possible.

20. Circulation and Vorticity. Circulation in fluid mechanics is designated by Γ (gamma) and defined by a *line integral* around a closed curve fixed in the flow. In the two-dimensional flowfield of Fig. 41 it is observed that each streamline will intersect the closed curve as some angle α and thus the component of velocity along the closed curve at the point of intersection is $V \cos \alpha$. An element of circulation $d\Gamma$ is defined as the product of the velocity component and the element ds of the closed curve. Thus

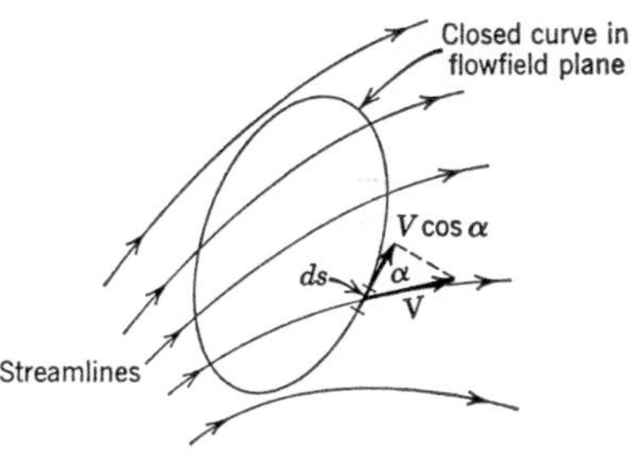

FIG. 41

$$d\Gamma = (V \cos \alpha)\, ds$$

The sum of the elements $d\Gamma$ along the closed curve defines the circulation:

$$\Gamma = \oint d\Gamma = \oint (V \cos \alpha)\, ds \tag{61}$$

in which the circle on the integral sign indicates that the integral is to be taken once around the closed curve and that the direction convention is counterclockwise.

Although the calculation of circulation around an arbitrary curve in a flowfield is generally a tedious step-by-step integration, the principle may be applied easily and fruitfully to specific closed curves such as circles and squares. Calculation of the circulation around the basic square element of Fig. 39 will yield a concept having great general significance. Proceed from A counterclockwise around the boundary of the element, setting down the products of velocity component and distance in order; since the element is of differential size, the resulting circulation is also a differential quantity, $d\Gamma$.

$$d\Gamma = \begin{bmatrix}\text{Mean velocity}\\ \text{along } AB\end{bmatrix} dx + \begin{bmatrix}\text{Mean velocity}\\ \text{along } BC\end{bmatrix} dy - \begin{bmatrix}\text{Mean velocity}\\ \text{along } CD\end{bmatrix} dx - \begin{bmatrix}\text{Mean velocity}\\ \text{along } DA\end{bmatrix} dy$$

Substituting values from the table of Fig. 39, expanding the products, and retaining only the terms of highest order of magnitude, there results

$$d\Gamma = \left(\frac{\partial v}{\partial x} - \frac{\partial u}{\partial y}\right) dx\, dy$$

in which $dx\,dy$ is the area inside the closed boundary. The *vorticity*, ξ (xi), is defined as the differential circulation per unit of area enclosed, which becomes

$$\xi = \frac{d\Gamma}{dx\,dy} = \frac{\partial v}{\partial x} - \frac{\partial u}{\partial y} \tag{62}$$

For polar coordinates, by the same procedure,

$$\xi = \frac{\partial v_t}{\partial r} + \frac{v_t}{r} - \frac{\partial v_r}{r\,\partial \theta} \tag{63}$$

From the definition of circulation, Γ, and the methods used to calculate vorticity, ξ, the reader will sense[5] that the latter quantity is some measure of the rotational aspects of the fluid elements as they move through the flowfield. If this is accepted tentatively, some new terms and definitions may be introduced. If a flow possesses vorticity (i.e., if $\xi \neq 0$), it is said to be a *rotational* flow; if a flow possesses no vorticity ($\xi = 0$), it is termed *irrotational*.[6] Such definitions are adequate for present purposes but may be misleading in that they imply that whole flowfields are either rotational or irrotational. Actually flowfields may possess zones of both irrotational and rotational flows, the latter frequently being concentrated in *singular points;* the free vortex flow (Art. 109) is a classic example of this.

Illustrative Problem

Calculate the vorticities of the flows described in the illustrative problems of Art. 19.

Solutions. For the first problem,

$$\xi = \frac{\partial}{\partial x}(-4y) - \frac{\partial}{\partial y}(4x) = 0 - 0 = 0 \tag{62}$$

Since the vorticity is zero, the flowfield of the first problem is an irrotational (or potential) flow.

For the second problem,

$$\xi = \frac{\partial}{\partial r}(\omega r) + \frac{\omega r}{r} - \frac{\partial}{r\,\partial \theta}(0) = \omega + \omega - 0 = 2\omega \tag{63}$$

Evidently this is a rotational flow possessing a constant vorticity (over the whole flowfield) of 2ω; the *forced vortex* is thus a *rotational* flowfield.

[5] Proof is offered in Art. 109.

[6] In Art. 112 it will be shown that *irrotational* flows are also characterized by a *velocity potential* and are therefore known as *potential* flows.

PROBLEMS

237. Sketch the following flowfields and derive general expressions for their components of acceleration: (*a*) $u = 4$, $v = 3$; (*b*) $u = 4$, $v = 3x$; (*c*) $u = 4y$, $v = 0$; (*d*) $u = 4y$, $v = 3$; (*e*) $u = 4y$, $v = 3x$; (*f*) $u = 4y$, $v = 4x$; (*g*) $u = 4y$, $v = -4x$; (*h*) $u = 4$, $v = 0$; (*i*) $u = 4$, $v = -4x$; (*j*) $u = 4x$, $v = 0$; (*k*) $u = 4xy$, $v = 0$; (*l*) $v_r = c/r$, $v_t = 0$; (*m*) $v_r = 0$, $v_t = c/r$.

238. The mean velocity of water in a 4-in. pipe line is 7 fps. Calculate the rate of flow in cubic feet per second, gallons per minute, pounds per second, and slugs per second.

239. One hundred pounds of water per minute flow through a 6-in. pipe line. Calculate the mean velocity.

240. One hundred gallons per minute of glycerin flow in a 3-in. pipe line. Calculate the mean velocity.

241. A 12-in. by 18-in. rectangular air duct carries a flow of 15 cfs. Calculate the mean velocity in the duct. If the duct tapers to 6-in. by 18-in. size, what is the mean velocity in this section?

242. Water flows in a pipe line composed of 3-in. and 6-in. pipe. Calculate the mean velocity in the 3-in. pipe when that in the 6-in. pipe is 8 fps. What is its ratio to the mean velocity in the 6-in. pipe?

243. A smooth nozzle with a tip diameter of 2 in. terminates a 6-in. water line. Calculate the mean velocity of efflux from the nozzle when the velocity in the line is 10 fps.

244. At a point in a two-dimensional fluid flow, two streamlines are parallel and 3 in. apart. At another point these streamlines are parallel but only 1 in. apart. If the velocity at the first point is 10 fps, calculate the velocity at the second.

245. A 12-in. pipe line leaves a large tank through a square-edged hole in its vertical wall. The mean velocity in the pipe is 15 fps. Assuming that the fluid in the tank approaches the center of the pipe entrance radially, what is the velocity of the fluid 2, 4, and 6 ft from the pipe entrance?

246. Five hundred cubic feet per second of water flow in a rectangular open channel 20 ft wide and 8 ft deep. After passing through a transition structure into a trapezoidal canal of 5-ft base width and sides sloping at 30°, the velocity is 6 fps. Calculate the depth of the water in the canal.

247. Calculate the mean velocities at C and D in problem 321, assuming them to be radial.

248. Fluid passes through this set of thin closely spaced blades. What flowrate q is required for the velocity V to be 10 fps?

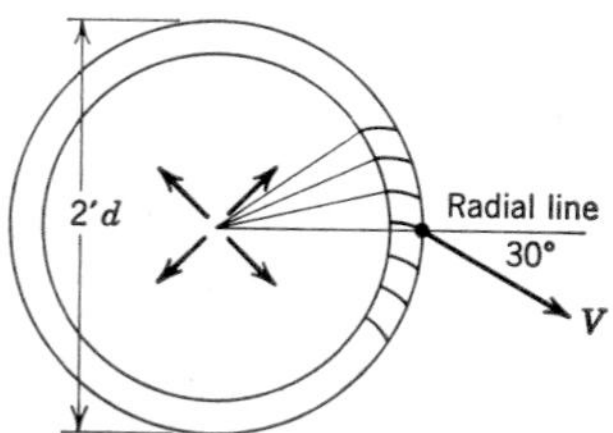

PROBLEM 248

249. A pipe line 12-in. in diameter divides at a Y into two branches 8-in. and 6-in. in diameter. If the flowrate in the main line is 10 cfs and the mean velocity in the 8-in. pipe is 8 fps, what is the flowrate in the 6-in. pipe?

250. A *manifold pipe* of 3-in. diameter has four openings in its walls spaced equally along the pipe and is closed at the downstream end. If the discharge from each opening is 0.50 cfs, what are the mean velocities in the pipe between the openings?

251. Calculate the mean velocities for these two-dimensional velocity profiles if $v_c = 10$ fps.

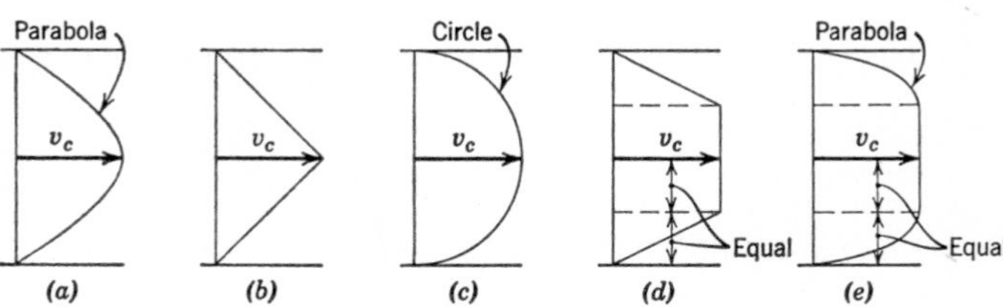

PROBLEM 251

252. Calculate the mean velocities in the preceding problem, assuming the velocity profiles to be axisymmetric in a cylindrical passage.

253. If the velocity profile in a passage of width $2R$ is given by the equation $v/v_c = (y/R)^{\frac{1}{n}}$, derive an expression for V/v_c in terms of n: (*a*) for a two-dimensional passage, (*b*) for a cylindrical passage.

254. Derive an expression for the flowrate q between two streamlines of radii r_1 and r_2 for the flowfields (*a*) $v_r = 0$, $v_t = c/r$ and (*b*) $v_r = 0$, $v_t = \omega r$.

255. For the flowfield $v_r = c/r$, $v_t = 0$ derive an expression for the flowrate q between any two streamlines.

256. Derive the equation of continuity in polar coordinates for the incompressible fluid.

257. Investigate the flowfields of problem 237 to see if they are physically possible (i.e., satisfy the equation of continuity).

258. Derive the equation for vorticity in polar coordinates.

259. For the flowfields of problem 237 derive expressions for vorticity and state whether the flowfield is rotational or irrotational.

chapter 4

Flow of an Incompressible Ideal Fluid

An insight into the basic laws of fluid flow can best be obtained from a study of the flow of a hypothetical *ideal fluid*. An ideal fluid is a fluid assumed to be inviscid, or devoid of viscosity.[1] In such a fluid, therefore, there can be no frictional effects between moving fluid layers or between these layers and boundary walls, and thus no cause for eddy formation or energy dissipation due to friction. The assumption of an ideal fluid allows a fluid to be treated as an aggregation of small particles which will support pressure forces normal to their surfaces but will slide over one another without resistance. Thus the motion of these ideal fluid particles is analogous to the motion of a solid body on a resistanceless plane; from this it may be concluded that unbalanced forces existing on particles of an ideal fluid will result in the acceleration of these particles according to Newton's second law.

By assuming frictionless motion, equations are considerably simplified and more easily assimilated by the beginner in the field. In many cases these simplified equations allow solution of engineering problems to an accuracy entirely adequate for practical use. The beginner should not jump to the conclusion that the assumption of frictionless flow will lead him to a useless abstraction which is always far from

[1] Compare with the definition of a perfect gas, Art. 3.

reality. In real situations, where friction is small, the frictionless assumption will give good results; where friction is large, it obviously will not. The identification of these situations is part of the art of fluid mechanics.

The further assumption of an incompressible (i.e., constant-density) fluid restricts the present chapter to the flow of liquids and of gases which undergo small changes of pressure and temperature. The flow of gases with large density changes will be discussed in Chapter 5, but there are numerous practical engineering problems which involve fluids whose densities may safely be considered constant; thus this assumption proves to be not only a practical one but also a useful simplification in the introduction to fluid flow, since it usually permits thermodynamic effects to be disregarded.

ONE-DIMENSIONAL FLOW

21. Euler's Equation. In 1750 Leonhard Euler first applied Newton's second law to the motion of fluid particles and thus laid the groundwork for an analytical approach to fluid dynamics.

Consider a streamline (or streamtube of differential cross section) as shown in Fig. 42. The forces tending to accelerate the cylindrical fluid mass are: pressure forces on the ends of the element, $p\,dA - (p + dp)\,dA = -dp\,dA$, and the component of weight in the direction of motion, $-\rho g\,ds\,dA\,(dz/ds) = -\rho g\,dA\,dz$. The differential mass being accelerated by the action of these differential forces is $dM =$

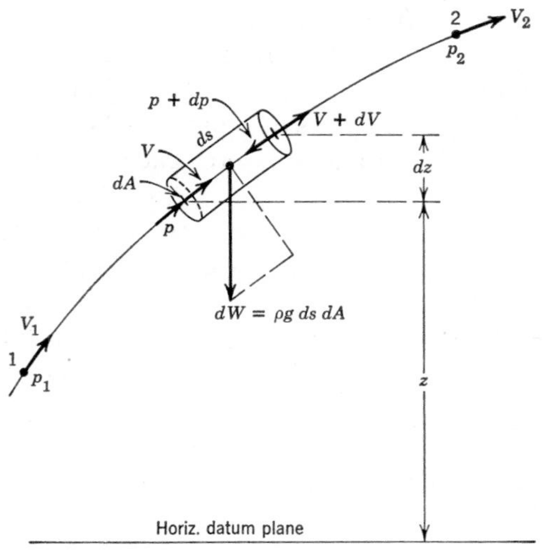

FIG. 42

$\rho\, ds\, dA$. Applying Newton's second law $dF = (dM)\, a$, using the one-dimensional expression for acceleration (equation 43),

$$-dp\, dA - \rho g\, dA\, dz = (\rho\, ds\, dA)\, V \frac{dV}{ds}$$

Dividing by $\rho\, dA$, the one-dimensional Euler equation results; this is

$$\frac{dp}{\rho} + V\, dV + g\, dz = 0$$

For incompressible flow this equation is usually divided by g and written

$$\frac{dp}{\gamma} + d\left(\frac{V^2}{2g}\right) + dz = 0$$

22. Bernoulli's Equation. For incompressible flow, the one-dimensional Euler equation may be easily integrated (since γ and g are both constant):

$$\int \frac{dp}{\gamma} + \int d\left(\frac{V^2}{2g}\right) + \int dz = \text{Constant}$$

to obtain Bernoulli's equation,

$$\frac{p}{\gamma} + \frac{V^2}{2g} + z = H \text{ (termed total head)} \qquad (64)$$

which will apply to all points on the streamline and thus provide a useful relationship between pressure p, velocity V, and height above datum, z.

Examination of the *Bernoulli terms* of equation 64 reveals that p/γ and z are, respectively, the pressure (either gage or absolute) and potential heads encountered in Art. 8 and hence may be visualized as vertical distances. Pitot's experiments (1732) showed that the sum of velocity head $V^2/2g$ and pressure head p/γ could be measured by placing a tiny open tube (now known as a *pitot tube*) in the flow with its open end upstream. Thus the Bernoulli equation may be visualized for liquids as in Fig. 43, the sum of the terms (total head) being the constant distance between the horizontal datum plane and the *total head line* or *energy line* [2] (E.L.). The *piezometric head line* or *hydraulic grade line* (H.G.L.) drawn through the tops of the piezom-

[2] The term *energy line*, although widely (and conveniently) used in engineering problems of liquid flow, is somewhat arbitrary and thus subject to some restrictions. Its position above datum depends upon pressure energy (see Art. 23), kinetic energy, and potential energy only—thus considerations of heat and internal energy are excluded; these restrictions are seldom critical in problems of incompressible flow.

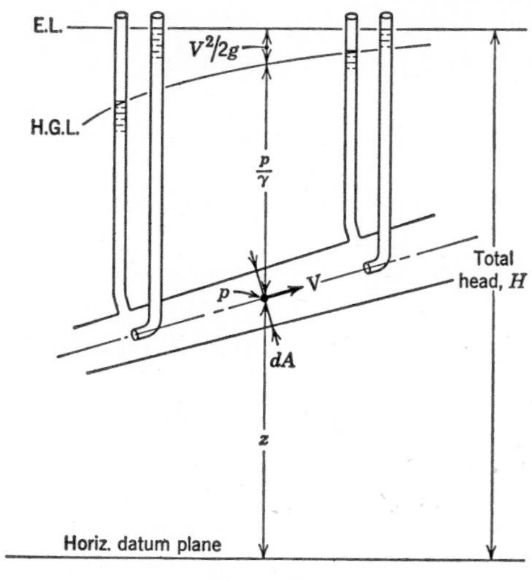

Fig. 43

eter columns gives a picture of the pressure variation in the flow; evidently (1) its distance from the streamtube is a direct measure of the static pressure in the flow, and (2) its distance below the energy line is proportional to the square of the velocity. Complete familiarity with these lines is essential because of their wide use in engineering practice and their great utility in problem solutions.

23. Bernoulli's Equation as an Energy Equation. Bernoulli's equation may also be derived from energy considerations to shed more light on its significance and utility. The kinetic energy per pound of fluid may be written directly from considerations of mechanics: the general expression for kinetic energy of translation is $MV^2/2$; for unit weight of fluid, however, $M = 1/g$, so the kinetic energy becomes $V^2/2g$ ft-lb/lb. Similarly the potential energy of a weight W at a vertical distance z above datum is (relative to the datum) Wz ft-lb; for a unit weight the potential energy is simply z ft-lb/lb. Restricting the derivation to steady flow and considering the continuity principle, each unit weight of fluid entering the boundary 1221 of Fig. 44 through section 1 must displace a unit weight of fluid which will move out of the boundary across section 2. Therefore

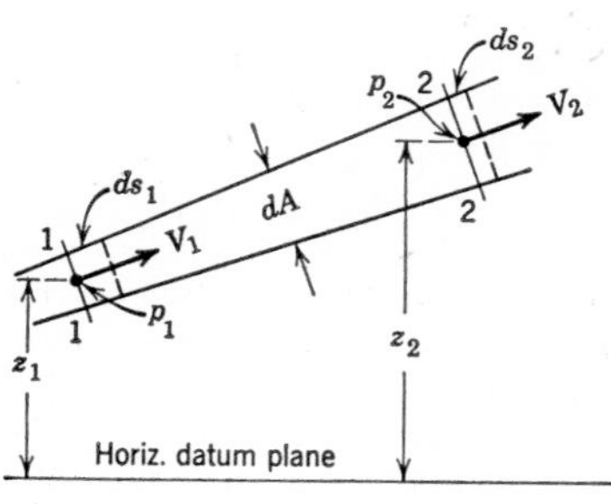

Fig. 44

$$\begin{bmatrix}\text{Energy carried in by} \\ \text{each unit weight of fluid}\end{bmatrix} = \frac{V_1^2}{2g} + z_1$$

$$\begin{bmatrix}\text{Energy carried out by} \\ \text{each unit weight of fluid}\end{bmatrix} = \frac{V_2^2}{2g} + z_2$$

Consider now the work done on the fluid within the boundary by a weight of fluid entering the boundary, and that done by the fluid within the boundary on the same weight of fluid leaving. Suppose that in a certain time a weight of fluid $\gamma\, ds_1\, dA_1$ moves into the boundary across section 1 and that in this time a weight of fluid $\gamma\, ds_2\, dA_2$ leaves the boundary across section 2. The work associated with these displacements can be computed from the products of the forces $p_1\, dA_1$ and $p_2\, dA_2$ and displacements ds_1 and ds_2 which give $p_1\, dA_1\, ds_1$ and $p_2\, dA_2\, ds_2$, respectively. For convenient inclusion in a work-energy equation these work terms must be written per unit weight of fluid flowing; this may be done by dividing by the total weight of fluid ($\gamma\, ds_1\, dA_1$ and $\gamma\, ds_2\, dA_2$) doing the work; the results are p_1/γ and p_2/γ, respectively. The energy equation may now be written

$$\begin{bmatrix}\text{Energy entering} \\ \text{boundary 1221}\end{bmatrix} + \begin{bmatrix}\text{Net work done on fluid} \\ \text{within boundary 1221}\end{bmatrix} = \begin{bmatrix}\text{Energy leaving} \\ \text{boundary 1221}\end{bmatrix}$$

or

$$\left[\frac{V_1^2}{2g} + z_1\right] + \left[\frac{p_1}{\gamma} - \frac{p_2}{\gamma}\right] = \left[\frac{V_2^2}{2g} + z_2\right]$$

which when rearranged is recognized as Bernoulli's equation,

$$\frac{p_1}{\gamma} + \frac{V_1^2}{2g} + z_1 = \frac{p_2}{\gamma} + \frac{V_2^2}{2g} + z_2 \qquad (64)$$

From the form of Bernoulli's equation it is natural that the flow work terms p/γ become associated with the energy terms, $V^2/2g$ and z; because of this, p/γ is generally (and conveniently) treated as the *pressure energy* of the flow. Although the foregoing development has shown that p/γ is not strictly a pressure energy, its use leads to no practical difficulty in fluid flow problems.[3]

24. The One-Dimensional Assumption for Streamtubes of Finite Cross Section. The foregoing development of Bernoulli's equation has

[3] In nonflow systems the flow work term does not appear, so p/γ cannot be interpreted as *pressure energy* in such problems.

been carried out for a single streamline or infinitesimal streamtube across which there is no variation of p, V, and z. However, the engineer may apply this equation easily and fruitfully to large streamtubes such as pipes and canals once its limitations are understood.

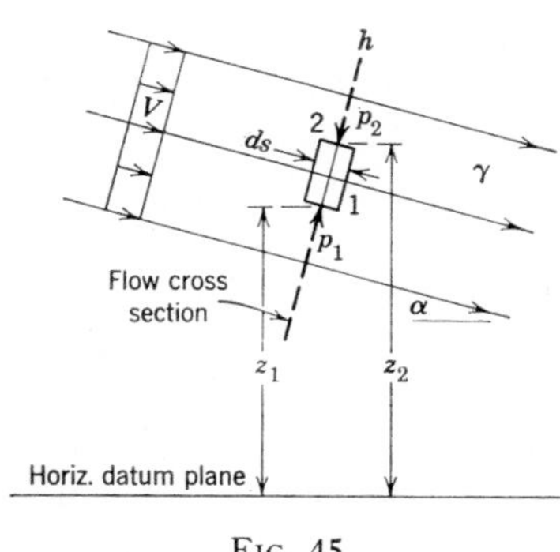

FIG. 45

Consider a cross section of a large flow (open or closed but not a free jet) through which *all streamlines are precisely straight and parallel* Fig. 45). The forces, normal to the streamlines, on the element of fluid are $(p_1 - p_2)\,ds$ and the component of the weight of the element, $\gamma h\ ds\ \cos\alpha$, in which $\cos\alpha = (z_2 - z_1)/h$. It is apparent that if (and only if) the streamlines are straight and parallel the acceleration toward the boundary is zero. This means that the forces defined above are in equilibrium:

$$(p_1 - p_2)\,ds = \gamma(z_2 - z_1)\,ds$$

yielding

$$\frac{p_1}{\gamma} + z_1 = \frac{p_2}{\gamma} + z_2 \tag{21}$$

and demonstrating that the quantity $(z + p/\gamma)$ is constant over the flow cross section when the streamlines are straight and parallel. This means that the Bernoulli equation of the single streamline may be extended to apply to two- and three-dimensional flows since at a given flow cross section $z + p/\gamma$ is the same for all streamlines as it is for the central streamline; stated in another way, it means that a single hydraulic grade line applies to all the streamlines of a flow, providing these streamlines are straight and parallel. In practice the engineer seldom if ever encounters flows containing precisely straight and parallel streamlines; however, because in pipes, ducts, and prismatic channels such lines are *essentially* straight and parallel, the approximation may be used for many practical calculations.

In ideal flows, because of the absence of friction, the distribution of velocity over a cross section of a flow containing straight and parallel streamlines is necessarily uniform; that is, all fluid particles pass a given cross section at the same mean velocity V. Accordingly no adjustment of the $V^2/2g$ term of the Bernoulli equation is to be expected in extending the equation from the infinitesimal to the finite streamtube; later when friction is considered it will be shown that a correction must be applied to $V^2/2g$ before the Bernoulli equation may be used for a streamtube of finite cross section.

Illustrative Problem

Water flows through this section of cylindrical pipe. If the static pressure at point C is 5.0 psi, what are the static pressures at A and B, and where is the hydraulic grade line at this flow cross section?

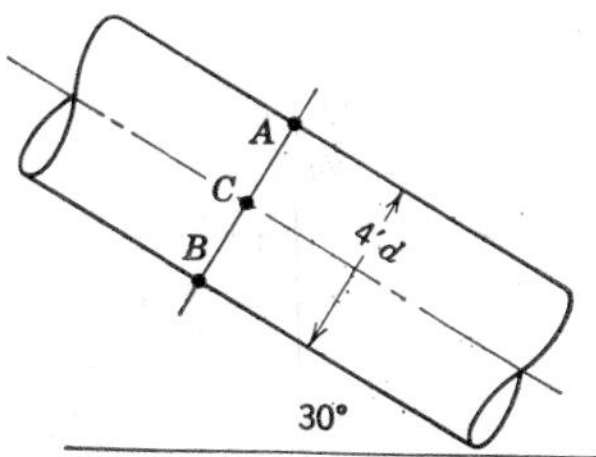

Solution. Using equation 21,

$$p_A = 5.0 - (62.4/144)(0.866)2 = 4.25 \text{ psi}$$

$$p_B = 5.0 + (62.4/144)(0.866)2 = 5.75 \text{ psi}$$

The hydraulic grade line is $(5.0 \times 144)/62.4 = 11.55$ ft (vertically) above C.

25. Applications of Bernoulli's Equation. Before proceeding to some engineering applications of Bernoulli's equation, it should first be noted that this equation gives further aid in the interpretation of streamline pictures, equation 64 indicating that, when velocity increases, the sum $(p/\gamma + z)$ of pressure and potential head must decrease. In many flow problems, the potential head z may vary little, allowing the approximate general statement: where velocity is high, pressure is low. Regions of closely spaced streamlines have been shown (Art. 18) to indicate regions of relatively high velocity, and now from the Bernoulli equation these are seen also to be regions of relatively low pressure.

In 1643, Torricelli showed that the velocity of efflux of an ideal fluid from a small orifice under a static head varies with the square root of the head. Today Torricelli's theorem is written

$$V = \sqrt{2gh}$$

the velocity being (ideally) equal to that attained by a solid body falling from rest through a height h. Torricelli's theorem is now recognized as a special case of the Bernoulli equation involving cer-

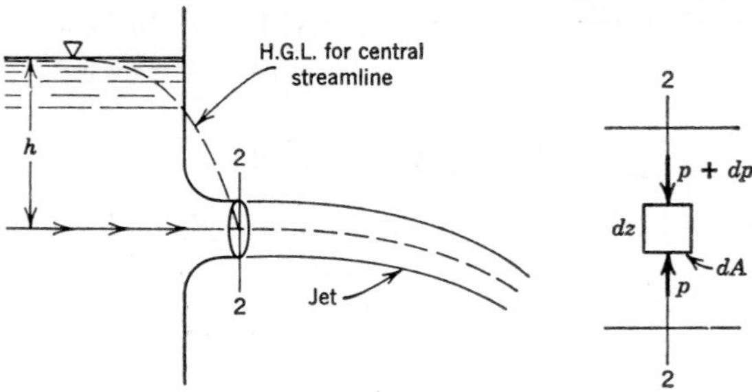

FIG. 46

tain conditions appearing in many engineering problems. Torricelli's equation may be easily derived by applying Bernoulli's equation from the reservoir to the tip of the nozzle in Fig. 46. The reservoir being assumed to be very large (compared to the nozzle), velocities may be neglected at all points in the reservoir except near the nozzle. Taking the datum plane at the center of the nozzle, $z + p/\gamma$ in the reservoir is seen to be equal to h. Writing Bernoulli's equation between the reservoir and the tip of the nozzle,

$$\frac{p_1}{\gamma} + \frac{V_1^2}{2g} + z_1 = h = \frac{p_2}{\gamma} + \frac{V_2^2}{2g} + 0$$

it is seen that Torricelli's equation results if $p_2 = 0$. From the validity of the Bernoulli and Torricelli equations it may be deduced that the pressure throughout the jet in the plane of the nozzle must be zero; however, analytical proof of this from Newton's law may be more convincing.

Consider the streamlines through section 2 of Fig. 46 to be essentially straight and parallel.[4] The pressure surrounding the jet is zero (gage) and the vertical acceleration of an elemental fluid mass $\rho\, dA\, dz$ is equal to g. The force causing the acceleration results from the pressure difference between top and bottom of the element, and the weight of the element. Writing Newton's second law in the vertical direction,

$$-(p + dp)\, dA + p\, dA - \gamma\, dA\, dz = -(\rho\, dA\, dz)g$$

from which it may be concluded that $dp = 0$. Thus there can be no pressure gradient across the jet at section 2, and with the pressures

[4] Actually the streamlines are curved and convergent (recall the trajectory theory of particle dynamics), but such convergence and curvature may be neglected for large h and high jet velocity.

zero at the jet boundaries it follows that the pressure throughout the jet must be zero. Downstream from section 2 it is customary to *assume* that the pressures throughout the free jet are also zero; this is equivalent to assuming that each fluid element follows a free trajectory streamline unaffected by adjacent fluid elements. This is an adequate approximation in many engineering problems, but it should not be taken to be scientifically exact.

Illustrative Problem

Calculate the flowrate through this pipe line and nozzle. Calculate the pressures at points 1, 2, 3, and 4 in the pipe and the elevation of the top of the jet's trajectory.

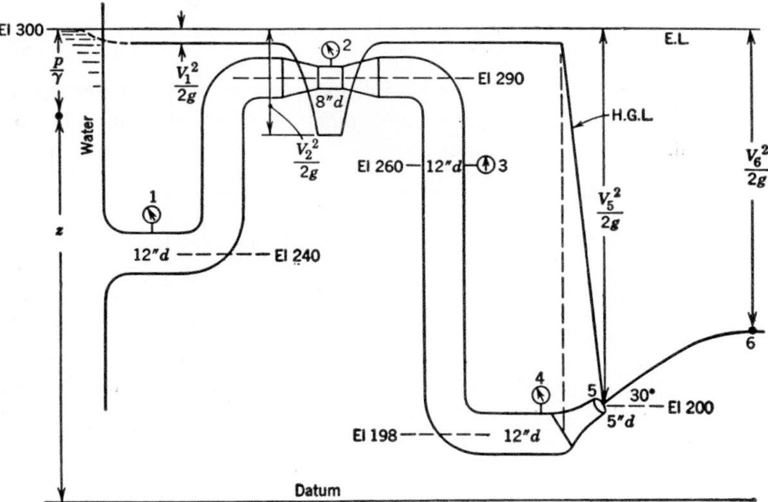

Solution. First sketch the energy line; it is evident that, at all points in the reservoir where the velocity is negligible, $(z + p/\gamma)$ will be the same. Thus the energy line has the same elevation as the water surface. Next sketch the hydraulic grade line; this is coincident with the energy line in the reservoir where velocity is negligible but drops below the energy line over the pipe entrance where velocity is gained. The velocity in the 12-in. pipe is everywhere the same, so the hydraulic grade line must be horizontal until the flow encounters the constriction upstream from section 2. Here, as velocity increases, the hydraulic grade line must fall (possibly to a level below the constriction). Downstream from the constriction the hydraulic grade line must rise to the original level over the 12-in. pipe and continue at this level to a point over the *base of the nozzle* at section 4. Over the nozzle the hydraulic grade line must fall to the nozzle tip and after that follow the jet, because the pressure in the jet is everywhere zero.

Since the vertical distance between the energy line and the hydraulic grade *line* at any point is the velocity head at that point, it is evident that

$$\frac{V_5^2}{2g} = 100 \quad \text{and therefore} \quad V_5 = 80.2 \text{ fps}$$

Thus

$$Q = 80.2 \times \frac{\pi}{4}\left(\frac{5}{12}\right)^2 = 10.95 \text{ cfs} \tag{53}$$

From continuity considerations,

$$V_1 = (\tfrac{5}{12})^2 V_5$$

and thus

$$\frac{V_1^2}{2g} = \left(\frac{5}{12}\right)^4 \frac{V_5^2}{2g} = \left(\frac{5}{12}\right)^4 100 = 3.0 \text{ ft}$$

Similarly

$$\frac{V_2^2}{2g} = \left(\frac{5}{8}\right)^4 100 = 15.3 \text{ ft}$$

Since the pressure heads are the vertical distances between the pipe center line and hydraulic grade line, the pressures in pipe and constriction may be computed as follows:

$$p_1/\gamma = 60 - 3.0 = 57.0 \text{ ft}; \quad p_1 = 24.7 \text{ psi}$$

$$p_2/\gamma = 10 - 15.3 = -5.3 \text{ ft}; \quad p_2 = 4.7 \text{ in. Hg vacuum}$$

$$p_3/\gamma = 40 - 3.0 = 37.0 \text{ ft}; \quad p_3 = 16.0 \text{ psi}$$

$$p_4/\gamma = 102 - 3.0 = 99.0 \text{ ft}; \quad p_4 = 42.8 \text{ psi}$$

The velocity of the jet at the top of its trajectory (where there is no vertical component of velocity) is given by

$$V_6 = 80.2 \cos 30° = 69.5 \text{ fps}$$

and the elevation here is $300 - (69.5)^2/2g = 225.0$ ft.

At this point it should be noted that, with increase of velocity or potential head, the pressure within a flowing fluid can drop no farther than to the absolute zero of pressure, thus placing a physical restriction upon the Bernoulli equation. Such a condition is not possible in gases, however, owing to their capacity for expansion with reduction of pressure, but it frequently assumes great importance in the flow of liquids. In liquids the absolute pressure can drop only to the vapor pressure of the liquid, whereupon vaporization takes place and *cavitation* [5] may occur with accompanying vibration, destructive action, and other deleterious effects.

[5] For a description of cavitation phenomena, see Appendix VII.

Illustrative Problem

The barometric pressure is 14.0 psia. What diameter of constriction can be expected to produce incipient cavitation at the throat of the constriction?

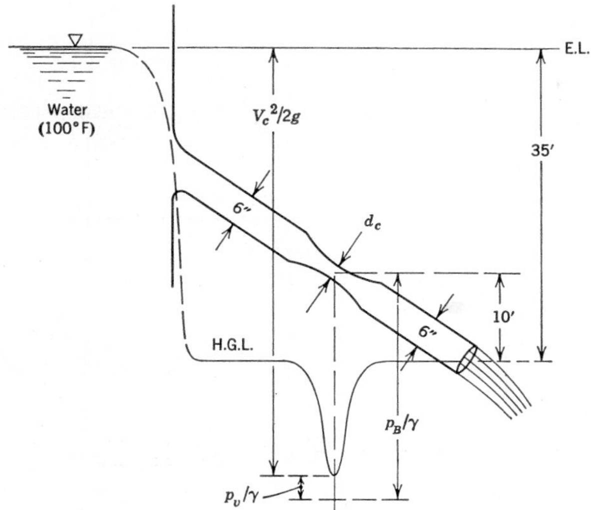

Solution. From Appendix II, $\gamma = 62.0$ lb/cuft and $p_v/\gamma = 2.2$ ft.

$$p_B/\gamma = 14.0 \times 144/62.0 = 32.5 \text{ ft}$$

Construct E.L. and H.G.L. as indicated on the sketch. Evidently the velocity head at the constriction is given by

$$v_c^2/2g = 35.0 - 10.0 + 32.5 - 2.2 = 55.3 \text{ ft}$$

However,

$$V_6^2/2g = 35.0 \text{ ft}$$

Since, from continuity considerations,

$$\left(\frac{d_c}{6}\right)^2 = \frac{V_6}{V_c}, \quad \left(\frac{d_c}{6}\right)^4 = \left(\frac{V_6}{V_c}\right)^2 = \frac{35.0}{55.3}$$

yielding $d_c = 5.35$ in. *Incipient* cavitation must be assumed in such ideal fluid flow problems for losses of head to be negligible; also with small cavitation there is more likelihood of the pipe flowing full at its exit. For a larger cavitation zone the same result will be obtained if losses are neglected and pipes assumed full at exit; however, the low point of the H.G.L. will be considerably flattened. See Appendix VII.

The Bernoulli equation is frequently written in terms of pressure rather than head and may be obtained in this form by multiplying equation 64 by γ and substituting ρ for γ/g; this results in

$$p_1 + \tfrac{1}{2}\rho V_1^2 + \rho g z_1 = p_2 + \tfrac{1}{2}\rho V_2^2 + \rho g z_2 \tag{65}$$

Here the Bernoulli terms p, $\rho V^2/2$, and $\rho g z$ are called static pressure, velocity pressure, and potential pressure, respectively. The *stagnation (or total) pressure*, p_s is defined by

$$p_s = p_o + \tfrac{1}{2}\rho V_o^2 \tag{66}$$

and from Fig. 47 it can be seen that this is the local pressure at the tip of a pitot tube or, more generally, the pressure at a point on

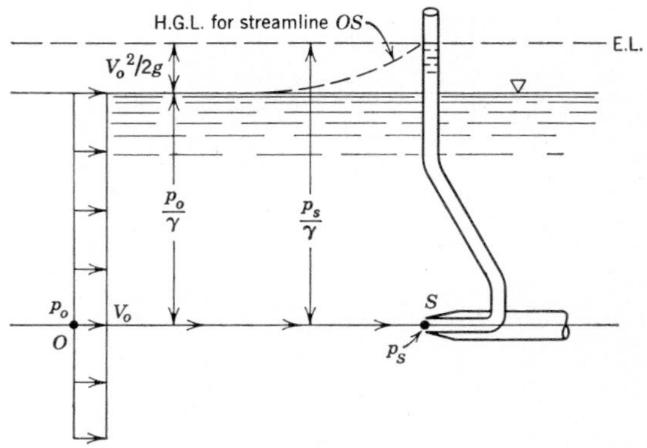

FIG. 47

the nose of any solid object in a flow. This point is called appropriately a *stagnation point* because here the flow momentarily stops, or *stagnates*. The variation of pressure and velocity along the central streamline to the nose of a solid object is shown in Fig. 47. Note that the pressure rises rather abruptly (but not discontinuously) from p_o to p_s just in front of the object, while the velocity decreases from V_o to zero. With stagnation pressure at s easily measurable, and pressure p_o known or measurable, the velocity V_o of the undisturbed stream can be computed from equation 66; this is the essence of the pitot tube principle which finds wide application in many velocity-measuring devices.

ILLUSTRATIVE PROBLEM

This pitot-static tube is carefully aligned with an air stream of density 0.00238 slug/cuft. If the attached differential manometer shows a reading of 6 in. of water, what is the velocity of the air stream?

Solution. Stagnation pressure will be found at the tip of the pitot tube. Assuming that the holes in the barrel of the static tube will collect the static

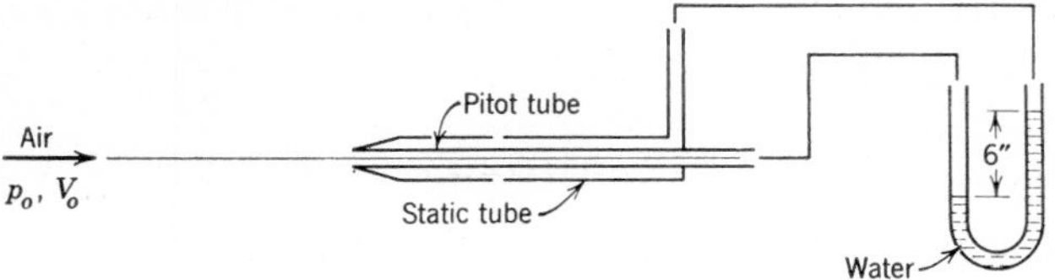

pressure p_o in the undisturbed air stream, the manometer will measure $(p_s - p_o)$. Applying equation 66,

$$p_s - p_o = (p_o + \tfrac{1}{2}\rho V_o^2) - p_o = \tfrac{1}{2}\rho V_o^2$$

Therefore

$$\tfrac{6}{12}(62.4 - 0.00238 \times 32.2) = \tfrac{1}{2}(0.00238)V_o^2; \quad V_o = 162 \text{ fps}$$

A constriction (Fig. 48) in a streamtube or pipe line is frequently used as a device for metering fluid flow. Simultaneous application of the continuity and Bernoulli principles to such a constriction will allow direct calculation of the flowrate when certain variables are measured.

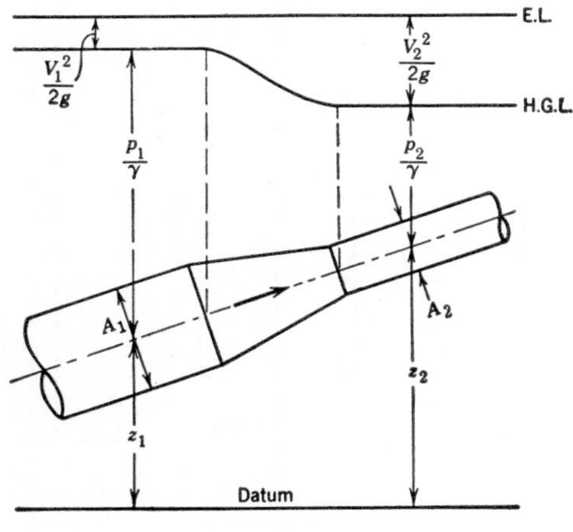

FIG. 48

The equations are

$$Q = A_1 V_1 = A_2 V_2 \tag{53}$$

$$\frac{p_1}{\gamma} + \frac{V_1^2}{2g} + z_1 = \frac{p_2}{\gamma} + \frac{V_2^2}{2g} + z_2 \tag{64}$$

By simultaneous solution of these equations the flowrate through a constriction may be easily computed if cross-sectional areas of pipe and constriction are known and pressures and heights above datum are measured, either separately or in the combination indicated.

Illustrative Problem

Calculate the flowrate of gasoline (s.g. = 0.82) through this pipe line, using first the gage readings and then the manometer reading.

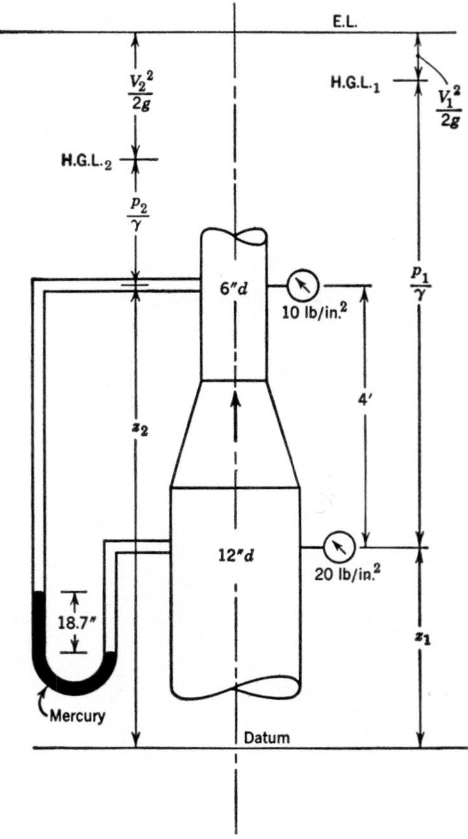

Solution. With sizes of pipe and constriction given, the problem centers around the computation of the value of the quantity $(p_1/\gamma + z_1 - p_2/\gamma - z_2)$ of the Bernoulli equation.

Taking datum at the lower gage and using the gage readings,

$$\left(\frac{p_1}{\gamma} + z_1 - \frac{p_2}{\gamma} - z_2\right) = \frac{20 \times 144}{0.82 \times 62.4} - \frac{10 \times 144}{0.82 \times 62.4} - 4 = 24.2 \text{ ft of gasoline}$$

To use the manometer reading, construct the hydraulic grade line levels at 1 and 2. It is evident at once that the difference in these levels is the quantity

$(p_1/\gamma + z_1 - p_2/\gamma - z_2)$ and, visualizing p_1/γ and p_2/γ as liquid columns, it is also apparent that the difference in the hydraulic grade line levels is equivalent to the manometer reading. Therefore

$$\left(\frac{p_1}{\gamma} + z_1 - \frac{p_2}{\gamma} - z_2\right) = \frac{18.7}{12}\left(\frac{13.55 - 0.82}{0.82}\right) = 24.2 \text{ ft of gasoline}$$

Insertion of 24.2 ft into the Bernoulli equation (64) followed by its simultaneous solution with the continuity equation (53) yields a flowrate Q of 8.03 cfs.

The Bernoulli principle may be applied to problems of *open flow* such as the overflow structure of Fig. 49. Such problems are featured by

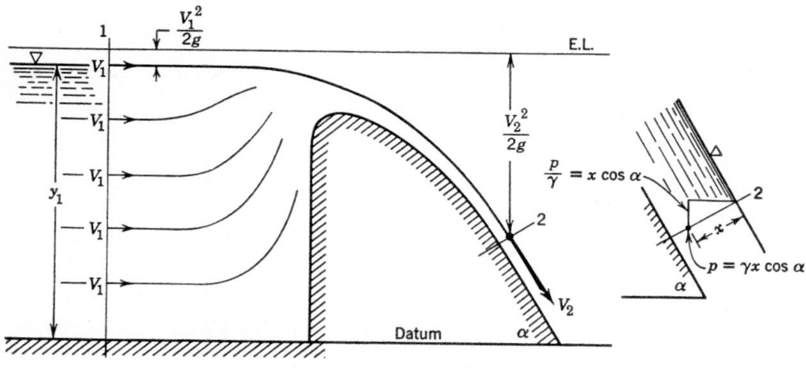

FIG. 49

a moving liquid surface in contact with the atmosphere and flow pictures dominated by gravitational action. A short distance upstream from the structure, the streamlines will be straight and parallel and the velocity distribution will be uniform. In this region the quantity $z + p/\gamma$ will be constant, the pressure distribution hydrostatic, and the hydraulic grade line (for all streamtubes) in the liquid surface; the energy line will be horizontal and located $V_1^2/2g$ above the liquid surface. With atmospheric pressure on the liquid surface the streamtube in the liquid surface behaves as a free jet allowing all surface velocities to be computed from the positions of liquid surface and energy line. The prediction of velocities elsewhere in the flowfield where streamtubes are severely convergent or curved is outside the province of one-dimensional flow; suffice it to say that all such velocities are interdependent and the whole flowfield must be established before any single velocity can be computed. At section 2, however (if the streamlines there are assumed straight and parallel), the pressures and velocities may be computed from the one-dimensional assumption.

Illustrative Problem

Refer to Fig. 49. At section 2 the water surface is at elevation 100, and the 60° spillway surface at elevation 98.5. The velocity in the water surface V_{S2} at section 2 is 20 fps. Calculate the pressure and velocity on the spillway face at section 2. If the bottom of the approach channel is at elevation 95, calculate the depth and velocity in the approach channel.

Solution.

Thickness of sheet of water at section 2 $= (100 - 98.5)/\cos 60° = 3.0$ ft
Pressure on spillway face at section 2 $= 3 \times 62.4 \times \cos 60° = 93.5$ psf
Elevation of energy line $= 100 + (20)^2/2g = 106.22$ ft

$$106.22 = 93.5/62.4 + V_{F2}^2/2g + 98.5; \quad V_{F2} = 20 \text{ fps}$$

which is to be expected from the one-dimensional assumption. Evidently all velocities through section 2 are 20 fps, so

$$q = 3 \times 1 \times 20 = 60 \text{ cfs per foot of spillway length}$$

At section 1, $y_1 = z_1 + p_1/\gamma$; and, applying the Bernoulli equation,

$$y_1 + V_1^2/2g = y_1 + (1/2g)(60/y_1)^2 = 11.22 \text{ ft} \tag{64}$$

Solving this (cubic) equation by trial, the roots are $y_1 = 10.74$, 2.54, and -2.05 ft. Obviously the second and third roots are invalid here, so the depth in the approach channel will be 10.74 ft. The velocity V_1 may be computed from

$$V_1^2/2g = 11.22 - 10.74$$

or from

$$V_1 = 60/10.74 \tag{55}$$

both of which give $V_1 = 5.6$ fps to slide rule accuracy.

The addition of mechanical energy (E_P ft-lb/lb) to a fluid flow by a pump or its extraction (E_T ft-lb/lb) by a turbine will alter the basic Bernoulli equation (64) to

$$\frac{p_1}{\gamma} + \frac{V_1^2}{2g} + z_1 + E_P = \frac{p_2}{\gamma} + \frac{V_2^2}{2g} + z_2 + E_T \tag{67}$$

in which the quantities E_P and E_T will appear as abrupt rises or falls of the energy line over the respective machines. Usually the engineer requires the total power of such machines, which may be computed from the product of weight flowrate (G lb/sec) and (E_P or E_T ft-lb/lb), yielding total power in ft-lb/sec. In terms of horsepower,

$$\text{Horsepower of machine} = \frac{G(E_P \text{ or } E_T)}{550} = \frac{Q\gamma(E_P \text{ or } E_T)}{550} \tag{68}$$

This equation may also be used as a general equation for converting any unit energy, or head, to the corresponding total power.

Illustrative Problem

How many horsepower must be supplied for the pump to maintain readings of 10 in. of mercury vacuum and 40 psi on gages 1 and 2, respectively, while delivering a flowrate of 5.0 cfs of water?

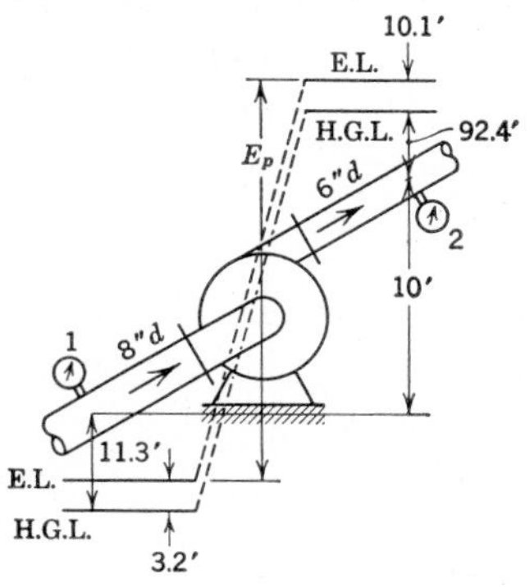

Solution. Converting the gage readings to feet of water, $p_1/\gamma = -11.3$ ft and $p_2/\gamma = +92.4$ ft. Thus the hydraulic grade lines are 11.3 ft below and 92.4 ft above points 1 and 2, respectively. The velocity heads at these points will be found to be 3.2 ft and 10.1 ft, respectively, so the energy lines will be these distances above the respective hydraulic grade lines. The rise in the energy line through the pump represents the energy supplied by the pump to each pound of fluid; from the sketch this is

$$E_P = 10.1 + 92.4 + 10.0 + 11.3 - 3.2 = 120.6 \text{ ft-lb/lb} \qquad (67)$$

$$\text{Pump horsepower} = 5.0 \times 62.4 \times 120.6/550 = 68.4 \text{ hp} \qquad (68)$$

It should be noted that the indicated H.G.L.'s are for the pipes only and do not include the pump passages, where the flow is not one-dimensional. The positions of these H.G.L.'s give no assurance that the pump will run cavitation-free, since local velocities in the pump passages may be considerably larger than the mean velocities in the pipes.

TWO-DIMENSIONAL FLOW

The solution of flowfield problems is vastly more complex than those of one-dimensional flow. Partial differential equations are invariably required for a formal mathematical approach to such problems, and usually these equations cannot be formally solved but must be handled by various trial-and-error techniques. The objective of the remainder of this chapter is merely to present an introduction to certain essentials and practical problems of vital importance to engineering students. To stress the similarities and differences of one- and two-dimensional flows, the subject is developed in parallel with the preceding treatment of one-dimensional flow. Coverage of advanced mathematical operations or broad generalizations is not attempted at this point;[6] the emphasis is on giving the beginner some appreciation of the intricacies of flowfield problems as compared to the relative simplicity of those of one-dimensional flow.

26. Euler's Equations. Euler's equations for a vertical two-dimensional flowfield may be derived by applying Newton's second law to a basic differential element of fluid of dimensions dx by dz (Fig. 50).

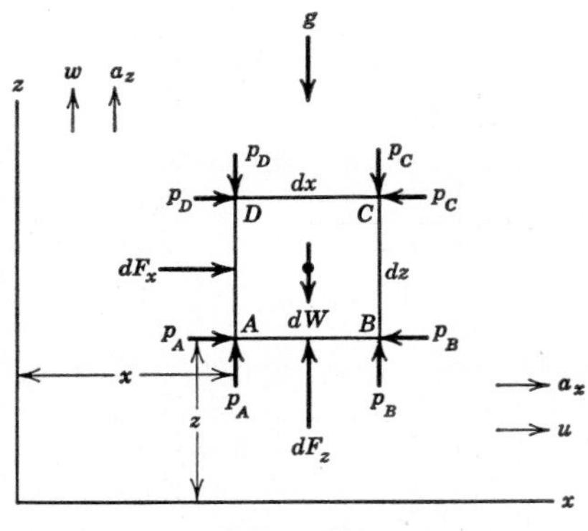

FIG. 50

The forces dF_x and dF_z on such an element have been identified in equations 16 and 17 of Art. 8, and with substitution of appropriate pressures they reduce to

$$dF_x = -\frac{\partial p}{\partial x}\,dx\,dz$$

$$dF_z = -\frac{\partial p}{\partial z}\,dx\,dz - \rho g\,dx\,dz$$

The accelerations of the element have been derived (equation 46, Art. 17) as

$$a_x = u\frac{\partial u}{\partial x} + w\frac{\partial u}{\partial z}$$

$$a_z = u\frac{\partial w}{\partial x} + w\frac{\partial w}{\partial z}$$

[6] Some of these will be found in Chapter 12.

Applying Newton's second law by equating the differential forces to the products of the mass of the element and respective accelerations,

$$-\frac{\partial p}{\partial x}\,dx\,dz = \rho\,dx\,dz\left(u\frac{\partial u}{\partial x} + w\frac{\partial u}{\partial z}\right)$$

$$-\frac{\partial p}{\partial z}\,dx\,dz - \rho g\,dx\,dz = \rho\,dx\,dz\left(u\frac{\partial w}{\partial x} + w\frac{\partial w}{\partial z}\right)$$

which by cancellation of $dx\,dz$ and slight rearrangement reduce to the Euler equations of two-dimensional flow in a vertical plane:

$$-\frac{1}{\rho}\frac{\partial p}{\partial x} = u\frac{\partial u}{\partial x} + w\frac{\partial u}{\partial z} \tag{69}$$

$$-\frac{1}{\rho}\frac{\partial p}{\partial z} = u\frac{\partial w}{\partial x} + w\frac{\partial w}{\partial z} + g \tag{70}$$

Accompanied by the equation of continuity,

$$\frac{\partial u}{\partial x} + \frac{\partial w}{\partial z} = 0 \tag{58}$$

the Euler equations form a set of three simultaneous partial differential equations which are basic to the solution of two-dimensional flowfield problems; complete solution of these equations will yield p, u, and w as functions of x and z, allowing prediction of pressure and velocity at any point in the flowfield.

27. Bernoulli's Equation. Bernoulli's equation may be derived by integrating the Euler equations, as was done for one-dimensional flow (Art. 22). Multiplying the first equation by dx and the second by dz and adding them,

$$-\frac{1}{\rho}\left[\frac{\partial p}{\partial x}dx + \frac{\partial p}{\partial z}dz\right] = u\frac{\partial u}{\partial x}dx + w\frac{\partial u}{\partial z}dx + u\frac{\partial w}{\partial x}dz + w\frac{\partial w}{\partial z}dz + g\,dz$$

The terms $w(\partial w/\partial x)\,dx$ and $u(\partial u/\partial z)\,dz$ are added to and subtracted from the right-hand side of the equation, and terms then collected in the following pattern:

$$-\frac{1}{\rho}\left[\frac{\partial p}{\partial x}dx + \frac{\partial p}{\partial z}dz\right] = \left[u\frac{\partial u}{\partial x}dx + w\frac{\partial w}{\partial x}dx\right] + \left[u\frac{\partial u}{\partial z}dz + w\frac{\partial w}{\partial z}dz\right] + (u\,dz - w\,dx)\left[\frac{\partial w}{\partial x} - \frac{\partial u}{\partial z}\right] + g\,dz$$

The bracket on the left-hand side of this equation will be recognized (see Appendix V) as the total differential dp. The sum of the first two brackets on the right-hand side is easily shown to be $d(u^2 + w^2)/2$, and the third bracket is identified as the vorticity, ξ (Art. 20). Reducing the equation accordingly and dividing it by g, it may be written

$$-\frac{dp}{\gamma} = \frac{d(u^2 + w^2)}{2g} + \frac{1}{g}(u\,dz - w\,dx)\xi + dz$$

and integrated to

$$\frac{p}{\gamma} + \frac{u^2 + w^2}{2g} + z = H - \frac{1}{g}\int \xi(u\,dz - w\,dx)$$

in which H is the constant of integration. Since the resultant velocity V at any point in the flowfield is related to its components u and w by $V^2 = u^2 + w^2$, the equation further simplifies to

$$\frac{p}{\gamma} + \frac{V^2}{2g} + z = H - \frac{1}{g}\int \xi(u\,dz - w\,dx) \tag{71}$$

This equation thus shows that the sum of the Bernoulli terms at any point in a flowfield will be a constant H if the vorticity ξ is zero, i.e., *if the flowfield is an irrotational* (or potential) one. Thus for irrotational flow the same constant applies to all the streamlines of the flowfield, or, in terms of the energy line, all fluid masses in an irrotational flowfield possess the same unit energy.

28. Application of Bernoulli's Equation. For irrotational flow of an ideal incompressible fluid the Bernoulli equation may be applied over the flowfield with a single (horizontal) energy line completely describing the energy situation. Figure 51 depicts this for two representative points A and B. From the position of the points (above datum) the quantity $(p/\gamma + V^2/2g)$ may be determined from the position of the energy line, but the pressures p_A, p_B, etc., cannot be calculated until the corresponding velocities V_A, V_B, etc., are known. However, in a flowfield all the velocities are interdependent and are determined by the streamline definition and by the differential equation (58) of continuity; until methods[7] are described for solving these equations, the pressures in the flowfield cannot be accurately

[7] Some of these methods will be cited in Chapter 12.

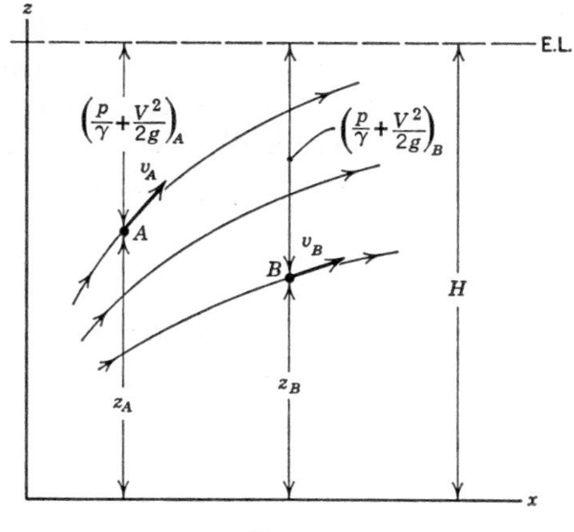

FIG. 51

predicted. However, the lack of formal mathematical solutions describing the entire velocity field need not deter the engineer from making a semiquantitative approach to such problems; indeed, when no formal solutions exist (as frequently happens) this is the only alternative.

One very useful tool in such approaches to flowfield problems is the effect of flow curvature on the pressure variation across the flow. For any element of streamline (Fig. 52) having radius of curvature r, the normal component of acceleration a_r is directed toward the center of curvature and equal to V^2/r (equation 44). Newton's second law applied to such a fluid element yields a general and useful result in the analysis and interpretation of flowfields. In the radial direction the components of force on the element are

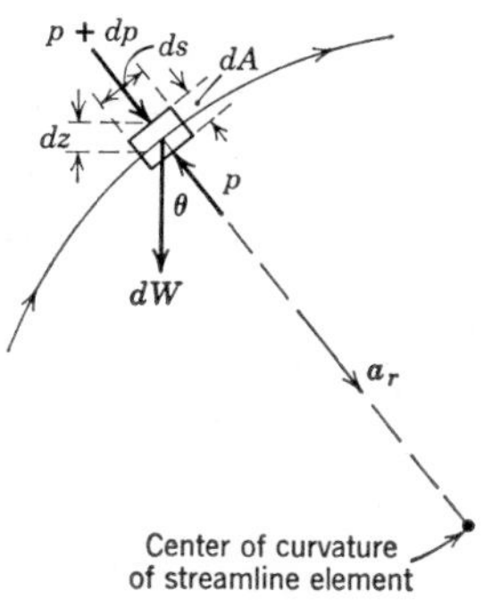

FIG. 52

$$(p + dp)\, ds - p\, ds + dW \cos\theta = \frac{(\rho\, dr\, ds)\, V^2}{r}$$

From the geometry of the element, $\cos\theta = dz/dr$ and $dW = \gamma\, dr\, ds$. Substituting these values in the equation above, dividing by $\gamma\, ds$, and rearranging,

$$\frac{d}{dr}\left(\frac{p}{\gamma} + z\right) = \frac{V^2}{gr} \tag{72}$$

from which it is seen that the gradient of $(p/\gamma + z)$ along r is always positive, or that an increase of $(p/\gamma + z)$ is to be expected along a direction outward from the center of curvature; conversely a drop of $(p/\gamma + z)$ will be expected along a direction toward the center of curvature. Although this equation is formally integrable only for vortex motion (where there is a single center of curvature), it is nevertheless of great value to the engineer in the "reading" of streamline pictures to distinguish regions of high and low $(p/\gamma + z)$. The reader will sense that this development, which predicts the variation of $(p/\gamma + z)$ in the radial direction, will also determine (through the Bernoulli equation) the variation of velocity with radial distance. This variation may be discovered by taking the derivative (in respect to r) of the Bernoulli equation of irrotational flow:

$$\frac{d}{dr}\left(\frac{p}{\gamma} + \frac{V^2}{2g} + z = H\right) = \frac{d}{dr}\left(\frac{p}{\gamma} + z\right) + \frac{2V}{2g}\frac{dV}{dr} = 0$$

or

$$\frac{d}{dr}\left(\frac{p}{\gamma} + z\right) = -\frac{V\,dV}{g\,dr}$$

which may be equated to the expression (equation 72) obtained from Newton's second law. Thus

$$\frac{V^2}{gr} = -\frac{V}{g}\frac{dV}{dr} \quad \text{or} \quad V\,dr + r\,dV = 0$$

which when integrated (for a flowfield having a single center of curvature) yields

$$Vr = \text{Constant} \tag{73}$$

Again it is emphasized that the equation cannot be generally integrated over a whole flowfield because of the numerous centers of curvature of the streamline elements, but it may nevertheless be used to conclude that generally a *decrease* [8] of velocity is to be expected with *increase* of distance from center of curvature in irrotational flowfields. The reader may be familiar with this type of velocity distribution from casual experience with vortices in the atmosphere,

[8] The beginner, schooled in the mechanics of solid-body rotation where tangential velocity varies linearly with radius, is frequently surprised to discover that this law does not apply to fluid flow. He should not be surprised, because the mobility of fluid particles in a curved flow is infinitely greater than that of solid particles "locked" together in a rotating object.

tornadoes, or the "drainhole vortex" which frequently develops when a tank is drained through an orifice in the bottom.

In Fig. 53 is shown a curved flow occurring through a passage in a vertical plane. Distortion of the velocity profile from a uniform distribution is shown; pitot tube and piezometer columns also indicate the variation of pressure and velocity throughout the flow. If

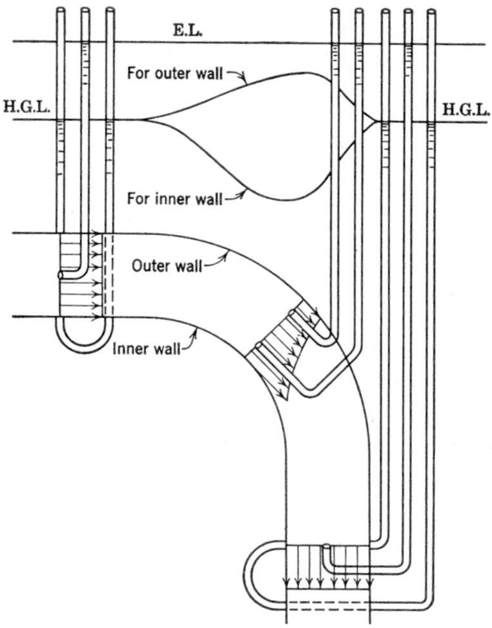

Fig. 53

these facts are supplemented by a streamline picture, streamlines which are uniformly spaced in the straight passages will be crowded together toward the inner wall of the curved passage and widely spaced toward its outer wall; *thus they cannot be concentric circular arcs.*

Another example of the effect of streamline curvature in a flowfield is seen in the convergent-divergent passage of Fig. 54. The streamlines AA along the walls are most sharply curved, whereas the central streamline possesses no curvature, and the streamlines in the region between walls and centerline are of intermediate curvature. Accordingly it may be deduced that the velocity profile at section 2 features higher velocity at the walls than on the centerline, and the relative position of the hydraulic grade lines for these streamlines

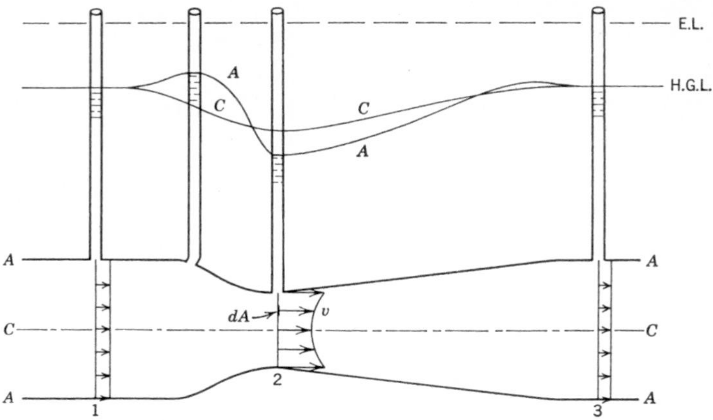

FIG. 54

is as shown. In passing, it is of interest to note that if incipient cavitation occurred at section 2 it would be expected to appear on the *upper wall* of the passage; for both walls at that cross section $(p/\gamma + z)$ is the same but, with z larger for the upper wall, the pressure will be less there. This problem has been treated by one-dimensional methods in Art 25; it is clearly a one-dimensional problem at sections 1 and 3, but between these sections there is a flowfield which can be treated as one-dimensional only as an approximation. The application of the continuity equation to such problems is basic and instructive. Clearly

$$Q = A_1 V_1 = A_2 V_2 = A_3 V_3 \tag{53}$$

in which V_1 and V_3 are the respective mean velocities at sections 1 and 3 and also the velocities of individual particles as they pass these sections. At section 2, however, the velocities of fluid particles are very different from the mean velocity, and a relation between these is to be sought. From equation 56 the mean velocity V_2 is given by

$$V_2 = \frac{1}{A_2} \int^{A_2} v \, dA$$

It is noted that, although V_2 can be easily calculated from Q, this yields no information on the distribution of velocity, which is determined by the shape and curvature of the passage walls.

Rigorous application of the Bernoulli equation to the flow between sections 1 and 2 is complicated by the nonuniform velocity profile at section 2. Although the same energy line applies to all the streamlines of the flowfield, the engineer is concerned with the separate terms p, z, and v, all of which vary across the flow at section 2. Here the Bernoulli equation must be written in terms of total power rather than unit energy; following the pattern of equation 68,

$$\begin{bmatrix}\text{Total power (ft-lb/sec)}\\ \text{at section 1}\end{bmatrix} = Q\gamma\left(\frac{p_1}{\gamma} + z_1 + \frac{V_1^2}{2g}\right)$$

the quantities $(p/\gamma + z)$ and $V^2/2g$ being constant across section 1. At section 2 the total power of the flow must be expressed as an integral, following the same pattern. Using $dQ = v\,dA$,

$$\begin{bmatrix}\text{Total power (ft-lb/sec)}\\ \text{at section 2}\end{bmatrix} = \int^{A_2} (v\,dA)\,\gamma\left(\frac{p}{\gamma} + z + \frac{v^2}{2g}\right)$$

Equating the expressions for total power, the "Bernoulli equation" becomes

$$Q\left(\frac{p_1}{\gamma} + z_1 + \frac{V_1^2}{2g}\right) = \int^{A_2} v\left(\frac{p}{\gamma} + z\right) dA + \int^{A_2} \frac{v^3}{2g}\,dA \qquad (74)$$

for which a method of solution is suggested in the following illustrative problem.

Illustrative Problem

Ideal fluid flows through this symmetrical constriction in the two-dimensional passage. Show how the flowrate may be calculated.

Solution. Draw the energy line from which $V_1^2/2g$ and $v_A^2/2g$ are identified. Calculate an approximate flowrate by the one-dimensional methods of Art. 25, assuming v_A to be the mean velocity at section 2. Since $v_A > V_2$, the approximate flowrate will be larger than the true flowrate; therefore assume a flowrate smaller than this for the first trial calculation. From the assumed Q, V_1 may be calculated (from the continuity principle), allowing the position of the energy line and the left-hand side of equation 74 to be tentatively established. Find a velocity profile of the form indicated which will satisfy the continuity equation for the assumed Q. From this velocity profile calculate the distribution of $(p/\gamma + z)$ across the flow. Finally, carry out the integrations (graphically if necessary) of equation 74 to see if this equation is satisfied; on the first trial it will not be, so assume another Q, repeating the procedure until the equation is satisfied. As a guide in the process, $d/dr\,(p/\gamma + z)$ at the walls may be computed from v_A^2/gr for assistance in establishing a suitable distribution of $(p/\gamma + z)$.

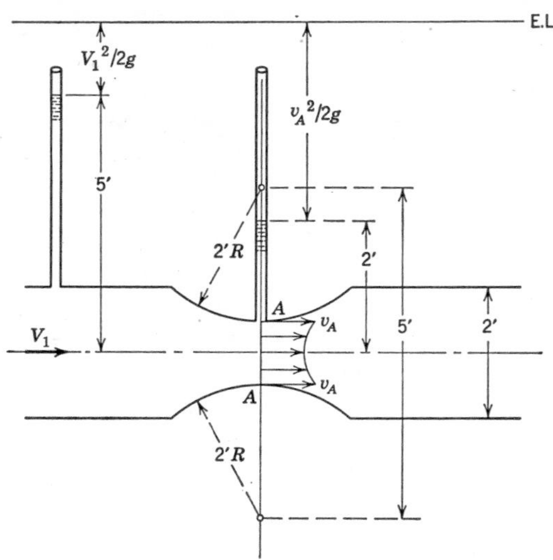

Another trial and error technique for the solution of such problems is that of the *flownet* described in Art. 114.

The identification of stagnation points in flowfields is another useful adjunct to an understanding of the flow process. In ideal flow these may be expected wherever a streamline is forced to turn a sharp corner; such points will be expected only on solid boundaries. Two stagnation points are shown in the external and internal flowfields of Fig. 55. At each stagnation point the velocity is locally zero, so the vertical distance between stagnation point and energy line is a direct measure of the pressure at the stagnation point. With this knowledge, the velocities at other points in the flowfield (if irrotational) may be calculated from static pressure measurements.

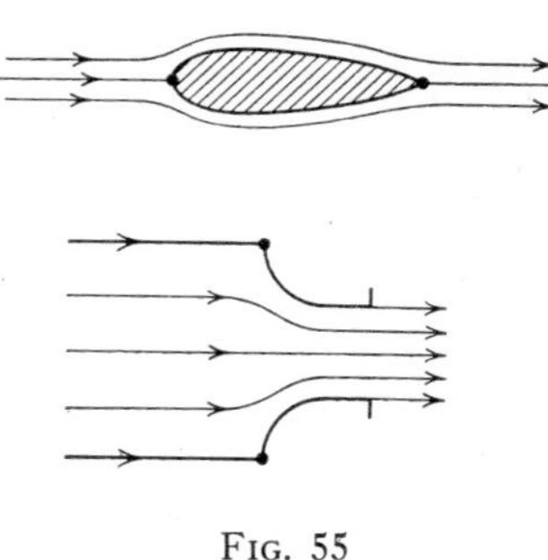

Fig. 55

Illustrative Problem

Ideal fluid of specific weight 50 lb/cuft flows down this pipe and out into the atmosphere through the "end cap" orifice. Gage *B* reads 6.0 psi and gage *A*, 2.0 psi. Calculate the mean velocity in the pipe.

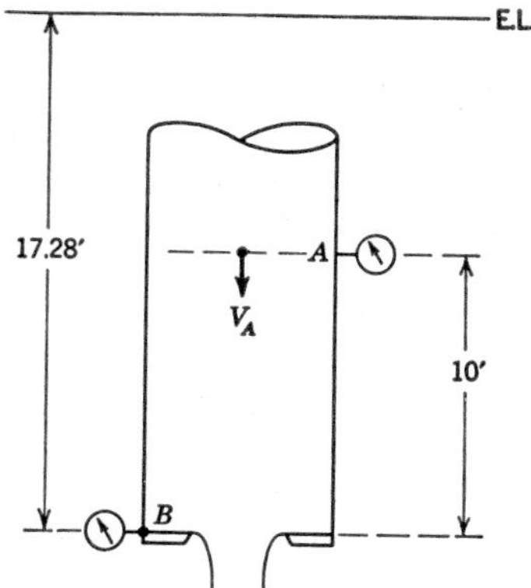

Solution. Identify point B as a stagnation point. The energy line will then be established $6 \times 144/50 = 17.28$ ft above B. Gage A measures the static pressure in the flow; p_A/γ is thus $2.0 \times 144/50 = 5.76$ ft. Therefore

$$\frac{V_A^2}{2g} = 17.28 - 10 - 5.76 = 1.52 \text{ ft}; \quad V_A = 9.90 \text{ fps}$$

A typical problem of two-dimensional irrotational open flow is that of the sharp-crested weir shown in Fig. 56. A short distance

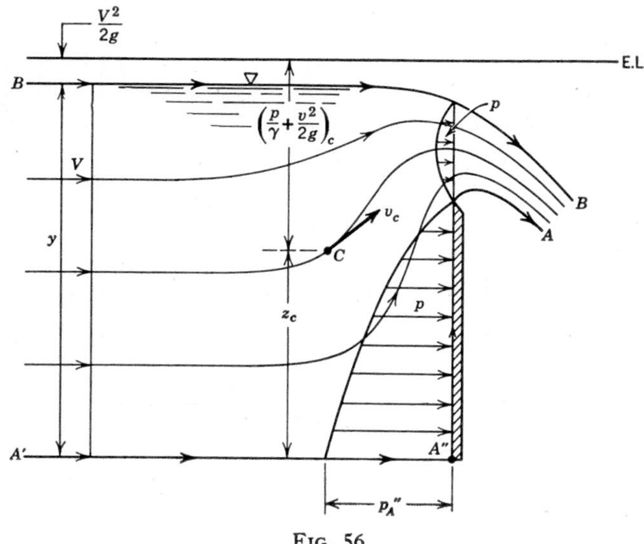

FIG. 56

upstream from such a structure the streamlines will be essentially straight and parallel, and a one-dimensional situation will therefore exist. Between this section and some point in the falling sheet of liquid where free fall begins, a flowfield occurs about as shown. Because of the velocity (V) of approach to the weir, the energy line may be visualized above the flow picture as indicated. The boundary streamlines BB and AA (downstream from the weir crest) are called *free streamlines*, their precise position in space being unknown, but the pressure on them being everywhere constant, in this case zero (gage); once their position is established (by analysis or experiment), the velocity at any point on them may be calculated since the vertical distance between any point and the energy line will be the velocity head ($v^2/2g$) at the point. The pressure distribution in the flow at section 1 is hydrostatic, the bottom pressure at A' being simply γy. The only stagnation point in the flow is noted at A'', at which the pressure will be $\gamma(y + V^2/2g)$. At any other point (C) in the flowfield, $(v_c^2/2g + p_c/\gamma)$ may be computed from the positions of point and energy line; however, the pressure there is not calculable without the velocity v_c, which is interrelated with all other velocities and not generally predictable until complete details of the whole flowfield are known. The pressure distribution in the plane of the weir plate is qualitatively predictable since the pressures at both boundaries (free streamlines) will be zero; with the streamlines of sharpest curvature nearest to the weir crest, $d/dr\ (p/\gamma + z)$ will be largest here, producing a positive pressure in the flow as shown. Downstream from this cross section the pressure within the falling sheet will diminish, becoming essentially zero as the streamlines become straighter and more parallel.

Illustrative Problem

Calculate the flowrate through this two-dimensional nozzle discharging to the atmosphere and designate any stagnation points in the flow.

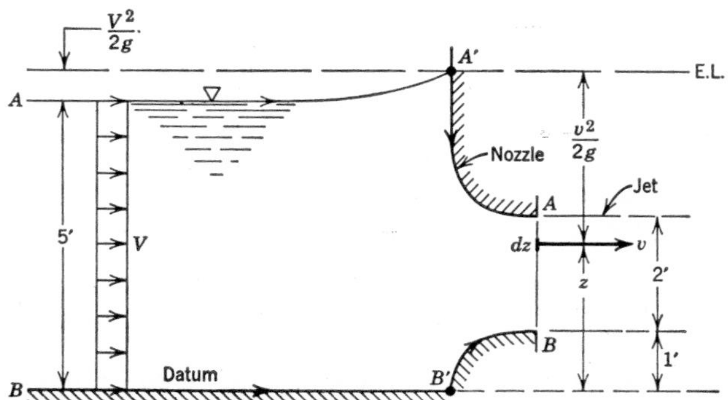

Solution. From the energy line drawn a distance $V^2/2g$ above the horizontal liquid surface, and with the pressure assumed zero throughout the free jet at the nozzle exit,

$$5 + V^2/2g = z + v^2/2g; \quad v = \sqrt{2g(5 + V^2/2g - z)} \tag{64}$$

The flowrate q may then be expressed

$$q = 5V = \int_1^3 v\,dz = \int_1^3 \sqrt{2g(5 + V^2/2g - z)}\,dz \tag{56}$$

which may be solved by trial to yield $q = 30.25$ cfs/ft.

Stagnation points on the boundary streamlines AA and BB are to be expected. The one at B' needs no comment. At some point on the plane $A'B'$ there must be a stagnation point on the top boundary streamline; assume that this is somewhere below A'. Such a point could not be a stagnation point since its distance below the energy line would indicate a velocity head and thus a velocity there. Accordingly it is concluded that the stagnation point must be at A' and the liquid surface must rise to this point.

Frequently it is possible to obtain the complete kinematics of a flowfield by mathematical methods which yield specific equations for the streamlines and velocities, from which accelerations and pressure variations may be predicted. A classic and useful example of this is the (irrotational) flowfield about a cylinder of radius R in a rectilinear flow of velocity U, which is shown in Fig. 57. The radial

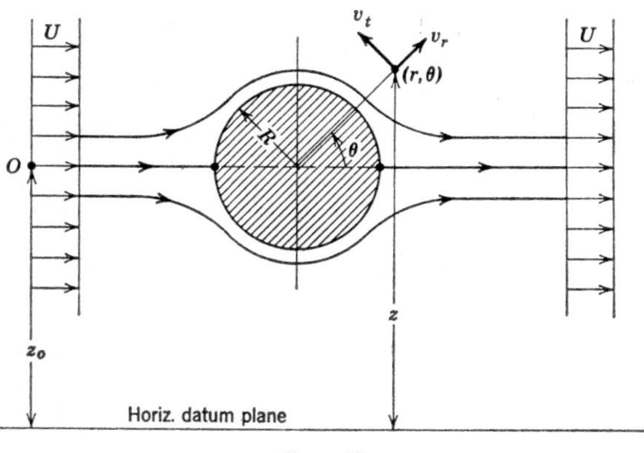

FIG. 57

and tangential components of velocity anywhere in the flowfield may may be shown [9] to be

$$v_r = U\left(1 - \frac{R^2}{r^2}\right)\cos\theta \quad \text{and} \quad v_t = -U\left(1 + \frac{R^2}{r^2}\right)\sin\theta$$

[9] See equations 298 and 299, Art. 111.

In such problems the velocity along the surface of the body is of greatest interest; here $r = R$, so $v_r = 0$ and $v_t = -2U \sin \theta$. For $\theta = 0$ and $\theta = \pi$, v_t will be zero, thus confirming the expected stagnation points at the head and tail of the body. Applying the Bernoulli equation between the undisturbed flow and any point on the body contour,

$$\frac{p_o}{\gamma} + \frac{U^2}{2g} + z_o = \left(\frac{p}{\gamma} + z\right) + \frac{(-2U \sin \theta)^2}{2g}$$

it is seen that $(p/\gamma + z)$ at any point on the cylinder may be predicted from the properties (p_o, z_o, and U) of the undisturbed flow; from the cylinder size, z is determined and thus computation of the pressures on the cylindrical surface may be accomplished. By similar methods any pressure in the flowfield may be computed.

Illustrative Problem

A cylinder of 6-in. diameter extends horizontally across the test section of large open-throat wind tunnel through which the air flows at a velocity of 100 fps, pressure 0 psi, and specific weight 0.08 lb/cuft. Calculate the velocity and pressure in the flowfield at the point ($\theta = 60°$, $r = 6$ in.).

Solution.

$$v_r = 100[1 - (\tfrac{3}{6})^2]0.5 = 37.5 \text{ fps} \qquad (298)$$

$$v_t = -100[1 + (\tfrac{3}{6})^2]0.866 = -108.1 \text{ fps} \qquad (299)$$

$$V = \sqrt{(37.5)^2 + (108.1)^2} = 114.4 \text{ fps}$$

Applying the Bernoulli equation, taking the horizontal datum plane through the center of the cylinder,

$$0 + (100)^2/2g + 0 = p/0.08 + (114.4)^2/2g + 0.5 \times 0.866 \qquad (64)$$

$$p/0.08 = -47.9 - 0.43; \quad p = -3.87 \text{ psf}$$

In gas flow problems it is customary to neglect the z terms in the Bernoulli equation; had this been done here, the 0.43 would not have appeared and the pressure would have been -3.83 psi, less than 1 percent from that calculated. In problems with larger velocities the z terms are of even less importance.

The flow through a sharp-edged opening (Fig. 58) produces a flowfield which has many engineering applications. Jet contraction and flow curvature are produced by the (approximately) radial approach of fluid to the orifice, the streamlines becoming essentially straight and parallel at a section (termed the vena contracta) [10] a short dis-

[10] Using advanced analytical methods, Kirchhoff showed the width of the vena contracta to be $(\pi/\pi + 2) \times$ (width of opening) for discharge at high velocity.

tance downstream from the opening. Here and at other sections downstream the pressure through the jet is essentially zero, as explained in Art. 25. Elsewhere the pressure is zero only on the free streamlines which bound the jet, but, with the centers of curvature of streamlines A, B, and C in the vicinity of O and the pressure increasing away from the center of curvature, it is apparent that, in the plane of the opening, pressures will increase and velocities will decrease toward the centerline; thus within the curved portion of the jet the pressures will be expected to be larger than zero. In engineering practice this problem is usually treated by one-dimensional methods, which are entirely adequate at the vena contracta; applied to the flow cross section in the plane of the opening they are quite meaningless and lead only to contradictions.

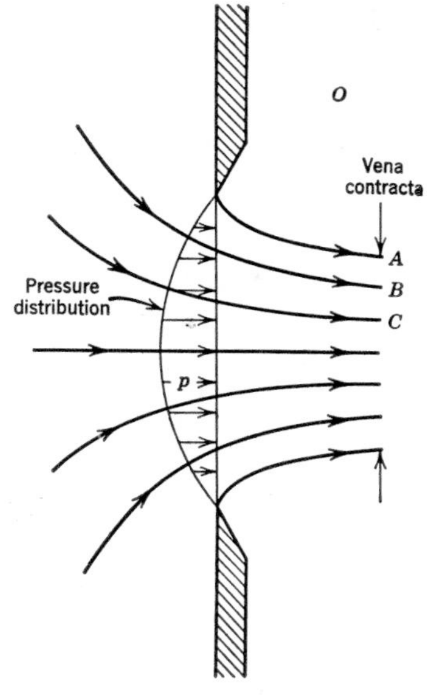

Fig. 58

Illustrative Problem

A two-dimensional flow of liquid discharges from a large reservoir through the sharp-edged opening; a pitot tube at the center of the vena contracta produces the reading indicated. Calculate the velocities at points A, B, C, and D, and the flowrate.

Solution. The pitot tube reading determines the position of the energy line (and also that of the free surface in the reservoir). Since the pressures at points A, B, C, and D are all zero, the respective velocity heads are determined by the vertical distances between the points and the energy line. The velocities at A, B, C, and D are 15.28 fps, 16.78 fps, 15.50 fps, and 16.50 fps, respectively. The flowrate may be computed by integrating the product $v\,dA$ over the flow cross section CD, in which $v = \sqrt{2gh}$:

$$q = \int \sqrt{2gh}\,dh = \sqrt{2g}\,\tfrac{2}{3}\,h^{3/2}\Big|_{3.771}^{4.229} = 7.35 \text{ cfs/ft} \tag{56}$$

The limits of integration are the vertical distances (in feet) from energy line to points D and C, respectively. Assuming that the mean velocity at section CD

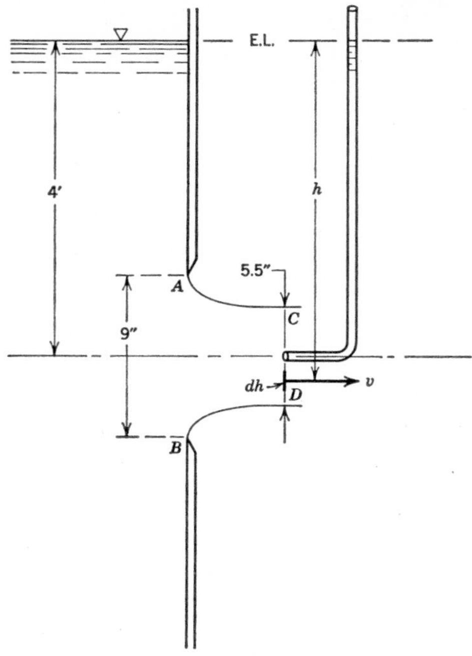

is at the center and thus measured by the pitot tube, $V = \sqrt{2g \times 4} = 16.04$ fps, the (approximate) flowrate, q, may be computed from $q = 16.04 \times 5.5/12 = 7.35$ cfs/ft, giving the same result as the more refined calculation. Actually the center velocity is greater than the mean, but for large ratios of head to orifice opening there is a negligible difference between them. For small ratio of head to orifice opening the flowfield is greatly distorted by drooping of the jet, producing sharply curved streamlines, and ill-defined vena contracta; for such conditions the foregoing methods of calculation may be used only as crude approximations.

PROBLEMS

260. On an overflow structure of 50° slope the water depth (measured normal to the surface of the structure) is 4 ft and the streamlines are essentially straight and parallel. Calculate the pressure on the surface of the structure.

261. Water flows in a pipe line. At a point in the line where the diameter is 7 in., the velocity is 12 fps and the pressure is 50 psi. At a point 40 ft away the diameter reduces to 3 in. Calculate the pressure here when the pipe is (*a*) horizontal, (*b*) vertical with flow downward.

262. A horizontal 6-in. pipe, in which 1000 gpm of carbon tetrachloride are flowing, contains a pressure of 30 psi. If this pipe reduces to 4-in. diameter, calculate the pressure in the 4-in. pipe.

263. In a pipe 1 ft in diameter, 10 cfs of water are pumped up a hill. On the hilltop (elevation 160) the line reduces to 8-in. diameter. If the pump maintains a pressure of 100 psi at elevation 70, calculate the pressure in the pipe on the hilltop.

264. In a 3-in. horizontal pipe line containing a pressure of 8 psi, 100 gpm of water flow. If the pipe line reduces to 1-in. diameter, calculate the pressure in the 1-in. section.

265. If benzene flows through this pipe line and its velocity at A is 8 fps, where is the benzene level in the open tube C?

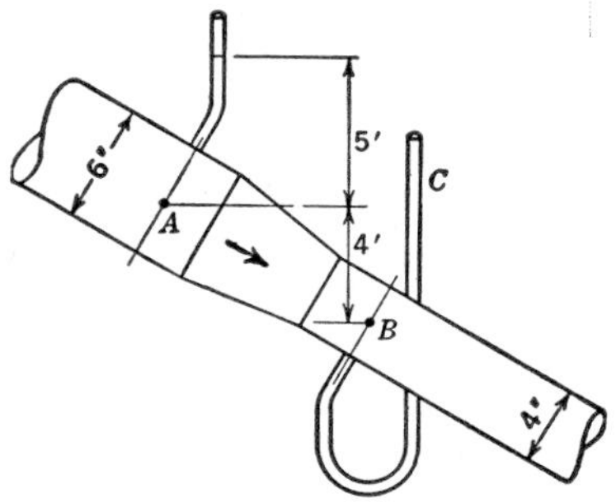

PROBLEM 265

266. Air of specific weight 0.08 lb/cuft flows through a 4-in. constriction in an 8-in. pipe line. When the flowrate is 2.5 lb/sec, the pressure in the constriction is 4 in. of mercury vacuum. Calculate the pressure in the 8-in. section, neglecting compressibility.

267. If 8 cfs of water flow through this pipe line, calculate the pressure at point A.

268. A pump draws water from a reservoir through a 12-in. pipe. When 12 cfs are being pumped, what is the pressure in the pipe at a point 8 ft above the reservoir surface, in psi? in feet of water?

269. If the pressure in the 12-in. pipe of problem 249 is 10 psi, what pressures exist in the branches, assuming all pipes are in the same horizontal plane? Water is flowing.

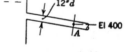

PROBLEM 267

270. If each gage shows the same reading for a flowrate of 1.00 cfs, what is the diameter of the constriction?

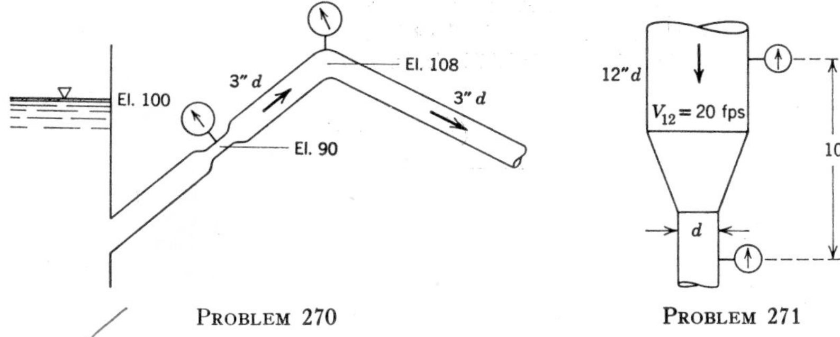

PROBLEM 270 PROBLEM 271

271. Water is flowing. Calculate the required pipe diameter, d, for the two gages to read the same.

272. For a flowrate of 75 cfs of air ($\gamma = 0.0763$ lb/cuft) what is the largest A_2 which will cause water to be drawn up to the piezometer opening? Neglect compressibility effects.

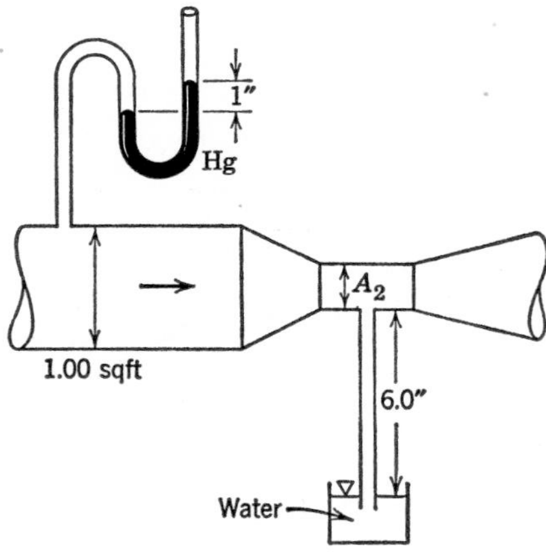

PROBLEM 272

273. When the head of water on a 2-in. diameter smooth nozzle is 10 ft, calculate the rate of flow therefrom.

274. A smooth nozzle of 2-in. diameter is connected to a water tank. Connected to the tank at the same elevation is an open U-tube manometer containing mercury and showing a reading of 25 in. The lower mercury surface is 20 in. below the tank connection. What flowrate will be obtained from the nozzle? Water fills the tube between tank and mercury.

275. Water discharges from a tank through a nozzle of 2-in. diameter into an air tank where the pressure is 5 psi. If the nozzle is 15 ft below the water surface, calculate the flowrate.

276. Water discharges through a nozzle of 1-in. diameter under a 20-ft head into a tank of air in which a vacuum of 10 in. of mercury is maintained. Calculate the rate of flow.

277. Air is pumped through the tank as shown. Neglecting effects of compressibility, compute the velocity of air in the 4-in. pipe. Atmospheric pressure is 13.2 psia and the specific weight of air 0.07 lb/cuft.

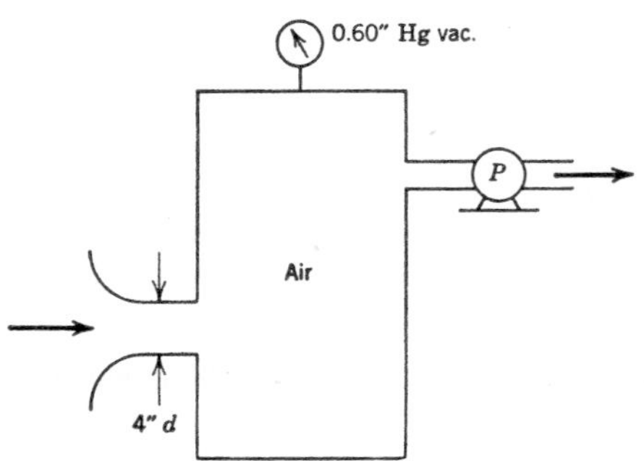

PROBLEM 277

278. A closed tank contains water with air above it. The air is maintained at a pressure of 15 psi, and 10 ft below the water surface a nozzle discharges into the atmosphere. At what velocity will water emerge from the nozzle?

279. Calculate the rate of flow through this pipe line and the pressures at A, B, C, and D.

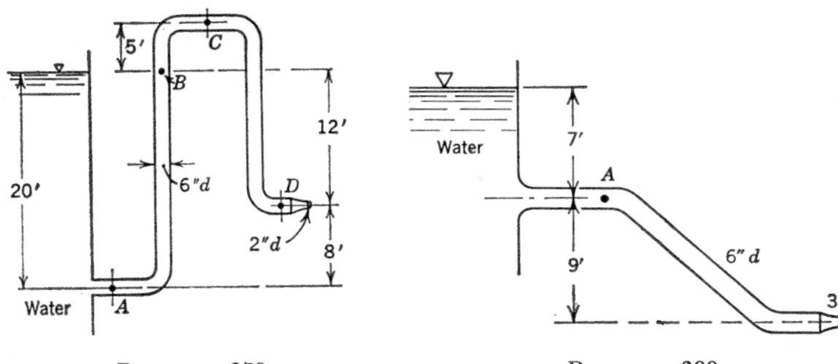

PROBLEM 279 PROBLEM 280

280. Calculate the pressure in the flow at A: (a) for the system shown and (b) for the pipe without the nozzle.

281. A siphon consisting of a 1-in. hose is used to drain water from a tank. The outlet end of the hose is 8 ft below the water surface, and the bend in the hose is 3 ft above the water surface. Calculate the pressure in the bend and the flowrate.

282. A 3-in. horizontal pipe is connected to a tank of water 5 ft below the water surface. The pipe is gradually enlarged to 3.5-in. diameter and discharges freely into the atmosphere. Calculate the flowrate and the pressure in the 3-in. pipe.

283. Calculate the pressure head in the $\frac{3}{4}$-in. section when $h = 0.5$ ft. Calculate the largest h at which the divergent tube can be expected to flow full.

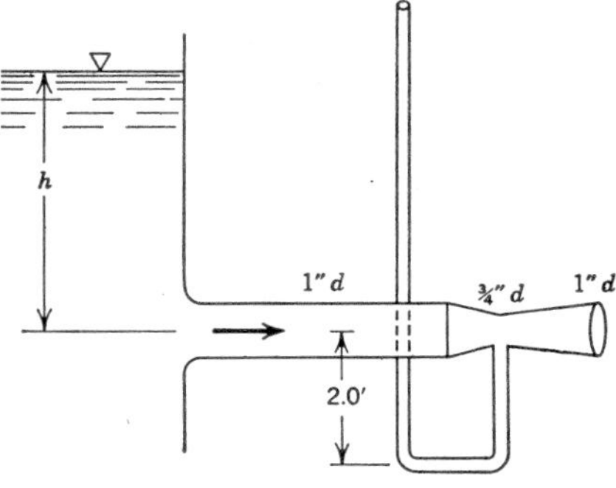

PROBLEM 283

284. If the nozzle of problem 273 is directed upward at angles of (a) 30°, (b) 45°, (c) 60°, (d) 90°, how high above the nozzle will the jet rise, and how far from the nozzle will the jet pass through the horizontal plane in which the nozzle lies? What is the diameter of the jet at the top of the trajectory?

285. A fire hose nozzle which discharges water at 150 fps is to throw a stream through a window 100 ft above and 100 ft horizontally from the nozzle. What is the minimum angle of elevation of the nozzle that will accomplish this?

286. The jet passes through point A. Calculate the flowrate.

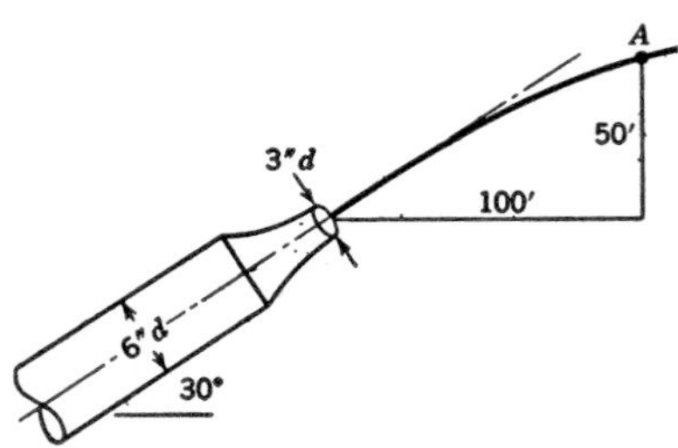

PROBLEM 286

287. A nozzle of 2-in. tip diameter discharges a flowrate of 0.40 cfs vertically downward. Calculate the velocity of the jet 10 ft below the nozzle tip.

288. A nozzle of 1-in. diameter on a horizontal 3-in. pipe discharges a stream of water with a velocity of 60 fps. Calculate the pressure in the pipe and the velocity of the jet at a point 20 ft below the nozzle.

289. Calculate the minimum flowrate which will pass over the wall.

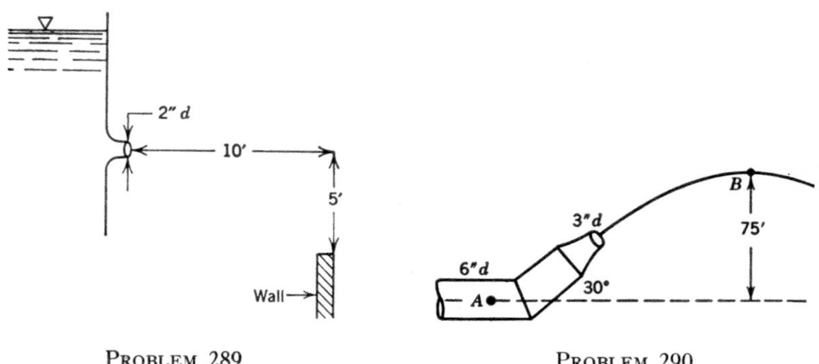

PROBLEM 289 PROBLEM 290

290. If the velocity at B is 50 fps, what is the pressure head at A?

291. From Bernoulli's equation derive a relation between the velocities at any two points on the trajectory of a free jet in terms of the vertical distance between the points.

292. A jet of water falling vertically has a velocity of 80 fps and a diameter of 2 in. at elevation 40 ft; calculate its velocity at elevation 15 ft.

293. Applying free trajectory theory to the centerline of a free jet discharging from a horizontal nozzle under a head h, show that the radius of curvature of the jet at the tip of the nozzle is $2h$.

294. For the two orifices discharging as shown, prove that $h_1y_1 = h_2y_2$.

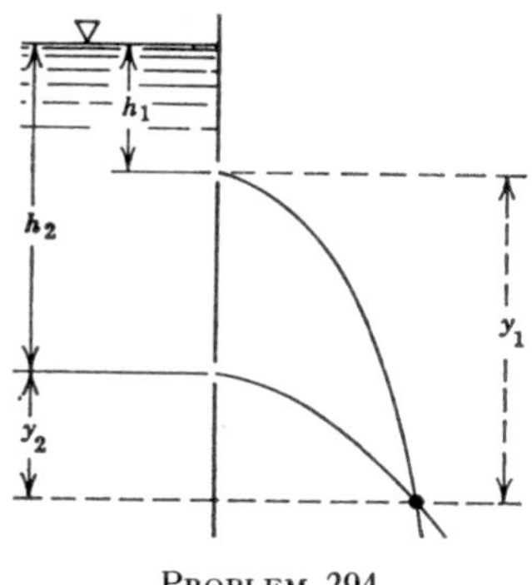

PROBLEM 294

295. A constriction of 6-in. diameter occurs in a horizontal 10-in. water line. It is stated that the pressure in the 10-in. pipe is 5 in. of mercury vacuum when the flowrate is 6.0 cfs. Is this possible? Why or why not?

296. Water flows through a 1-in. constriction in a horizontal 3-in. pipe line. If the water temperature is 150°F and the pressure in the line is maintained at 40 psi, what is the maximum flowrate which may occur? Barometric pressure is 14.7 psia.

297. The liquid has specific gravity 1.60 and negligible vapor pressure. Calculate the flowrate for incipient cavitation in the 3-in. section, assuming that the tube flows full. Barometric pressure is 14.0 psia.

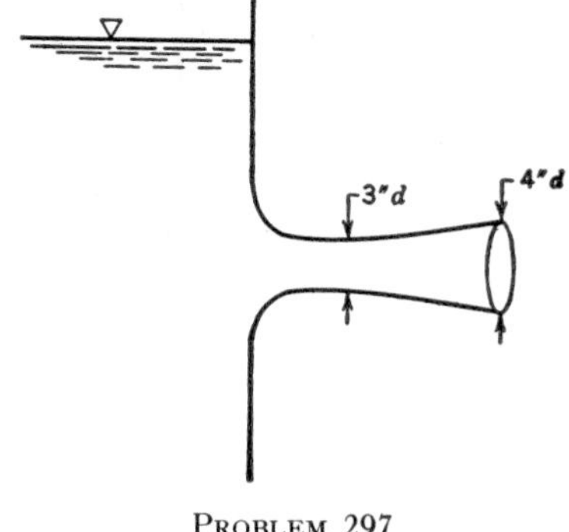

PROBLEM 297

298. Water flows between two open reservoirs. Barometric pressure is 13.4 psia. Vapor pressure is 6.4 psia. Above what value of h will cavitation be expected in the 2-in. constriction?

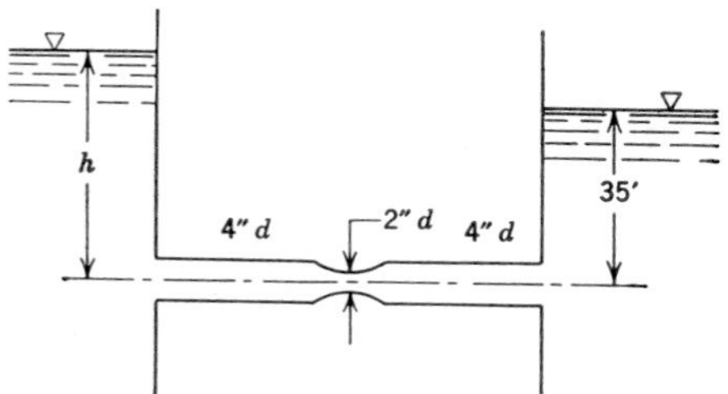

PROBLEM 298

299. Barometric pressure is 14.7 psia. For $h > 2$ ft, cavitation is observed at the 2-in. section. If the pipe is horizontal and flows full throughout, what is the vapor pressure of the water?

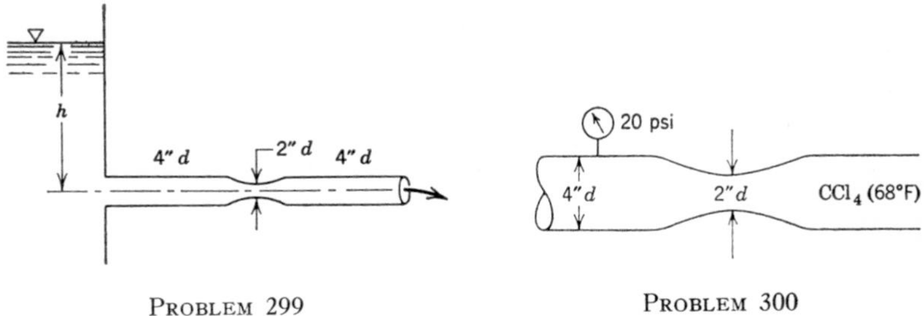

PROBLEM 299

PROBLEM 300

300. If cavitation is observed in the 2-in. section, what is the flowrate? Barometric pressure is 14.2 psia.

301. Barometric pressure is 14.0 psia. What is the maximum flowrate which can be obtained by opening the valve?

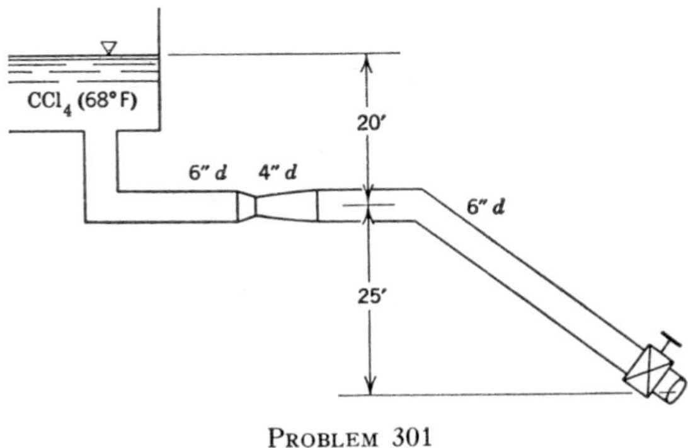

PROBLEM 301

302. A variety of combinations of d and h will allow maximum possible flowrate to occur through this system. Derive a relationship between d and h which will always produce this. Barometric pressure is 14.5 psia.

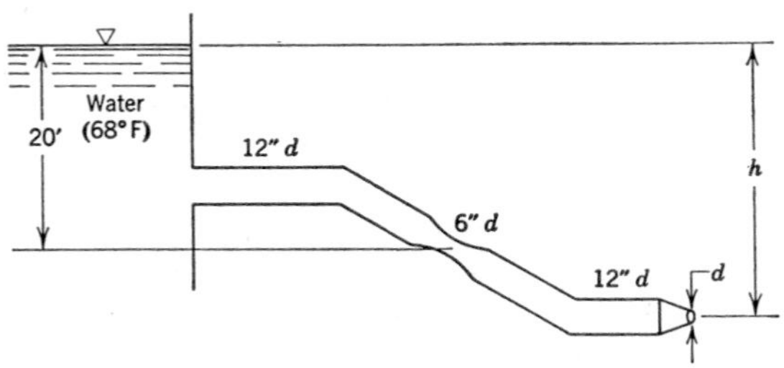

PROBLEM 302

303. The pressure in this closed tank is gradually increased by pumping. Calculate the gage reading at which cavitation will appear in the 1-in. constriction. The barometric pressure is 13.6 psia.

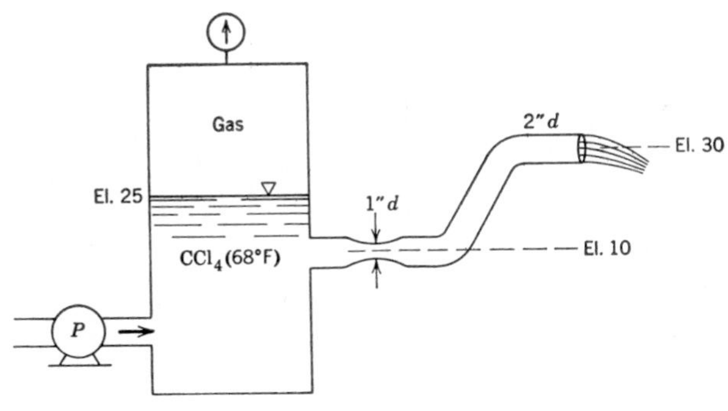

PROBLEM 303

304. Calculate the head h above which cavitation at the throat of the tube is to be expected. Barometric pressure is 14.3 psia.

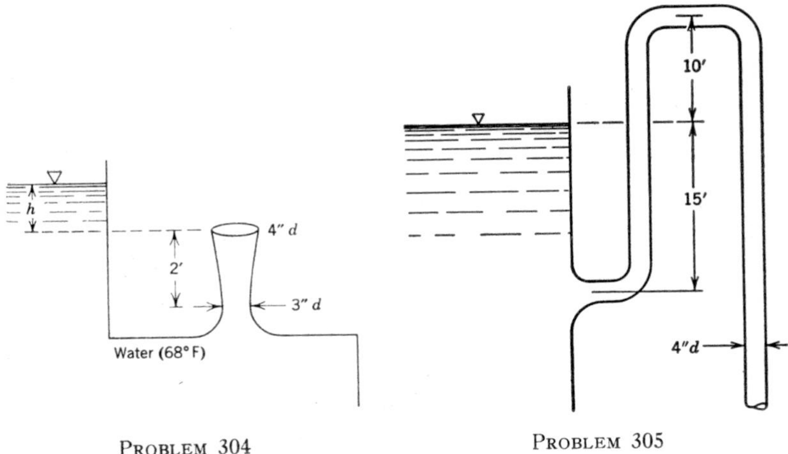

PROBLEM 304 PROBLEM 305

305. Calculate the maximum flowrate which can occur through this water line, assuming barometric pressure 13.5 psia and vapor pressure 0.5 psia.

306. Water at 150°F is being siphoned from a tank through a hose in which the velocity is 15 fps. What is the maximum theoretical height of the high point ("crown") of the siphon above the water surface which will allow this flow to occur? Assume standard barometric pressure.

307. Barometric pressure is 14.7 psia and vapor pressure of the water is 4.7 psia. Calculate the elevation of end B for maximum flowrate through this pipe line.

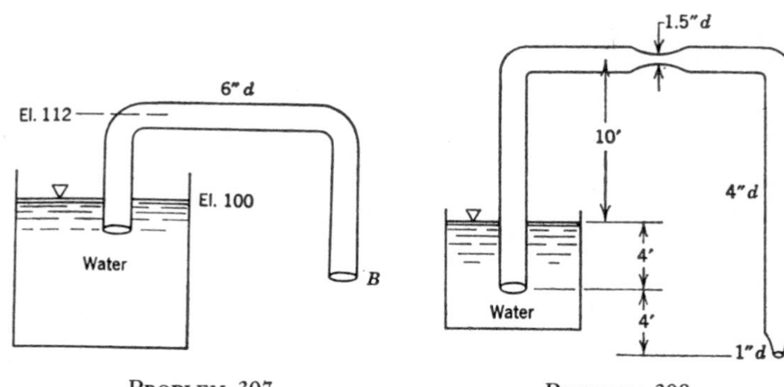

PROBLEM 307 PROBLEM 308

308. Barometric pressure is 14.8 psia and vapor pressure 7.0 psia. Calculate the pressure in the constriction.

309. The pressure in the test section of a wind tunnel is -1.07 in. of water when the velocity is 60 mph. Calculate the pressure on the nose of an object when placed in the test section. Assume the specific weight for air 0.0763 lb/cuft.

310. The pressure in a 4-in. pipe line carrying 1000 gpm of fluid weighing 70 lb/cuft is 20 psi. Calculate the pressure on the upstream end of a small object placed in this pipe line.

311. An airship flies through still air at 50 mph. What is the pressure on the nose of the ship if the air temperature is 40°F and pressure is 13.0 psia?

312. A submarine moves at 10 knots (1 knot = 6080 ft/hr) through salt water (s.g. = 1.025) at a depth of 50 ft. Calculate the pressure on the nose of the submarine.

313. Reach AB is the highest and smallest (diameter) part of this pipe line. Predict the largest possible manometer reading to be expected when the barometric pressure is 14.25 psia.

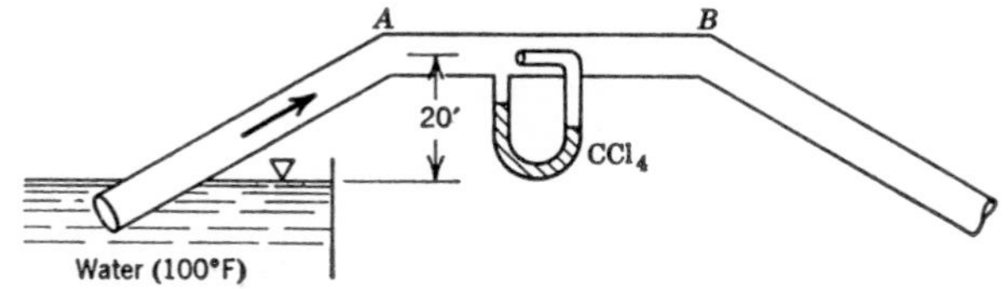

PROBLEM 313

314. Calculate the flowrate of water in this pipe line.

315. Ten cubic feet per second of water are flowing. Calculate the manometer reading (a) using the sketch as shown, (b) when the pitot tube is at section 2 and the static pressure connection is at section 1.

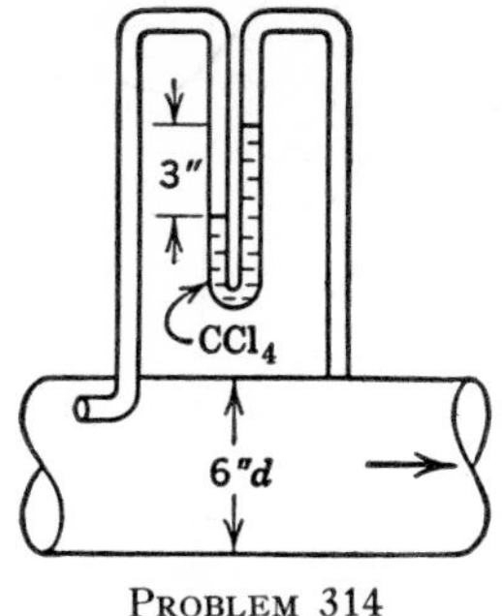

PROBLEM 314

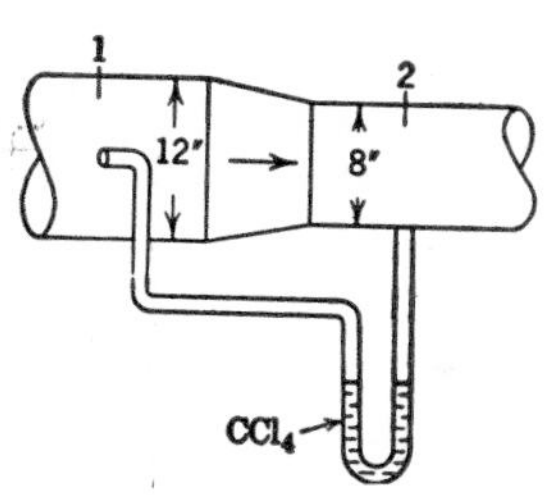

PROBLEM 315

316. Calculate the flowrate through this pipe line.

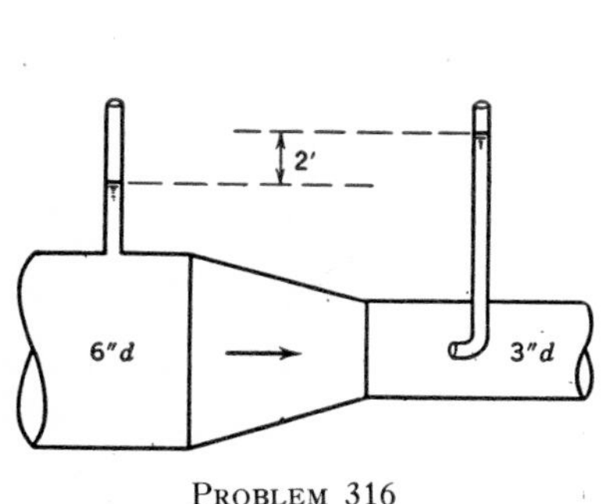

PROBLEM 316

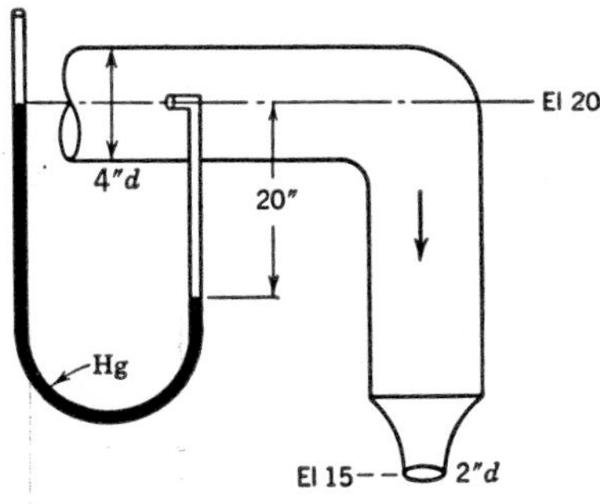

PROBLEM 317

317. Water is flowing. Calculate the flowrate.

318. Calculate the gage reading.

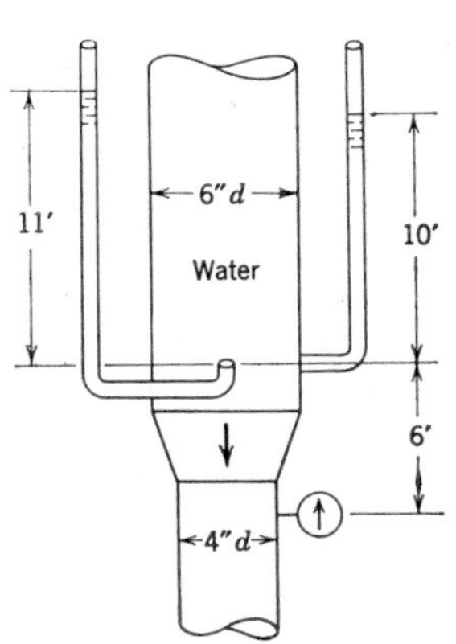

PROBLEM 318

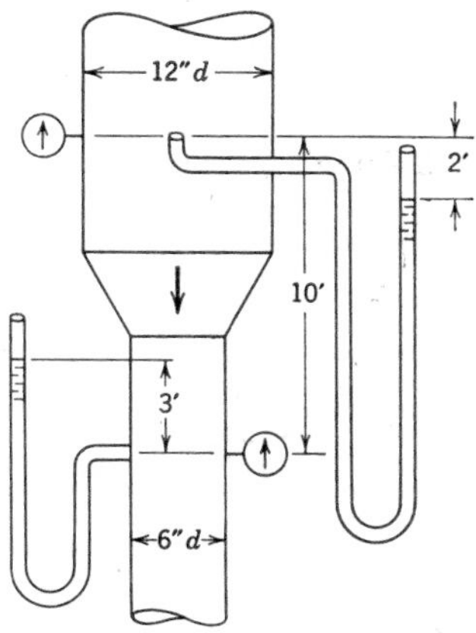

PROBLEM 319

319. Gasoline (s.g. 0.85) is flowing. Calculate gage readings and flowrate.

320. Calculate the velocity head in the 3-in. constriction.

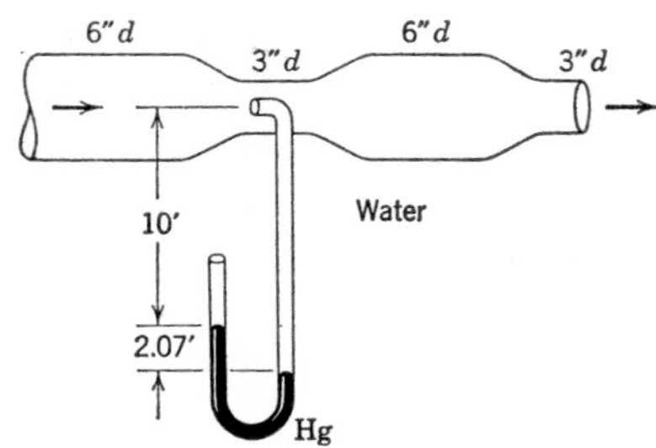

PROBLEM 320

321. Water is flowing. Assume the flow between the disks to be radial and calculate the pressures at A, B, C, and D. The flow discharges to the atmosphere.

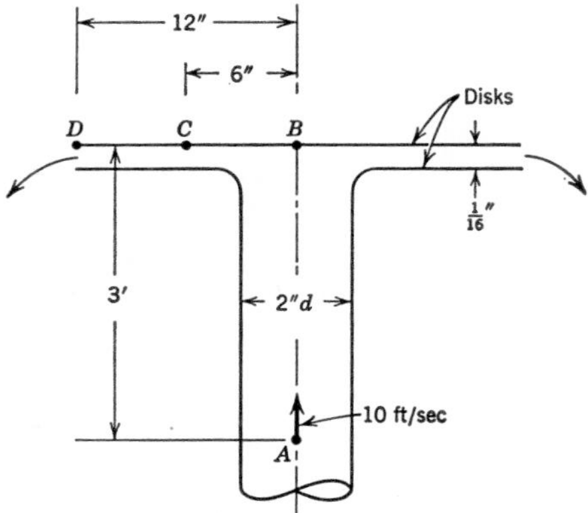

PROBLEM 321

322. Calculate the gage reading.

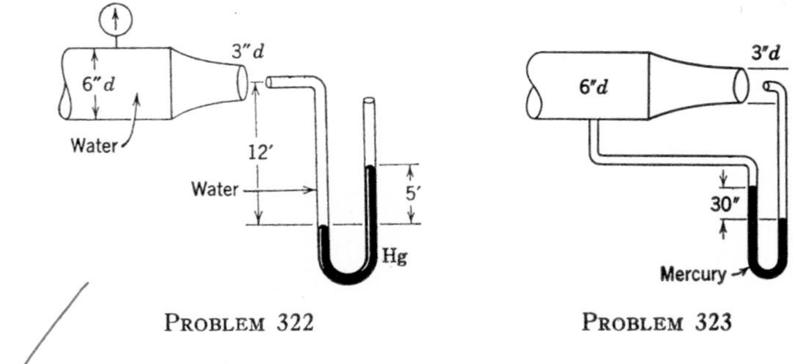

PROBLEM 322 PROBLEM 323

323. Calculate the flowrate of water through this nozzle.

324. Water is flowing. Calculate the flowrate and gage reading.

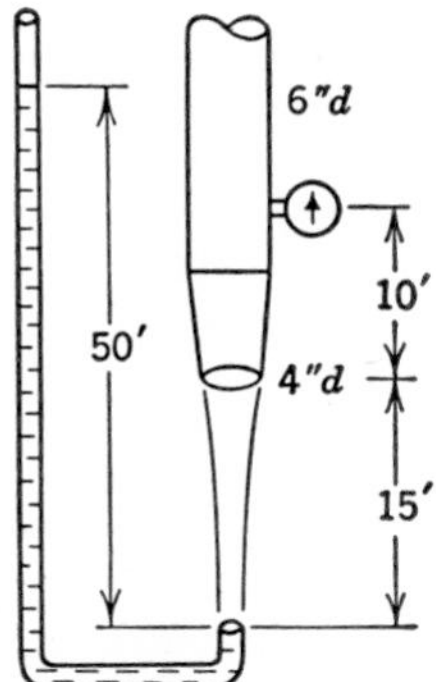

PROBLEM 324

325. A flowrate of 0.50 cfs of water discharges downward from a vertical 2-in. pipe line and strikes a horizontal surface 10 ft below the end of the pipe. What is the pressure at the intersection of jet axis and surface?

326. The tip of the pitot tube is at the top of the jet. Calculate the flowrate and the angle θ.

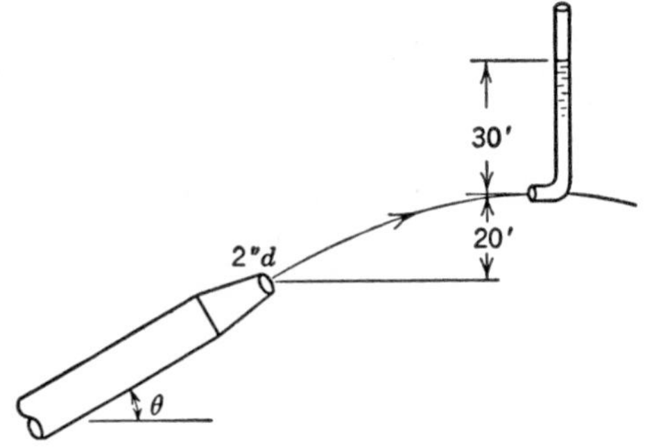

PROBLEM 326

327. Water is flowing. Calculate the pressure gage reading.

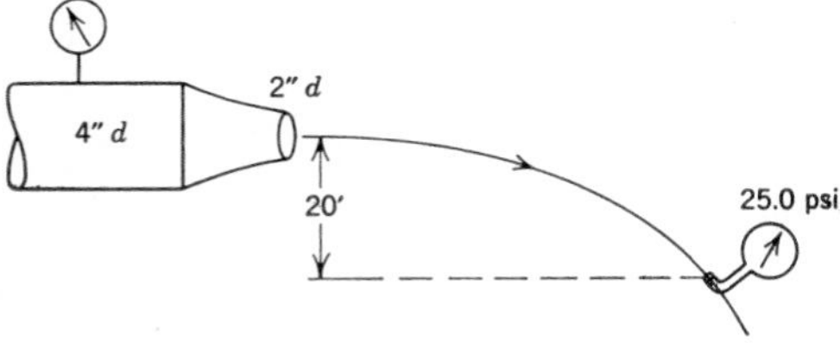

PROBLEM 327

328. A 1.5-in. smooth nozzle is connected to the end of a 6-in. water line. The pressure in the pipe behind the nozzle is 40 psi. Calculate the flowrate.

329. Air discharges from a duct of 12-in. diameter through a 4-in. nozzle into the atmosphere. The pressure in the duct is found by a draft gage to be 1.00 in. of water. Assuming the specific weight of air to be constant at 0.075 lb/cuft, what is the flowrate?

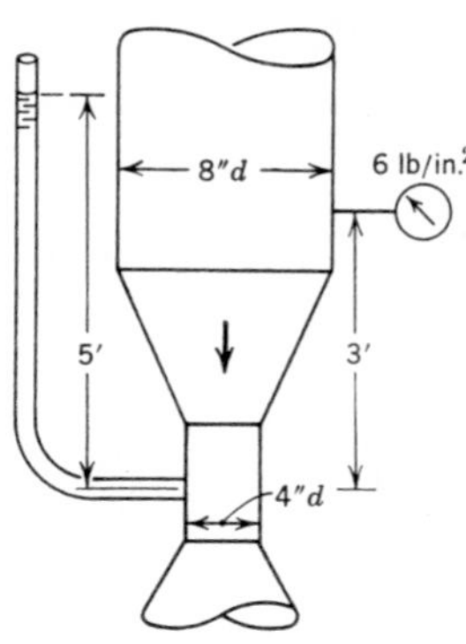

PROBLEM 333

330. If in problem 265 the vertical distance between the liquid surfaces of the piezometer tubes is 2 ft, what is the flowrate?

331. A smooth 2-in. nozzle terminates a 4-in. pipe line and discharges water vertically upward. If a pressure gage on the pipe 6 ft below the nozzle tip reads 50 psi, calculate the flowrate.

332. Water flows through a 3-in. constriction in a horizontal 6-in. pipe. If the pressure in the 6-in. section is 50 psi and that in the constriction 30 psi, calculate the velocity in the constriction and the flowrate.

333. Water is flowing. Compute the flowrate.

334. Ten cubic feet per second of liquid of specific weight 50 lb/cuft are flowing. Calculate the manometer reading.

335. In the preceding problem calculate the flowrate when the manometer reading is 10 in.

336. Water flows upward through a 6-in. constriction in a vertical 12-in. pipe. In the constriction there is a vacuum of 8 in. of mercury, and at a point on the pipe 5 ft below the constriction a pressure gage reads 30 psi. Calculate the flowrate.

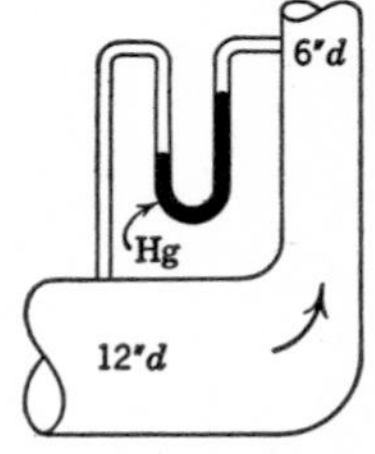

PROBLEM 334

337. Carbon tetrachloride flows downward through a 2-in. constriction in a 3-in. vertical pipe line. If a differential manometer containing mercury is connected to the constriction and to a point in the pipe 4 ft above the constriction and this manometer reads 14 in., calculate the flowrate. (Carbon tetrachloride fills manometer tubes to the mercury surfaces.)

338. Calculate the flowrate.

339. If a free jet of fluid strikes a circular disk and produces the flow picture shown, what is the flowrate?

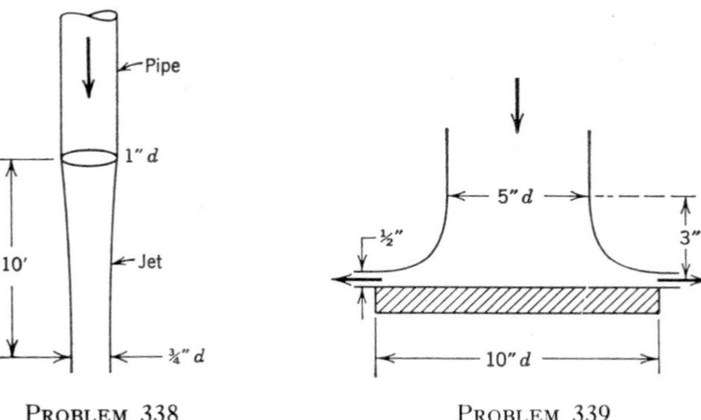

PROBLEM 338 PROBLEM 339

340. Through a transition structure between two rectangular open channels the width narrows from 8 ft to 7 ft, the depth decreases from 5 ft to 3.5 ft, and the bottom rises 1 ft. Calculate the flowrate.

341. Through a transition structure between two rectangular open channels the width increases from 10 ft to 12 ft while the water surface remains horizontal. If the depth in the upstream channel is 5 ft, what is the depth in the downstream channel?

342. If the two-dimensional flowrate over this sharp crested weir is 10 cfs/ft, what is the thickness of the sheet of falling water at a point 3 ft below the weir crest?

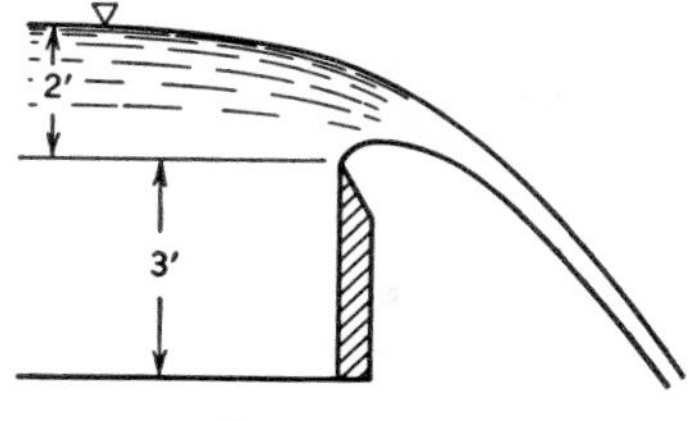

PROBLEM 342

343. Calculate the two-dimensional flowrate through this frictionless sluice gate when the depth h is 5 ft. Also calculate the depth h for a flowrate of 35 cfs/ft.

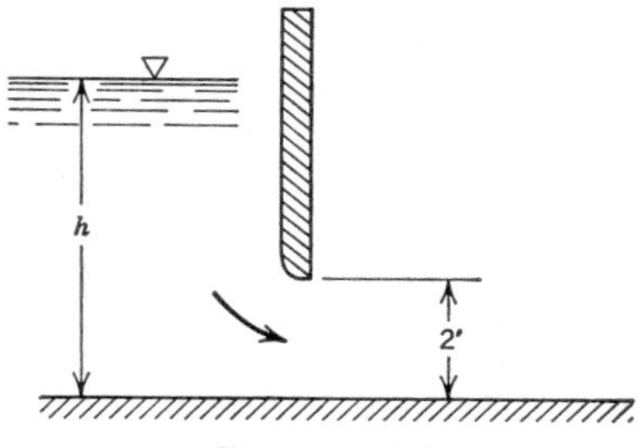

PROBLEM 343

344. A pump having a 4-in. suction pipe and 3-in. discharge pipe pumps 500 gpm of water. At a point on the suction pipe a vacuum gage reads 6 in. of mercury; on the discharge pipe 12 ft above this point a pressure gage reads 48 psi. Calculate the horsepower supplied by the pump.

345. A pump of what horsepower is theoretically required to raise 200 gpm of water from a reservoir of surface elevation 100 to one of surface elevation 250?

346. If 12 cfs of water are pumped over a hill through an 18-in. pipe line, and the hilltop is 200 ft above the surface of the reservoir from which the water is being taken, calculate the pump horsepower required to maintain a pressure of 25 psi on the hilltop.

347. A pump has 8-in. inlet and 6-in. outlet. If the hydraulic grade line rises 60 ft across the pump when the flowrate is 5.0 cfs of water, how many horsepower is the pump delivering to the fluid? For the same rise in hydraulic grade line, what flowrate will be maintained by an expenditure of 60 hp?

348. Water is flowing. Calculate the pump horsepower for a flowrate of 1.0 cfs.

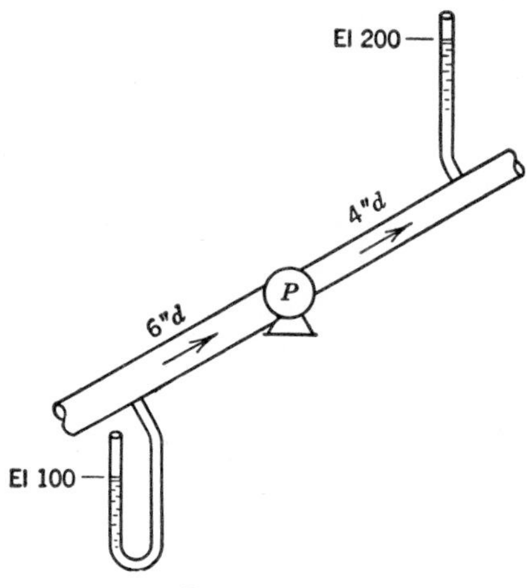

PROBLEM 348

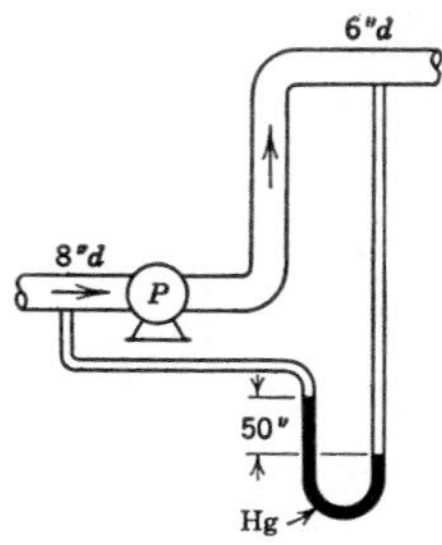

PROBLEM 349

349. Four cubic feet per second of water are flowing. Calculate the pump horsepower.

350. If pitot tubes replace the static pressure connections of the preceding problem and give the same manometer reading, what is the flowrate if the pump is supplying 10 hp to the fluid?

351. In this pump test the flowrate is 3 cfs of water. Calculate the pump horsepower.

352. This 7.0-hp pump maintains a flowrate of 2.0 cfs of water. Calculate the gage reading.

353. A pump takes water from a tank and discharges it into the atmosphere through a horizontal 2-in. nozzle. The nozzle is 15 ft above the water surface and is connected to the pump's 4-in. discharge pipe. What horsepower must the pump have to maintain a pressure of 40 psi just upstream from the nozzle?

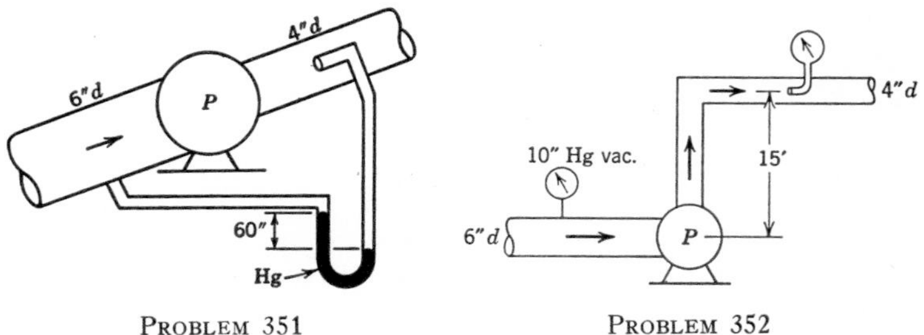

PROBLEM 351 PROBLEM 352

354. Calculate the pump horsepower, assuming that the diverging tube flows full.

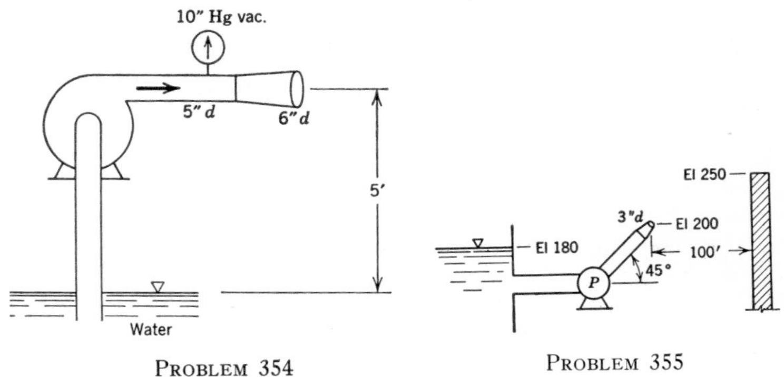

PROBLEM 354 PROBLEM 355

355. Calculate the minimum pump horsepower which will send the jet over the wall.

356. Calculate the pump horsepower required to maintain the flowrate indicated.

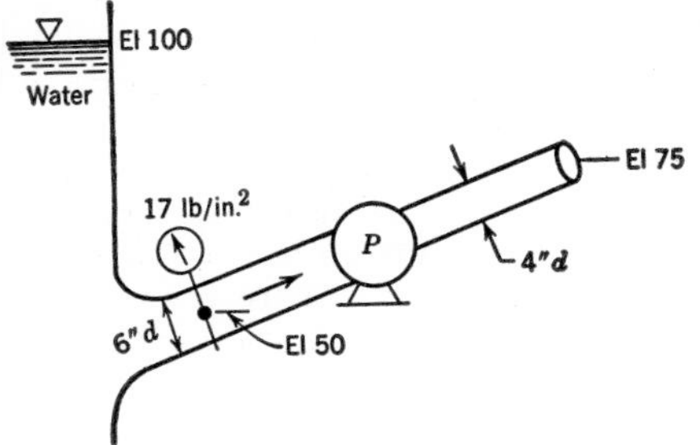

PROBLEM 356

357. Calculate the pump horsepower.

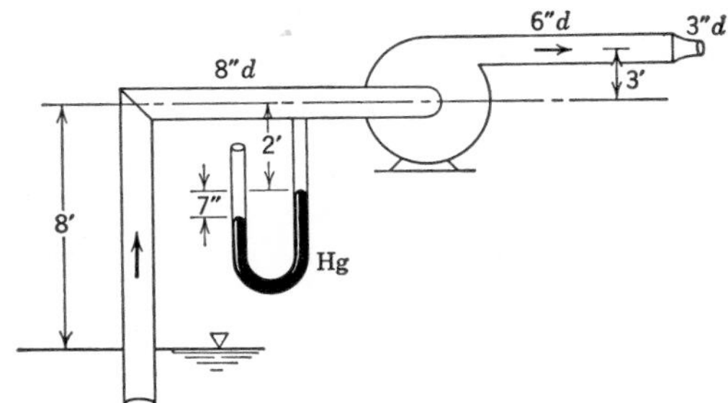

PROBLEM 357

358. Compute the pump horsepower required to maintain a flowrate of 4.0 cfs, if the barometric pressure is 14.3 psia and the vapor pressure is 1.0 psia. Calculate the maximum possible x for reliable operation.

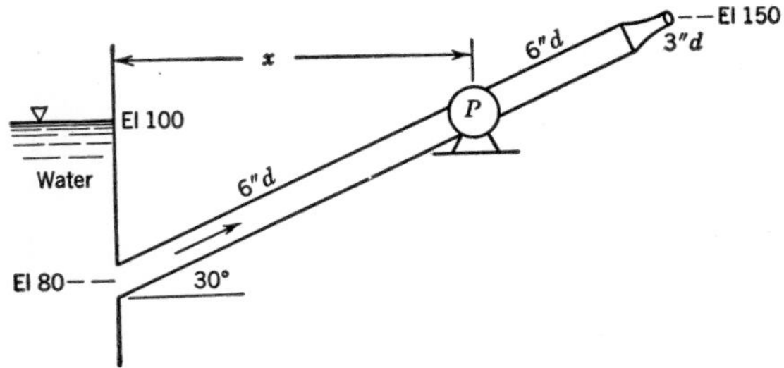

PROBLEM 358

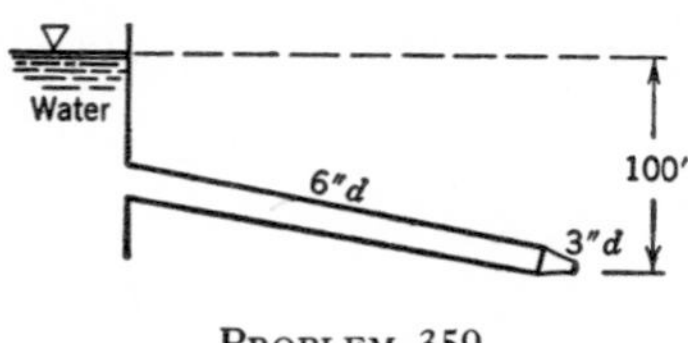

PROBLEM 359

359. A pump is to be installed to increase the flowrate through this pipe and nozzle by 20 percent. Calculate the required horsepower.

360. Through a 4-in. pipe, 1.0 cfs of water enters a small hydraulic motor and discharges through a 6-in. pipe. The inlet pipe is lower than the discharge pipe, and at a point on the inlet pipe a pressure gage reads 70 psi; 14 ft above this on the discharge pipe a pressure gage reads 40 psi. What horsepower is developed by the motor? For the same gage readings, what flowrate would be required to develop 10 hp?

361. A hydraulic turbine in a power plant takes 100 cfs of water from a reservoir of surface elevation 235 and discharges it into a river of surface elevation 70. What theoretical horsepower is available in this flow?

362. Calculate the h which will produce a flowrate of 3.0 cfs and a turbine output of 20 hp.

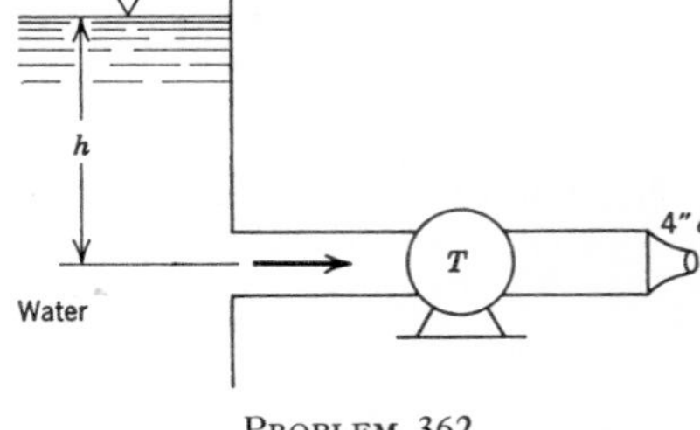

PROBLEM 362

363. Water flows from the rectangular channel into a 2-ft diameter pipe where the turbine extracts 10 ft-lb of energy from every pound of fluid. How far above the pipe outlet will the water surface be in the column connected to the pitot tube?

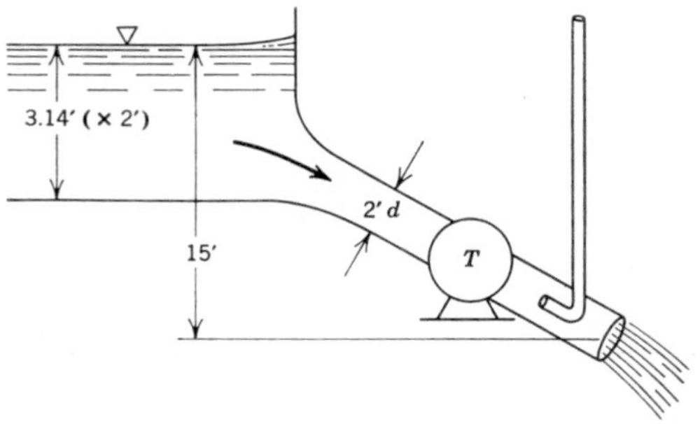

PROBLEM 363

364. Water is flowing. Calculate the horsepower of the turbine.

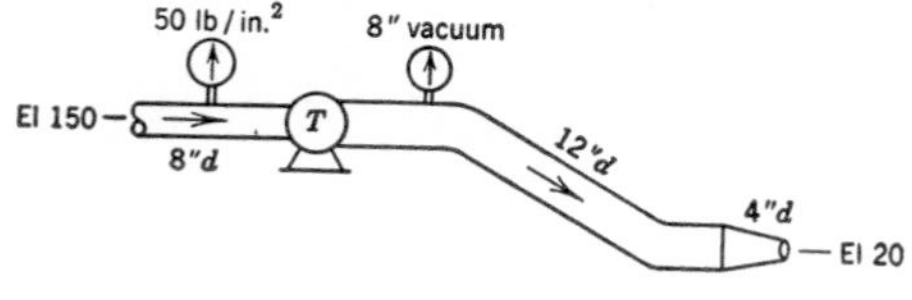

PROBLEM 364

365. Calculate the horsepower output of this turbine.

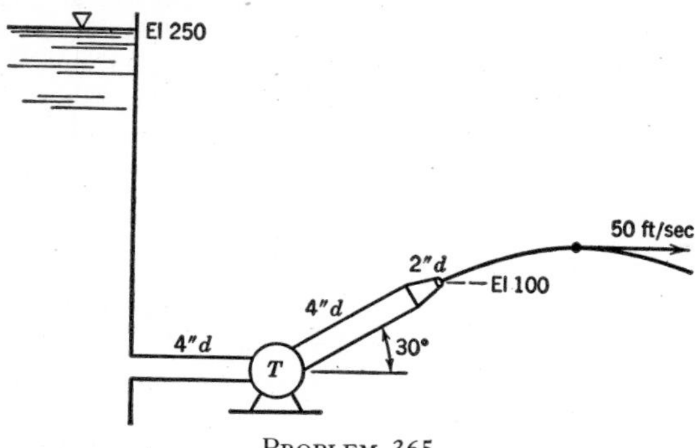

PROBLEM 365

366. What is the maximum horsepower the turbine can extract from the flow before cavitation will occur at some point in the system? Barometric pressure is 14.8 psia and vapor pressure of the water is 0.5 psia.

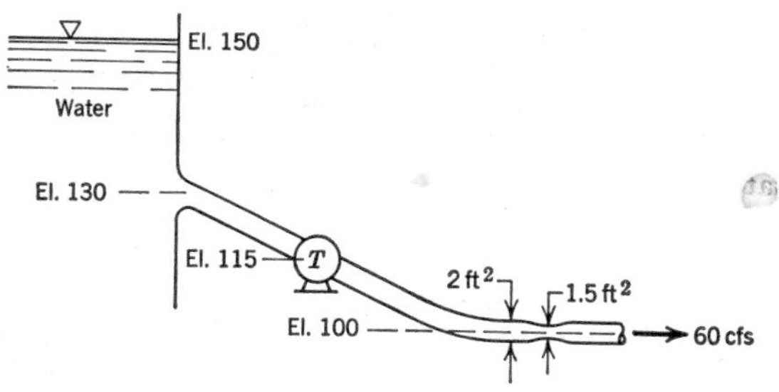

PROBLEM 366

367. Assuming a very large tank and density of air constant at 0.0025 slug/cuft, calculate (*a*) pressure just upstream from the turbine, (*b*) the turbine horsepower, and (*c*) the jet velocity when the turbine is removed.

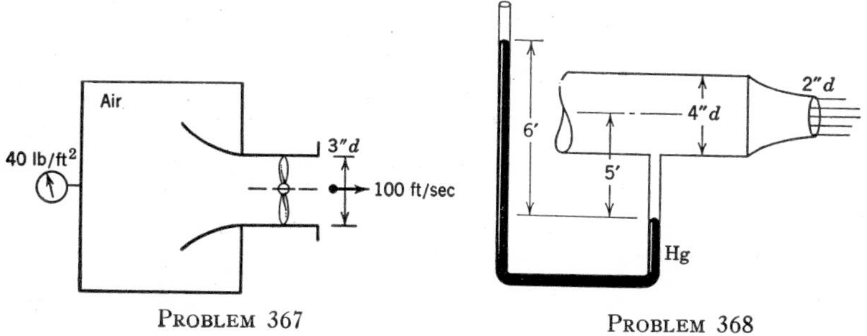

PROBLEM 367

PROBLEM 368

368. Calculate the horsepower available in the jet of water issuing from this nozzle.

369. If the irrotational flow of problem 237*f* is in a horizontal plane and the pressure head at the origin of coordinates is 10 ft, what is the pressure head at point (2, 2)? What is this pressure head when the flowfield is in a vertical plane?

370. If the irrotational flowfields of problem 237*l* and 237*m* are in a horizontal plane and the velocity and pressure head at a radius of 2 ft are 15 fps and 10 ft, respectively, what pressure head exists at a radius of 3 ft? When these flowfields are in vertical planes and the pressure head at ($\theta = 90°$, $r = 2$) is 10 ft, what is the pressure head at ($\theta = 0°$, $r = 3$)?

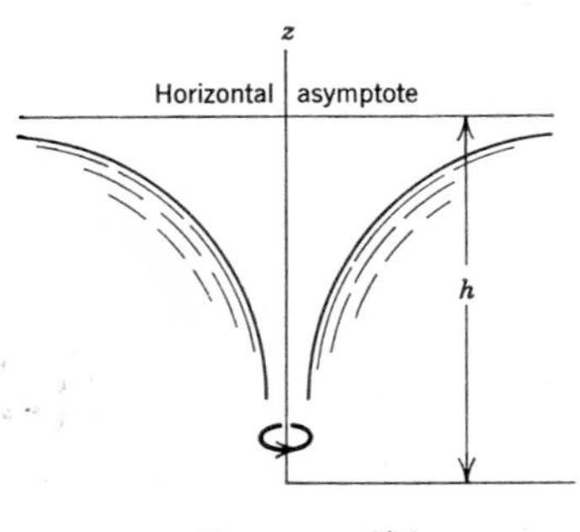

PROBLEM 371

371. If the irrotational flowfield of a free vortex is described by the equations $v_t = c/r$ and $v_r = 0$, derive an equation for the profile of the liquid surface.

372. A 180° circular bend occurs in a horizontal plane in a passage 3 ft wide. The radius of the inner wall is 1 ft and that of the outer wall 4 ft. If the velocity (of the water) halfway around the bend on the outer wall is 10 fps and the pressure there is 5 psi, what will be the velocity and pressure at the corresponding point on the inner wall (using the relation of equation 73)? Note that this equation is an approximation because the centers of curvature of all streamlines are not coincident with the center of curvature of the bend. Also calculate the approximate flowrate in the passage.

373. If the bend of the preceding problem is the top of a siphon, and thus in a vertical plane, what are the velocity and pressure? What is the approximate flowrate?

374. Estimate the peripheral velocity 100 ft from the center of a tornado if the peripheral velocity 400 ft from the center is 50 mph. If the barometer reads 28.0 in. of mercury at the latter point, what reading can be expected at the former? Assume air density 0.0025 slug/cuft.

375. Predict the flowrate through this two-dimensional outflow structure for a water depth of 5 ft in the canal. Also calculate the water depth on the upstream face of the structure and the pressure at *A*.

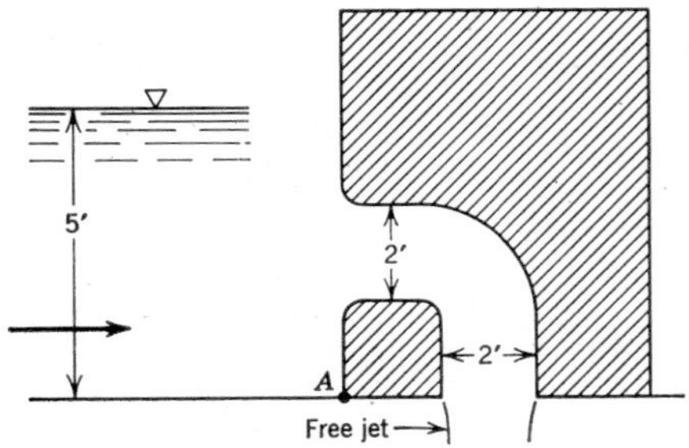

PROBLEM 375

376. The flowrate is 25.2 cfs. The gage pressure in the corner at A is 12.0 psi. Calculate the gage pressure at B and the horsepower of the turbine.

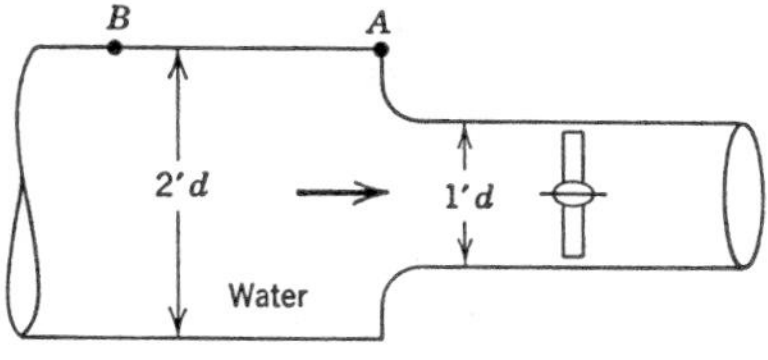

PROBLEM 376

377. A small circular cylinder spans the test section of a large water tunnel. A water and mercury differential manometer is connected to two small openings in the cylinder, one facing directly into the approaching flow the other at 90° to this point. For a manometer reading of 10 in. calculate the velocity approaching the cylinder.

378. For the flowfield described by equations 298 and 299 derive the equation of a line at all points of which the pressure is p_o. See Fig. 57.

379. If the depth at the vena contracta caused by this sluice gate is 2 ft, calculate the two-dimensional flowrate and the water depth on the upstream face of the gate.

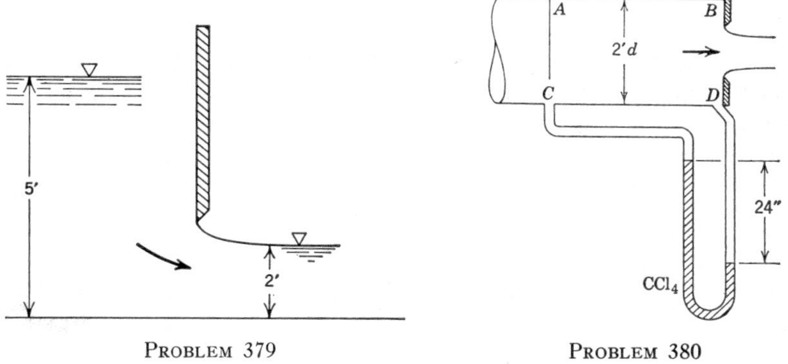

PROBLEM 379 PROBLEM 380

380. Water is flowing. The pressure at A is 1.75 psi. Calculate the flowrate, diameter of vena contracta, and pressures at points B, C, and D.

chapter 5

Flow of a Compressible Ideal Fluid

The development of the laws of motion for a fluid with large density changes will parallel that of Chapter 4 for an incompressible fluid; however, the development of these laws will prove more difficult for the compressible fluid because of (1) their dependence on thermodynamic considerations; (2) the appearance of new physical phenomena; and (3) more unwieldy equations. The first of these is the most troublesome because the author assumes that the reader has not yet had formal training in thermodynamics and sees no hope of including it here. Accordingly some thermodynamic equations are used with explanations which, although unsatisfactory for comprehensive understanding, are adequate for pursuit of the subject at hand. The beginner must realize that a complete understanding of the problems of compressible fluid flow cannot be acquired from fluid mechanics alone but depends upon successful synthesis of fluid mechanics with thermodynamics. The intent of the following treatment is to give the beginner some feeling for the subject, an understanding of some of its difficulties, and an appreciation of the striking differences between compressible and incompressible flow.

The assumption of an ideal (inviscid) fluid restricts the discussion to *frictionless* flow processes, as in Chapter 4 for the incompressible fluid. A further restriction here is the assumption of no heat transfer to or from the fluid, which is the definition of an *adiabatic* process.

In thermodynamics a *frictionless* adiabatic process is called an *isentropic* one since it is accompanied by no change of entropy. Such processes are closely approximated in practice if they occur with small friction and with such rapidity that there is little opportunity for heat transfer; thus the assumption of isentropic flow and the necessity that the results be of some engineering value confine developments of this chapter to high-velocity gas flow (*gas dynamics*) over short distances (e.g., nozzles rather than pipes) where in real problems friction and heat transfer are small.

High-velocity gas flow is associated in general with large changes of pressure, temperature, and density, but changes of these variables are of course much smaller in low-velocity flow; therefore many of the latter problems may be treated approximately yet satisfactorily by the methods of Chapter 4. However, there is no precise boundary between high-velocity and low-velocity motion, and whether a gas may be treated as an incompressible [1] fluid depends upon the accuracy of the results required. Only experience with the equations of Chapters 4 and 5 will provide an answer to this question.

ONE-DIMENSIONAL FLOW

29. Euler's Equation. Development of the Euler equation for one-dimensional flow of a compressible fluid is the same as that for an incompressible one (Art. 21). Although fluid density will vary along the streamline (Fig. 59), the mean density ρ of the differential element of fluid differs negligibly from that at its ends and thus may be taken as constant throughout the element.[2]

Euler's equation is therefore (as before)

$$\frac{dp}{\rho} + V\,dV + g\,dz = 0$$

For compressible flow, however, the term $g\,dz$ is usually dropped and the Euler equation written

$$\frac{dp}{\rho} + V\,dV = 0 \quad \text{or} \quad \frac{dp}{\gamma} + \frac{V\,dV}{g} = 0 \tag{75}$$

This simplification is justified by the fact that compressible-flow problems are usually concerned with gases of light weight and with flows of small vertical extent in which changes of pressure and

[1] For example, in the aerodynamics of low-speed aircraft it is a sufficiently accurate approximation to consider the air incompressible; for military aircraft, missiles, jet engines, etc., such an assumption would obviously be unsatisfactory.

[2] A similar situation is discussed in Art. 8.

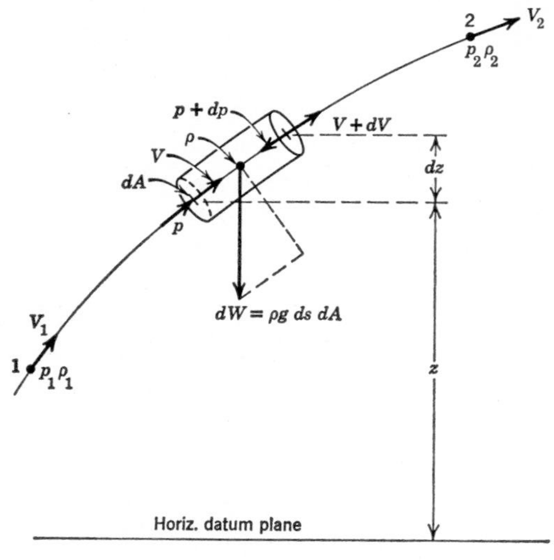

FIG. 59

velocity are predominant and changes of elevation negligible[3] by comparison; another way of justifying this simplification is to assume g zero, which is equivalent to neglecting the weight (but not the mass) of the fluid.

30. The Energy Equation. The energy equation for the flow of a compressible fluid in a small streamtube (Fig. 60) may be written as an extension of the energy equation (64) of Art. 23. For the compressible fluid, internal energy (I) must be included and $(z_1 - z_2)$ may be neglected. The result is

$$I_1 + \frac{p_1}{\gamma_1} + \frac{V_1^2}{2g} = I_2 + \frac{p_2}{\gamma_2} + \frac{V_2^2}{2g} \tag{76}$$

in which internal, pressure, and kinetic energies (ft-lb/lb) are set down in order. Internal energy is the energy associated with kinetic energy of the molecules and the forces between them and thus manifests itself in temperature,[4] high or low temperature implying high or low internal energy, respectively. Thermodynamics defines the *enthalpy* H as $I + p/\gamma$, which allows equation 76 to be written

$$H_1 + \frac{V_1^2}{2g} = H_2 + \frac{V_2^2}{2g} \tag{77}$$

[3] See the fourth illustrative problem of Art. 28.

[4] For perfect gases, thermodynamics shows I to be a function of temperature only and change of internal energy related to change of temperature by $I_1 - I_2 = c_v(T_1 - T_2)$.

For application to perfect gases these equations may be written in other useful forms. Since $I = c_v T$ and $p/\gamma = RT$, $H = (c_v + R)T$; however, from Art. 4, $c_v + R = c_p$, which gives $H = c_p T$. Substituting this into equation 77, it becomes

$$c_p T_1 + \frac{V_1^2}{2g} = c_p T_2 + \frac{V_2^2}{2g} \tag{78}$$

The foregoing equations apply equally well to frictional and frictionless flow, but only for the adiabatic case in which no heat is added to or extracted from the fluid; also they provide only for the situation where no mechanical energy is added to or extracted from the fluid by pump or turbine. For the remainder of this chapter their application will be confined to frictionless adiabatic (termed *isentropic*) flow.

Comparison of the energy and Euler equations is easily accomplished by writing the former in differential form:

$$dI + p\,d\left(\frac{1}{\gamma}\right) + \frac{dp}{\gamma} + \frac{V\,dV}{g} = 0$$

However, for the isentropic process, thermodynamics shows that

$$dI + p\,d\left(\frac{1}{\gamma}\right) = 0$$

through which the energy equation reduces to

$$\frac{dp}{\gamma} + \frac{V\,dV}{g} = 0$$

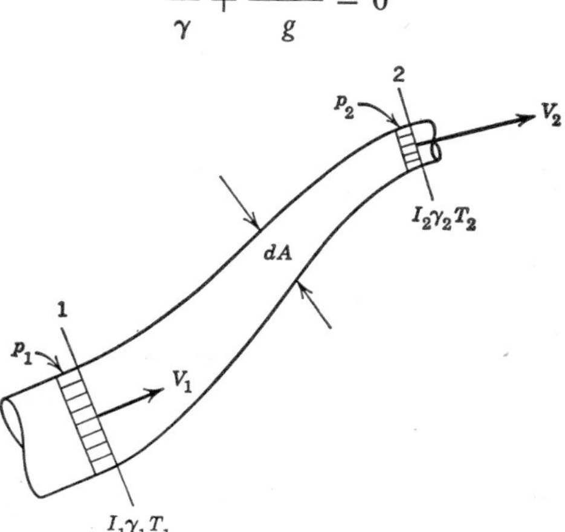

FIG. 60

which is recognized as the Euler equation and thus proves the energy and Euler equations to be identical for isentropic flow.

31. Integration of the Euler Equation. When the Euler equation is integrated along the streamline (or small streamtube) for isentropic flow of perfect gases, it becomes

$$\frac{V_2^2 - V_1^2}{2g} = \int_{p_2}^{p_1} \frac{dp}{\gamma} = \frac{p_1}{\gamma_1} \frac{k}{k-1} \left[1 - \left(\frac{p_2}{p_1}\right)^{\frac{k-1}{k}}\right]$$

$$\text{or} \quad \frac{p_2}{\gamma_2} \frac{k}{k-1} \left[\left(\frac{p_1}{p_2}\right)^{\frac{k-1}{k}} - 1\right] \qquad (79)$$

in which the integration has been performed by substitution of the isentropic relation $p_1/\gamma_1^k = p_2/\gamma_2^k$. The same result may also be obtained from the energy equation (78) by rearranging it and substituting [5] $kp/\gamma(k-1)$ for c_pT and $(p_2/p_1)^{\frac{1}{k}}$ for γ_2/γ_1. From equations 78 and 79, the variation of pressure and temperature (and thus density) with velocity may be predicted along the streamline of Fig. 59 or the small streamtube of Fig. 60.

Equation 78, which shows $(c_pT + V^2/2g)$ to be constant along any streamline in adiabatic flow, is frequently written

$$c_pT_1 + \frac{V_1^2}{2g} = c_pT_2 + \frac{V_2^2}{2g} = c_pT_s \qquad (78)$$

At a stagnation point where V is zero, T is the *stagnation temperature* T_s, which is seen to be constant [6] for all points on the streamline.

For the isentropic flow of vapors, equation 77 may be used directly; the enthalpy H is a function of pressure and temperature and may be obtained from appropriate tables and diagrams; for isentropic flow, values of H_1, and H_2 for the same entropy must be used.

Illustrative Problems

At one point on a streamline in an air flow the velocity, pressure, and temperature are 100 fps, 50 psia, and 300°F, respectively. At a second point on the same streamline the velocity is 500 fps. Calculate the pressure and temperature there.

[5] See Art. 4.

[6] Thus stagnation temperature in adiabatic flow is analogous to the total head of ideal incompressible flow.

Solution.

$$\frac{(500)^2 - (100)^2}{2g} = 186.5(T_1 - T_2) \tag{78}$$

$$T_1 - T_2 = 20°\text{F}; \quad t_2 = 280°\text{F}$$

Using equation 79 (with RT_1 substituted for p_1/γ_1),

$$\frac{(500)^2 - (100)^2}{2g} = 53.3(300 + 460)\frac{1.4}{0.4}\left[1 - \left(\frac{p_2}{p_1}\right)^{0.286}\right] \tag{79}$$

$$p_2/p_1 = 0.911, \quad p_2 = 45.5 \text{ psia}$$

Steam in a large tank is at pressure and temperature of 300 psia and 900°F, respectively. The steam flows from the tank into a passage, and at a point in the passage the pressure is observed to be 200 psia. Determine the temperature and velocity at this point.

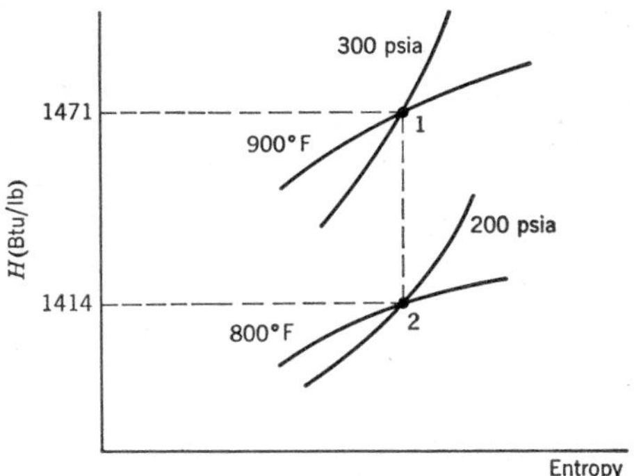

Solution. Refer to the *Mollier diagram* contained in any text on thermodynamics; this is a plot of enthalpy H against entropy for various pressures and temperatures. At the intersection of the 300 psia and 900°F lines on the diagram, H_1 is found to be 1471 Btu/lb. Now drop vertically down the chart (at constant entropy) to the 200 psia line. From the temperature line passing through point 2 the temperature may be read; it is 780°F. For this point the enthalpy H_2 is found to be 1414 Btu/lb. Substituting H_1 and H_2 into equation 77, noting that the velocity V_1 in the large tank will be zero,

$$V_2^2/2g = 778(1471 - 1414); \quad V_2 = 1660 \text{ fps} \tag{77}$$

32. The Stagnation Point. In modern gas dynamics, equation 79 is usually expressed in terms of Mach number, **M**. Using the first form of the equation, this may be easily accomplished by recalling (equation 10) that the acoustic (sonic) velocity a is given by $\sqrt{kp/\rho}$ and thus $a_1^2 = gkp_1/\gamma_1$. Substituting this into equation 79 and rearranging, there results

$$\frac{V_2^2}{a_1^2} = \mathbf{M}_1^2 + \frac{2}{k-1}\left[1 - \left(\frac{p_2}{p_1}\right)^{\frac{k-1}{k}}\right] \tag{80}$$

Now consider the application of this equation to a stagnation point (s) in a compressible flow (Fig. 61). With the fluid compressible, the rise of pressure at the stagnation point will cause compression of the fluid, producing a higher density (ρ_s) and temperature (T_s) there. Evidently, the extent of these compression effects will depend primarily upon the magnitude of the stream velocity V_o; they will be large at high velocities and small (often negligible) at low velocities. At the stagnation point, $V_2 = 0$ and $p_2 = p_s$; substituting these values and $p_1 = p_o$ and $\mathbf{M}_1 = \mathbf{M}_o$ into equation 80, the stagnation pressure p_s is given by

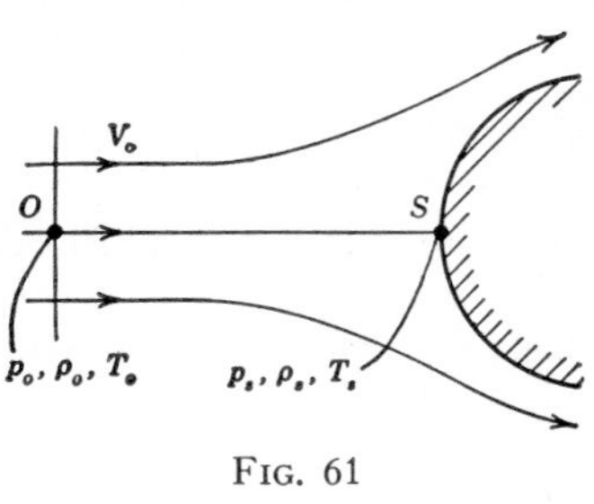

FIG. 61

$$\frac{p_s}{p_o} = \left[1 + \mathbf{M}_o^2\,\frac{k-1}{2}\right]^{\frac{k}{k-1}} \tag{81}$$[7]

If the right-hand side of this equation is expanded by the binomial theorem there results (retaining the first three terms)

$$p_s = p_o + \tfrac{1}{2}\rho_o V_o^2\,[1 + \tfrac{1}{4}\mathbf{M}_o^2 + \cdots] \tag{82}$$[7]

Comparison of this equation with equation 66 of Art. 25 shows that the effects of compressibility have been isolated in the bracketed quantity and that these effects depend only upon the Mach number. The bracketed quantity may thus be considered a "compressibility correction factor" and the effect of compressibility on $(p_s - p_o)$ calculated with fair precision.[8]

From equation 81 it will be observed that measurements of p_s and p_o will allow calculation of the "free stream" Mach number $\mathbf{M}_o$ of the undisturbed flow. However, to obtain the velocity V_o a temperature measurement is also required, and in practice the temperature T_s is measured at the stagnation point. From T_s, p_s, and p_o, the velocity V_o may be handily calculated by using the second form of

[7] The use of equations 81, 82, and 83 is restricted to $\mathbf{M}_o < 1$. For $\mathbf{M}_o > 1$ a shock wave exists across the streamline between points o and s and the flow is no longer isentropic. See Arts. 46 and 94.

[8] The precision is excellent for small $\mathbf{M}_o$ but decreases with increasing $\mathbf{M}_o$ as the neglected terms of the binomial expansion become significant.

equation 79 between points o and s; this equation becomes (RT_s having been substituted for p_s/γ_s)

$$\frac{V_o^2}{2g} = c_p T_s \left[1 - \left(\frac{p_o}{p_s}\right)^{\frac{k-1}{k}}\right] \tag{83}$$ [7]

whence V_o may be calculated directly.

ILLUSTRATIVE PROBLEM

An airplane flies at 400 mph (586 fps) through still air at 13.0 psia and 0°F. Calculate pressure, temperature, and air density at the stagnation points (on nose of fuselage and leading edges of wings).

Solution.

$$\rho_o = \frac{13.0 \times 144}{32.2 \times 53.3 \times 460} = 0.00237 \text{ slug/cuft} \qquad (1 \text{ and } 3)$$

$$a_o = \sqrt{1.4 \times 32.2 \times 53.3 \times 460} = 1052 \text{ fps} \qquad (11)$$

$$\mathbf{M}_o = 586/1052 = 0.557$$

$$(\text{approx.})\ p_s = 13.0 + \frac{1}{2}\frac{0.00237 \times (586)^2}{144}\left[1 + \tfrac{1}{4}(0.557)^2\right] = 16.05 \text{ psia} \qquad (82)$$

$$(\text{exact})\ p_s = 13.0\left[1 + (0.557)^2 \frac{1.4 - 1}{2}\right]^{3.5} = 16.18 \text{ psia} \qquad (81)$$

$$(586)^2/2g = 186.5(T_s - 460); \quad T_s = 488.5°\text{R}; \quad t_s = 28.5°\text{F} \qquad (78)$$

$$\rho_s = \frac{16.18 \times 144}{488.5 \times 32.2 \times 53.3} = 0.00278 \text{ slug/cuft} \qquad (1 \text{ and } 3)$$

33. The One-Dimensional Assumption. The foregoing equations of this chapter have been developed for flow along a streamline or through an infinitesimal streamtube across which there is no variation of velocity, pressure, temperature, and density; however (as in incompressible flow), they may be applied successfully to passages of finite cross section when the streamlines are essentially straight and parallel. Consistent with neglecting the difference in the z terms in the Euler and energy equations, the pressure is taken to be constant [9] over the flow cross section; the absence of friction permits no variation of velocity. With pressure and velocity constant throughout the flow cross section, constancy of temperature and density will follow from equations 78 and 79.

[9] For the incompressible fluid $(p/\gamma + z)$ is constant over the flow cross section; see Art. 24.

When streamlines are not essentially straight and parallel, variations of pressure, velocity, temperature, and density are to be expected. As with incompressible flow,[10] increase of pressure and decrease of velocity will be found with increasing distance from center of curvature; in compressible flow, such variations of pressure and velocity will produce variations of temperature and density as well. Although mean values of the variables may be visualized and computed, their use in the equations of one-dimensional flow is not recommended except for approximate calculations.

34. Subsonic and Supersonic Velocities. Combination of the continuity and Euler equations will yield information on the superficial shapes of passages required to produce changes of flow velocity when such velocities are subsonic or supersonic. These equations are

$$\frac{dA}{A} + \frac{d\rho}{\rho} + \frac{dV}{V} = 0 \tag{51}$$

$$\frac{dp}{\rho} + V\,dV = 0 \tag{75}$$

Multiplying the first term of equation 75 by $d\rho/d\rho$, a^2 is recognized (equation 9) as $dp/d\rho$, and $d\rho/\rho$ is obtained as $-V\,dV/a^2$ or $-V^2\,dV/a^2V$. Substituting this in equation 51, identifying V/a as the Mach number $\mathbf{M}$, and rearranging,

$$\frac{dA}{A} = \frac{dV}{V}(\mathbf{M}^2 - 1) \tag{84}$$

From this equation may be deduced some far-reaching and somewhat surprising conclusions. Analysis of the equation for $dV/V > 0$ shows that: for $\mathbf{M} < 1$, $dA/A < 0$; for $\mathbf{M} = 1$, $dA/A = 0$; for $\mathbf{M} > 1$, $dA/A > 0$. This means that, for subsonic flow ($\mathbf{M} < 1$), a reduction of cross-sectional area (Fig. 62) is required for an increase of velocity; however, for supersonic flow ($\mathbf{M} > 1$), *an increase of area is required to produce an increase of velocity.* For flow at sonic speed ($\mathbf{M} = 1$) the rate of change of area must be zero; that is, this might occur at a maximum or minimum cross section of the streamtube; use of the

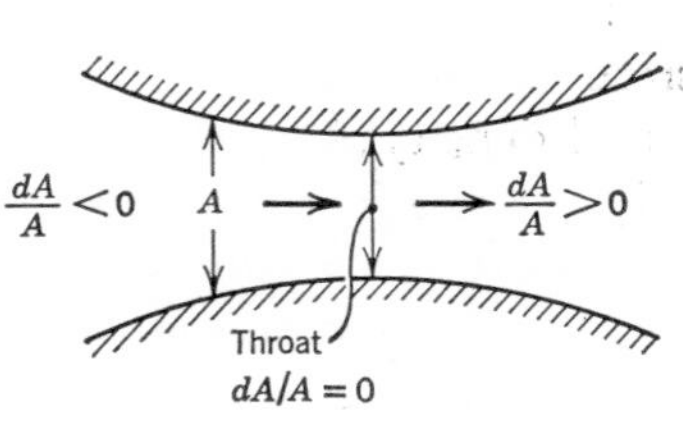

Fig. 62

[10] See Art. 28

preceding conclusions will show this to be restricted to a minimum section (*a throat*) only. However, equation 84 does *not* allow the conclusion that a throat will always produce a flow of sonic velocity since $\mathbf{M}$ is *not necessarily one* when $dA/A = 0$. If, at a throat ($dA/A = 0$) $\mathbf{M} \gtreqless 1$, it follows that $dV/V = 0$, implying a maximum or minimum velocity there. Upstream from the throat $dA/A < 0$, so here $dV/V < 0$ for $\mathbf{M} > 1$ and $dV/V > 0$ for $\mathbf{M} < 1$. Accordingly it may be concluded that if the throat velocity is not sonic it will be a maximum in subsonic flow and a minimum in supersonic flow.

35. The Convergent Nozzle. Consider now the frictionless flow of a gas from a large tank ($A_1 \sim \infty$, $V_1 \sim 0$) through a convergent

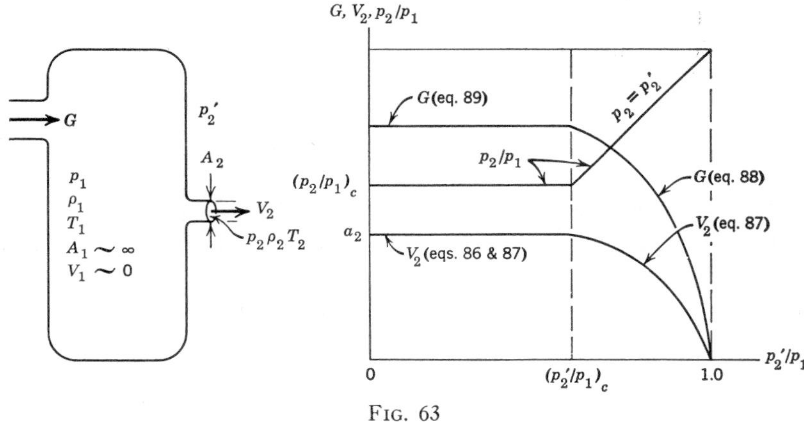

FIG. 63

nozzle (Fig. 63) into a region of pressure, p_2'. Using the second form of equation 79 and substituting a_2^2 for kgp_2/γ_2 and $\mathbf{M}_2$ for V_2/a_2,

$$\mathbf{M}_2^2 = \frac{2}{k-1}\left[\left(\frac{p_1}{p_2}\right)^{\frac{k-1}{k}} - 1\right] \tag{85}$$

In view of the preceding development, supersonic velocities will not be expected in this problem, since the fluid starts from rest and there are no divergent passages; accordingly $\mathbf{M}_2 \leqq 1$. If the pressure difference ($p_1 - p_2'$) is large enough to produce sonic velocity, this velocity must exist at the throat of the nozzle and $\mathbf{M}_2 = 1$; placing this value in equation 85 and solving for p_2/p_1, the result is designated as the critical pressure ratio $(p_2/p_1)_c$:

$$\left(\frac{p_2}{p_1}\right)_c = \left(\frac{2}{k+1}\right)^{\frac{k}{k-1}} \tag{86}$$

Thus, if the sonic velocity is attained by the fluid, the absolute pressure, p_2, in the minimum section is in fixed ratio to the absolute pressure, p_1, in the tank and *therefore independent of the pressure* p_2'. It follows that in a free jet moving at sonic velocity *the pressure within the jet at the nozzle exit is never less, and is usually more, than the pressure which surrounds it;* the outside pressure, p_2', tends to penetrate the jet at the sonic speed but cannot distribute itself over the whole jet cross section at the nozzle exit because the fluid there is moving at this same high velocity.

Of course, the sonic velocity is not attained unless the pressure drop between the inside and outside of the tank is large enough. For small pressure drops (i.e., for pressure ratios, p_2/p_1, above the critical) the pressures p_2' and p_2 are the same, and the velocity at the nozzle exit may be computed (from equation 79) by

$$\frac{V_2^2}{2g} = \frac{p_1}{\gamma_1}\frac{k}{k-1}\left[1 - \left(\frac{p_2}{p_1}\right)^{\frac{k-1}{k}}\right] \tag{87}$$

A graphic summary of these facts is given in Fig. 63.

In many problems (especially those of fluid metering) the flowrate is the most important quantity to be computed in flow through nozzles. The computation may be easily made by combining some of the foregoing equations.

If the pressure ratio p_2'/p_1 is more than the critical, the weight flowrate may be calculated from $G = A_2\gamma_2 V_2$, using V_2 from equation 87 and the isentropic relation $p_2/p_1 = (\gamma_2/\gamma_1)^k$. The result is

$$G = A_2\sqrt{\frac{2gk}{k-1}p_1\gamma_1\left[\left(\frac{p_2}{p_1}\right)^{\frac{2}{k}} - \left(\frac{p_2}{p_1}\right)^{\frac{k+1}{k}}\right]} \tag{88}$$

If the pressure ratio p_2'/p_1 is less than the critical, $p_2/p_1 = [2/(k+1)]^{k/(k-1)}$, and substituting this in equation 88, along with p_1/RT_1 for γ_1, yields

$$G = \sqrt{\frac{gk}{R}\left(\frac{2}{k+1}\right)^{\frac{k+1}{k-1}}}\;\frac{A_2 p_1}{\sqrt{T_1}} \tag{89}$$

in which the first square root is obviously a characteristic constant of the gas, dependent upon R, k, and g; this allows a simple calculation for flowrate and is good reason for selecting metering nozzles of such proportions that sonic velocities are produced.

ILLUSTRATIVE PROBLEM

Air discharges from a large tank, in which the pressure is 100 psi and temperature 100°F, through a convergent nozzle of 1-in. tip diameter. Calculate the flowrates when the pressure outside the jet is (*a*) 30 psi, and (*b*) 80 psi, and the barometric pressure is 14.7 psia. Also calculate the pressure, temperature, velocity, and sonic velocity at the nozzle tip for these flowrates.

Solution. First calculate specific weight of the air in the tank and the critical pressure ratio.

$$\gamma_1 = 114.7 \times 144/53.3 \times 560 = 0.553 \text{ lb/cuft} \qquad (3)$$

$$(p_2/p_1)_c = (2/2.4)^{3.5} = 0.528 \qquad (86)$$

(*a*)
$$p_2'/p_1 = 44.7/114.7 = 0.390 < 0.528$$

so

$$p_2 = 0.528 \times 114.7 = 60.6 \text{ psia}$$

and substituting $g = 32.2$, $k = 1.40$, $R = 53.3$, $A_2 = 0.005454$, $p_1 = 114.7 \times 144$, and $T_1 = 560$ into equation 89, $G = 2.03$ lb/sec. Using the first form of equation 79, with $V_1 = 0$ and $p_2/p_1 = 0.528$, $V_2 = 1060$ fps, which is also the sonic velocity. Using equation 78 with $V_1 = 0$, $V_2 = 1060$ fps, and $T_1 = 560°\text{R}$: $T_2 = 466°\text{R}$.

(*b*)
$$p_2'/p_1 = 94.7/114.7 = 0.825 > 0.528$$

so

$$p_2 = 94.7 \text{ psia}$$

and, substituting $A_2 = 0.005454$, $g = 32.2$, $k = 1.40$, $p_1 = 114.7 \times 144$, $\gamma_1 = 0.553$, and $p_2/p_1 = 0.825$ into equation 88: $G = 1.59$ lb/sec. Using the first form of equation 79 with $V_1 = 0$ and $p_2/p_1 = 0.825$: $V_2 = 604$ fps. Using equation 78 with $V_1 = 0$, $V_2 = 604$ fps, and $T_1 = 560°\text{R}$: $T_2 = 530°\text{R}$. Calculating the sonic velocity from equation 11, $a_2 = 1130$ fps, giving $\mathbf{M}_2 = 0.534$.

36. Constriction in Streamtube. When a compressible fluid flows through a constriction in a streamtube or pipe line (Fig. 64), the equations of Art. 35 are not applicable unless the constriction is very

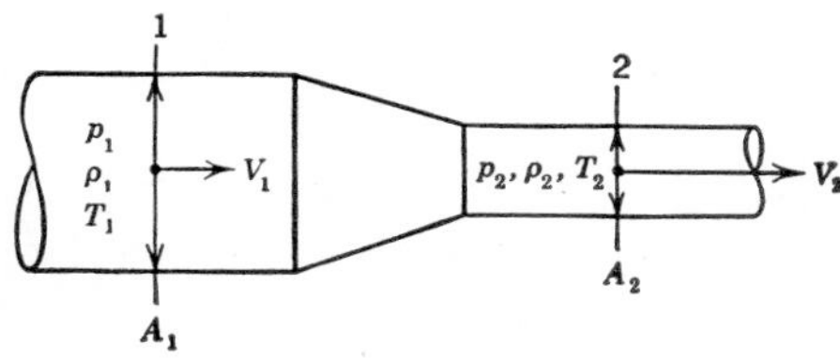

FIG. 64

small (compared to the pipe). When the constriction is larger, the velocity V_1 is no longer negligible compared to V_2 (see equation 79), and adjustments must be made in the foregoing equations to account for this. However, when sonic velocities are attained in the constriction, it has usually been chosen small enough that V_1 is negligible and thus adjustment of equation 89 is not usually necessary and will not be discussed here.

For flow of gases well above the critical pressure ratio, however, V_1 usually cannot be neglected. For this case an equation for flow-rate may be derived by simultaneous solution of

$$\frac{V_2^2 - V_1^2}{2g} = \frac{p_1}{\gamma_1}\frac{k}{k-1}\left[1 - \left(\frac{p_2}{p_1}\right)^{\frac{k-1}{k}}\right] \tag{79}$$

$$G = A_1\gamma_1 V_1 = A_2\gamma_2 V_2 \tag{52}$$

$$p_2/p_1 = (\gamma_2/\gamma_1)^k \tag{7}$$

This yields

$$G = \frac{A_2}{\sqrt{1 - \left(\frac{p_2}{p_1}\right)^{\frac{2}{k}}\left(\frac{A_2}{A_1}\right)^2}}\sqrt{\frac{2gk}{k-1}\,p_1\gamma_1\left[\left(\frac{p_2}{p_1}\right)^{\frac{2}{k}} - \left(\frac{p_2}{p_1}\right)^{\frac{k+1}{k}}\right]} \tag{90}$$

Comparison of this equation with equation 88 indicates that the effect of including V_1 is concentrated in the first square root, which can be seen to approach 1.00 rapidly as the area ratio, A_2/A_1, decreases.

Equation 90, because of its unwieldy form, is sometimes solved by the use of an *expansion factor*, Y, applied to the simpler and analogous solution for incompressible flow,[11] which is

$$Q = \frac{A_2}{\sqrt{1 - \left(\frac{A_2}{A_1}\right)^2}}\sqrt{2g\left(\frac{p_1 - p_2}{\gamma}\right)} \tag{91}$$

Y is defined as the factor which, when multiplied into the product of γ_1 and equation 91, will yield equation 90. Thus

$$G = \frac{YA_2\gamma_1}{\sqrt{1 - \left(\frac{A_2}{A_1}\right)^2}}\sqrt{2g\left(\frac{p_1 - p_2}{\gamma_1}\right)} \tag{92}$$

[11] See Art. 25.

An expression for Y may be derived by equating equations 90 and 92; it is found to be a function of the three variables p_2/p_1, A_2/A_1, and k, and thus may be computed once for all and presented in tables or plots. The equation for Y and its tabulated values will be found in Appendix VIII.

Illustrative Problem

Air flows through a 1-in. constriction in a 1.50-in. pipe line. In the pipe the pressure and temperature of the air are 100 psi and 100°F, respectively. Calculate the flowrate if the pressure in the constriction is 80 psi. Barometric pressure is 14.7 psia.

Solution. $p_2/p_1 = 0.825$, $p_1 = 16{,}520$ psfa, $\gamma_1 = 0.553$ lb/cuft, as in the preceding illustrative problem. Substituting these quantities in equation 90 gives a flowrate of 1.71 lb/sec. (Compare this with the flowrate of the preceding illustrative problem.)

The flowrate may also be calculated by interpolating a value of Y (0.874) from Appendix VIII and using it in equation 92; then

$$G = \frac{0.874 \times \frac{\pi}{4}\left(\frac{1}{12}\right)^2 \times 0.553}{\sqrt{1-(2/3)^4}} \sqrt{2g\,\frac{20 \times 144}{0.553}} = 1.71 \text{ lb/sec} \qquad (92)$$

37. The Convergent-Divergent Nozzle. From the study of the flow of compressible fluid through a convergent-divergent passage (Laval nozzle) much may be learned of basic phenomena and engineering application. For simplicity consider the discharge from a large reservoir through such a passage with pressure p_1 and temperature T_1 in the reservoir (Fig. 65). If sonic velocity exists at the throat, the flowrate G is determined by equation 89, the throat area A_2 being known. Assumption of the pressure p_3 in the jet at the nozzle exit allows computation of the exit area A_3 from equation 88, using A_3 for A_2 and p_3 for p_2; however, it will be found that *two very different* p_3*'s will yield the same area* A_3 *and flowrate* G. The higher of these pressures (p_3'') will cause subsonic velocity in the diverging part of the passage, the lower one (p_3''') will produce supersonic velocity there. The variations of pressure along the passage for these two conditions are shown in Fig. 65.

It is of fundamental importance to examine such a nozzle at the same reservoir conditions (p_1 and T_1) but with "back pressure" p_3' other than those used for the determination of A_3. Reduction of p_3' below p_3''' cannot affect conditions at the throat, so no change of flowrate is produced and no change of nozzle performance is to be expected (except that the compressed gas passing the nozzle exit

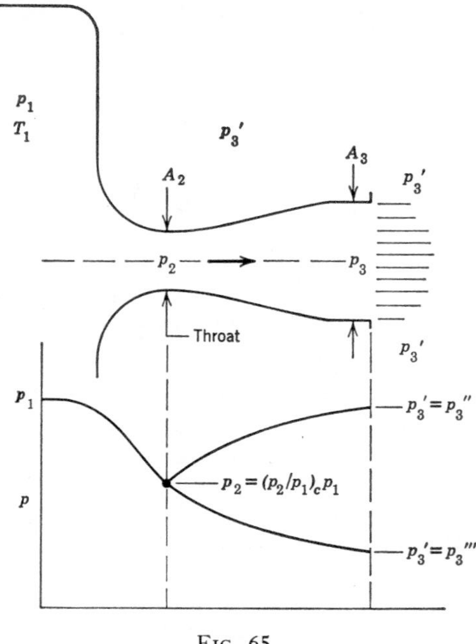

FIG. 65

will expand rapidly upon emergence into a region of lower pressure). Raising the back pressure above p_3'' shown on Fig. 66 will cause a reduction of flowrate, velocities throughout the nozzle will become subsonic, and the upper pressure distribution will exist in the nozzle.

Between p_3'' and p_3''' there are an infinite number of "back pressures," none of which can satisfy the equations of isentropic flow; this is due to the formation of a *normal shock wave* [12] (with considerable internal friction and increase of entropy) in the divergent passage. Through such a wave the velocity drops abruptly from supersonic to subsonic and the pressure jumps abruptly about as shown on curve A. For the curve A and for others similar to it, the flow will emerge from the nozzle at subsonic velocity and the pressure p_3 will equal p_3'. At some lower value of p_3' (curve B) the shock wave will occur at the nozzle exit, which is the limiting case to be treated by one-dimensional methods; for pressures p_3' between p_B and p_3''' the jet will emerge from the nozzle with $p_3 = p_3'''$ and the shock wave will be in the flowfield downstream from the nozzle exit. Since such flowfields are either two- or three-dimensional, they cannot be described by the foregoing one-dimensional equations.

[12] See Art. 46.

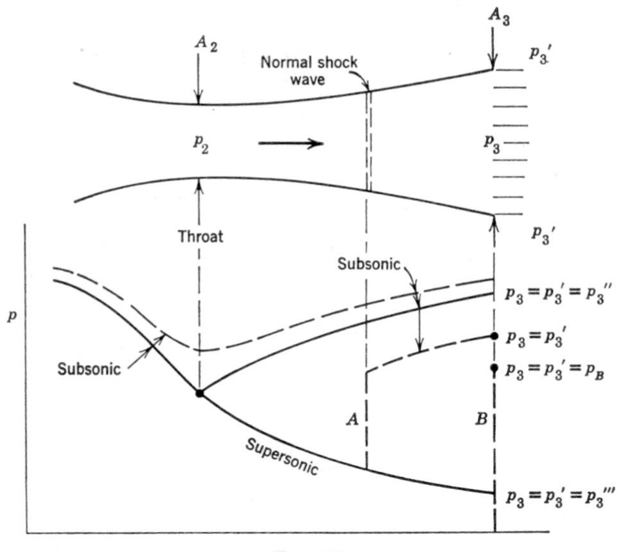

Fig. 66

Illustrative Problem

Air discharges through a convergent-divergent passage (of 1-in. throat diameter) into the atmosphere. The pressure and temperature in the reservoir are 100 psi and 100°F, respectively; the barometric pressure is 14.7 psia. Calculate the nozzle tip diameter required for $p_3 = 14.7$ psia. Calculate the flow velocity, sonic velocity, and Mach number at the nozzle exit. Determine the pressure p_3'' which will yield the same flowrate, and the pressure p_B which will produce a normal shock wave at the nozzle exit.

Solution. Since supersonic velocities will occur in the divergent portion of the nozzle, sonic velocity will be expected at the throat. As these are the conditions of the illustrative problem of Art. 35, the flowrate G is 2.03 lb/sec as before.

$$\frac{V_3^2}{2g} = \frac{1.4}{0.4} \times \frac{114.7 \times 144}{0.553}\left[1 - \left(\frac{14.7}{114.7}\right)^{0.286}\right]; \quad V_3 = 1732 \text{ fps} \qquad (87)$$

$$(1732)^2/2g = 186.5(T_1 - T_3); \quad T_1 - T_3 = 250°\text{F}, \quad T_3 = 310°\text{R} \qquad (78)$$

$$a_3 = \sqrt{1.4 \times 32.2 \times 53.3 \times 310} = 862 \text{ fps}; \quad \mathbf{M}_3 = 1732/862 = 2.01$$

$$\gamma_3 = 14.7 \times 144/53.3 \times 310 = 0.128 \text{ lb/cuft} \qquad (3)$$

$$A_3 = 2.03/0.128 \times 1732 = 0.00915 \text{ sqft}; \quad d_3 = 1.295 \text{ in.} \qquad (52)$$

$$2.03 = 0.00915\sqrt{\frac{64.4 \times 1.4}{0.4} 114.7 \times 144 \times 0.553\left[\left(\frac{p_3''}{p_1}\right)^{1.43} - \left(\frac{p_3''}{p_1}\right)^{1.715}\right]} \qquad (88)$$

Solving by trial,

$$p_3''/p_1 = 0.90; \quad p_3'' = 0.90 \times 114.7 = 103.3 \text{ psia}$$

$$p_B = 14.7\left[1 + \frac{2 \times 1.4}{1.4 + 1}(\overline{2.01^2} - 1)\right] = 66.8 \text{ psia} \tag{115}$$

TWO-DIMENSIONAL FLOW

The study of two-dimensional fields of compressible flow presents the same difficulties as those of incompressible flow (Arts. 26–28) and a further complication: variation of density over the flowfield—which means another variable in the equations and increased difficulty of solution. Although formal mathematical solution of such problems is not possible, special techniques have been invented for the solution of certain problems of engineering interest; however, review of these methods is outside the scope of an elementary text. The intent of the following treatment is merely to develop the basic equations, to describe certain flowfields, and to discuss the applicability and limitations of the equations.

38. Euler's Equations and Their Integration. Euler's equations for the two-dimensional flow of an ideal compressible fluid are the same as those for the incompressible fluid except for the neglect of g, which was justified in Art. 29. This neglection allows the flowfield to be considered to be in a horizontal (x–y) plane and the Euler equations to be written as

$$-\frac{1}{\rho}\frac{\partial p}{\partial x} = u\frac{\partial u}{\partial x} + v\frac{\partial u}{\partial y} \tag{69}$$

$$-\frac{1}{\rho}\frac{\partial p}{\partial y} = u\frac{\partial v}{\partial x} + v\frac{\partial v}{\partial y} \tag{70}$$

They may be rearranged and combined in the same pattern as that of Art. 26 to give

$$-\frac{dp}{\rho} = d\left(\frac{u^2 + v^2}{2}\right) + (u\,dy - v\,dx)\left(\frac{\partial v}{\partial x} - \frac{\partial u}{\partial y}\right)$$

in which $(\partial v/\partial x - \partial u/\partial y)$ is recognized as the vorticity ξ, and integrated to yield

$$\int\frac{dp}{\rho} + \frac{V^2}{2} = C + \int \xi(u\,dy - v\,dx)$$

which reduces, *for irrotational flow* ($\xi = 0$), to

$$\int \frac{dp}{\rho} + \frac{V^2}{2} = C$$

in which C is a constant for all points in the flowfield (Fig. 67). For an isentropic flowfield this equation may be integrated by using the isentropic relation between p and ρ of equation 7; the result is

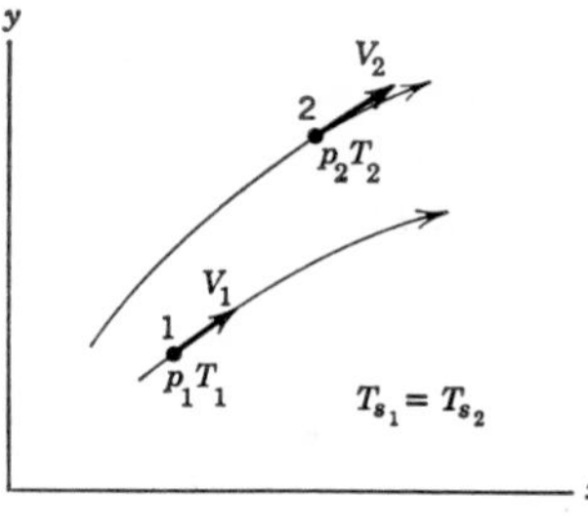

FIG. 67

$$\frac{V_2^2 - V_1^2}{2g} = \frac{k}{k-1}\frac{p_1}{\gamma_1}\left[1 - \left(\frac{p_2}{p_1}\right)^{\frac{k-1}{k}}\right] \tag{93}$$

Although equations 79 and 93 are identical, application of the former is restricted to points on the same streamline, whereas in equation 93 points 1 and 2 may be any points in the flowfield. Use of the isentropic relation between pressure and temperature in equation 93, along with $p_1/\gamma_1 = RT_1$, will yield

$$c_pT_1 + \frac{V_1^2}{2g} = c_pT_2 + \frac{V_2^2}{2g} = c_pT_s \tag{94}$$

which is identical with equation 78 but shows that stagnation temperature is constant not only along single streamlines but also at all points in an isentropic flowfield. The limitations of equations 93 and 94 deserve emphasis: they have been derived for frictionless and adiabatic (i.e., isentropic) irrotational flow and hence cannot be expected to apply to real flowfields with heat transfer, boundary friction, and shock waves. They do, however, provide a useful method of approach to many compressible flow problems.

Consider now (for comparison with the foregoing) a particular compressible flowfield which is *nonisentropic* but throughout which the stagnation temperature is constant. With no change of stagnation temperature along the streamlines, equation 78 may be applied to conclude that there is no exchange of heat energy between adjacent streamtubes; however, frictional processes may vary from streamtube to streamtube with accompanying variability of entropy between them. This situation is closely approximated downstream from a curved shock wave. *Crocco's theorem*,[13] a classic synthesis of

[13] Development and discussion of Crocco's theorem will be found in the references at the end of this chapter.

thermodynamic and fluid mechanics principles, shows that such a nonisentropic flowfield cannot be irrotational.

39. Application of the Equations. Application of equation 93 and 94 to a flowfield is straightforward enough if all the velocities are known, along with the temperature and pressure at one point in the flowfield; from these, all pressures and temperatures throughout the field may be computed, and from pressure and temperature the fluid density at any point may be predicted.

Illustrative Problem

Air approaches this streamlined object at the speed, pressure, and temperature shown. Calculate the pressures, temperatures, and Mach numbers at points A and B, where the velocities are 800 and 900 fps, respectively.

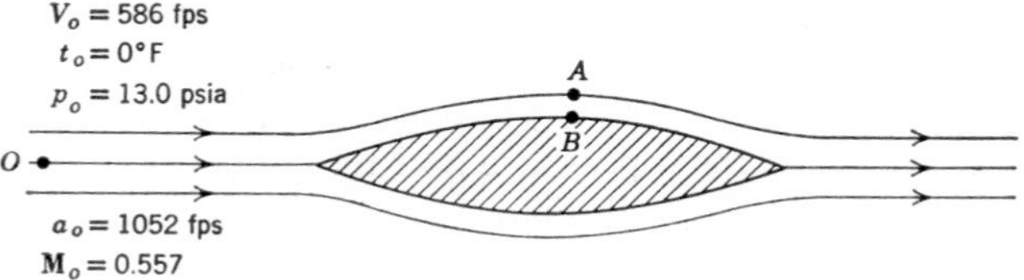

Solution. Refer to the illustrative problem of Art. 33 for some preliminary calculations and for pressure and temperature at the stagnation point. Then

$$(800)^2/2g = 186.5(488.5 - T_A); \quad T_A = 435.3°\text{R} \tag{94}$$

$$(900)^2/2g = 186.5(488.5 - T_B); \quad T_B = 421.0°\text{R} \tag{94}$$

From the isentropic relationship between pressure and temperature (equation 7),

$$\frac{p_A}{16.18} = \left(\frac{435.3}{488.5}\right)^{3.5}, \quad \frac{p_B}{16.18} = \left(\frac{421.0}{488.5}\right)^{3.5}; \quad p_A = 10.85 \text{ psia}, \quad p_B = 9.65 \text{ psia}$$

$$a_A = \sqrt{1.4 \times 32.2 \times 53.3 \times 435.3} = 1021 \text{ fps} \tag{11}$$

$$a_B = \sqrt{1.4 \times 32.2 \times 53.3 \times 421.0} = 983 \text{ fps} \tag{11}$$

$$\mathbf{M}_A = 800/1021 = 0.783, \quad \mathbf{M}_B = 900/983 = 0.915$$

Formal mathematical solution of the inverse of the foregoing problem, when boundary conditions are specified and the flowfield through a passage or about an object is to be determined (i.e., velocities, pressures, etc., predicted), is impossible; answers to such problems are obtained by approximations, linearizations, trial and error, graphical techniques, etc., most of which are beyond the scope of this elementary book. However, it is useful to consider such a problem and some of the methods and limitations of its solution.

The flowfield around the simple streamlined object of Fig. 68 is to be predicted. The shape and orientation of the object are known, and the *boundary condition* of velocity, pressure, and density of the

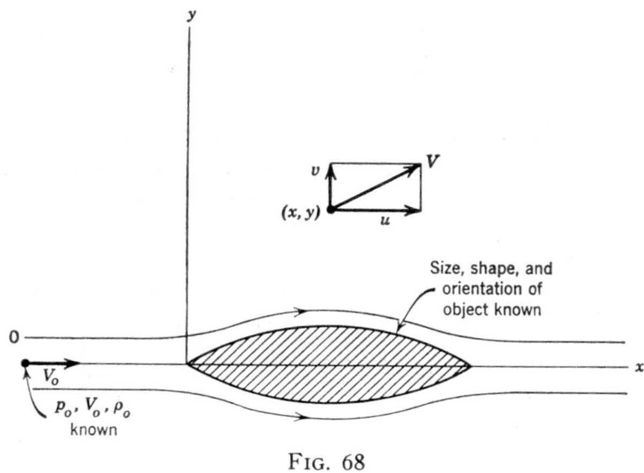

FIG. 68

fluid approaching the object is also known. The available independent equations are

$$\frac{\partial}{\partial x}(\rho u) + \frac{\partial}{\partial y}(\rho v) = 0 \tag{57}$$

$$-\frac{1}{\rho}\frac{\partial p}{\partial x} = u\frac{\partial u}{\partial x} + v\frac{\partial u}{\partial y} \tag{69}$$

$$-\frac{1}{\rho}\frac{\partial p}{\partial y} = u\frac{\partial v}{\partial x} + v\frac{\partial v}{\partial y} \tag{70}$$

$$\frac{p}{\rho^k} = \text{Constant} \tag{7}$$

The objective in solving these equations is to obtain u and v as functions of x and y. If such a solution can be obtained, pressures, temperatures, and fluid densities anywhere in the flowfield may also be predicted. The form and orientation of the object, of course, affects the flow picture; this enters the solution of the problem as a *boundary condition* expressing the fact that the velocity V along the surface of the body is everywhere tangent to the body. Mathematically, this means

$$\left[\frac{dy}{dx} = \frac{v}{u}\right]_{\text{at surface of body}}$$

The unknowns in the problem are seen to be u, v, x, y, ρ, and p, and with four equations and two boundary conditions the unknowns are seen to be determinable; however, integration of such equations is not yet possible, so a formal mathematical solution of the problem cannot be obtained. Although the solution of such problems by trial and error may have little appeal, it is possible through the exercise of considerable time and patience; with this approach an incompressible flow solution would be sought as a first trial, after which appropriate adjustments would be made until the foregoing equations for compressible flow are satisfied at all points in the flowfield. This type of solution is most efficiently handled by a modification of the *flownet* technique described in Chapter 12.

The isentropic treatment of such compressible flowfields around solid objects is limited to the case where velocities are everywhere subsonic. If the approaching and leaving velocities (Fig. 68) are both subsonic and at a point on the body (near its midsection) the velocity becomes supersonic, a shock wave is to be expected downstream from this point where the velocity becomes subsonic again. Analysis of such problems is exceedingly complex because of the unknown position and extent of the shock wave, and its nonisentropic and discontinuous nature. Upstream from and outboard of the shock wave the flow of the ideal compressible fluid is (whether subsonic or supersonic) irrotational and isentropic, but downstream from such waves (in general) the flow is rotational and nonisentropic;[14] however, if such shock waves are straight (or essentially so), the flowfield downstream from them may be shown to be irrotational and isentropic; this simplifies the problem somewhat.

Flowfields within passages present the same difficulties as the external flowfields described above, so further examples need not be cited here. In general, such internal flowfields are less amenable to the use of the ideal fluid because of the pervasiveness of wall frictional effects which render the isentropic (irrotational) assumption invalid. However, such methods may be effective for short passages, duct inlets, etc., where wall friction is of small importance.

[14] See Art. 38.

REFERENCES

G. I. Taylor and J. W. Maccoll, *The Mechanics of Compressible Fluids*, Part H of Vol. III of *Aerodynamic Theory* (W. F. Durand, Ed.), Julius Springer, Berlin, 1935.

R. Sauer, *Introduction to Theoretical Gas Dynamics*, J. W. Edwards, 1947.

L. Howarth (Ed.), *Modern Developments in Fluid Dynamics; High Speed Flow*, Vols. I and II, Oxford University Press, 1953.

A. H. Shapiro, *The Dynamics and Thermodynamics of Compressible Fluid Flow*, Vol. I, Ronald Press Co., 1953.

K. Oswatitsch, *Gas Dynamics*, Academic Press, 1956.

H. W. Liepmann and A. Roshko, *Elements of Gasdynamics*, John Wiley and Sons, 1957.

A. B. Cambel and B. H. Jennings, *Gas Dynamics*, McGraw-Hill Book Co., 1958.

H. W. Emmons (Ed.), *Fundamentals of Gas Dynamics*, Vol. III of *High Speed Aircraft and Jet Propulsion*, Princeton University Press, 1958.

R. von Mises, *Mathematical Theory of Compressible Fluid Flow*, Academic Press, 1958.

S.-I Pai, *Introduction to the Theory of Compressible Flow*, D. Van Nostrand Co., 1959.

PROBLEMS

381. The velocity and temperature at a point in an isentropic flow of helium are 300 fps and 200°F respectively. Predict the temperature on the same streamline where the velocity is 600 fps. What is the ratio between the pressures at the two points?

382. At a point in an adiabatic flow of nitrogen the velocity is 700 fps. Between this point and another one on the same streamline the rise of temperature is 20°F. Calculate the velocity at the second point.

383. The velocity and temperature are measured at two points on the same streamline in a flow of carbon dioxide and found to be 200 fps, 100°F and 400 fps, 90°F. This flow is *not* adiabatic. How much heat energy (ft-lb/lb) has been added to or extracted from the fluid between the two points?

384. Derive equation 82 from equation 81.

385. Revise equation 82 by including the fourth term in the binomial expansion.

386. Carbon dioxide flows in a duct at a velocity of 300 fps, pressure 20 psia, and temperature 200°F. Calculate pressure and temperature on the nose of a small object placed in this flow.

387. If nitrogen at 60°F is flowing and the stagnation temperature on the nose of a small object in the flow is measured as 100°F, what is the velocity in the pipe?

388. Oxygen flows in a passage at a pressure of 25 psia. The pressure and temperature on the nose of a small object in the flow are 28 psia and 150°F, respectively. What is the velocity in the passage?

389. Calculate the stagnation pressure in an air stream of pressure, temperature, and velocity, 14.7 psia, 60°F, and 1050 fps, respectively, using (*a*) equation 82 and (*b*) equation 81. Compare results.

390. What is the pressure on the nose of a bullet moving through standard sea level air at 900 fps, according to (*a*) equation 82 and (*b*) equation 81? Compare results.

391. With equation 81 invalid for $\mathbf{M}_o > 1$ derive an expression for the minimum allowable value of p_o/p_s which may be used for velocity computation in equation 83.

392. Derive equations (*a*) 88 and (*b*) 89.

393. Nitrogen flows from a large tank, through a convergent nozzle of 2-in. tip diameter, into the atmosphere. The temperature in the tank is 200°F. Calculate pressure, velocity, temperature, and sonic velocity in the jet; and calculate the flowrate when the tank pressure is (*a*) 30 psia and (*b*) 25 psia. Barometric pressure is 15.0 psia. What is the lowest tank pressure that will produce sonic velocity in the jet? What is this velocity and what is the flowrate?

394. Air flows from the atmosphere into an evacuated tank through a convergent nozzle of 1.50-in. tip diameter. If atmospheric pressure and temperature are 14.7 psia and 59°F, respectively, what vacuum must be maintained in the tank to produce sonic velocity in the jet? What is the flowrate? What is the flowrate when the vacuum is 10 in. of mercury?

395. Oxygen discharges from a tank through a convergent nozzle. The temperature and velocity in the jet are 0°F and 900 fps, respectively. What is the temperature in the tank? What is the temperature on the nose of a small object in the jet?

396. Carbon dioxide discharges from a tank through a convergent nozzle into the atmosphere. If the tank temperature and pressure are 100°F and 20 psi, respectively, what jet temperature, pressure, and velocity can be expected? Barometric pressure is 14.7 psia.

397. In the illustrative problem of Art. 35, calculate the pressure, temperature, velocity, and sonic velocity at a point in the nozzle where the diameter is 2 in.

398. Air (at 100°F and 100 psia) in a large tank flows into a 6-in. pipe, whence it discharges to the atmosphere (15.0 psia) through a convergent nozzle of 4-in. tip diameter. Calculate pressure, temperature, and velocity in the pipe.

399. Carbon dioxide flows through a convergent nozzle in a wall between two large tanks in which the pressures are 65 and 30 psia. Calculate the jet velocity if the jet temperature is 50°F. What is the temperature in the upstream tank?

400. Calculate the required diameter of a convergent nozzle to discharge 5.0 lb/sec of air from a large tank (in which the temperature is 100°F) to the atmosphere (14.7 psia) if the pressure in the tank is: (*a*) 25.0 psia, (*b*) 30.0 psia.

401. Air flows from a 6-in. pipe, through a 1-in. convergent nozzle, into the atmosphere. Calculate the absolute pressure in the pipe when the pressure in the jet is 20 psia.

402. Derive equation 90.

403. Carbon dioxide flows through a 4-in. constriction in a 6-in. pipe. The pressures in pipe and constriction are 40 and 35 psia, respectively. The temperature in the pipe is 100°F. Calculate (*a*) flowrate, (*b*) temperature in the constriction, (*c*) sonic velocity in the constriction, and (*d*) velocities in pipe and constriction.

404. Nitrogen discharges from a 4-in. pipe through a 2-in. nozzle into the atmosphere. Barometric pressure is 14.7 psia, and the pressure in the pipe is 1 ft of water. The temperature in the pipe is 60°F. Calculate the flowrate, (*a*) neglecting expansion of the gas and (*b*) including expansion. What percent error is induced by neglecting expansion?

405. Five pounds of air per second discharge from a tank through a convergent-divergent nozzle into another tank where a vacuum of 10 in. of mercury is maintained. If the pressure and temperature in the upstream tank are 100 in. of mercury abs and 100°F, respectively, what nozzle-exit diameter must be provided for full expansion? What throat diameter is required? Calculate pressure, temperature, velocity, and sonic velocity in throat and nozzle exit. Barometric pressure is 30 in. of mercury.

406. Carbon dioxide flows from a tank through a convergent-divergent nozzle of 1-in. throat and 2-in. exit diameter. The pressure and temperature in the tank are 35 psia and 100°F, respectively. Calculate the flowrate when the pressure p_3' is (*a*) 25 psia and (*b*) 32 psia.

407. If the nozzle of the illustrative problem of Art. 37 is a convergent one of 1.295-in. tip diameter, calculate velocity and sonic velocity in the jet, and the flowrate. Compare results, and note the effect of changing the shape of a nozzle without changing its exit diameter.

408. If the nozzle in the illustrative problem of Art. 37 has a divergent passage 4 in. long, calculate the required passage diameter, at three equally spaced points between throat and nozzle tip, for a linear pressure drop through this part of the passage.

409. A convergent-divergent nozzle of 2-in. tip diameter discharges to the atmosphere (15.0 psia) from a tank in which air is maintained at a pressure and temperature of 100 psia and 100°F, respectively. What is the maximum flowrate which can occur through this nozzle? What throat diameter must be provided to produce this flowrate?

410. Atmospheric air (at 14.3 psia and 68°F) is drawn into a vacuum tank through a convergent-divergent nozzle of 2-in. throat diameter and 3-in. exit diameter. Calculate the largest flowrate which can be drawn through this nozzle under these conditions.

411. The exit section of a convergent-divergent nozzle is to be used for the test section of a supersonic wind tunnel. If the pressure in the test section is to be 20 psia, what pressure is required in the reservoir to produce a Mach number of 5 in the test section? For the air temperature to be 0°F in the test section, what temperature is required in the reservoir? What ratio of throat area to test section area is required to meet these conditions?

412. In the nozzle of problem 405, what range of pressures in the tank will cause a normal shock wave in the nozzle?

413. A perfect gas (k = 2.0, and R = 116.8 ft-lb/lb °R) discharges from a tank through a convergent-divergent nozzle into the atmosphere (barometric pressure 28 in. of mercury). The gage pressure in the tank is 50 psi, and the temperature is 100°F. The throat and exit diameters of the nozzle are 1.00 in. and 1.05 in., respectively. Show calculations to prove whether a normal shock wave is to be expected in this nozzle.

414. Air discharges through a convergent-divergent nozzle attached to a large reservoir. At a point in the nozzle a normal shock wave is detected across which the pressure jumps from 10 to 30 psia. Calculate the pressures in the throat of the nozzle and in the reservoir.

chapter 6

The Impulse–Momentum Principle

The impulse-momentum principle, along with the continuity and energy principles, provides a third basic tool for the solution of fluid flow problems. Sometimes its application leads to the solution of problems which cannot be solved by the energy principle alone; more often it is used in conjunction with the energy principle to obtain more comprehensive solutions of engineering problems.

The following treatment of the subject is limited to steady one-dimensional flow; its objective is to develop the principle and demonstrate its application to a variety of practical problems. Extensive treatment of many of these problems is far beyond the scope of this book; the reader will find textbooks and other abundant engineering literature dealing with these subjects, for which references are provided at the end of the chapter.

40. Development of the Principle. Consider the flow in the streamtube of Fig. 69*a* in which the streamlines are essentially straight and parallel at sections 1 and 2. Obviously a change of momentum is occurring between these sections since net changes in magnitude and direction of mean velocity are evident. By constructing a *control surface* around this section of streamtube the forces producing this change of momentum may be identified. Within the control volume the "internal" pressure and shearing forces [1] (Fig. 69*a*) existing at

[1] In ideal-fluid problems such shearing forces are zero; in most engineering problems with real fluids they are usually negligible compared to the pressure forces; the shearing forces will be neglected throughout this development.

the surfaces of adjacent elements are seen to cancel, so that a summation of such forces will yield only those exerted on the fluid at the control surface $ABCD$; the summation of the *body forces* (dW) will, of course, yield the total weight of fluid (W) within the control volume. The separate forces on the control surface may now be identified (Fig. 69*b*). For streamlines essentially straight and parallel at sections 1 and 2, the forces F_1 and F_2 will result from hydrostatic pressure distribution in the flowing fluid. The weight W is the total weight of fluid within the control volume and thus dependent only upon specific weight of fluid and control volume geometry.[2] The force F is the resultant of the pressure distributions along the curved surfaces AB and CD; it will be shown that, although these pressure distributions are usually unknown in detail, the impulse-momentum principle still allows prediction of their resultant. The summation of these forces may be carried out by the well-known methods of statics,[3] using Fig. 69*b*. The resulting equations are

$$\Sigma F_x = F_{1x} - F_{2x} - F_x \tag{95}$$

$$\Sigma F_z = F_{1z} - F_{2z} + F_z - W \tag{96}$$

$$\overset{+\curvearrowleft}{\Sigma M_o} = (F_{2x})z_2' - (F_{1x})z_1' - (F_{2z})x_2' + (F_{1z})x_1' + (F)r - (W)x_W \tag{97}$$

Assuming that ΣF_x, ΣF_z, and ΣM_o may be obtained independently from change of momentum considerations and that x_1', x_2', z_1', z_2', F_{1x}, F_{1z}, F_{2x}, F_{2z}, x_W, and W are all known, the magnitude, direction, and position of F may be determined.

The change of momentum of the fluid as it passes through the control volume may now be determined for a time interval dt as shown in Fig. 69*c*. Consider that the fluid mass moves from position $A'B'C'D'$ to $A''B''C''D''$ in time dt. Since the mass $A''B'C'D''$ experiences no change of momentum in this time, the differential change of momentum (a vector quantity) must be

$$d(MV) = (Q_2\rho_2\,dt)\,\overrightarrow{V_2} - (Q_1\rho_1\,dt)\,\overrightarrow{V_1}$$

[2] Frequently in engineering problems this force is small compared to the forces on the surfaces of the control volume, but decision to neglect it requires either preliminary calculations or considerable experience. It obviously plays no part if flow is in a horizontal plane, since its direction is perpendicular to this plane.

[3] The points (x_1', z_1') and (x_2', z_2') identify the *centers of pressure* (not the centroids) of sections 1 and 2. However, these will be coincident for uniform pressure distributions. See Art. 11.

which may be written

$$\frac{d}{dt}(MV) = (\overrightarrow{Q\rho V})_2 - (\overrightarrow{Q\rho V})_1 \tag{98}$$

in which the quantities $Q\rho V$ (termed *momentum flux*) may be considered to be fictitious forces (Fig. 69*d*) having magnitude, direction, and position. The magnitude is $Q\rho V$, the direction is the same as that of V, and the location (for symmetrical velocity profiles) is at

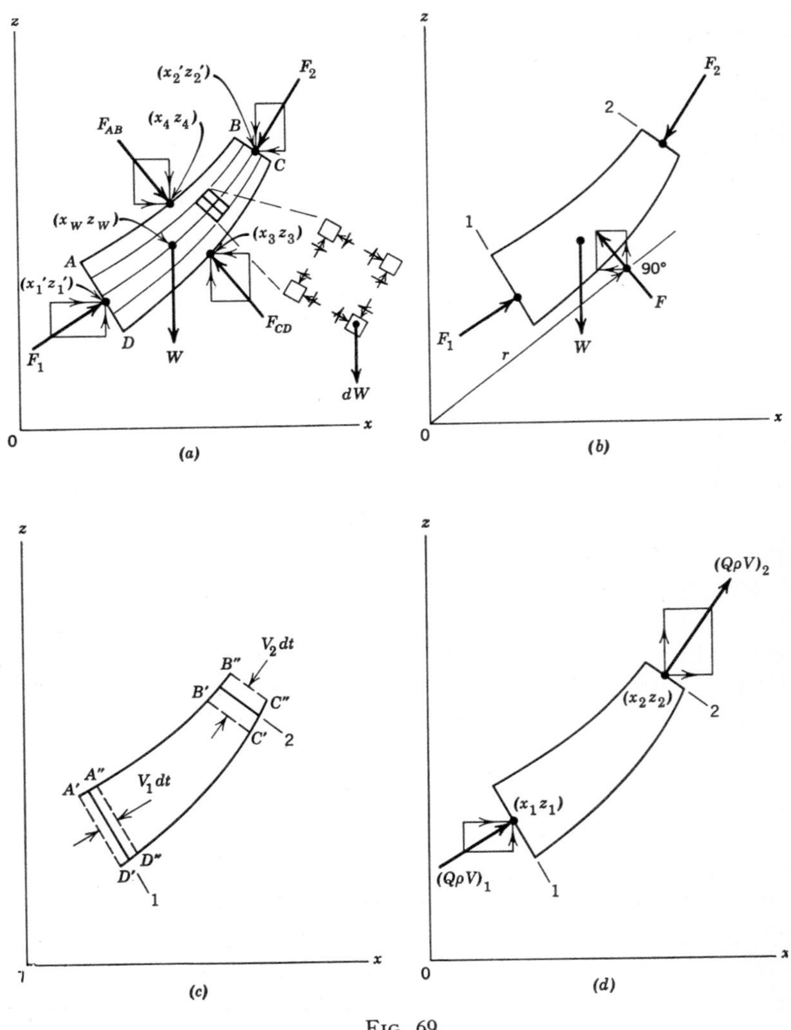

FIG. 69

the center of the flow cross section. The rate of change of momentum in the x- and z-directions and the rate of change of moment-of-momentum about 0 may be expressed

$$\frac{d}{dt}(MV)_x = (Q\rho V)_{2x} - (Q\rho V)_{1x} \tag{99}$$

$$\frac{d}{dt}(MV)_z = (Q\rho V)_{2z} - (Q\rho V)_{1z} \tag{100}$$

$$\frac{d}{dt}(r\overset{+}{\curvearrowleft}{MV})_o = [(Q\rho V)_{2z}x_2 - (Q\rho V)_{2x}z_2] - [(Q\rho V)_{1z}x_1 - (Q\rho V)_{1x}z_1] \tag{101}$$

Finally, the impulse-momentum principle states that

$$\Sigma F_x = \frac{d}{dt}(MV)_x \tag{102}$$

$$\Sigma F_z = \frac{d}{dt}(MV)_z \tag{103}$$

$$\Sigma \overset{+}{\curvearrowleft}{M}_o = \frac{d}{dt}(r\overset{+}{\curvearrowleft}{MV})_o \tag{104}$$

so equations 95, 96, and 97 may be equated respectively to equations 99, 100, and 101, allowing solution of a variety of engineering problems. The inherent advantage of the impulse-momentum principle is seen in the equations which show that only the flow conditions at inlet and exit of the control volume are needed for successful application; the detailed (and often complex) flow processes within the control volume need not be known to apply the principle. It hardly needs emphasis that efficient application of the principle depends greatly upon the judicious selection of a convenient control volume with streamlines essentially straight and parallel at inlet and exit. Although the development of the impulse-momentum principle may be generalized so that the control volume may be drawn arbitrarily, such a generalization is beyond the scope of an elementary textbook.

41. Pipe Bends, Enlargements, and Contractions. The force exerted by a flowing fluid on a bend, enlargement, or contraction in a pipe line may be readily computed by application of the impulse-momentum principle without detailed information on shape of passage or pressure distribution therein.

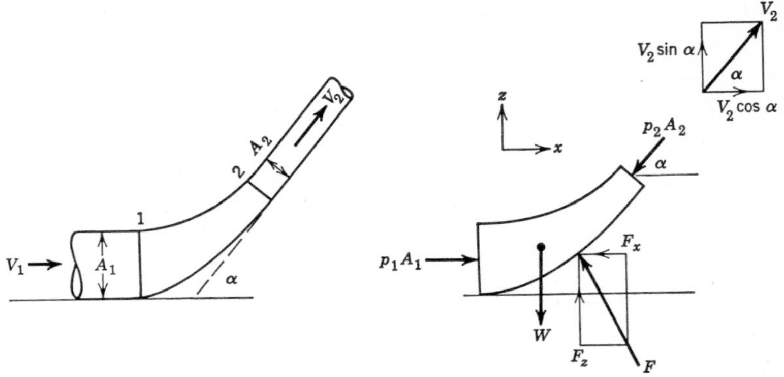

FIG. 70

The reducing pipe bend of Fig. 70 is typical of many of the problems encountered in fluid mechanics; the flowrate, velocities, and superficial flow geometry are known, and a force is to be calculated. With mean pressures p_1 and p_2 at sections 1 and 2, $F_1 = p_1 A_1$ and $F_2 = p_2 A_2$. F is the force exerted *by the bend on the fluid* and is the resultant of the pressure distribution over the curved surface of the bend; the force exerted *by the fluid on the bend* is the equal and opposite of this. When equations 102 and 103 are applied,

$$\Sigma F_x = p_1 A_1 - p_2 A_2 \cos \alpha - F_x = Q\rho(V_2 \cos \alpha - V_1)$$

$$\Sigma F_z = F_z - W - p_2 A_2 \sin \alpha = Q\rho(V_2 \sin \alpha - 0)$$

from which F_x and F_z can be obtained, and the magnitude and direction of F readily established. The position of F depends upon the location of W acting at the center of gravity of the fluid within the bend; this in turn will depend upon the geometry of the bend. Use of equation 104 will then allow prediction of the location of F. The following illustrative problem demonstrates the method.

ILLUSTRATIVE PROBLEM

When 10 cfs of water flow through this vertical 12-in. by 8-in. pipe bend, the pressure at the entrance is 10 psi. Calculate the force on the bend if the volume of the bend is 3 cuft.

Solution. From the continuity principle $V_1 = 12.72$ fps, $V_2 = 28.6$ fps, and from the Bernoulli equation $p_2 = 3.43$ psi.

Now, for the free-body diagram, the forces F_1 and F_2 may be computed

$$F_1 = \frac{\pi}{4}(12)^2 10 = 1132 \text{ lb}, \quad F_2 = \frac{\pi}{4}(8)^2 3.43 = 173 \text{ lb}$$

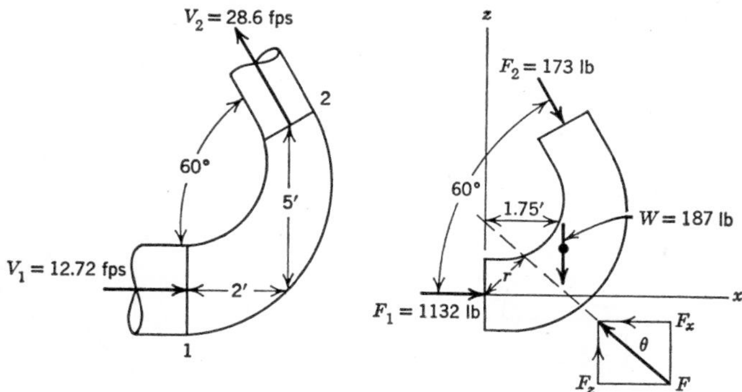

With this and the velocity diagram, equations 102 and 103 may be applied:

$$1132 - F_x + (173)0.5 = 10 \times 1.935(-28.6 \times 0.5 - 12.72) \quad (102)$$

$$F_z - (173)0.866 - 187 = 10 \times 1.935(28.6 \times 0.866 - 0) \quad (103)$$

$$F_x = +1742 \text{ lb}, \quad F_z = +817 \text{ lb}, \quad F = 1920 \text{ lb}, \quad \theta = 25°$$

The plus signs confirm the direction assumptions for F_x and F_z. Therefore the force *on the bend* is 1920 lb downward to the right at 25° with the horizontal.

Now, assuming that the bend is of such shape that the center of gravity of the fluid therein is 1.75 ft to the right of section 1 and that F_1 and F_2 act at the centroids of sections 1 and 2, respectively, equation 104 may be used to find the location of F. Taking moments about the center of section 1, $(1920)r - (187)1.75 - (173)0.866 \times 2 - (173)0.5 \times 5 = (10 \times 1.935 \times 28.6 \times 0.5)5 + (10 \times 1.935 \times 28.6 \times 0.866)2$; $r = 1.77$ ft

42. Structures in Open Flow. The forces exerted by flowing water on overflow or underflow structures (weirs or gates) may frequently be calculated from the impulse-momentum principle, although the location of such forces depends upon the flowfield produced and cannot be obtained from one-dimensional methods. The sluice gate of Fig. 71 is an example of this. From the depths y_1 and y_2 simultaneous application of the Bernoulli and continuity principles will yield the flowrate and the velocities V_1 and V_2. Selecting a control volume bounded by sections 1 and 2, and applying equation 102,

$$\Sigma F_x = F_1 - F_2 - F_x = \gamma y_1^2/2 - \gamma y_2^2/2 - F_x = q\rho(V_2 - V_1)$$

in which F_1 and F_2 have been computed from hydrostatics principles and unit width (normal to the plane of the paper) has been assumed. For a plane gate surface and an ideal fluid, F is normal to the gate

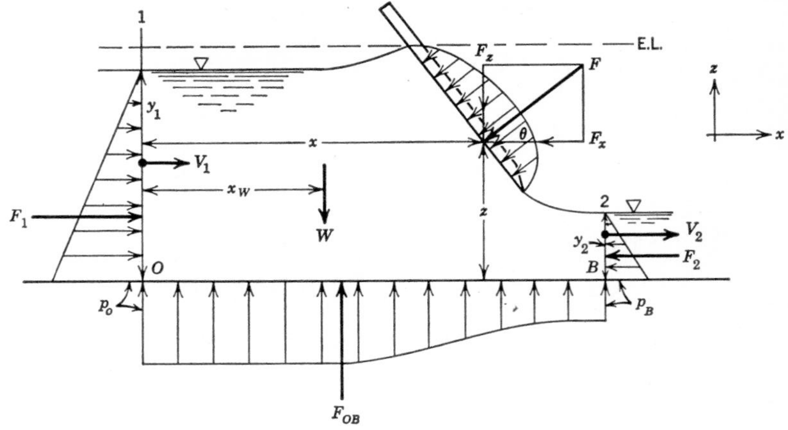

FIG. 71

and thus may be computed directly from $F = F_x/\cos\theta$; similarly, $F_z = F_x \tan\theta$. Application of equation 103 gives

$$\Sigma F_z = F_{OB} - W - F_z$$

in which W is the weight of fluid between sections 1 and 2, and F_{OB} the force exerted by the surface OB on the fluid. With the magnitude and location of W dependent upon the detailed form of the control volume and the location of F dependent upon its (unknown) pressure distribution, it is evident that computations can be carried no further until the complete geometry of the flowfield is known. However, equation 104 may still be applied to demonstrate the method. Taking moments about O,

$$(F_{OB})x_3 + (F_x)z - (F_z)x + (F_2)y_2/3 - (F_1)y_1/3 - (W)x_W$$
$$= -(q\rho V_2)y_2/2 + (q\rho V_1)y_1/2$$

ILLUSTRATIVE PROBLEM

This two-dimensional overflow structure (shape and size unknown) produces the flowfield shown. Calculate the magnitude and direction of the horizontal component of force on the structure.

Solution. Assuming ideal fluid, construct a horizontal energy line above the liquid surface, whence

$$V_1^2/2g + 5 = V_2^2/2g + 2$$

from continuity

$$q = 5V_1 = 2V_2$$

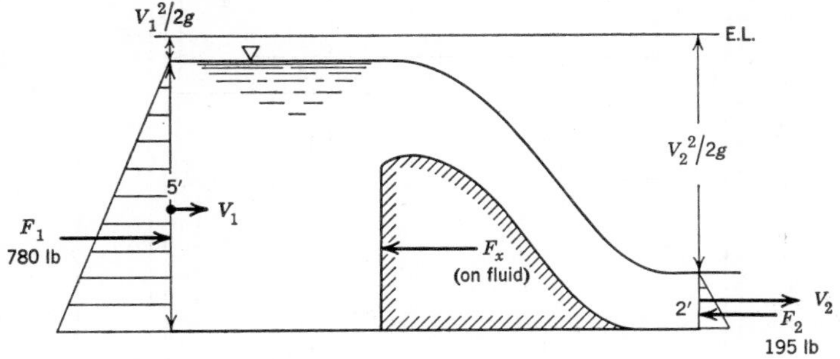

Solving these simultaneous equations,

$$V_1 = 6.07 \text{ fps}, \quad V_2 = 15.2 \text{ fps}, \quad q = 30.4 \text{ cfs/ft}$$

Next construct a control volume of fluid between sections of 5-ft and 2-ft depth, through which streamlines are straight and parallel.

$$F_1 = 62.4(5)^2 = 780 \text{ lb/ft}, \quad F_2 = 62.4(2)^2/2 = 195 \text{ lb/ft}$$

Now, applying equation 102,

$$\Sigma F_x = 780 - F_x - 195 = 30.4 \times 1.935(15.2 - 6.07)$$

whence F_x, the force component desired (and exerted by the structure on the fluid), is +47 lb/ft. The direction of this force component *on the structure* is therefore in a *downstream* direction. Since this force is the integral of the horizontal components of an *unknown* pressure distribution over a structure of *undefined* form, the effectiveness of the impulse-momentum principle is again demonstrated.

43. Abrupt Enlargement in a Closed Passage. The impulse-momentum principle may be employed to predict the fall of the energy line caused by an abrupt axisymmetric enlargement in a passage (Fig. 72). Consider the control surface $ABCD$, drawn to enclose the zone of momentum change. The flowrate Q through the control volume is

$$Q = A_1 V_1 = A_2 V_2 \tag{53}$$

The rate of change of momentum is given by

$$\frac{d}{dt}(MV) = \frac{Q\gamma}{g}(V_2 - V_1) = \frac{A_2 \gamma V_2}{g}(V_2 - V_1) \tag{99}$$

The forces producing this change of momentum act on the surfaces

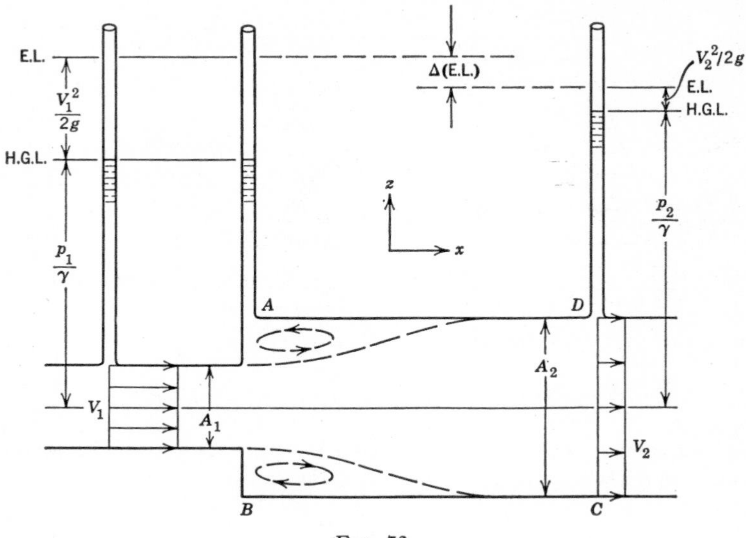

FIG. 72

AB and CD; if these are assumed to result from hydrostatic [4] pressure distributions over the areas, they may be calculated as

$$\Sigma F_x = p_1 A_2 - p_2 A_2 = (p_1 - p_2) A_2 \tag{95}$$

Applying equation 102,

$$\frac{p_1 - p_2}{\gamma} = \frac{V_2(V_2 - V_1)}{g}$$

However, from the energy line,

$$\frac{p_1 - p_2}{\gamma} = \frac{V_2^2}{2g} - \frac{V_1^2}{2g} + \Delta \text{ (E.L.)}$$

Equating these expressions for $(p_1 - p_2)/\gamma$ and solving for Δ (E.L.),

$$\Delta \text{ (E.L.)} = \frac{(V_1 - V_2)^2}{2g} \tag{105}$$

This analysis is a classic one of early analytical hydraulics; Δ (E.L.) is frequently termed the Borda-Carnot *head loss*, after those who contributed to its original development.

[4] This is a good assumption for area CD since the conditions of one-dimensional flow are satisfied. For area AB it is an approximation because of the dynamics of the eddies in the "dead water" zone. Accordingly the result of the analysis will be approximate and will require experimental verification. See Art. 76.

44. The Hydraulic Jump. Frequently in open channel flow, when liquid at high velocity discharges into a zone of lower velocity, a rather abrupt rise (a standing wave) occurs in the liquid surface which is accompanied by violent turbulence, eddying, air entrainment, and surface undulations; such a wave is known as a *hydraulic jump*.[5] In spite of the foregoing complications and the consequential large head loss, application of the impulse-momentum principle will give results very close to those observed in field and laboratory; the engineering problem is to find the relation between the depths for a given flowrate. Assuming a two-dimensional situation with flowrate

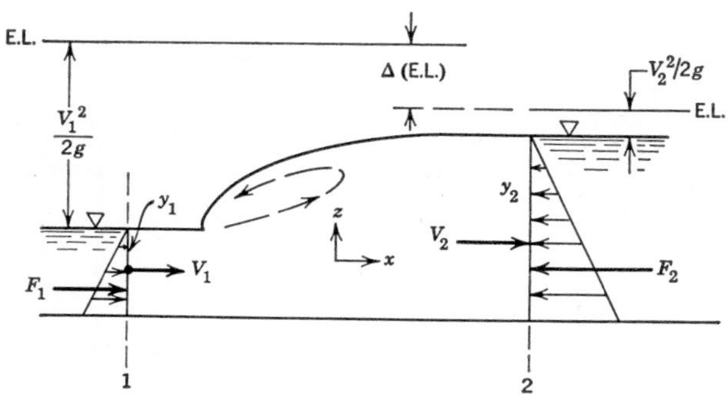

FIG. 73

q and constructing a control volume (Fig. 73) enclosing the jump, the only horizontal forces are seen to be (neglecting friction) the hydrostatic ones of equation 27. Applying equation 102,

$$\Sigma F_x = F_1 - F_2 = \frac{\gamma y_1^2}{2} - \frac{\gamma y_2^2}{2} = \frac{q\gamma}{g}(V_2 - V_1)$$

Substituting the continuity relations $V_2 = q/y_2$ and $V_1 = q/y_1$ and rearranging give the desired relationship between y_1, y_2, and q:

$$\frac{q^2}{gy_1} + \frac{y_1^2}{2} = \frac{q^2}{gy_2} + \frac{y_2^2}{2}$$

and solving for y_2/y_1 yields

$$\frac{y_2}{y_1} = \frac{1}{2}\left[-1 + \sqrt{1 + \frac{8q^2}{gy_1^3}}\right] = \frac{1}{2}\left[-1 + \sqrt{1 + \frac{8V_1^2}{gy_1}}\right] \quad (106)$$

[5] Further discussion of the hydraulic jump will be found in Art. 88.

From this equation it will be seen that: (*a*) for $V_1^2/gy_1 = 1$, $y_2/y_1 = 1$; (*b*) for $V_1^2/gy_1 > 1$, $y_2/y_1 > 1$; and (*c*) for $V_1^2/gy_1 < 1$, $y_2/y_1 < 1$.

Condition (*c*), that of a *fall of the liquid surface*, although satisfying the impulse-momentum and continuity equations, will be found to produce a rise of the energy line through the jump, and is thus physically impossible. Accordingly it may be concluded that, for a hydraulic jump to occur, the upstream conditions must be such that $V_1^2/gy_1 > 1$. Later (Art. 61) V_1^2/gy_1 will be shown to be the square of the *Froude number*, the ratio of flow velocity to wave velocity, and thus analogous to the Mach number of compressible flow.

Illustrative Problem

Water flows in a horizontal open channel at a depth of 2 ft; the flowrate is 40 cfs/ft of width. If a hydraulic jump is possible, calculate the depth just downstream from the jump and the power dissipated in it.

Solution.

$$V_1 = 40/2 = 20 \text{ fps} \tag{55}$$

$$V_1^2/gy_1 = 400/32.2 \times 2 = 6.22$$

which is greater than 1, so a jump is possible.

$$y_2 = \tfrac{2}{2}[-1 + \sqrt{1 + 8 \times 6.22}] = 6.11 \text{ ft} \tag{106}$$

$$V_2 = 40/6.11 = 6.55 \text{ fps}$$

$$\Delta \text{ (E.L.)} = 2 + (20)^2/2g - 6.11 - (6.55)^2/2g = 1.44 \text{ ft}$$

Horsepower dissipated (per foot of channel width):

$$40 \times 62.4 \times 1.44/550 = 65.3 \text{ hp} \tag{68}$$

which shows the hydraulic jump to be an excellent energy dissipator; frequently it is used for this purpose in engineering designs.

45. The Oblique Standing Wave. Under certain circumstances in open flow with $V_1 > \sqrt{gy_1}$, a diagonal standing wave will form whose properties may be analyzed by the impulse-momentum principle. A wedge-shaped pier producing such a wave is shown in Fig. 74. Upstream from the wave front the liquid depth is y_1, and down stream it is y_2. By isolating a control volume $ABCD$ of length 1 ft parallel to the wave front, application of the impulse-momentum and continuity principles is readily made.

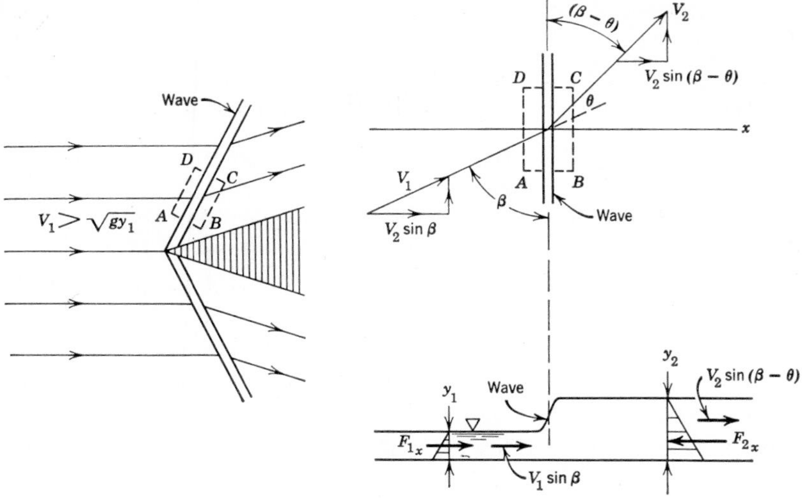

FIG. 74

Applying the continuity principle by noting that, since (from symmetry) the flowrate across AB is equal to that across CD, the flowrates across AD and BC are also equal. Accordingly,

$$q = y_1 V_1 \sin \beta = y_2 V_2 \sin (\beta - \theta) \tag{107}$$

The impulse-momentum principle applied along the y-direction shows (from symmetry) that the forces on the surfaces AB and CD are equal and thus the y-component of velocity upstream and down stream from the wave must also be equal; therefore

$$V_1 \cos \beta = V_2 \cos (\beta - \theta)$$

Applying the impulse-momentum principle along x, the difference in the hydrostatic forces must account for the rate of change of momentum per unit time. This is written

$$\Sigma F_x = (F_1 - F_2)_x = \frac{\gamma {y_1}^2}{2} - \frac{\gamma {y_2}^2}{2}$$

$$= \frac{q\gamma}{g} [V_2 \sin (\beta - \theta) - V_1 \sin \beta] \tag{108}$$

and, since $F_{2x} > F_{1x}$, $V_2 \sin (\beta - \theta) < V_1 \sin \beta$ and the indicated direction of deflection of the streamlines through the wave is therefore justified.

Combination of these equations will yield

$$\frac{\tan^2(\beta - \theta)}{\tan^2\beta} = \frac{1 + 2(V_2^2/gy_2)\sin^2(\beta - \theta)}{1 + 2(V_1^2/gy_1)\sin^2\beta} \tag{109}$$

which will show that (for $V_1^2/gy_1 > 1$): for small values of β, $V_2^2/gy_2 > 1$ and, for large values of β, $V_2^2/gy_2 < 1$. Thus the oblique standing wave resembles the hydraulic jump of Art. 44 for large values of β but is quite unlike it for small values of β.

A relationship[6] analogous to that of equation 106 for the hydraulic jump may be set down directly by comparing the development leading to equation 106 with that leading to equations 107 and 108; these are identical except for the terms sin β and sin $(\beta - \theta)$ which accompany the velocities V_1 and V_2, respectively. Hence, for the oblique standing wave, equation 106 may be adjusted to read

$$\frac{y_2}{y_1} = \frac{1}{2}\left[-1 + \sqrt{1 + \frac{8V_1^2\sin^2\beta}{gy_1}}\right] \tag{110}$$

Illustrative Problem

Water flows in an open channel at a depth of 2 ft and velocity 20 fps. A wedged-shaped pier is to be placed in the flow to produce a standing wave of angle (β) 40°. Calculate the required wedge angle (2θ) and the depth just downstream from the wave.

Solution.

$$V_1^2/gy_1 = (20)^2/32.2 \times 2 = 6.22$$

$$y_2 = \tfrac{2}{2}[-1 + \sqrt{1 + 8(6.22)(0.642)^2}] = 3.64 \text{ ft} \tag{110}$$

$$V_2 \sin(\beta - \theta) = (2/3.64) \times 20 \times 0.642 = 7.06 \text{ fps} \tag{107}$$

$$\tan^2(\beta - \theta) = 0.839\left[\frac{1 + 2(7.06)^2/32.2 \times 3.64}{1 + 2(6.22) \times (0.642)^2}\right] \tag{109}$$

$$\tan(\beta - \theta) = 0.461, \quad (\beta - \theta) = 25°, \quad \theta = 15°, \quad 2\theta = 30°$$

46. The Normal Shock Wave. The compression shock wave mentioned in Art. 37 is analogous to the hydraulic jump of Art. 44. The normal shock wave occurs in gases with abrupt reduction from

[6] For other relationships see A. T. Ippen, Chapter VIII of *Engineering Hydraulics* (H. Rouse, Ed.), John Wiley and Sons, 1950.

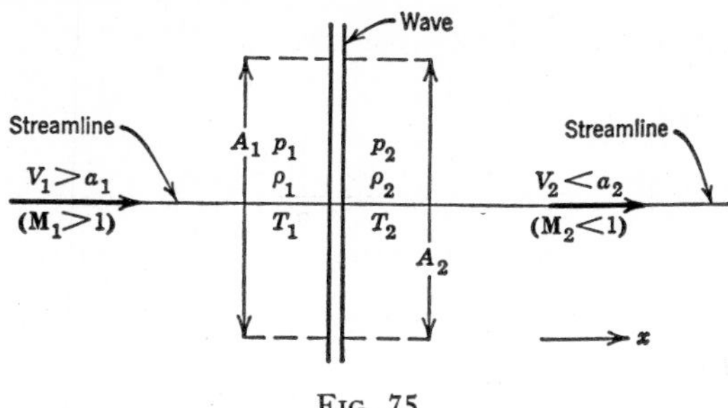

FIG. 75

supersonic to subsonic velocity and is accompanied by large and abrupt[7] rises of pressure, density, temperature, and entropy.

Applying the continuity principle to the normal shock wave of Fig. 75,

$$G = A_1\gamma_1 V_1 = A_2\gamma_2 V_2 \tag{111}$$

The impulse-momentum principle, applied as in the hydraulic jump, gives

$$\Sigma F_x = p_1A_1 - p_2A_2 = \frac{G}{g}(V_2 - V_1)$$

Eliminating G and noting that $A_1 = A_2$, combination of these equations yields

$$p_2 - p_1 = \frac{1}{g}(\gamma_1 V_1^2 - \gamma_2 V_2^2) \tag{112}$$

which will allow the pressure jump $(p_2 - p_1)$ across the wave to be computed. However, this will require the use of another equation to obtain V_2 from p_1, γ_1, and V_1. The energy equation

$$\frac{V_2^2}{2g} - \frac{V_1^2}{2g} = H_1 - H_2 \tag{77}$$

may be used, since internal friction (with increase of entropy) is to be expected in the shock wave without flow of heat to or from the fluid. For perfect gases, $H_1 - H_2 = c_p(T_1 - T_2)$ (see equation 78); with $c_p = Rk/(k-1)$ and $RT = p/\gamma$, equation 77 may be written

$$\frac{V_2^2}{2g} - \frac{V_1^2}{2g} = \frac{k}{k-1}\left(\frac{p_1}{\gamma_1} - \frac{p_2}{\gamma_2}\right) \tag{113}$$

[7] The shock wave is so thin that, for computations outside the wave, it may be considered to be a single line, i.e., a discontinuity in the flow.

and solved simultaneously with equation 112 to yield,[8] after some algebraic manipulation, a relationship between the Mach numbers $\mathbf{M}_1$ and $\mathbf{M}_2$. This is

$$\mathbf{M}_2^2 = \frac{1 + \dfrac{k-1}{2}\mathbf{M}_1^2}{k\mathbf{M}_1^2 - \dfrac{k-1}{2}} \qquad (114)^8$$

from which $\mathbf{M}_2$ may be computed from $\mathbf{M}_1$. Since this equation will be satisfied when $\mathbf{M}_1 = \mathbf{M}_2 = 1$, all other solutions must have the property: for $\mathbf{M}_1 > 1$, $\mathbf{M}_2 < 1$ and for $\mathbf{M}_1 < 1$, $\mathbf{M}_2 > 1$; however, it can be shown that only the first of these is physically possible—the second solution implies a *loss* of entropy through the wave and thus violates the second law of thermodynamics. Accordingly it may be concluded that through a normal shock wave the velocity must fall from supersonic to subsonic.

The pressure jump through the shock wave may also be computed in terms of the *shock strength* $(p_2 - p_1)/p_1$ to yield

$$\frac{p_2 - p_1}{p_1} = \frac{2k}{k+1}(\mathbf{M}_1^2 - 1) \qquad (115)^8$$

and the velocity ratio, V_2/V_1,

$$\frac{V_2}{V_1} = \frac{(k-1)\mathbf{M}_1^2 + 2}{(k+1)\mathbf{M}_1^2} \qquad (116)^8$$

and, from p_2, V_2, and $\mathbf{M}_2$, temperature, density, and sonic velocity downstream from the wave may be readily obtained.

Illustrative Problem

Upstream from a normal shock wave in an air flow the pressure, velocity, and sonic velocity are 14.7 psia, 1732 fps, and 862 fps, respectively. Calculate these quantities just downstream from the wave, and the rise in temperature through the wave.

Solution.

$$\mathbf{M}_1 = 1732/862 = 2.01$$

$$\mathbf{M}_2^2 = \frac{1 + 0.4(2.01)^2/2}{1.4(2.01)^2 - 0.4/2} = 0.271; \quad \mathbf{M}_2 = 0.52 \qquad (114)$$

[8] For detailed development of these equations refer to any text on the flow of compressible fluids. See references at the end of Chapter 5.

$$p_2 = 14.7 \left[1 + \frac{2(1.4)}{2.4}(\overline{2.01^2} - 1)\right] = 66.8 \text{ psia} \tag{115}$$

$$V_2 = 1732 \left[\frac{0.4(2.01)^2 + 2}{2.4(2.01)^2}\right] = 645 \text{ fps} \tag{116}$$

$$a_2 = 645/0.52 = 1240 \text{ fps}$$

$$1240 = \sqrt{1.4 \times 32.2 \times 53.3 \times T_2}; \quad T_2 = 640°\text{R} \tag{11}$$

$$862 = \sqrt{1.4 \times 32.2 \times 53.3 \times T_1}; \quad T_1 = 310°\text{R} \tag{11}$$

Rise of temperature through wave, $T_2 - T_1 = 330°\text{F}$

The conditions which will produce this particular shock wave in a nozzle are shown in the illustrative problem of Art. 37.

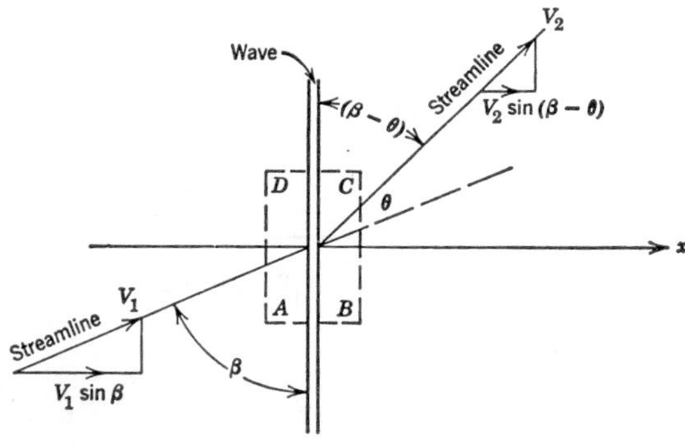

FIG. 76

47. The Oblique Shock Wave. In an age of high-speed flight the reader has had casual acquaintance with the shock waves produced by objects traveling at supersonic speeds.[9] The geometry of such two-dimensional waves may be studied by application of the continuity, impulse-momentum, and energy principles in the following manner.

Construct a control surface $ABCD$ as indicated in Fig. 76, taking $BC = AD = 1$ ft, and also consider the control volume to be 1 ft deep perpendicular to the plane of the paper. From symmetry it is seen that the flowrates across AB and CD are equal; from this it follows that the flowrates, G, across BC and AD are also equal. These are

$$G = \gamma_1 V_1 \sin \beta = \gamma_2 V_2 \sin(\beta - \theta) \tag{117}$$

[9] See Art. 122 for photograph and analytical justification for such waves.

Since the forces caused by the pressures on surfaces AB and CD are equal and opposite, the impulse-momentum equation applied in the y-direction discloses that the components of velocity parallel to the wave front are equal, giving

$$V_1 \cos \beta = V_2 \cos (\beta - \theta) \tag{118}$$

The impulse-momentum principle applied in the x-direction gives

$$p_1 - p_2 = \frac{G}{g}[V_2 \sin (\beta - \theta) - V_1 \sin \beta] \tag{119}$$

The energy equation (without addition or extraction of heat) applied across the wave is (as in Art. 46)

$$\frac{V_2^2}{2g} - \frac{V_1^2}{2g} = \frac{k}{k-1}\left(\frac{p_1}{\gamma_1} - \frac{p_2}{\gamma_2}\right) \tag{113}$$

These equations may be combined and manipulated to yield a variety of useful relationships, one of which is

$$\mathbf{M}_2^2 \sin^2 (\beta - \theta) = \frac{1 + \dfrac{k-1}{2}\mathbf{M}_1^2 \sin^2 \beta}{k\mathbf{M}_1^2 \sin^2 \beta - \dfrac{k-1}{2}} \tag{120}$$ [10]

This equation reveals that (for $\mathbf{M}_1 > 1$): $\mathbf{M}_2 < 1$ for large β and $\mathbf{M}_2 > 1$ for small β. The so-called *strong shock* (featured by large β) is similar to the normal shock of Art. 46 in that the flow changes from supersonic to subsonic in passing through the wave, whereas for small β (*weak shock*) the flow is supersonic on both sides of the wave. Through all such waves, however, there is internal friction with increase of entropy, the flow is rotational (see Art. 38), and $\mathbf{M}_2 < \mathbf{M}_1$.

Illustrative Problem

Upstream from an oblique shock wave in an air flow, the pressure, velocity, and sonic velocity are 14.7 psia, 1732 fps, and 862 fps, respectively. The wave angle (β) is 40°. Calculate angle (θ) required to produce this wave, the pressure, velocity, and sonic velocity just downstream from the wave, and the rise of temperature through the wave.

[10] This result and others for V_2/V_1 and $(p_2 - p_1)/p_1$ may be derived directly from equations 114, 115, and 116 in the following manner. Compare simultaneous equations 111, 112, and 113 with equations 117, 119, and 113. These are identical except for $\sin \beta$ and $\sin (\beta - \theta)$. However, these quantities are multiplied into V_1 and V_2, respectively, and hence will appear in derived results in this pattern and also multiplied into $\mathbf{M}_1$ and $\mathbf{M}_2$, respectively.

Solution.

$$\mathbf{M}_1 = 1732/862 = 2.01$$

$$862 = \sqrt{1.4 \times 32.2 \times 53.3 \times T_1}; \quad T_1 = 310°\text{R} \qquad (11)$$

Applying equation 115, appropriately adjusted with $\sin \beta$,

$$p_2 = 14.7 \left[1 + \frac{2(1.4)}{2.4}\left(\overline{2.01}^2 \times \overline{0.642}^2 - 1\right)\right] = 26.0 \text{ psia}$$

Applying equation 116, appropriately adjusted with $\sin \beta$ and $\sin (\beta - \theta)$,

$$\frac{V_2 \sin (\beta - \theta)}{V_1 \sin \beta} = \frac{(k - 1)\mathbf{M}_1^2 \sin^2 \beta + 2}{(k + 1)\mathbf{M}_1^2 \sin^2 \beta}$$

and, using $V_2/V_1 = \cos \beta / \cos (\beta - \theta)$ from equation 118,

$$\frac{\tan (\beta - \theta)}{\tan 40°} = \frac{0.4(2.01 \times 0.642)^2 + 2}{2.4(2.01 \times 0.642)^2}; \quad \beta - \theta = 29.3°, \quad \theta = 10.7°$$

$$\mathbf{M}_2^2 = \frac{1}{(0.489)^2} \frac{1 + (0.4/2)(2.01 \times 0.642)^2 + 2}{1.4(2.01 \times 0.642)^2 - 0.4/2} = 2.635; \quad \mathbf{M}_2 = 1.62 \qquad (120)$$

$$V_2 = 1732 \cos 40°/\cos 29.3° = 1523 \text{ fps} \qquad (118)$$

$$a_2 = 1523/1.62 = 940 \text{ fps}$$

$$940 = \sqrt{1.4 \times 32.2 \times 53.3 \times T_2}; \quad T_2 = 367.5°\text{R} \qquad (11)$$

or

$$\frac{(1732)^2 - (1523)^2}{2g} = 186.5(T_2 - T_1); \quad T_2 - T_1 = 57.5°, \quad T_2 = 367.5°\text{R} \qquad (78)$$

48. Jet Propulsion. In a modern aircraft jet-propulsion system (Fig. 77), air is drawn in at the upstream end and its pressure,

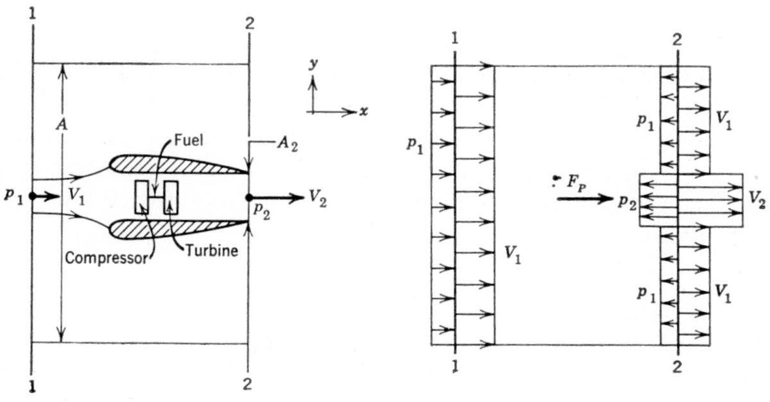

FIG. 77

density, and temperature are raised by a compressor. Just downstream from the compressor, fuel is injected and burned, adding energy to the fluid. The mixture then expands and passes through a gas turbine (which drives the compressor) on its way to the nozzle, whence it emerges into the atmosphere at high velocity and at a pressure equal to or greater than that surrounding the nozzle.

An impulse-momentum analysis of such a unit will allow its propulsive force to be computed. Here a control volume of indefinite vertical extent may be used as shown in Fig. 77, section 1 located well upstream from the unit and section 2 in the plane of the nozzle exit. If it be assumed that air which does not pass through the unit experiences no net change of velocity or pressure, the pressures and velocities at all points of sections 1 and 2 will be p_1 and V_1, respectively, except in the nozzle exit, where they are p_2 and V_2. Taking the summation of forces on the fluid within the control volume,

$$\Sigma F_x = p_1 A - p_1(A - A_2) - p_2 A_2 + F_P = (p_1 - p_2)A_2 + F_P$$

in which F_P is the propulsive force exerted *by the unit on the fluid.* Between the point of fuel injection and nozzle exit the velocity of the mass flowrate (G_f/g) of fuel increases from zero to V_2; between sections 1 and 2 the velocity of the mass flowrate (G_a/g) of air increases from V_1 to V_2. Accordingly (equation 99)

$$\frac{d}{dt}(MV)_x = (G_f/g)V_2 + (G_a/g)(V_2 - V_1)$$

Equating ΣF_x to $d/dt\,(MV)_x$ (equation 102) yields the propulsive force, F_P, obtainable from the unit; it is

$$F_P = (G_a/g)(V_2 - V_1) + (G_f/g)V_2 + (p_2 - p_1)A_2 \tag{121}$$

Although this force is fundamental to preliminary design calculations, it yields no information about the optimum design of passages, compressor, turbine, etc., for which other principles must be used. It does, however, demonstrate again the strength of the impulse-momentum principle in the bypassing of internal complexities and obtaining useful results from upstream and downstream flow conditions alone.

Illustrative Problem

A jet-propulsion unit is to develop a propulsive force of 1000 lb when moving through the ICAO Standard Atmosphere (Appendix IV) at 30,000 ft altitude at a speed of 500 fps. The velocity, pressure, and area at the nozzle exit are 4000 fps, 5.0 psia, and 1 sqft, respectively. How much air must be drawn through the unit? Fuel may be neglected.

Solution.

$$1000 = (G_a/32.2)(4000 - 500) + (5.00 - 4.37)144 \times 1; \quad G_a = 8.4 \text{ lb/sec} \tag{121}$$

A suitable compressor must then be designed to provide this flowrate.

49. Propellers and Windmills. Although the screw propellers of ships or aircraft cannot be analyzed with the impulse-momentum and energy principles alone, application of these principles to the problem will lead to some of the laws which characterize their design.

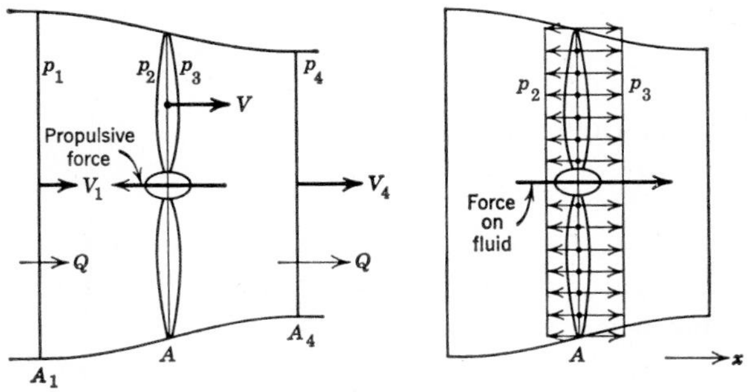

FIG. 78

A screw propeller with its slipstream is shown in Fig. 78. For such a propeller operating in an unconfined fluid, the pressures p_1 and p_4 at some distance ahead of and behind the propeller, and the pressures over the slipstream boundary, are the same. However, from the shape of the slipstream (using the continuity and Bernoulli principles) the mean pressure p_2 just upstream from the propeller is smaller than p_1, and the pressure p_3 just downstream from the propeller is larger than p_4.[11] When the fluid in the slipstream between sections 1 and 4 is isolated, it is observed that the only force acting is that exerted by the propeller on the fluid.[12] This may be computed either from the pressure difference $(p_3 - p_2)$ or from the gain in momentum flux between sections 1 and 4. Therefore

$$(p_3 - p_2)\,A = F = Q\rho(V_4 - V_1) = A\rho V(V_4 - V_1) \tag{122}$$

[11] Applying this, and the fact of the preceding sentence, at the propeller tip apparently results in two different pressures there. This is due to the assumption of a uniform pressure distribution; in reality the pressure is not uniform over the propeller disk because at the propeller tip it must equal that surrounding the slipstream.

[12] The same pressure distributed over sections 1, 4, and the slipstream boundary can produce no net force.

in which V is the mean velocity through the propeller disk; cancelling A's in this equation,

$$p_3 - p_2 = \rho V(V_4 - V_1) \tag{123}$$

Now, applying the Bernoulli principle between sections 1 and 2,

$$p_1 + \tfrac{1}{2}\rho V_1^2 = p_2 + \tfrac{1}{2}\rho V_2^2 \tag{65}$$

and between sections 3 and 4,

$$p_3 + \tfrac{1}{2}\rho V_3^2 = p_4 + \tfrac{1}{2}\rho V_4^2 \tag{65}$$

and using $p_1 = p_4$, another expression may be derived for $(p_3 - p_2)$; it is

$$p_3 - p_2 = \tfrac{1}{2}\rho(V_4^2 - V_1^2) \tag{124}$$

By equating equations 123 and 124,

$$V = (V_1 + V_4)/2 \tag{125}$$

which shows that the velocity through the propeller disk is the numerical average of the velocities at some distance ahead of and behind the propeller; in other words, there is the same increase of velocity ahead of the propeller as behind it. This result, modified slightly to allow for friction, rotational effects, etc., is one of the basic assumptions of propeller design.

The useful power output, P_o, derived from a propeller, is the thrust, F, multiplied by the velocity, V_1, at which the propeller is moving forward.

$$P_o = FV_1 = Q\rho(V_4 - V_1)V_1 \tag{126}$$

The power input, P_i, is that required to maintain continual increase of velocity of the slipstream from V_1 to V_4. From equation 68,

$$P_i = \frac{Q\rho}{2}(V_4^2 - V_1^2) = Q\rho(V_4 - V_1)\left(\frac{V_4 + V_1}{2}\right) = Q\rho(V_4 - V_1)V \tag{127}$$

The ideal efficiency, η, of the propeller is then

$$\eta = \frac{P_o}{P_i} = \frac{V_1}{V} \tag{128}$$

and, since V is always greater than V_1, it may be concluded that the efficiency of a propeller even in an ideal fluid can never be 100 percent.[13]

[13] When $V = V_1$, a propeller of 100 percent efficiency results, but such a propeller produces no propulsive force! (See equation 122.) Practical ship and airplane propellers may have efficiencies of 80 percent.

There are many similarities between propeller and windmill, but their purposes are quite different. The propeller is designed primarily to create a propulsive force or thrust, whereas the windmill is designed for extracting energy from the wind; indeed, it would be advantageous (for structural considerations) to have the axial force on a windmill a minimum. Because of the different objectives of windmill and propeller, their efficiencies are calculated differently. However, comparison of Figs. 78 and 79 will show the windmill to be, as far as the flow picture is concerned, the inverse of the propeller. In the windmill the "slipstream" widens as it passes the machine and the pressure p_2 is greater than pressure p_3. However, through applying the Bernoulli and impulse-momentum principles as before, the velocity through the windmill disk, as through the propeller, may be shown to be the numerical average of V_1 and V_4.

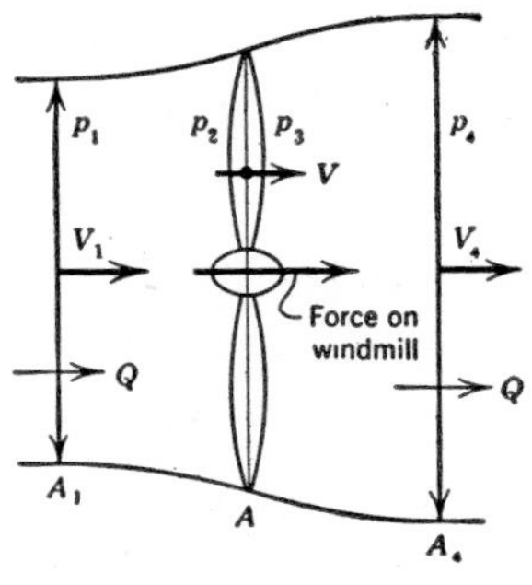

FIG. 79

In a frictionless machine, the power delivered to the windmill must be exactly that extracted from the air, which in turn is represented by the decrease of the kinetic energy of the slipstream between sections 1 and 4. This is the *output* of the machine and is given by

$$P_o = \frac{Q\rho}{2}(V_1^2 - V_4^2) \tag{129}$$

It is customary to define windmill efficiencies as the ratio of this power output to the total power *available* in a streamtube of cross-sectional area A and wind velocity V_1. Thus the efficiency of an ideal windmill is

$$\eta = \frac{P_o}{P_a} = \frac{(V_1^2 - V_4^2)AV\rho/2}{AV_1\rho V_1^2/2} = \frac{(V_1 + V_4)(V_1^2 - V_4^2)}{2V_1^3} \tag{130}$$

$(V_1 + V_4)/2$ having been substituted for V. The maximum efficiency is found by differentiating η in respect to V_4/V_1 and setting the result equal to zero. This gives a value of $V_4/V_1 = \frac{1}{3}$, which when substituted in equation 130 produces a maximum efficiency of $\frac{16}{27}$, or 59.3 percent. Because of friction and other losses this efficiency is, of course, not realized in practice: the highest possible efficiency for a real windmill appears to be around 50 percent; the traditional "Dutch windmill" with large sail-like blades operates at an efficiency around 5 percent.

Illustrative Problem

The engine of an airplane flying through still air (specific weight 0.0765 lb/cuft) at 200 mph (293 fps) delivers 1500 hp to an ideal propeller 10 ft in diameter. Calculate slipstream velocity, velocity through the propeller disk, and the diameter of the slipstream ahead of and behind the propeller. Also calculate thrust and efficiency.

Solution.

$$P_i = 1500 \times 550 = \frac{\pi}{4}(10)^2\left(\frac{V_4 + 293}{2}\right)\frac{0.0765}{32.2}\left(\frac{V_4^2 - \overline{293}^2}{2}\right);$$

$$V_4 = 338 \text{ fps} \tag{127}$$

$$V = (338 + 293)/2 = 315.5 \text{ fps}, \quad Q = 24{,}800 \text{ cfs} \tag{125}$$

$$A_1 = 24{,}800/293 = 84.7 \text{ sqft}, \quad d_1 = 10.4 \text{ ft}$$

$$A_4 = 24{,}800/338 = 73.3 \text{ sqft}, \quad d_4 = 9.67 \text{ ft}$$

$$F = 24{,}800(0.0765/32.2)(338 - 293) = 2610 \text{ lb} \tag{122}$$

$$\eta = \frac{2610 \times 293}{1500 \times 550} = 92.7\% \quad \text{or} \quad \eta = \frac{293}{315.5} = 92.7\% \tag{128}$$

50. Deflectors and Blades—The Impulse Turbine. When a free jet is deflected by a blade surface, a change of momentum occurs and a force is exerted on the blade. If the blade is allowed to move, this force will act through a distance, and power may be derived from the moving blade; this is the basic principle of the impulse turbine.

The jet of Fig. 80 is deflected by a fixed blade and may be assumed to be in a horizontal plane.[14] With the control surface drawn around the region of momentum change, it is seen at once that the force exerted by the blade is the only force acting on the fluid. Therefore, from equations 102 and 103,

$$\Sigma F_x = -F_x = Q\rho(V_2 \cos\beta - V_1) \tag{131}$$

$$\Sigma F_y = F_y = Q\rho(V_2 \sin\beta - 0) \tag{132}$$

If this blade now moves (Fig. 81) with velocity u in the same direction as the original jet, the jet is no longer deflected through an angle β since the leaving velocity, V_2, is now the resultant of the blade velocity and the velocity of the fluid over the blade. The velocity, v, of fluid relative to the blade is obviously $(V_1 - u)$, and if friction is neglected this relative velocity is the same at the entrance

[14] The difference in elevation between beginning and end of blade is usually negligible in practical blade problems.

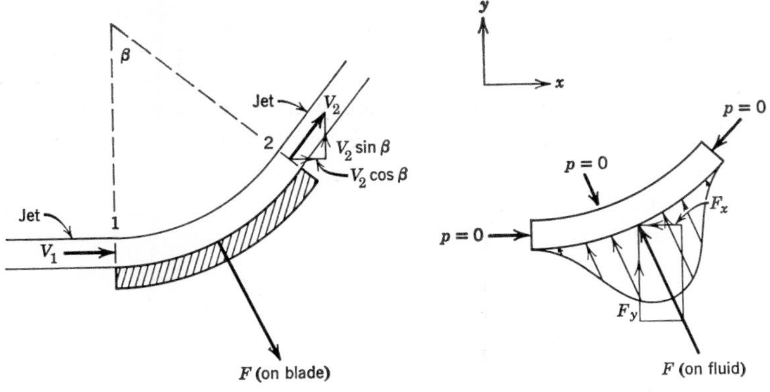

FIG. 80

and exit of the blade system; therefore, as the jet leaves the blade, it has an absolute velocity, V_2, equal to the vector sum of u and $(V_1 - u)$. Then, from Fig. 81 and equations 102 and 103,

$$\Sigma F_x = -F_x = Q\rho(V_{2x} - V_{1x}) = -Q\rho(V_1 - u)(1 - \cos \beta) \tag{133}$$

$$\Sigma F_y = F_y = Q\rho(V_{2y} - V_{1y}) = Q\rho(V_1 - u) \sin \beta \tag{134}$$

Engineers frequently prefer to treat such problems with the relative velocities v_1 and v_2, both of which are equal to $(V_1 - u)$, thus reducing the problem to that of the stopped blade of Fig. 80; the validity of this may be seen from the velocity triangles and by noting that the substitution of $(V_1 - u)$ for V_1 and V_2 in equations 131 and 132 will yield equations 133 and 134, respectively.

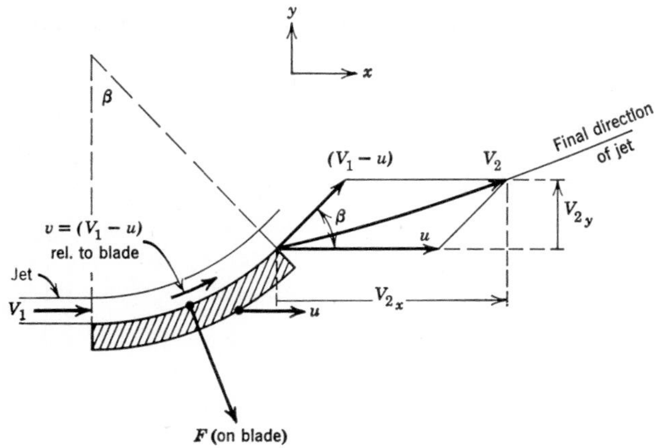

FIG. 81

For a single moving blade the flowrate, Q, in these equations will not be the flowrate in the jet, since less[15] fluid is deflected per second by a single blade than is flowing in the jet. However, in the practical application of this theory to a turbine, where a series of blades is used, the fluid deflected *by the blade system* per second is the same as that flowing.

Problems of this type should be recognized as unsteady-flow problems for which the equations have been written for time-average conditions. As a blade moves through the jet a time-varying force

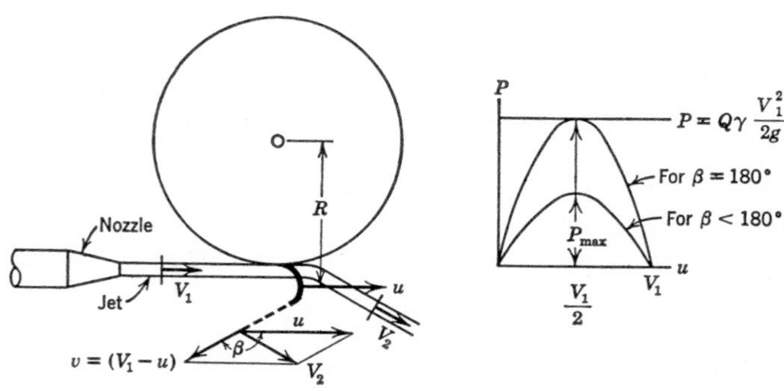

FIG. 82

will act on it, rising from zero and falling back to zero. From this it is clear that the foregoing equations can give no information about the maximum force exerted on the blade or the force on the blade at any instant of time, or the optimum spacing of blades to secure a required deflection of fluid; nevertheless the equations may be used effectively in preliminary design and performance calculations.

In an impulse turbine (Fig. 82) a series of blades of the type above is mounted on the periphery of a wheel. While a blade is in the jet it is moving in a direction approximately parallel to that of the jet; thus the equations above may be applied directly to find the power characteristics of the machine. The force component F_y does no work, since it acts through no distance; when the component F_x is multiplied by u, the power transferred from jet to turbine is obtained. In a frictionless machine this is the output and is given by

$$P = Q\rho(V_1 - u)(1 - \cos\beta)u \qquad (135)$$

From this it is evident (1) that no power is obtained from the machine when $u = 0$ (machine stopped) and when $u = V_1$ (runaway

[15] For a blade moving in the same direction as the jet.

speed) and (2) that, with Q, ρ, β, and V_1 constant, the relationship between P and u is a parabolic one. Taking $dP/du = 0$ to obtain the properties of maximum output shows that this will occur when $u = V_1/2$ and will be given by

$$P_{\max} = Q\rho \frac{V_1^2}{4}(1 - \cos\beta) \tag{136}$$

For a blade angle of 180°, this becomes

$$P_{\max} = \frac{Q\rho V_1^2}{2} = \frac{Q\gamma V_1^2}{2g} \tag{137}$$

Since this is exactly the power of the free jet (see equation 68), it may be concluded that all the jet power may be theoretically extracted and transferred to the machine (at 100 percent efficiency), if (1) the peripheral speed is one-half the jet speed and if (2) the blade angle is 180°. In practice, blade angles will be found to be around 165°, operating peripheral speeds to be about 48 percent of jet velocity, with resulting peak efficiencies near 90 percent.

Another and more general type of free-jet turbine (Fig. 83) may be successfully analyzed by using the energy and impulse-momentum principles. The power P transmitted from jet to blade system is (neglecting friction) exactly that lost from the jet (see equation 68):

$$P = Q\rho(V_1^2 - V_2^2)/2 \tag{138}$$

which shows that $V_2 < V_1$. This power is calculable from $P = F_y u$, in which F_y is the working component of force exerted by the jet

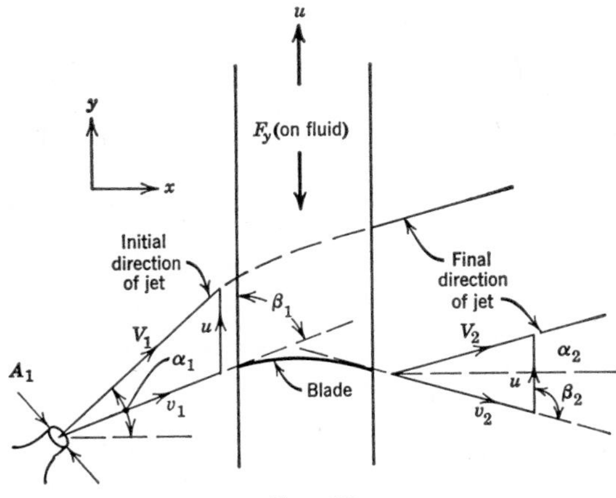

FIG. 83

on the moving blade system. From the impulse-momentum principle, F_y (on the fluid) is given by

$$\Sigma F_y = -F_y = Q\rho(V_2 \sin \alpha_2 - V_1 \sin \alpha_1) \tag{139}$$

showing that $V_2 \sin \alpha_2 < V_1 \sin \alpha_1$. These requirements (without regard to the details of the blade geometry) justify the relative magnitudes and positions of V_1 and V_2 shown in Fig. 83.

Suitable blades to produce these changes of velocity may now be designed by observing that the absolute velocity (V) of the fluid is the vector sum of relative velocity (v) and blade velocity (u), the relative velocity being everywhere tangent to the blade (i.e., the fluid does not separate from the blade) for good design. Accordingly the velocity triangles are as shown and the inlet and exit blade angles, β_1 and β_2, determined. A further requirement of a frictionless system is that the relative velocities v_1 and v_2 must be equal, as in Fig. 81.

The foregoing equations and requirements allow the preliminary design of free-jet turbines to produce a given power at a given flowrate and required speed; this is illustrated in the following problems.

Illustrative Problems

An impulse turbine of 6-ft diameter is driven by a water jet of 2-in. diameter moving at 200 fps. Calculate the force on the blades and the horsepower developed at 250 rpm. The blade angles are 150°.

Solution.

$$u = (250/60)2\pi \times 3 = 78.6 \text{ fps}$$

$$v_1 = v_2 = V_1 - u = 200 - 78.6 = 121.4 \text{ fps}$$

From the velocity diagram,

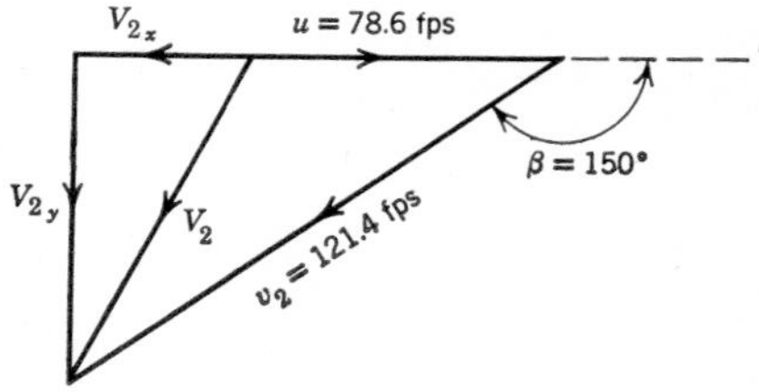

$$V_{2x} = -(121.4)\ 0.866 + 78.6$$

$$= -26.6 \text{ fps}$$

$$V_{2y} = (121.4)0.50 = 60.7 \text{ fps}$$

$$V_2 = 66.5 \text{ fps}$$

The flowrate is

$$Q = (\pi/4)(2/12)^2 200 = 4.36 \text{ cfs}$$

The working component of force on the fluid is

$$-F_x = 4.36 \times 1.935[-26.6 - 200] = -1915 \text{ lb} \tag{133}$$

The power developed may be computed in two ways:

From mechanics,

$$P = 1915 \times 78.6/550 = 274 \text{ hp}$$

From equation 138,

$$P = 4.36 \times 1.935(\overline{200}^2 - \overline{66.5}^2)/2 \times 550 = 274 \text{ hp}$$

A free-jet impulse turbine is to produce 100 hp at a blade speed of 75 fps. A water jet having $V_1 = 150$ fps, diameter $= 2$ in., and $\alpha_1 = 60°$ is used to drive the machine. Calculate the required blade angles β_1 and β_2.

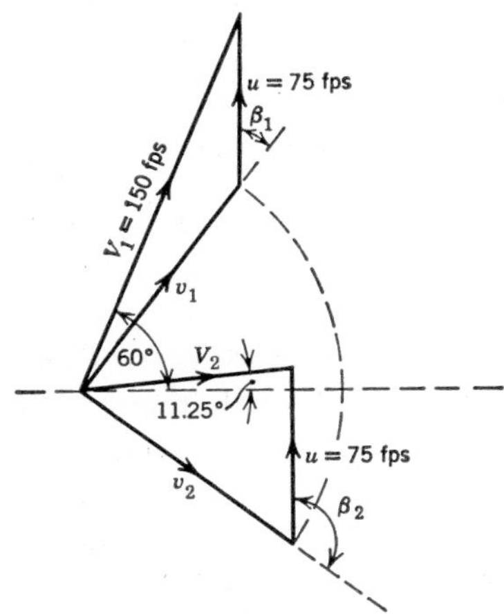

Solution.

$$Q = (\pi/4)(2/12)^2 150 = 3.27 \text{ cfs} \quad (53)$$

$$F_y = 550 \times 100/75 = 733 \text{ lb}$$

$$100 \times 550 = 3.27 \times 1.935(\overline{150}^2 - V_2^2)/2, \quad V_2 = 71.7 \text{ fps} \quad (138)$$

$$733 = 3.27 \times 1.935(150 \times 0.866 - 71.7 \sin \alpha_2); \quad \alpha_2 = 11.25° \quad (139)$$

From the geometry of the velocity triangles,

$$\beta_1 = 54°, \quad \beta_2 = 131°.$$

51. Reaction Turbine and Centrifugal Pump. One of the most important engineering applications of the impulse-momentum principle is the design of turbines, pumps, and other *turbomachines* such as fluid drives and torque converters, all of which involve very complex three-dimensional flowfields beyond the scope of this book. However, the principle may be applied to the two-dimensional flowfield of a simplified reaction turbine and centrifugal pump to demonstrate the method and to gain some basic understanding of the design and operation of such machines.

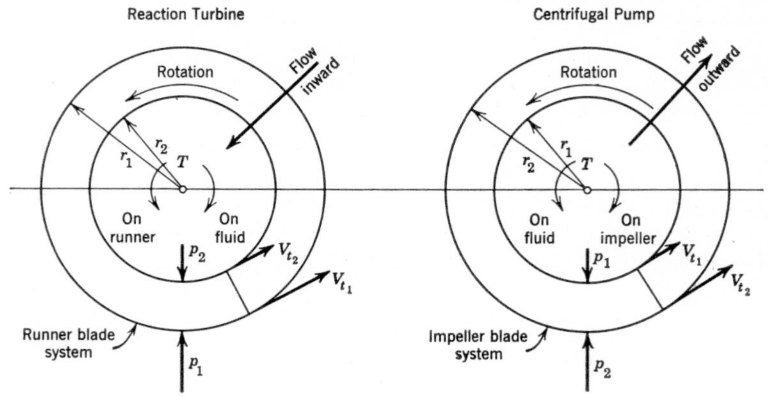

FIG. 84

Figure 84 is a definition sketch for both of these machines, which feature a moving blade system symmetrical about the axis of rotation. This blade system is the essential part of the rotating element of the machine, called the *runner* for the turbine and the *impeller* for the pump. As fluid flows through these blade systems, the tangential component of the absolute velocity changes, decreasing through the runner of the turbine, increasing through the impeller of the pump. Constructing control surfaces (concentric circles) around the blade systems, section 1 becomes the inlet and section 2 the outlet of the control volume for both machines. The impulse-momentum principle may now be applied. Resolving the velocities (V) into tangential and radial components, $d/dt\,(rMV)_o$ may be easily computed from equation 101 since the radial components of velocity pass through the center of moments; the result (for both machines) is

$$\frac{d}{dt}(rMV)_o = Q\rho V_{t_2}r_2 - Q\rho V_{t_1}r_1 \tag{140}$$

The forces exerted on the fluid within the control volume are produced by the pressures on sections 1 and 2, and by the blades of the moving blade system; as the former are wholly radial, they can have no moment about O, the radial components of the latter will also cancel for the same reason, leaving only the tangential components exerted by blade system on fluid to be considered. Although these forces are not identified as to magnitude and location, their total effect is a *torque, T, by the blade system on the fluid.* From equation 97, $\Sigma M_o = T$ and, applying equation 104,

$$\Sigma M_o = T = Q\rho\,[V_{t_2}r_2 - V_{t_1}r_1] \tag{141}$$

From the relative magnitude of $V_{t_1}r_1$ and $V_{t_2}r_2$ it is evident that T (*on the fluid*) is clockwise for the turbine and counterclockwise for the pump; the torques *on runner and impeller* will necessarily be in the opposite directions.

The change of unit energy, E, associated with either of these machines may be calculated from equation 141 by recalling from mechanics that $P = T\omega$ and from equation 68 that $P = Q\gamma E$. The result is

$$E = \frac{\omega}{g}[V_{t_2}r_2 - V_{t_1}r_1] \tag{142}$$

which gives some insight into the relation between the internal dynamics of the machine and the head, E_P or E_T, which was emphasized in Art. 25. For a pump, where $V_{t_2}r_2 > V_{t_1}r_1$, equation 142 is written

$$E_P = \frac{\omega}{g}[V_{t_2}r_2 - V_{t_1}r_1] \tag{143}$$

but for the turbine, where $V_{t_2}r_2 < V_{t_1}r_1$,

$$E_T = \frac{\omega}{g}[V_{t_1}r_1 - V_{t_2}r_2] \tag{144}$$

The alternative to this adjustment in the equation is to set up a special sign convention; this is convenient in more advanced analysis but hardly necessary here.

A two-dimensional reaction turbine is shown in Fig. 85 with accompanying velocity diagrams. Fixed guide vanes exert a torque on the fluid which gives it a tangential velocity component; since these guide vanes do not move, this torque does no work. The fluid then passes through the moving runner, through which the tangential component of velocity decreases, producing a torque according to equation 144. As in Art. 50, the blades of the runner must be designed for smooth flow to accomplish this. Blade angles are determined from the velocity diagrams as before, the size of such diagrams being dictated by the equation of continuity ($q = 2\pi r_1 V_{r_1} = 2\pi r_2 V_{r_2}$), the angular speed of the runner through $u = \omega r$, and the guide vane angle, α. Important features of the velocity triangles are: (1) the increase of V_r through the runner and (2) the same V_1 at guide vane exit and runner entrance. The latter indicates no abrupt change in magnitude and direction of the velocity as it passes from guide vanes to moving runner; because this is imperative for good design, runner blades are shaped to accomplish it.

A section through the impeller of a two-dimensional centrifugal pump is shown in Fig. 86. With no guide vanes upstream from the impeller, the tangential component of absolute velocity is zero at the inlet and it increases through the impeller as torque is exerted on the fluid. This power is transferred from impeller to fluid and the energy of the fluid increases; both pressure and kinetic energies of the fluid increase through the impeller. As in the turbine, blade angles required to produce specified heads and flowrates for specified rotational speeds may be deduced from the velocity triangles.

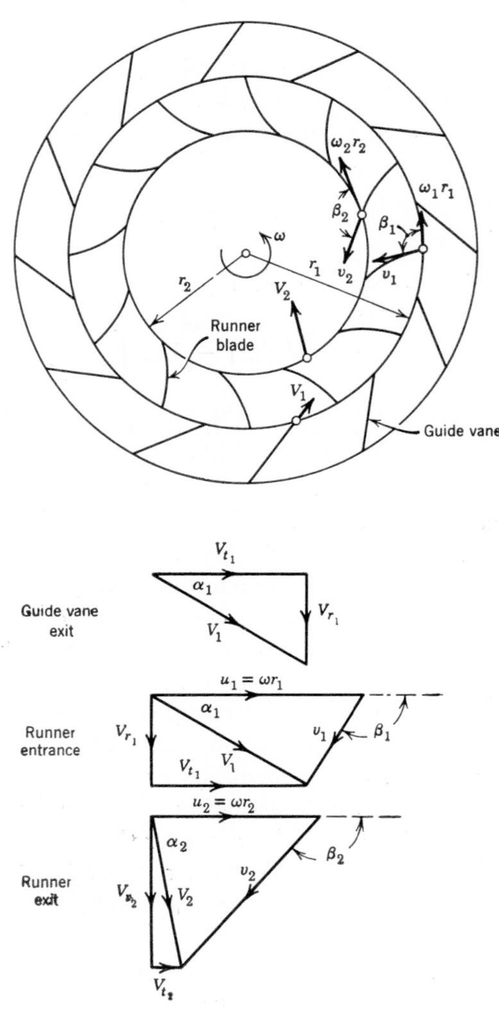

FIG. 85

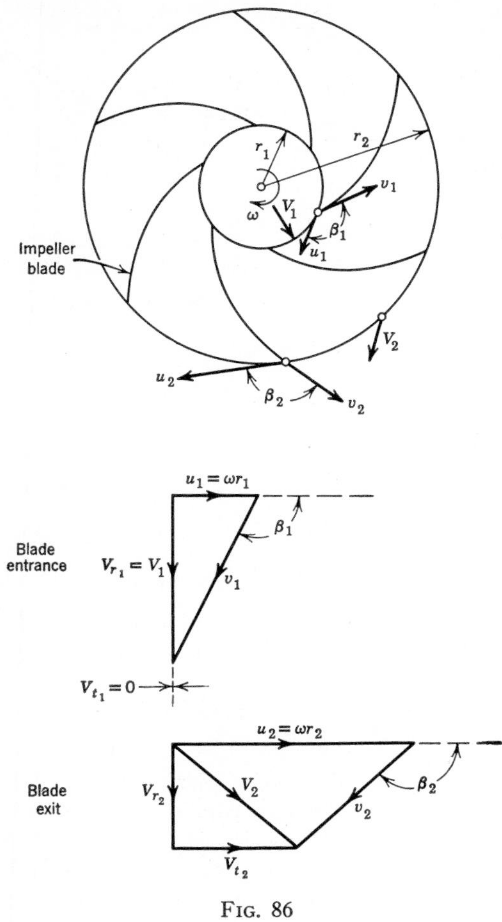

FIG. 86

ILLUSTRATIVE PROBLEMS

A two-dimensional reaction turbine has $r_1 = 5$ ft, $r_2 = 3.5$ ft, $\beta_1 = 60°$, $\beta_2 = 150°$, and thickness of 1 ft parallel to the axis of rotation. With a guide vane angle of 15° and a flowrate of 333 cfs of water, calculate the required speed of the runner for smooth flow at the inlet. For this condition calculate also the torque exerted on the runner, the horsepower developed, the energy extracted from each pound of fluid, and the pressure drop through the runner.

Solution.

$$V_{r_1} = 333/2\pi(5) = 10.6 \text{ fps}$$

$$V_{r_2} = 333/2\pi(3.5) = 15.15 \text{ fps}$$

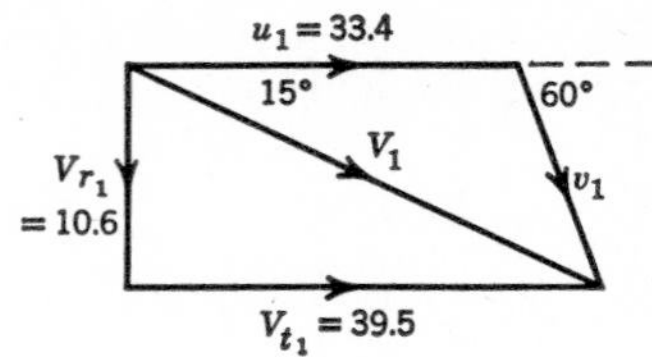

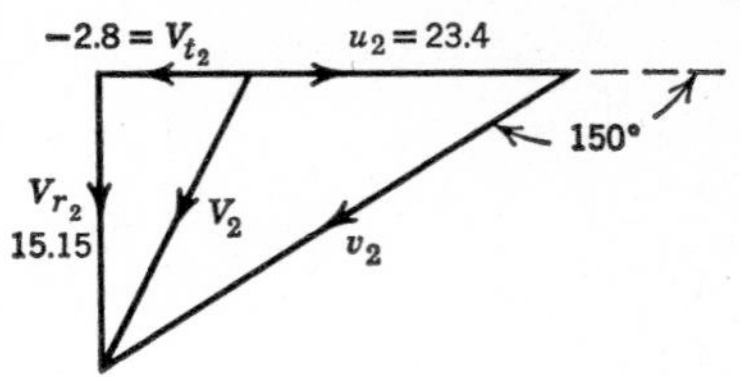

From the velocity triangles,

$$V_{t_1} = 10.6 \cot 15° = 39.5 \text{ fps}$$

$$u_1 = \omega r_1 = 39.5 - 10.6 \tan 30° = 33.4 \text{ fps}$$

$$\omega = 33.4/5 = 6.68 \text{ rad/sec} = 63.8 \text{ rpm}$$

$$u_2 = \omega r_2 = 6.68 \times 3.5 = 23.4 \text{ fps}$$

$$V_{t_2} = 23.4 - 15.15 \cot 30° = -2.8 \text{ fps}$$

$$T = 333 \times 1.935[39.5 \times 5 - (-2.8)3.5] = 133{,}300 \text{ ft-lb} \tag{141}$$

$$\text{Horsepower developed} = 133{,}300 \times 6.68/550 = 1620 \text{ hp}$$

$$333 \times 62.4 \times E_T/550 = 1620; \quad E_T = 42.9 \text{ ft-lb/lb} \tag{68}$$

From the diagrams, $V_1 = 40.8$ fps, $V_2 = 15.4$ fps, and, applying the Bernoulli equation between sections 1 and 2,

$$\frac{p_1}{\gamma} + \frac{(40.8)^2}{2g} = \frac{p_2}{\gamma} + \frac{(15.4)^2}{2g} + 42.9; \quad p_1 - p_2 = 9.0 \text{ psi} \tag{67}$$

A two-dimensional centrifugal pump impeller has $r_1 = 3$ in., $r_2 = 10$ in., $\beta_1 = 120°$, $\beta_2 = 135°$, and thickness of 1 in. parallel to the axis of rotation. If it delivers 4 cfs with no tangential velocity component at the entrance, what is its rotational speed? For this condition, calculate torque and horsepower of the machine, the energy given to each pound of water, and the pressure rise through the impeller.

Solution.

$$V_1 = V_{r_1} = 4 \times 144/2\pi \times 3 \times 1 = 30.6 \text{ fps}$$

$$V_{r_2} = 4 \times 144/2\pi \times 10 \times 1 = 9.18 \text{ fps}$$

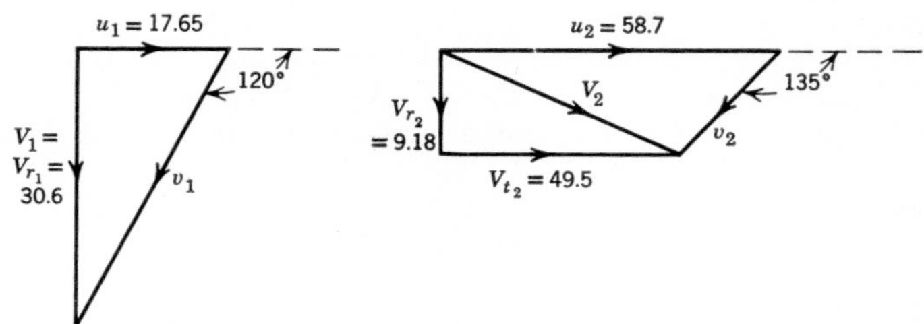

From the velocity triangles,

$$u_1 = \omega r_1 = 30.6 \tan 30^\circ = 17.65 \text{ fps}$$

$$\omega = 17.65/0.25 = 70.5 \text{ rad/sec} = 674 \text{ rpm}$$

$$u_2 = \omega r_2 = 70.5 \times 10/12 = 58.7 \text{ fps}$$

$$V_{t_2} = 58.7 - 9.18/\tan 45^\circ = 49.5 \text{ fps}$$

$$T = 4.0 \times 1.935(49.5 \times 10/12 - 0) = 319 \text{ ft-lb} \qquad (141)$$

$$P = 319 \times 70.5 = 22{,}500 \text{ ft-lb/sec} = 41.0 \text{ hp}$$

$$4.0 \times 62.4 \times E_P/550 = 41.0; \quad E_P = 90 \text{ ft} \qquad (68)$$

From the diagrams, $V_1 = 30.6$ fps, $V_2 = 50.3$ fps, and, applying Bernoulli's equation between sections 1 and 2,

$$\frac{p_1}{\gamma} + \frac{(30.6)^2}{2g} + 90.2 = \frac{p_2}{\gamma} + \frac{(50.3)^2}{2g}; \quad p_2 - p_1 = 28.3 \text{ psi} \qquad (67)$$

REFERENCES

Jet Propulsion

J. G. Keenan, *Elementary Theory of Gas Turbines and Jet Propulsion*, Oxford University Press, 1946.

V. C. Finch, *Jet Propulsion—Turbojets*, The National Press, 1948.

V. C. Finch, *Jet Propulsion—Turboprops*, The National Press, 1950.

F. P. Durham, *Aircraft Jet Power Plants*, Prentice-Hall, 1950.

G. G. Smith, *Gas Turbines and Jet Propulsion for Aircraft*, 5th Edition, Illisse and Son, 1950.

D. Küchemann and J. Weber, *Aerodynamics of Propulsion*, McGraw-Hill Book Co., 1953.

M. J. Zucrow, *Aircraft and Missile Propulsion*, John Wiley and Sons, 1958.

O. E. Lancaster (Ed.), *Jet Propulsion Engines*, Princeton University Press, 1959.

Propellers and Windmills

F. E. Weick, *Aircraft Propeller Design*, McGraw-Hill Book Co., 1930.

D. W. Taylor, *The Speed and Power of Ships*, Revised Edition, Ramsdell, 1933.

H. Glauert, "Airplane Propellers," Part L of Vol. IV of *Aerodynamic Theory* (W. F. Durand, Ed.), Julius Springer, Berlin, 1935.

H. Glauert, *Elements of Airfoil and Airscrew Theory*, 2nd Edition, The Macmillan Co., 1947.

F. T. Meacock, *Elements of Aircraft Propeller Design*, E. and F. N. Spon, 1947.

P. C. Putnam, *Power from the Wind*, D. Van Nostrand, 1948.

T. Theodorsen, *Theory of Propellers*, McGraw-Hill Book Co., 1948.

Turbines and Pumps

R. L. Daugherty, *Centrifugal Pumps*, McGraw-Hill Book Co., 1915.

R. L. Daugherty, *Hydraulic Turbines*, Third Edition, McGraw-Hill Book Co., 1920.

W. Spannhake, *Centrifugal Pumps, Turbines, and Propellers*, Technology Press, 1934.

A. Pfau, *Modern Hydraulic Prime Movers*, Allis-Chalmers Co., 1943.

A. H. Church, *Centrifugal Pumps and Blowers*, John Wiley and Sons, 1944.

G. F. Wislicenus, *Fluid Mechanics of Turbomachinery*, McGraw-Hill Book Co., 1947.

J. W. Daily, "Hydraulic Machinery," Chapter XIII of *Engineering Hydraulics* (H. Rouse, Ed.), John Wiley and Sons, 1950.

D. G. Shepherd, *Principles of Turbomachinery*, The Macmillan Co., 1956.

A. J. Stepanoff, *Centrifugal and Axial Flow Pumps*, 2nd Edition, John Wiley and Sons, 1957.

PROBLEMS

415. A horizontal 6-in. pipe, in which 2.2 cfs of water are flowing, contracts to a 3-in. diameter. If the pressure in the 6-in. pipe is 40 psi, calculate the magnitude and direction of the horizontal force exerted on the contraction.

416. A horizontal 2-in. pipe, in which 400 gpm of water are flowing, enlarges to a 4-in. diameter. If the pressure in the smaller pipe is 20 psi, calculate magnitude and direction of the horizontal force on the enlargement.

417. A conical enlargement in a vertical pipe line is 5 ft long and enlarges the pipe from 12-in. to 24-in. diameter. Calculate the magnitude and direction of the vertical force on this enlargement when 10 cfs of water flow upward through the line and the pressure at the smaller end of the enlargement is 30 psi.

418. A conical diverging tube is horizontal, 12 in. long, has 3-in. throat diameter, 4-in. exit diameter, and discharges 1.0 cfs of water into the atmosphere. Calculate the magnitude and direction of the force components exerted by the water on the tube.

419. A 4-in. nozzle is bolted (with 6 bolts) to the flange of a 12-in. horizontal pipe line and discharges water into the atmosphere. Calculate the tension load in each bolt when the pressure in the pipe is 85 psi. Neglect vertical forces.

420. Calculate the resultant force of the water on this orifice plate. Pipe is of 12-in. diameter, orifice 8 in., and vena contracta 6.5 in.

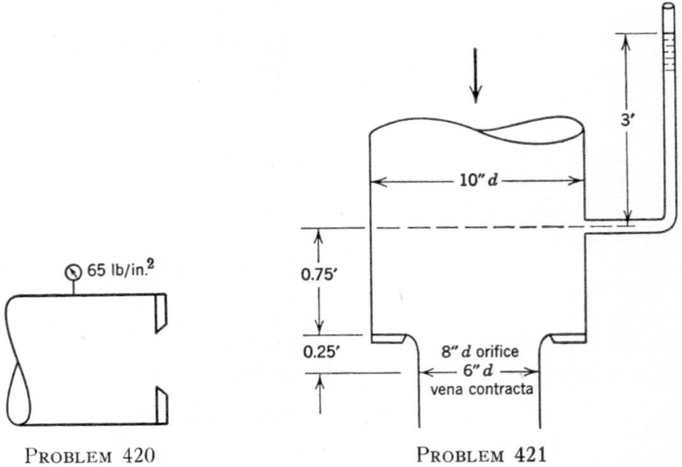

PROBLEM 420 PROBLEM 421

421. Calculate the force exerted by the water on this orifice plate. Assume that water in the jet between orifice plate and vena contracta weighs 4.0 lb.

422. The projectile partially fills the end of the 12-in. pipe. Calculate the force required to hold the projectile in position when the mean velocity in the pipe is 20 fps

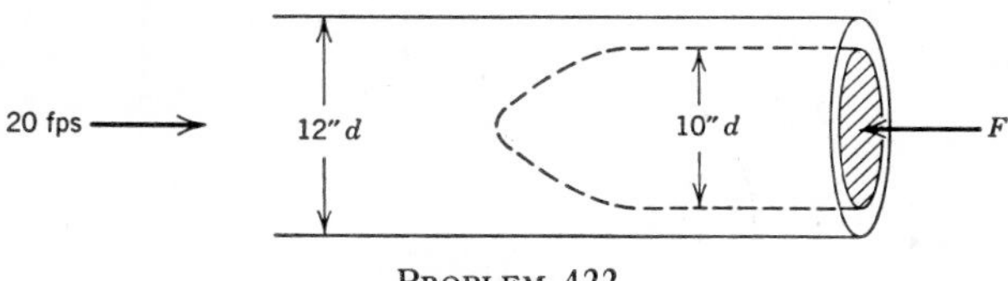

PROBLEM 422

423. Calculate the horizontal force of the water on the cone when the mean velocity in the 8-in. pipe is 10 fps.

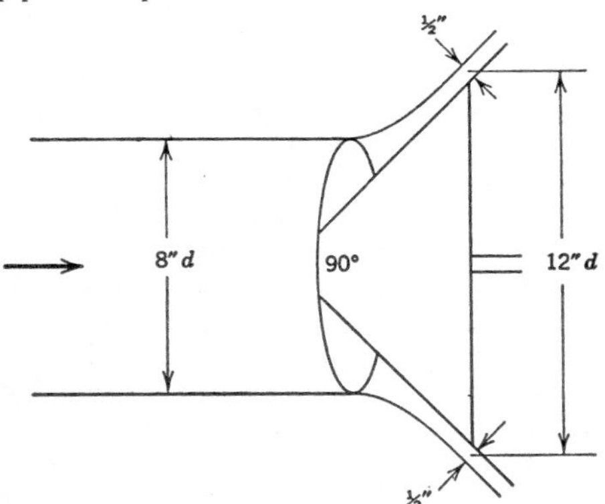

PROBLEM 423

424. This "needle nozzle" discharges a free jet of water at a velocity of 100 fps. The tension force in the stem is measured experimentally and found to be 1000 lb. Predict the horizontal force on the bolts.

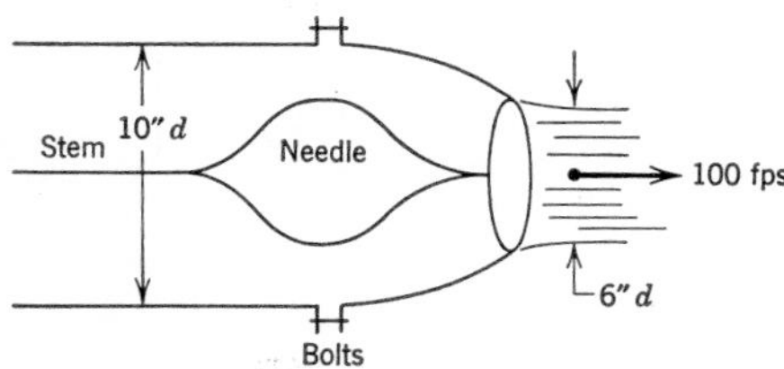

PROBLEM 424

425. A 90° bend occurs in a 12-in. horizontal pipe in which the pressure is 40 psi. Calculate the magnitude and direction of the horizontal force on the bend when 10 cfs of water flow therein.

426. A 6-in. horizontal pipe line bends through 90° and while bending changes its diameter to 3 in. The pressure in the 6-in. pipe is 30 psi. Calculate the magnitude and direction of the horizontal force on the bend when 2.0 cfs of water flow therein. Both pipes are in the same horizontal plane.

427. A 4-in. by 2-in. 180° pipe bend lies in a horizontal plane. Find the horizontal force of the water on the bend, when the pressures in the 4-in. and 2-in. pipes are 15.0 psi and 5.0 psi, respectively.

428. Calculate the force on the bolts. Water is flowing.

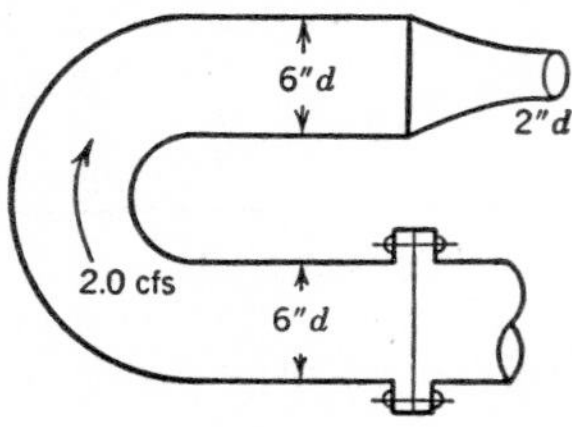

PROBLEM 428

429. Water flows through a tee in a horizontal pipe system. The velocity in the stem of the tee is 15 fps, and the diameter is 12 in. Each branch is of 6-in. diameter. If the pressure in the stem is 20 psi, calculate magnitude and direction of the force of the water on the tee if the flowrates in the branches are the same.

430. A double nozzle on the end of a 6-in. pipe line discharges into the atmosphere. One arm of the nozzle is of 2-in. diameter and discharges at an angle of 90° to the axis of the pipe. The other nozzle which is of 2-in. diameter discharges at 45° with the direction of flow in the pipe. The nozzles are on opposite sides of the pipe, and both nozzle jets have velocities of 50 fps. Calculate magnitude and direction of the horizontal force exerted by the water on the nozzle combination. Assume axes of pipe and nozzles to be in a horizontal plane.

431. A nozzle of 2-in. tip diameter discharges 0.66 cfs of water vertically upward. Calculate the volume of water in the jet between the nozzle tip and a section 12 ft above this point.

432. The block weighs 1 lb and is held up by the water jet issuing from the nozzle. Calculate H for a flowrate of 0.05454 cfs. Ignore the small quantity of water above plane AA.

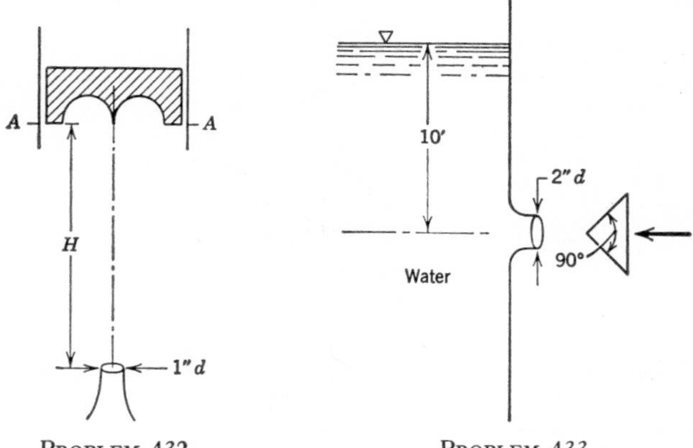

PROBLEM 432 PROBLEM 433

433. Calculate the force required to hold this weightless cone in the jet.

434. When round jets of the same velocity meet head on this flow picture results. Derive (ignoring gravity effects) an expression for θ in terms of d_1 and d_2.

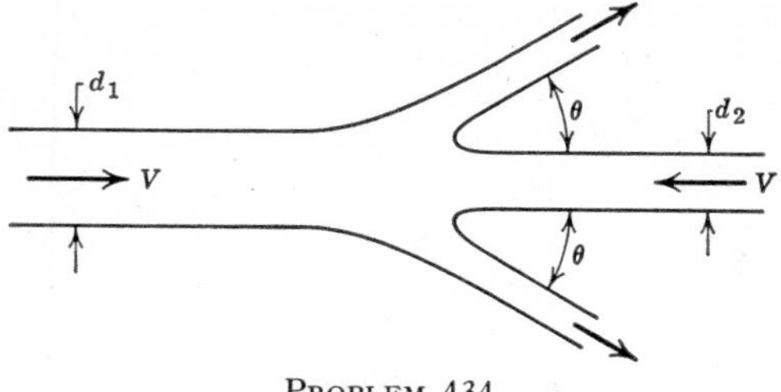

PROBLEM 434

435. The lower tank weighs 50.0 lb, and the water in it weighs 200.0 lb. If this tank is on a platform scales, what weight will register on the scale beam?

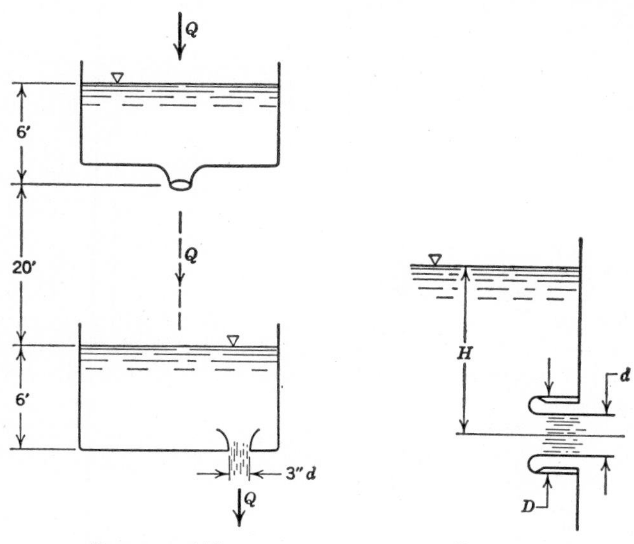

PROBLEM 435 PROBLEM 436

436. A free jet issues from this "Borda orifice" into the atmosphere. Calculate d/D.

437. The pressure difference results from head loss caused by eddies downstream from the orifice plate. Wall friction is negligible. Calculate the force exerted by the water on the orifice plate. The flowrate is 7.86 cfs.

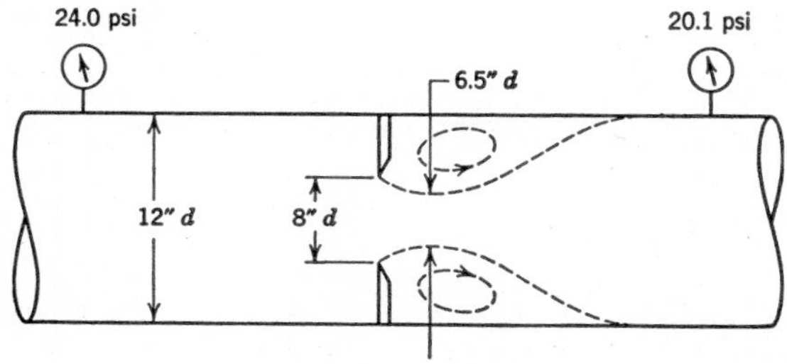

PROBLEM 437

438. The pump, suction pipe, discharge pipe, and nozzle are all welded together as a single unit. Calculate the horizontal component of force (magnitude and direction) exerted by the water on the unit when the pump is developing a head of 74 ft.

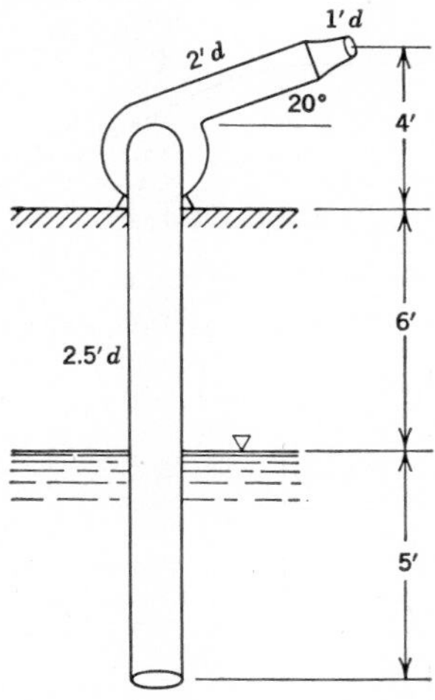

PROBLEM 438

439. When the pump is stopped (and offering no resistance to the flow) the force in the spring is 150 lb. With the pump running the force increases to 500 lb. How many horsepower is the pump delivering to the water? The water surface in the tank does not change.

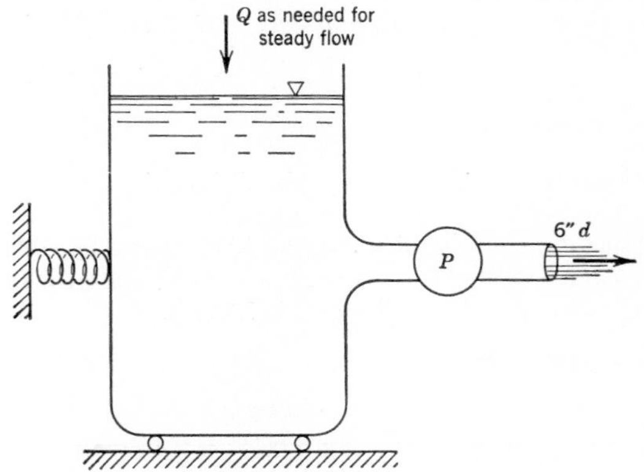

PROBLEM 439

440. When the pump is started strain gages at A and B indicate longitudinal tension forces in the pipes of 23 and 100 lb, respectively. Assuming a frictionless system, calculate flowrate and pump horsepower.

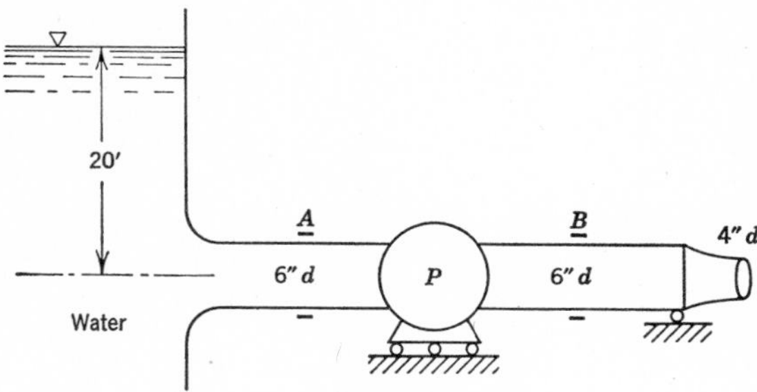

PROBLEM 440

441. The flowrate passing over this sharp-crested weir in a channel of 1 ft width is 3.5 cfs. Calculate the magnitude and direction of the force exerted by the water on the weir plate.

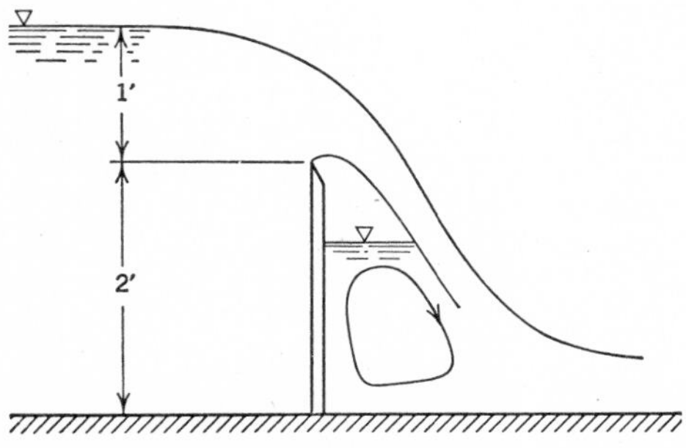

PROBLEM 441

442. The passage is 4 ft wide normal to the paper. What will be the horizontal component of force exerted by the water on the structure?

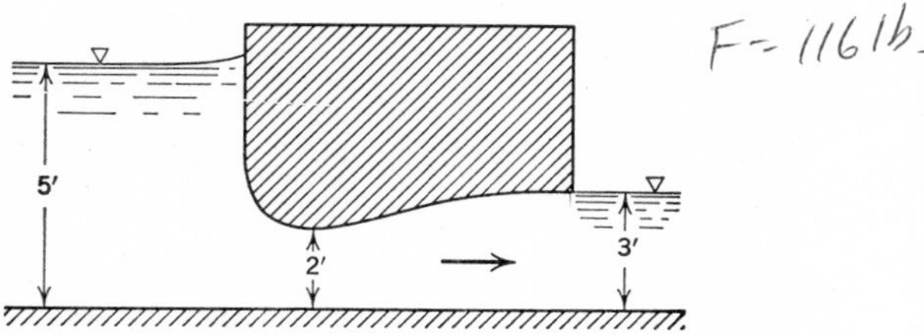

F = 116 lb.

PROBLEM 442

443. If the two-dimensional flowrate through this sluice gate is 50 cfs/ft, calculate the horizontal and vertical components of force on the gate, neglecting wall friction.

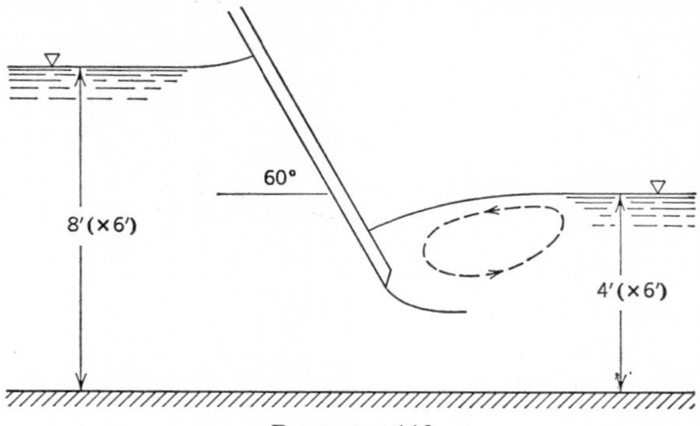

Problem 443

444. Calculate the magnitude and direction of the horizontal component of force exerted by the flowing water on this (hatched) outflow structure. Assume velocity distribution uniform where streamlines are straight and parallel.

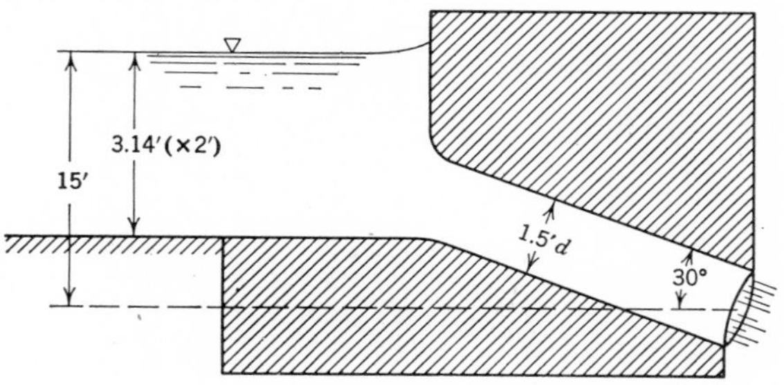

Problem 444

445. Calculate the magnitude and direction of the horizontal force exerted by the water on the (hatched) outflow structure. The flow is two-dimensional and the width of the flow picture (normal to the paper) is 6 ft.

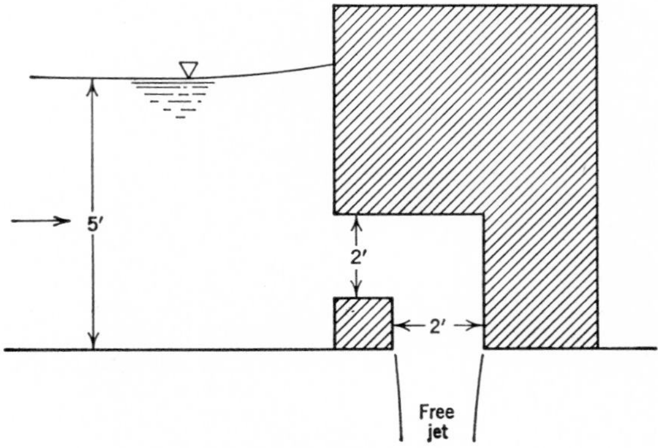

Problem 445

446. This two-dimensional structure (hatched) is used to divert part of the flow from the open channel. Calculate the magnitude and direction of the horizontal component of force exerted by the water on the structure, which is 5 ft wide normal to the paper.

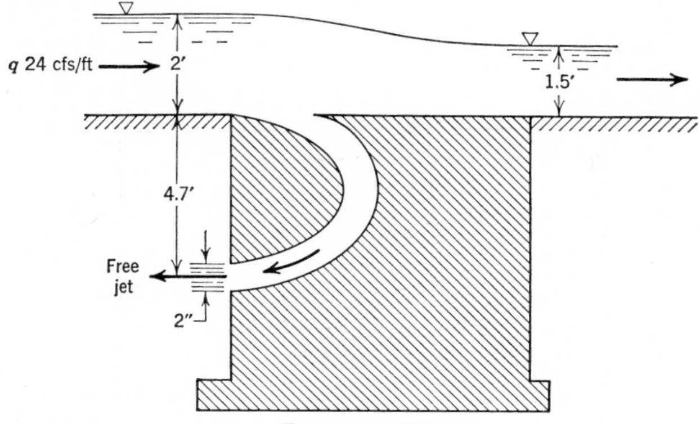

PROBLEM 446

447. Calculate the magnitude and direction of the horizontal force exerted by the water on the frictionless "drop structure" AB. Assume the structure to be 1 ft wide normal to the paper.

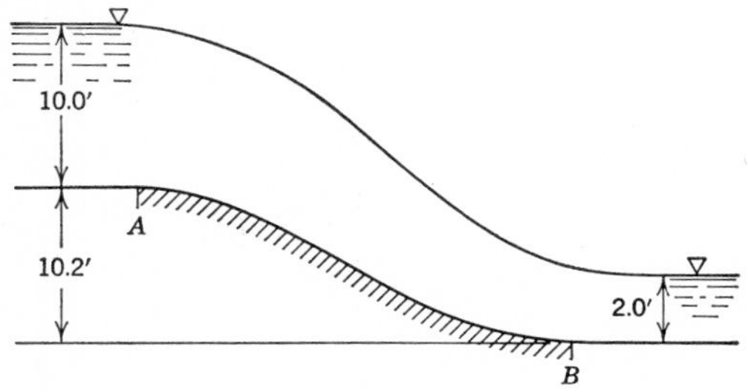

PROBLEM 447

448. Calculate the magnitude and direction of the horizontal and vertical components of resultant force exerted by the flowing water on the "flip bucket" AB. Assume that the water between sections A and B weighs 600 lb and that downstream from B the moving fluid may be considered to be a free jet.

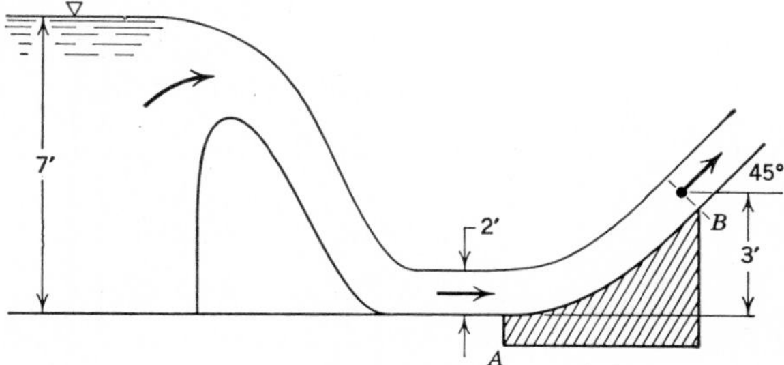

PROBLEM 448

449. Calculate F_x exerted by the water on the block which has been placed at the end of this horizontal open channel. The channel and block are 4 ft wide normal to the paper.

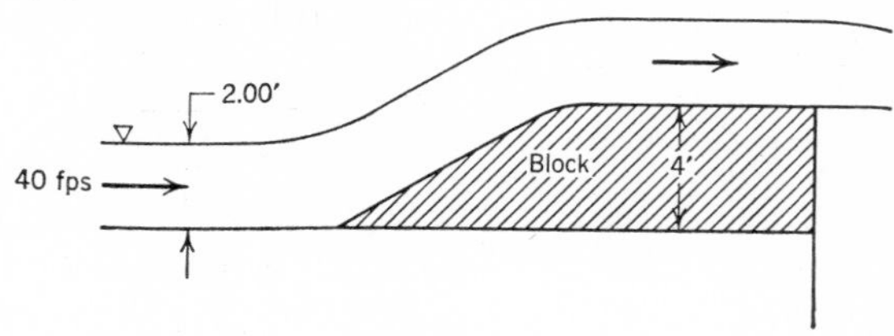

PROBLEM 449

450. A water scoop for a locomotive tender has dimensions as shown. Calculate the approximate horsepower expended in driving the scoop through the water at 30 mph. Assume the scoop and trough to be 1 ft wide.

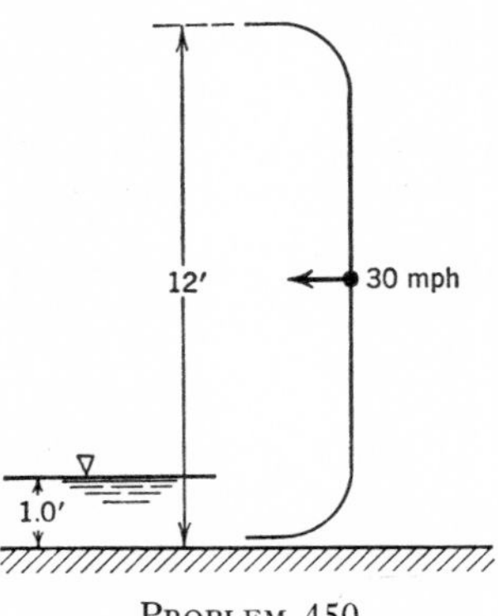

PROBLEM 450

451. Upstream from an axisymmetric abrupt enlargement in a horizontal passage the mean pressure is 20 psi and the diameter is 6 in. Downstream from the enlargement the mean pressure is 30 psi and the diameter is 12 in. Estimate the flowrate through the passage and the force on the enlargement.

452. At a point in a rectangular channel 20 ft wide just downstream from a hydraulic jump, the depth is observed to be 12 ft when the flowrate is 2000 cfs. What were the depth and velocity just upstream from the jump?

453. A hydraulic jump is observed in an open channel 10 ft wide. The approximate depths (such depths are very difficult to measure accurately) upstream and downstream from the jump are 2 ft and 5 ft, respectively. Estimate the flowrate in the channel.

454. In an open channel 5 ft wide the depth of water is 1 ft. Small standing waves (caused by imperfections in the sidewalls) are observed on the water surface at an angle of 30° with the walls. Estimate the flowrate in the channel.

455. Calculate the drop in the energy line (the head loss) produced by the oblique wave described in the illustrative problem of Art. 45.

456. In an open rectangular channel 10 ft wide the water depth is 1 ft. A standing wave 0.5 ft high is produced by narrowing the channel with a straight diagonal wall. The angle between the wave and the original channel wall is observed to be 60°. Estimate the flowrate in the channel. What must be the angle of the diagonal wall to produce such a wave?

457. A normal shock wave exists in an air flow. The pressure, velocity, and temperature just upstream from the wave are 30 psia, 2000 fps, and 0°F, respectively. Calculate the pressure, velocity, temperature, and sonic velocity just downstream from the shock wave.

458. If through a normal shock wave (in air) the pressure rises from 40 to 60 psia and the velocity diminishes from 1500 to 600 fps, what temperatures are to be expected upstream and downstream from the wave?

459. The stagnation temperature in an air flow is 300°F upstream and downstream from a normal shock wave. The stagnation pressure downstream from the shock wave is 23.0 psia. Through the wave the pressure rises from 15 to 20 psia. Determine the velocities upstream and downstream from the wave.

460. A wedge-shaped object in a supersonic wind tunnel is observed to produce an oblique shock wave of angle (β) 60°. The pressure through the wave rises from 30 to 40 psia and the temperature of the air upstream from the object is 100°F. Calculate the velocity of the air at this point.

461. The velocities and pressures upstream and downstream from an oblique shock wave in an air flow are found to be 2000 fps, 20 psia, and 1500 fps, 30 psia, respectively. Calculate the Mach numbers upstream and downstream from the wave if the wave angle (β) is 30°.

462. An airplane flies at twice the speed of sound in the ICAO Standard Atmosphere (Appendix IV) at an altitude of 30,000 ft. If the leading edge of the wing may be approximated by a wedge of angle (2θ) 20°, what pressure rise through the oblique shock wave is to be expected?

463. A smooth horizontal nozzle of end area A discharges from a tank under a head H. Calculate the net horizontal force exerted on the tank-and-nozzle combination in terms of the hydrostatic force exerted on the same area under the same head.

464. A horizontal convergent-divergent nozzle is bolted to a water tank from which it discharges under a 12-ft head. The throat and tip diameters of the nozzle are 5 in. and 6 in., respectively. Determine the net horizontal force exerted on the tank-and-nozzle combination by the flowing fluid.

465. Calculate the force exerted on the tank-and-nozzle by the air flow in the illustrative problem of Art. 35 if discharge is to the atmosphere.

466. Calculate the force exerted on the tank-and-nozzle in the illustrative problem of Art. 37 for flow without the shock wave.

467. The pump maintains a pressure of 10 psi at the gage. Calculate the tension force in the cable.

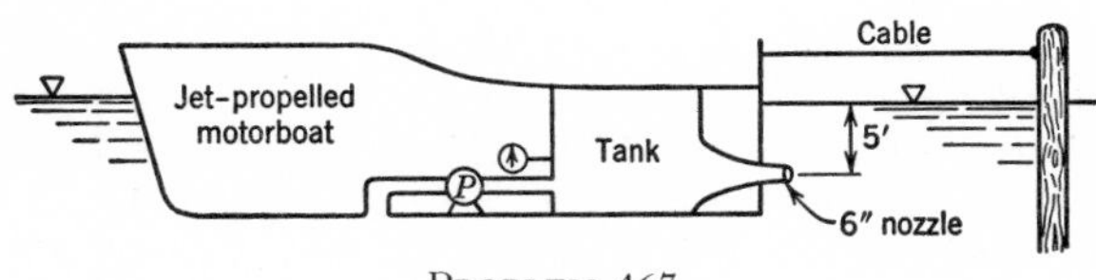

PROBLEM 467

468. A motorboat moves up a river at a speed of 30 fps (relative to the land). The river flows at a velocity of 5 fps. The boat is powered by a jet-propulsion unit which takes in water at the bow and discharges it (beneath the surface) at the stern. Measurements in the jet show its velocity (relative to the boat) to be 60 fps. For a flowrate through the unit of 5 cfs, calculate the propulsive force produced.

469. A head of 6 ft is maintained on the nozzle. What is the propulsive force: (*a*) on the car, (*b*) on the blade, (*c*) on the tank-and-nozzle.

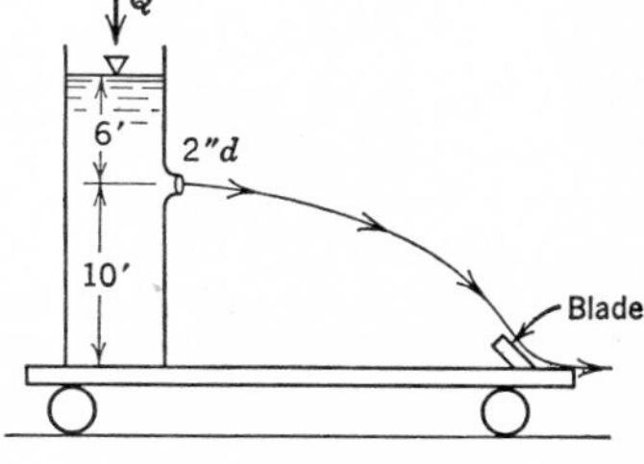

PROBLEM 469

470. A ship moves up a river at 20 mph relative to the shore. The river current has a velocity of 5 mph. The velocity of the water a short distance behind the propellers is 40 mph relative to the ship. If the velocity of 100 cfs of water is changed by the propeller, calculate its thrust.

471. An airplane flies at 120 mph through still air of specific weight 0.0765 lb/cuft. The propeller is 8 ft in diameter, and its slipstream has a velocity of 180 mph relative to the fuselage. Calculate: (*a*) the propeller efficiency, (*b*) the velocity through the plane of the propeller, (*c*) the horsepower input, (*d*) the horsepower output, (*e*) the thrust of the propeller, and (*f*) the pressure difference across the propeller disk.

472. A propeller must produce a thrust of 2000 lb to drive an airplane at 175 mph. An ideal propeller of what size must be provided if it is to operate at 90 percent efficiency? Assume air density 0.00238 slug/cuft.

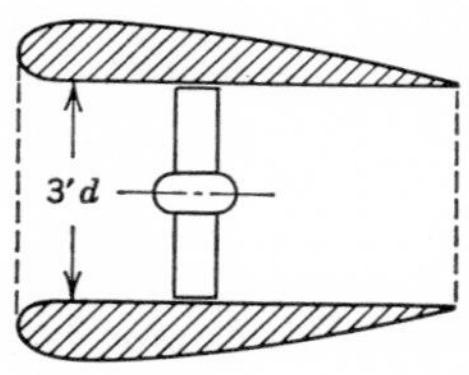

PROBLEM 473

473. This ducted propeller unit drives a ship through still water at a speed of 15 fps. Within the duct the mean velocity of the water (relative to the unit) is 50 fps. Calculate the propulsive force produced by the unit. Calculate the force exerted on the fluid by the propeller. Account for the difference between these forces.

474. A propeller type of turbine wheel is installed near the end of a horizontal pipe line discharging water to the atmosphere. Derive an expression for the horsepower to be expected from the turbine in terms of the gage pressure upstream from the turbine disk, and the flowrate.

475. What is the maximum power which can be expected from a windmill 100 ft in diameter in a wind of 30 mph? Assume air density 0.00238 slug/cuft.

476. If an ideal windmill is operating at best efficiency in a wind of 30 mph, what is the velocity through the disk and at some distance behind the windmill? What is the thrust on this windmill, assuming a diameter of 100 ft and an air density of 0.00238 slug/cuft? What are the mean pressures just ahead of and directly behind the windmill disk?

477. The flowrate through a 10-ft diameter windmill is 6000 cfs. The mean pressures just upstream and downstream from the windmill plane are 5.0 psf and −3.9 psf, respectively. Assuming the air density 0.0025 slug/cuft, calculate the wind velocity, the axial force on the windmill, and the horsepower of the machine.

478. A 250-gpm horizontal jet of water of 1-in. diameter strikes a stationary blade which deflects it 60° from its original direction. Calculate the vertical and horizontal components of force exerted by the liquid on the blade.

479. A 2-in. jet of water moving at 120 fps has its direction reversed by a smooth stationary deflector. Calculate the magnitude and direction of the force on the deflector.

480. The jet of the preceding problem strikes a stationary flat plate whose surface is normal to the jet. Calculate the magnitude and direction of the force on the plate.

481. Calculate the resultant force of the water on this plate. The flowrate is 1.50 cfs.

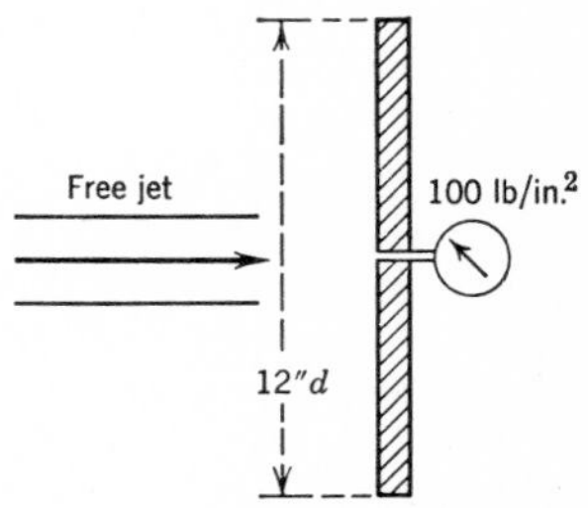

PROBLEM 481

482. The plate covers the 5-in. diameter hole. What is the maximum H which can be maintained without leaking?

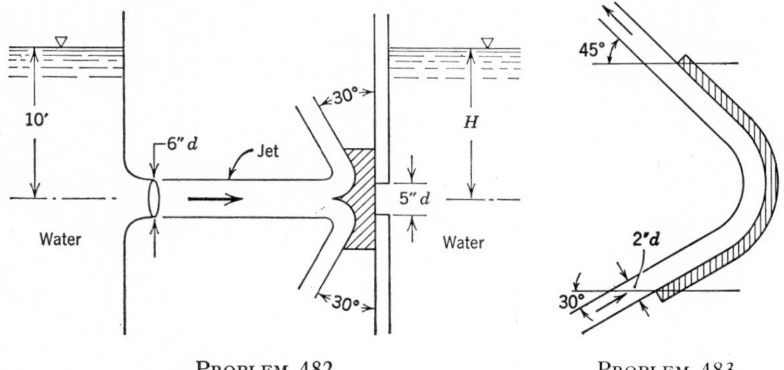

PROBLEM 482 PROBLEM 483

483. Calculate the magnitude and direction of the vertical and horizontal components and the total force exerted on this stationary blade by a 2-in. jet of water moving at 50 fps.

484. A smooth nozzle of 2-in. tip diameter is connected to the bottom of a large water tank and discharges vertically downward. The water surface in the tank is 5 ft above the tip of the nozzle. Ten feet below the nozzle tip there is a horizontal disk of 4-in. diameter surface. Calculate the force exerted by the water jet on the disk.

485. Water is flowing in a horizontal plane. Calculate F_x and F_y exerted on this stationary deflector which divides the flow equally.

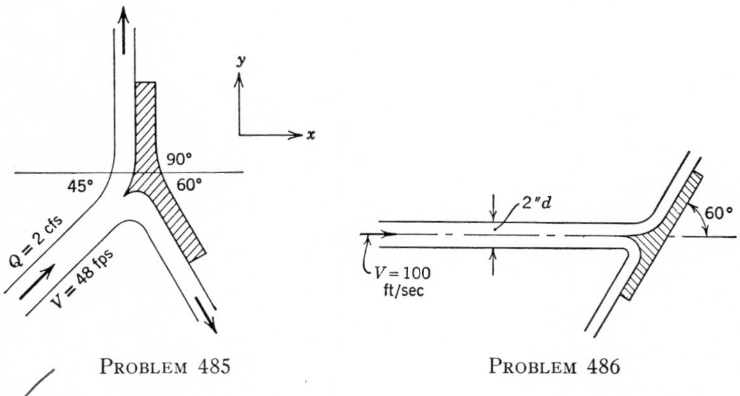

PROBLEM 485 PROBLEM 486

486. This water jet of 2-in. diameter moving at 100 fps is divided in half by a "splitter" on the stationary flat plate. Calculate the magnitude and direction of the force on the plate. Assume that flow is in a horizontal plane.

487. If the splitter is removed from the plate of the preceding problem and sidewalls are provided on the plate to keep the flow two-dimensional, how will the jet divide after striking the plate?

488. Calculate the gage reading when the plate is pushed horizontally to the left at a constant speed of 25 fps. Also calculate the force and power required to push the plate at this speed.

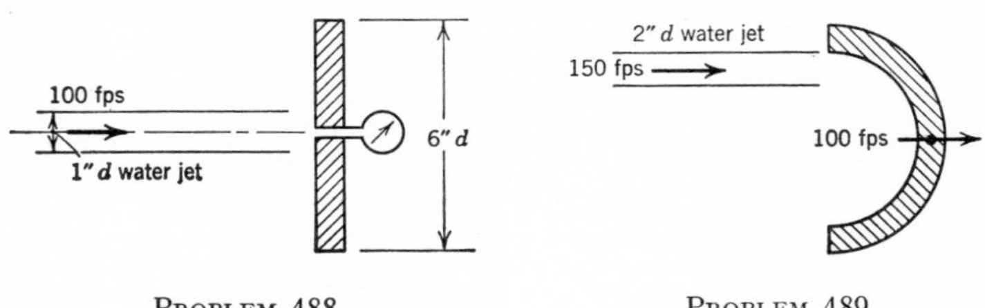

PROBLEM 488 PROBLEM 489

489. The blade is one of a series. Calculate the force exerted by the jet on the blade system.

490. This blade is one of a series. What force is required to move the series horizontally against the direction of the jet at a velocity of 50 fps? What horsepower is required to accomplish this motion?

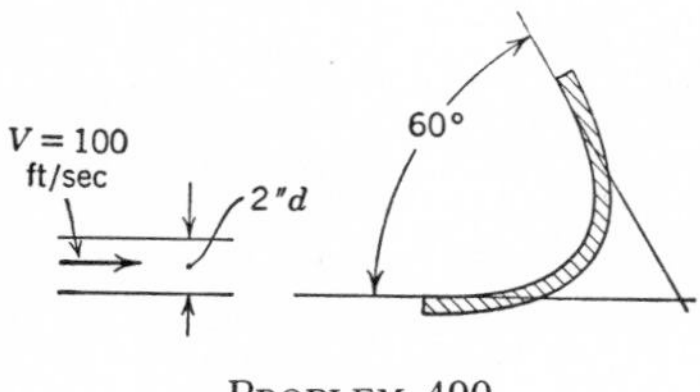

PROBLEM 490

491. From a water jet of 2-in. diameter, moving at 200 fps, 180 hp are to be transferred to a blade system (Fig. 81) which is moving in the direction of the jet at 50 fps. Calculate the required blade angle.

492. A series of blades (Fig. 81), moving in the same direction as a water jet of 1-in. diameter and of velocity 150 fps, deflects the jet 75° from its original direction. What relation between blade velocity and blade angle must exist to satisfy this condition? What is the force on the blade system?

493. If a system of blades (Fig. 81) is free to move in a direction parallel to that of a jet, prove that the direction of the force on the blade system is the same for all blade velocities.

494. A crude impulse turbine has flat radial blades and is in effect a "paddle wheel." If the 1-in. diameter water jet has a velocity of 100 fps and is tangent to the rim of the wheel, calculate the approximate horsepower of the machine when the blade velocity is 40 fps.

495. A 6-in. pipe line equipped with a 2-in. nozzle supplies water to an impulse turbine 6 ft in diameter having blade angles of 165°. Plot a curve of theoretical horsepower vs. rpm when the pressure behind the nozzle is 100 psi. What is the force on the blades when the maximum horsepower is being developed? Plot a curve of force on the blades vs. rpm.

496. When an air jet of 1-in. diameter strikes a series of blades on a turbine rotor the (absolute) velocities are as shown. If the air is assumed to have a constant specific weight of 0.08 lb/cuft, what is the force on the turbine rotor? How many horsepower are transferred to the rotor? What must be the velocity of the blade system?

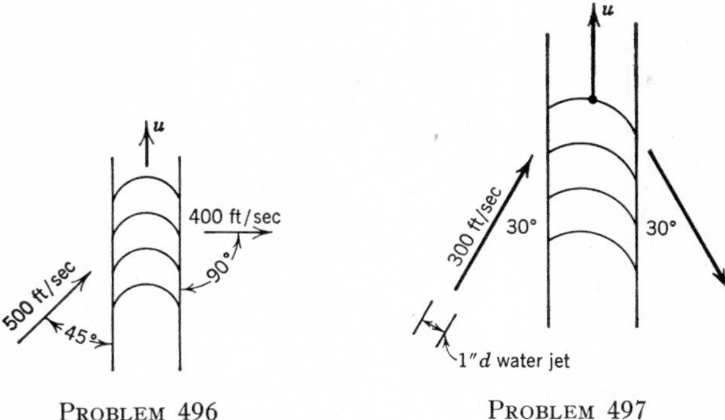

PROBLEM 496 PROBLEM 497

497. This system of blades develops 200 hp under the influence of the jet shown. Calculate the blade velocity.

498. If $\alpha_1 = \alpha_2$, calculate: α, F_x on the blade system, and the horsepower developed.

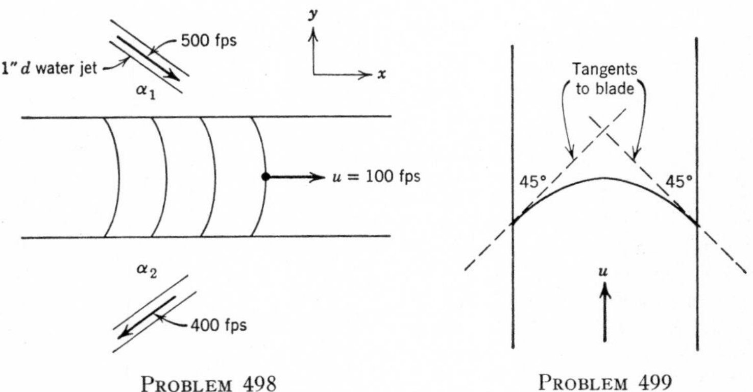

PROBLEM 498 PROBLEM 499

499. In passing through this blade system, the absolute jet velocity decreases from 136 to 73.8 fps. If the flowrate is 2.00 cfs of water, calculate the power transferred to the blade system and the vertical force component exerted on the blade system.

500. If $u = v_1 = v_2 = 100$ fps for the blade system of the preceding problem, calculate the absolute velocities of the jet entering and leaving the blade system. If the flowrate is 5 cfs, how many horsepower may be expected from the machine?

501. The flowrate is 2.0 cfs of water, $u = 30$ fps, and $v_1 = v_2 = 40$ fps. Calculate: the absolute velocity of the water entering and leaving the system, F_x and F_y (magnitude and direction) on the blade system, and the horsepower transferred from jet to blade system.

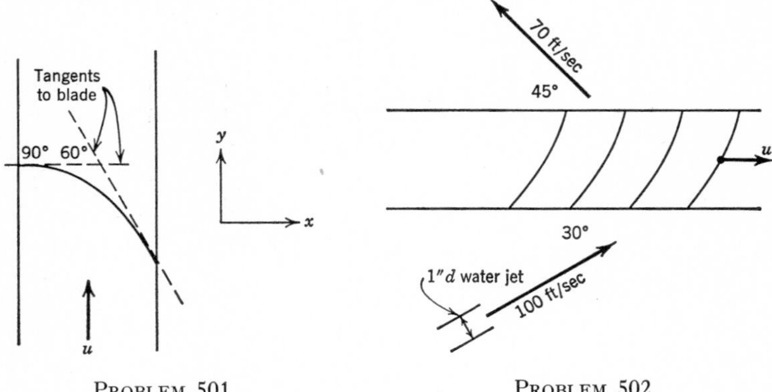

PROBLEM 501

PROBLEM 502

502. The (absolute) velocities and directions of the jets entering and leaving the blade system are as shown. Calculate the power transferred from jet to blade system and the blade angles required.

503. This stationary blade is pivoted at point O. Calculate the torque (about O) exerted thereon when a 2-in. water jet moving at 100 fps passes over it as shown.

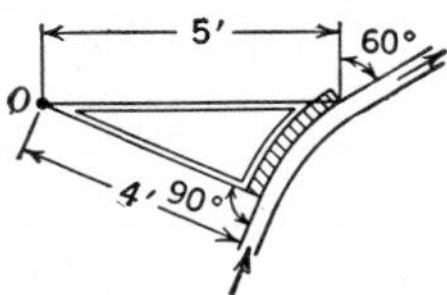

PROBLEM 503

504. Eight thousand cubic feet per second of water flow into a hydraulic turbine whose guide vanes, set at an angle of 15°, have an exit circle of 8-ft radius and are 6 ft high. In the draft tube the flow is observed to have no tangential component. What is the torque on the runner? If the runner is rotating at 150 rpm, how many horsepower are being delivered to the turbine and how much energy is being extracted from each pound of water?

505. A simple reaction turbine has $r_1 = 3$ ft, $r_2 = 2$ ft, and its flow cross section is 1 ft high. The guide vanes are set so that $\alpha_1 = 30°$. When 100 cfs of water flow through this turbine the angle α_2 is found to be 60°. Calculate the torque exerted on the turbine runner. If the angle β_2 is 150°, calculate the speed of rotation of the turbine runner and the angle β_1 necessary for smooth flow into the runner. Calculate the horsepower developed by the turbine at the above speed.

506. The nozzles are all of 1-in. diameter and each nozzle discharges 0.25 cfs of water. If the turbine rotates at 100 rpm, calculate the horsepower developed.

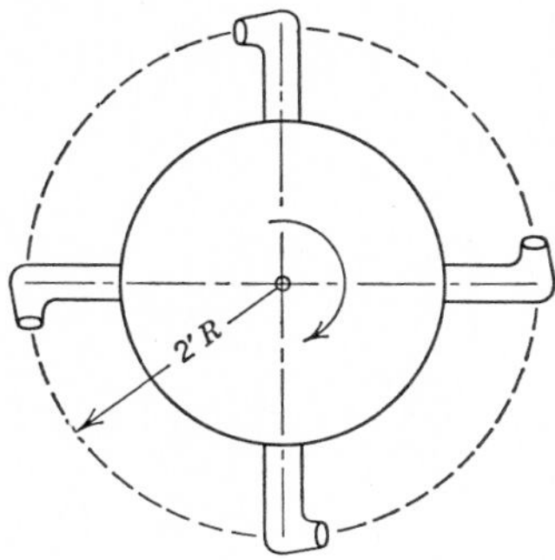

PROBLEM 506

507. A centrifugal pump impeller having $r_1 = 2$ in., $r_2 = 6$ in., and width $b =$ 1.5 in. is to pump 8 cfs of water and supply 40 ft-lb of energy to each pound of fluid. The impeller rotates at 1000 rpm. What blade angles are required? What horsepower is required to drive this pump? Assume smooth flow at the inlet of the impeller.

508. A centrifugal pump impeller having dimensions and angles as shown rotates at 500 rpm. Assuming a radial direction of velocity at the blade entrance, calculate the flowrate, the pressure difference between inlet and outlet of blades, and the torque and horsepower required to meet these conditions.

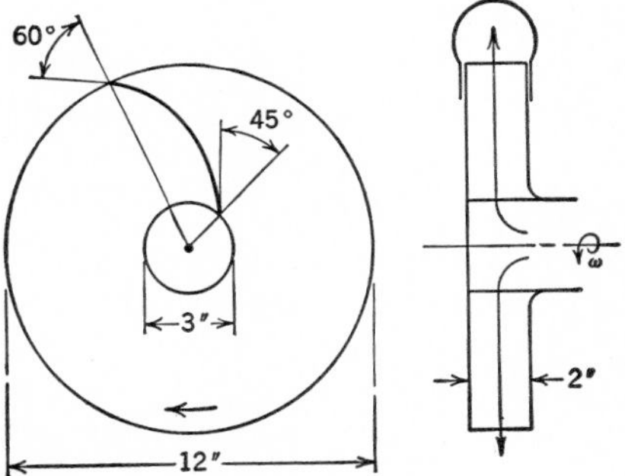

PROBLEM 508

509. If the impeller of the preceding problem rotates between horizontal planes of infinite extent and the flowrate is 0.86 cfs, what rise of pressure may be expected between one point having $r = 6$ in. and another having $r = 9$ in.?

510. At the outlet of a pump impeller of diameter 2 ft and width 6 in., the (absolute) velocity is observed to be 100 fps at an angle of 60° with a radial line. Calculate the torque exerted on the impeller.

chapter 7

Flow of a Real Fluid

The flow of a real fluid is vastly more complex than that of an ideal fluid, owing to phenomena caused by the existence of viscosity. Viscosity introduces resistance to motion by causing shear or friction forces between fluid particles and between these and boundary walls. For flow to take place, work must be done against these resistance forces, and in the process energy is converted into heat. The inclusion of viscosity also allows the possibility of two physically different flow regimes, and in addition (by causing separation and secondary flows) it frequently produces flow situations entirely different from those of the ideal fluid; its effects on the velocity profile also render invalid the assumption of uniform velocity distribution. Although the Euler equations may be altered to include the shear stresses of a real fluid, the result is a set of partial differential equations to which no general solution is known. In view of these complexities the engineer must resort to experimental results and semi-empirical methods to solve problems which (at present) offer no hope of analytical solution. This requires a good basic understanding of a variety of physical phenomena, which will be described in this chapter and which are the basic features of fluid friction. Numerous applications, quantitative and qualitative, will be found for these features.

52. Reynolds' Experiments and Their Significance. The effects of viscosity cause the flow of a real fluid to occur under two very different conditions, or regimes: that of *laminar* flow and that of *turbulent* flow. The characteristics of these regimes were first demonstrated by Reynolds,[1] with an apparatus similar to that of Fig 87. Water flows from a tank through a bell-mouthed glass pipe, the flow being controlled by the valve A. A thin tube, B, leading from a reservoir of dye, C, has its opening within the entrance of the glass pipe. Reynolds discovered that, for low velocities of flow in the glass pipe, a thin filament of dye issuing from the tube did not diffuse but was maintained intact throughout the pipe, forming a thin straight line parallel to the axis of the pipe. As the valve was opened, however, and greater velocities were attained, the dye filament wavered and broke, diffusing through the flowing water in the pipe. Reynolds

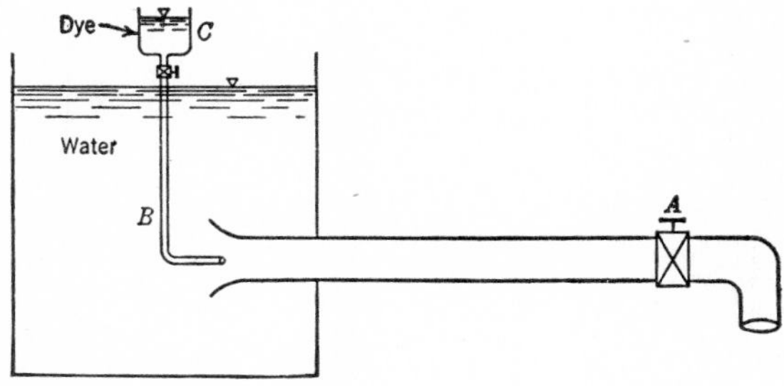

FIG. 87

found that the mean velocity at which the filament of dye began to break up (termed *critical velocity*) was dependent upon the degree of quiescence of the water in the tank, higher critical velocities being obtainable with increased quiescence. He also discovered that if the dye filament had once diffused it became necessary to decrease the velocity in order to restore it, and that the restoration always occurred at approximately the same mean velocity in the pipe.

Since intermingling of fluid particles during flow would cause diffusion of the dye filament, Reynolds deduced from his experiments that at low velocities this intermingling was absent and that the fluid particles moved in parallel layers, or laminae, sliding past adjacent laminae but not mixing with them;[2] this is the regime of *laminar flow*. Since at higher velocities the dye filament diffused through the pipe,

[1] O. Reynolds, "An Experimental Investigation of the Circumstances Which Determine Whether the Motion of Water Shall Be Direct or Sinuous and of the Law of Resistance in Parallel Channels," *Phil. Trans. Roy. Soc.*, Vol. 174, Part III, p. 935, 1883.

[2] This is the premise upon which viscosity is defined; see Fig. 4 of Art. 5. It is to be noted, however, that intermingling of fluid particles does occur in laminar flow—but only on a molecular scale; indeed, it is the effect of such molecular activity which contributes materially to viscous action. See Art. 5.

it was apparent that intermingling of fluid particles was occurring, or, in other words, the flow was *turbulent.* Laminar flow broke down into turbulent flow at some critical velocity above that at which turbulent flow was restored to the laminar condition; the former velocity is called the *upper critical velocity*, and the latter, the *lower critical velocity.*

Other evidence of the existence of two flow regimes may be deduced from the following simple experiment of Fig. 88. Here the fall of pressure between two points in a section of a long straight pipe is measured by the manometer reading h and correlated with the mean

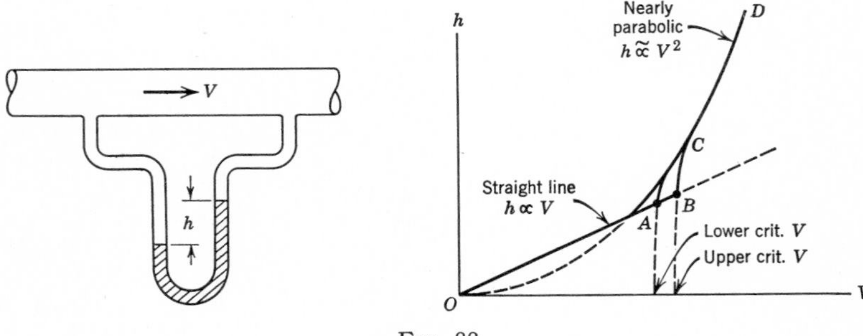

Fig. 88

velocity V. For the small values of V a plot of h against V will be found to yield a straight line ($h \propto V$), but at higher values of V a nearly parabolic curve ($h \mathrel{\tilde{\propto}} V^2$) will result; evidently the flow is laminar in the first case, turbulent in the second. Between the two flow regimes lies an interesting *transition region*; as V is increased, data will follow the line $OABCD$ but with diminishing V will follow $DCAO$. From these results and Reynolds' observations it may be deduced that points A and B define the lower and upper critical velocities, respectively.

Reynolds was able to generalize his conclusions from his dye stream experiments by the introduction of a dimensionless term $\mathbf{R}$, later called the Reynolds number, which was defined by

$$\mathbf{R} = Vd\rho/\mu \quad \text{or} \quad Vd/\nu \tag{145}$$

in which V is the mean velocity of the fluid in the pipe, d the diameter of the pipe, and ρ and μ the density and viscosity of the fluid flowing therein. Reynolds found that certain critical values of the Reynolds number, $\mathbf{R}_c$, defined the upper and lower critical velocities for all fluids flowing in all sizes of pipes, and thus he deduced the fact that single numbers define the limits of laminar and turbulent pipe flow *for all fluids.*

The upper limit of laminar flow was found by Reynolds to correspond to $12{,}000 < \mathbf{R}_c < 14{,}000$, but unfortunately this upper critical Reynolds number is indefinite, being dependent upon several incidental conditions such as: (1) initial quiescence of the fluid,[3] (2) shape of pipe entrance, and (3) roughness of pipe. However, these high values are of little practical interest, and the engineer may take the upper limit of laminar flow to be defined by $2700 < \mathbf{R}_c < 4000$.

The lower limit of turbulent flow, defined by the lower critical Reynolds number, is of greater engineering importance; it defines a condition below which all turbulence entering the flow from any source will eventually be damped out by viscosity. This lower critical Reynolds number thus sets a limit below which laminar flow will always occur; many experiments have indicated the lower critical Reynolds number to have a value of approximately 2100. Between Reynolds numbers 2100 and 4000 a region of uncertainty exists, and the engineer must make conservative selection in this region of the variables which depend upon the Reynolds number.

The concept of a critical Reynolds number delineating the regimes of laminar and turbulent flow is indeed a useful one in promoting concise generalization of certain flow phenomena. Applying this concept to the flow of *any fluid in cylindrical pipes*, the engineer may predict that the flow will be laminar if $\mathbf{R} < 2100$ and turbulent if $\mathbf{R} > 4000$. However, it is to be emphasized that critical Reynolds number is very much a function of boundary geometry. For flow between parallel walls (using mean velocity V, and spacing d) $\mathbf{R}_c \cong 1000$; for flow in a wide open channel (using mean velocity V and depth d) $\mathbf{R}_c \cong 500$; for flow about a sphere (using approach velocity V and diameter d) $\mathbf{R}_c \cong 1$. Also noteworthy is the fact that such critical Reynolds numbers must be determined experimentally; because of the obscure origins of turbulence, analytical methods for predicting critical Reynolds numbers have yet to be developed.

Illustrative Problem

Water of kinematic viscosity 0.00001 sqft/sec flows in a cylindrical pipe of 1-in. diameter. Calculate the largest flowrate for which laminar flow can be expected.

Solution. Taking $\mathbf{R}_c = 2100$ as the conservative upper limit of laminar flow,

$$2100 = V(1/12)/0.00001; \quad V = 0.252 \text{ fps} \tag{145}$$

$$Q = 0.252 \times (\pi/4)(1/12)^2 = 0.001375 \text{ cfs} \tag{53}$$

[3] Ekman, working in 1910 with Reynolds' original apparatus, was able to obtain laminar flow up to a Reynolds number of 50,000 by quieting the water for many days before running his tests.

53. Laminar and Turbulent Flow. In laminar flow, agitation of fluid particles is of molecular nature only, and these particles are constrained to motion in essentially parallel paths by the action of viscosity. The shearing stress between adjacent moving layers is determined in laminar flow by the viscosity and is completely defined by the differential equation (Art. 5),

$$\tau = \mu \frac{dv}{dy} \tag{12}$$

the stress being the product of viscosity and velocity gradient (Fig. 89).

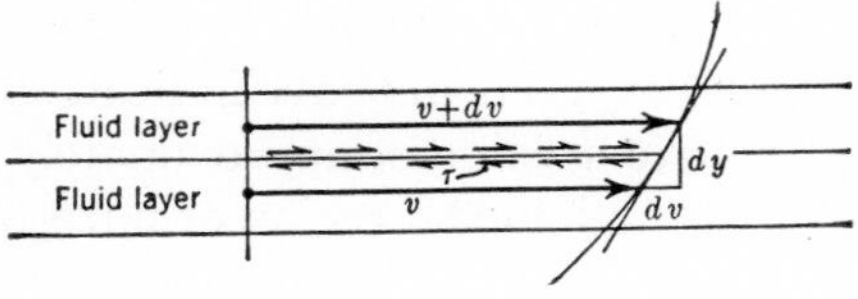

FIG. 89

In turbulent flow, fluid masses are not retained in layers but move in heterogeneous fashion through the flow, sliding past other masses and colliding with others in an entirely haphazard manner, causing mixing of the fluid as flow occurs. These small, randomly moving fluid masses cause at any point in the flow a rapid and irregular pulsation of velocity. This may be visualized as in Fig. 90, where v is the temporal mean velocity [4] and v' is the instantaneous velocity, necessarily a function of time. The instantaneous velocity v' may be considered to be composed of the vector sum of the temporal mean velocity and the components of the pulsations v_x and v_y, both functions of time; by defining and isolating v_x and v_y in this manner, certain essentials of turbulence can be fruitfully studied. Measurements of v_x and v_y by hot-wire anemometer (Art. 95) yield a record similar to that of Fig. 90, which because of the random nature of turbulence discloses no regular period or amplitude; nevertheless such records allow the definition of certain turbulence characteristics. The root-mean-square (rms) value of v_x or v_y is a measure of the violence of the turbulent fluctuations, i.e., the magnitude of the departure of v' from v; this value is widely known as the *intensity* of turbulence. The mean time interval [5] between reversals in the sign of v_x (or v_y) is a measure of the *scale* of the turbulence since it is a measure of the size of the turbulent eddies passing the point. In general, the

[4] Measurable with a small pitot tube.

[5] This would correspond to the half-period of a simple harmonic vibration.

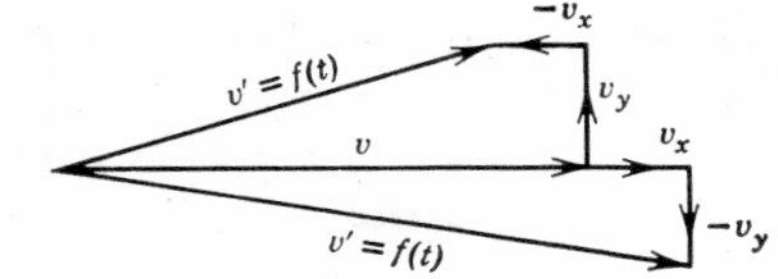

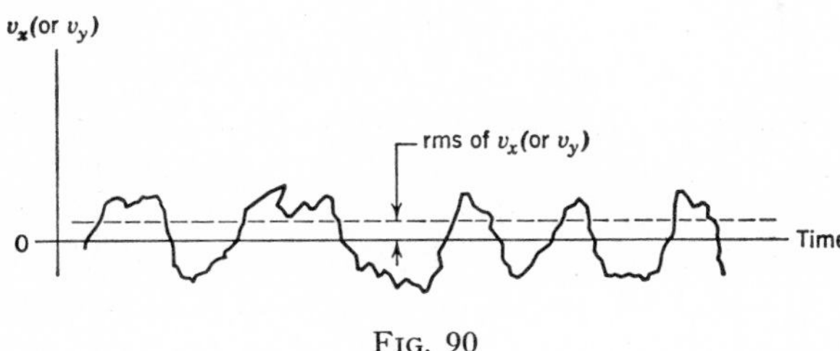

FIG. 90

intensity of turbulence increases with velocity, and scale increases with boundary dimensions. The former is easily imagined from the more rapid diffusion of the dye filament in a Reynolds apparatus with increased velocity; the latter can be visualized from the expectation that turbulent eddies will be larger in a large canal than in a small pipe for the same mean velocity.

Since turbulence is an entirely chaotic motion of small fluid masses through short distances in every direction as flow takes place, the motion of individual fluid particles is impossible to trace and characterize mathematically, but mathematical relationships may be obtained by considering the average motion of aggregations of fluid particles or by statistical methods.

Shearing stress in turbulent flow may be visualized by considering two adjacent points in a flow cross section (Fig. 91); at one of these points the temporal mean velocity is v, at the other, $v + \Delta v$. If the small transverse distance between the points is l, a velocity gradient dv/dy (of temporal mean velocity) is implied. If v and $v + \Delta v$ are now taken to be the mean velocities of (fictitious) fluid layers, the

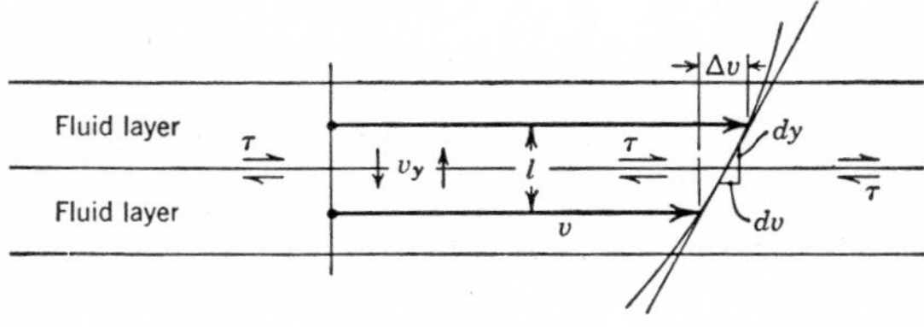

FIG. 91

turbulence velocity v_y implies a transverse motion of small fluid masses between layers, such mass being transferred in one direction or in the other. However, before transfer these fluid masses have velocities (v and $v + \Delta v$) which after transfer become $v + \Delta v$ and v, respectively; this means that their *momentum is changed* during the transfer process—tending to speed up the slower layer and slow down the faster one—just as if there were a shearing stress between them. Thus the existence of shearing stress in turbulent flow is seen to be deducible from momentum considerations; however, in the sense that it is deducible but not measurable, it is to this extent fictitious, but it is nevertheless an extremely useful concept for engineering treatment of turbulent flow.

For nearly a century scientists have grappled with the problem of developing a useful and accurate expression (in terms of velocity gradient) for turbulent shear stress; considerable progress has been made, but perfection has not yet been attained. Following is a brief summary [6] of progress to the present time.

The first attempt to express turbulent shear stress in mathematical form was made by Boussinesq,[7] who followed the pattern of the laminar flow equation and wrote

$$\tau = \epsilon \frac{dv}{dy} \tag{146}$$

in which ϵ, called the *eddy viscosity*, was a property of the flow (not of the fluid alone) which depended primarily upon the structure of the turbulence. From its definition the eddy viscosity can be seen to have the disadvantageous feature of varying from point to point throughout the flow. Nevertheless this first expression for turbulent shear is used frequently today because of the comparison between μ and ϵ. Now the equation is usually written

$$\tau = (\mu + \epsilon)\frac{dv}{dy}$$

to cover the combined situation where both viscous action and turbulent action are present in a flow. For the limiting conditions where the flow is entirely laminar or entirely turbulent, ϵ or μ is taken to be zero, respectively, and the foregoing equation reverts to equation 12 or 146.

[6] Although considerable progress is now being made through the use of statistical methods, this cannot be included in an elementary treatment of the subject.

[7] J. Boussinesq, "Essay on the Theory of Flowing Water," French Academy of Sciences, 1877.

Taking the velocities of fluid particles due to turbulence as v_y and v_x, respectively, normal to and along the direction of general motion, it is evident that, if homogeneous turbulence is assumed,

$$v_x = v_y$$

Using this assumption, Reynolds[8] showed that the shearing stress between moving fluid layers in turbulent flow is given by

$$\tau = \rho\overline{v_x v_y}$$

in which $\overline{v_x v_y}$ is the mean value of the product of v_x and v_y. Prandtl[9] succeeded in relating the above velocities of turbulence to the general flow characteristics by proposing that small aggregations of fluid particles are transported by turbulence a certain mean distance,[10] l, from regions of one velocity to regions of another and in so doing suffer changes in their general velocities of motion. Prandtl termed the distance l the *mixing length* and suggested that the change in velocity, Δv, incurred by a fluid particle moving through the distance l was proportional to v_x and to v_y.

From this he suggested

$$\tau = \rho l^2 \left(\frac{dv}{dy}\right)^2 \tag{147}$$

as a valid equation for shearing stress in turbulent flow. This expression, although satisfactory in many respects, had the disadvantage that l was a function of y, the mixing length becoming larger at greater distance from the boundary wall. This disadvantage was partially overcome by von Kármán,[11] who proposed, after comparing the properties of turbulent velocity profiles and the accompanying variation of l with y, that

$$\tau = \rho\kappa^2 \frac{(dv/dy)^4}{(d^2v/dy^2)^2} \tag{148}$$

in which κ is a (dimensionless) constant of the turbulence. However, experimental measurements have demonstrated that κ is not precisely constant, implying some imperfections in the theory and leading to the expectation of some inaccuracies in results obtained from

[8] O. Reynolds, "On the Dynamical Theory of Incompressible Viscous Fluids and the Determination of the Criterion," *Phil. Trans. Roy. Soc.*, A1, Vol. 186, p. 123, 1895.

[9] L. Prandtl, "Ueber die ausgebildete Turbulenz," *Proc. 2nd Intern. Congr., Appl. Mech.*, Zurich, p. 62, 1926.

[10] This concept is analogous to that of the mean free path in molecular theory.

[11] Th. von Kármán, "Turbulence and Skin Friction," *Jour. Aero. Sci.*, Vol. 1, No. 1, p. 1, 1936.

the use of equation 148. Nevertheless this equation is an adequate and relatively simple expression for turbulent shear stress which finds considerable engineering application; it is known as the Prandtl-Kármán equation.

ILLUSTRATIVE PROBLEMS

Show that, if the velocity profile in laminar flow is parabolic, the shear stress profile must be a straight line.

Solution. For a parabolic relation between v and y, $v = c_1 y^2 + c_2$; therefore $dv/dy = 2c_1 y$. Since $\tau = \mu\, dv/dy$, $\tau \propto y$.

A turbulent flow of water occurs in a pipe of 2-ft diameter. The velocity profile is measured experimentally and found to be closely approximated by the equation $v = 10 + \ln y$, in which v is in feet per second and y (the distance from the pipe wall) is in feet. The shearing stress in the fluid at a point 4 in. from the wall is calculated analytically from measurements of pressure drop (see Art. 57) to be 0.2 psf. Calculate the eddy viscosity, mixing length, and turbulence constant at this point.

Solution.

$$\frac{dv}{dy} = \frac{d}{dy}(10 + \ln y) = \frac{1}{y}$$

$$\frac{d^2v}{dy^2} = \frac{d}{dy}\left(\frac{1}{y}\right) = -\frac{1}{y^2}$$

$$0.2 = \epsilon(1/0.33); \quad \epsilon = 0.067 \text{ lb-sec/sqft} \tag{146}$$

$$0.2 = 1.94 l^2(1/0.33)^2; \quad l = 0.107 \text{ ft} \tag{147}$$

$$0.2 = 1.94\kappa^2(1/0.33)^4/(-1/\overline{0.33^2})^2; \quad \kappa = 0.32 \tag{148}$$

The magnitude of the eddy viscosity ϵ when compared with the viscosity μ (approximately 0.00002 lb-sec/sqft) is of special interest in that it provides a direct comparison between the (large) turbulent shear and (small) laminar shear for the same velocity gradient. The mixing length, l, when compared with the pipe radius is found to be about 10 percent of the latter dimension; this is a nominal value of correct order of magnitude, as is the turbulence constant, κ.

54. Fluid Flow Past Solid Boundaries. A knowledge of flow phenomena near a solid boundary is of great value in engineering problems because in practice flow is always affected to some extent by the solid boundaries over which it passes; for example, the classic aeronautical problem is the flow of fluid over the surfaces of an object such as a wing or fuselage, and in many other branches of engineering the problem of flow *between* solid boundaries, as in pipes and channels, is of paramount importance.

Although a layer of ideal fluid adjacent to a solid surface may move relatively to it, it is not possible for a real fluid to do this: experimental evidence shows that the velocity of the layer adjacent to the surface is zero (relative to the surface). This means that a conventional velocity profile must show a velocity of zero at the boundary. In visualizing the flow over a boundary surface it is well to imagine a very thin layer of fluid, possibly having the thickness of but a few molecules, adhering to the surface with a continuous increase of velocity of the fluid layers farther away from the surface,

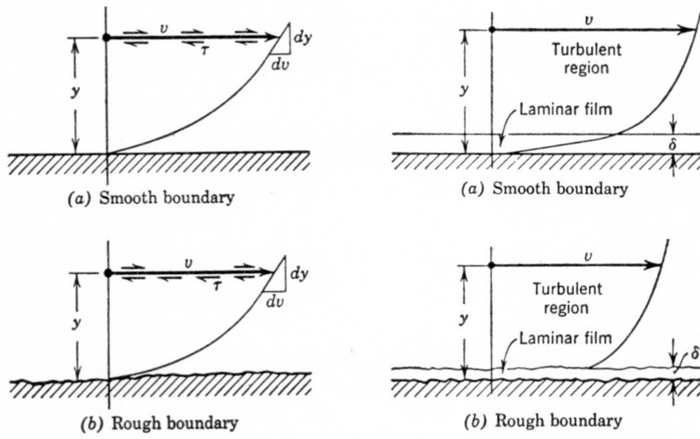

FIG. 92. Laminar flow.

FIG. 93. Turbulent flow.

the magnitude of the velocity gradient dependent upon the shear in the fluid. For rough surfaces this simple picture is somewhat compromised since small eddies tend to form between the roughness projections, causing local unsteadiness of the flow.

Laminar flow occurring over smooth [12] or rough boundaries (Fig. 92) possesses essentially the same properties, the velocity being zero at the boundary surface and the shear stress throughout the flow being given by equation 12. Thus, in laminar flow, *surface roughness has no effect* on the flow picture.

In turbulent flow, however, the roughness of the boundary surface will affect the physical properties of the fluid motion. When turbulent flow occurs over *smooth* solid boundaries, it is always separated from the boundary by a *film* (or *sublayer*) of laminar flow (Fig. 93). This laminar film has been observed experimentally, and its existence may be justified theoretically by the following simple reasoning: The

[12] In the dynamics of solids, a "smooth" surface is often assumed to be a frictionless one; this concept is irrelevant here.

presence of a boundary in a turbulent flow will curtail the freedom of the turbulent mixing process by reducing the available mixing length, and, in a region very close to the boundary, the available mixing length is reduced to zero (i.e., the turbulence is completely extinguished) and a film of laminar flow over the boundary results. Thus, in problems of turbulent flow over surfaces, the engineer must be prepared to apply principles and equations of both laminar and turbulent flow.

In the laminar film the shear stress, τ, is given by equation 12, and at a distance from the boundary where turbulence is completely developed, by equation 148. Between the latter region and the laminar film lies a transition zone in which shear stress results from a complex combination of both turbulent and viscous action, turbulent mixing being inhibited by the viscous effects due to the proximity of the wall. The fact that there is a transition from fully developed turbulence to no turbulence at the boundary surface shows that the laminar film, although given (for convenience) an arbitrary thickness,[13] δ (Fig. 93), does *not* imply a sharp line of demarcation between the laminar and turbulent regions.

As stated above, the roughness of boundary surfaces will affect the physical properties of turbulent flow, and the effect of this roughness is dependent upon the relative size of roughness and laminar film. A boundary surface is said to be *smooth* if its projections or protuberances are so completely submerged in the laminar film (Fig. 93*b*) that they have no effect on the structure of the turbulence. However, experiments have shown that roughness heights larger than about one-fourth of film thickness will augment the turbulence and have some effect upon the flow. Thus the thickness of the laminar film is the criterion of effective roughness, and, because the thickness of this film also depends upon certain properties of the flow, it is quite possible for the same boundary surface to behave as a smooth one or a rough one, depending upon the size of the Reynolds number and of the laminar film which tends to form over it.[14]

Since surface roughness increases the turbulence in a flowing fluid and thus decreases the effect of viscous action, some prediction may

[13] Intensive research on the laminar film has shown that its thickness varies with time, showing the film flow to be unsteady, this unsteadiness being associated with eddy formation adjacent to the surface. As these eddies are carried away from the wall their presence accounts for the increased turbulence observed in the transition zone.

[14] It will be shown later that film thickness decreases with increasing Reynolds number. Usually in practice the change of a smooth surface to a rough one results from an increase of Reynolds number brought about by an increase of velocity.

be made about the effect of roughness on energy losses. In turbulent flow over rough surfaces, energy is dissipated by the work done in the continual generation of turbulence by the roughness protuberances; the energy involved in this turbulence is composed of the kinetic energy of fluid masses, which is known to be proportional to the squares of their velocities. Since these velocities are in turn proportional to the velocities of general motion, it may be expected that energy dissipation and resistance caused by rough surfaces should vary with the squares of velocities.

As turbulent flow takes place over smooth surfaces, work is done at the expense of available fluid energy against the shear stress due to viscous action in the laminar film. No predictions will be made as to the relation between energy conversions and velocities in this case where a combination of turbulent and viscous action exists, but many experiments have indicated that, for turbulent flow over smooth surfaces, energy dissipation and resistance will vary with the 1.75 to 1.85 power of velocities. In laminar flow over boundary surfaces, energy dissipation and resistance will subsequently be shown to be directly proportional to velocities.

55. Velocity Distribution and Its Significance. The shearing stresses of laminar and turbulent flow have been seen to produce velocity distributions characterized by reduced velocities near the boundary surfaces. These deviations from the uniform velocity distribution of ideal fluid flow will necessitate alterations[15] in the methods for calculation of velocity head and momentum flux. The effect of nonuniform velocity distribution on the computation of flowrate has been indicated in Art. 18.[16]

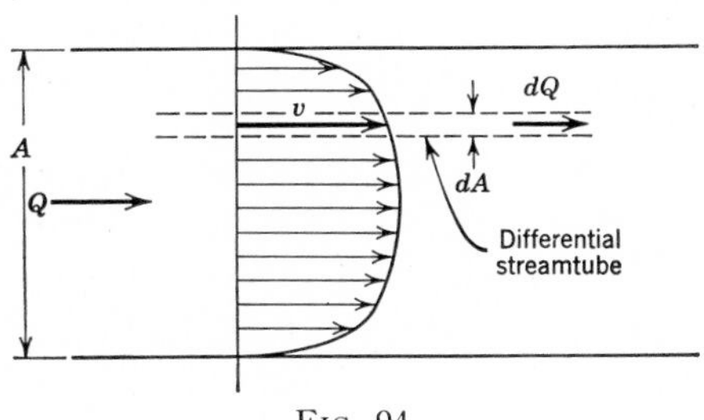

FIG. 94

The kinetic energy of fluid moving in the differentially small streamtube of Fig. 94 will be (from equation 68) $(dQ)\gamma v^2/2g$ or

[15] In many practical problems, however, these alterations are so small that they may be neglected.

[16] Art. 18 should be restudied at this point.

$\rho v^3\, dA/2$. The momentum flux (Art. 40) will be $(dQ)\rho v$ or $\rho v^2\, dA$. The total kinetic energy and momentum flux will be the respective integrals of these differential quantities; thus general expressions are

$$\text{Total kinetic energy (ft-lb/sec)} = \frac{\rho}{2}\int^A v^3\, dA \tag{149}$$

$$\text{Momentum flux (lb)} = \rho\int^A v^2\, dA \tag{150}$$

Although these quantities are computed as indicated in many engineering situations, they are also usefully expressed in terms of mean velocity V and total flowrate Q, using the same forms of the equations; thus

$$\text{Total kinetic energy (ft-lb/sec)} = \alpha Q\gamma\frac{V^2}{2g} = Q\gamma\left(\alpha\frac{V^2}{2g}\right) \tag{151}$$

$$\text{Momentum flux (lb)} = \beta Q\rho V \tag{152}$$

in which α and β are dimensionless and represent correction factors to the conventional velocity head $V^2/2g$ and momentum flux $Q\rho V$, respectively. For a uniform velocity distribution no such correction factors appear, so $\alpha = \beta = 1$; for a nonuniform velocity profile $\alpha > \beta > 1$. Expressions for α and β may be derived by equating the equations 149 and 151, equations 150 and 152, and substituting $\int^A v\, dA$ for Q, yielding

$$\alpha = \frac{1}{V^2}\frac{\int^A v^3\, dA}{\int^A v\, dA} \quad \text{and} \quad \beta = \frac{1}{V}\frac{\int^A v^2\, dA}{\int^A v\, dA} \tag{153}$$

from which it may be easily seen that $\alpha = \beta = 1$ for uniform distribution of velocity. Comparing the rather pointed (far from uniform) velocity distributions of laminar flow (Fig. 92) with the rather flattened (nearer to uniform) ones of turbulent flow (Fig. 93), it may be concluded directly that high values of α and β are to be expected for the former, lower ones for the latter.

For real fluid flowing in a prismatic passage the analysis of Art. 24 applies,[17] and $(p/\gamma + z)$ is constant throughout any flow cross sec-

[17] Providing that the shearing stresses in the fluid perpendicular to the direction of motion may be neglected. These are safely negligible in engineering problems except for flows dominated by viscous action. Such flows occur at very low Reynolds numbers and low velocity; they are frequently termed *creeping motions.*

tion. From this it is immediately seen that $(z + p/\gamma + v^2/2g)$ cannot be constant throughout the flow cross section, and this in turn requires some re-examination of the energy line concept. Taking A and C as typical streamlines, it is noted from Fig. 95 that each streamline is associated with a different energy line; in other words, the flow along different streamlines will possess different amounts of total energy, and a comprehensive energy line picture would be a

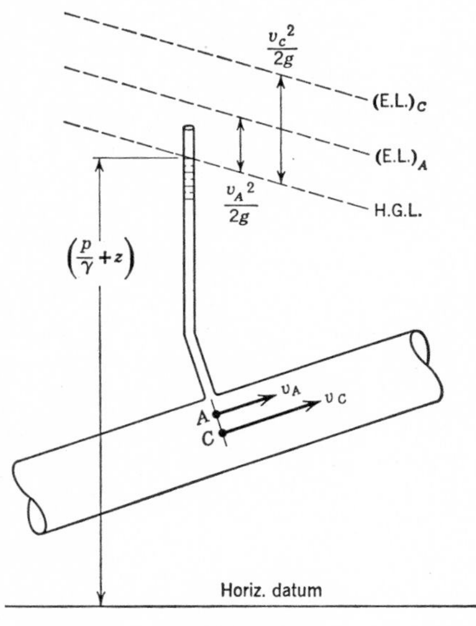

FIG. 95

"bundle of energy lines"—one for each streamline. However, for flow in a parallel-walled passage, such as a pipe, duct, or open channel, the properties of individual streamlines are seldom of interest, and the whole aggregation of streamlines (i.e., the whole flow) is characterized by a single effective energy line a distance $\alpha V^2/2g$ above the hydraulic grade line; thus alterations in the Bernoulli equation due to nonuniform velocity distribution are concentrated in the coefficient α alone. However, in problems involving flowfields each streamline will in general be associated with different amounts of total energy, a fact essential to the analysis and interpretation of such problems.

Another consequence of nonuniform velocity distribution can be shown in its simplest aspects by considering a flow through a short constriction (Fig. 96) in a passage where the coefficient α changes from section to section. The trend of change of α from section 1 to section 2 may be established by noting that all fluid particles in passing from section 1 to section 2 experience the same change in $(p/\gamma + z)$ or the same drop in the hydraulic grade line. The velocities at section 1 are relatively small, and those at section 2 relatively

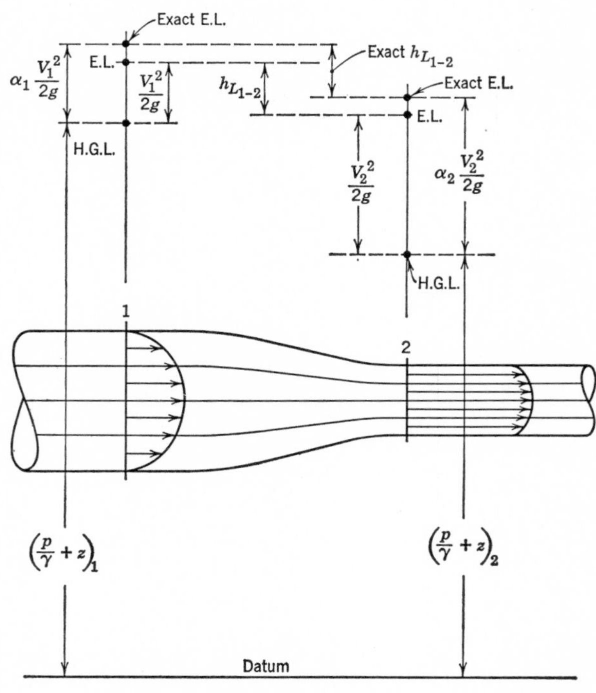

FIG. 96

large; with frictional effects in the constriction relatively small, the increase in the velocity of all particles from section 1 to section 2 will tend to be about the same; the velocity profile at section 2 will therefore be flatter than that at section 1, and α_2 will accordingly be less than α_1. The *exact head loss* between sections 1 and 2 is given by the drop in the energy line between these sections; from Fig. 96,

$$\text{Exact } h_{L_{1-2}} = \left(\alpha_1 \frac{V_1^2}{2g} + \frac{p_1}{\gamma} + z_1\right) - \left(\alpha_2 \frac{V_2^2}{2g} + \frac{p_2}{\gamma} + z_2\right)$$

However, the conventional head loss used in engineering computations is defined by ignoring the α terms in the Bernoulli equation and writing

$$h_{L_{1-2}} = \left(\frac{V_1^2}{2g} + \frac{p_1}{\gamma} + z_1\right) - \left(\frac{V_2^2}{2g} + \frac{p_2}{\gamma} + z_2\right)$$

Comparison of the two equations leads to

$$h_{L_{1-2}} = \text{Exact } h_{L_{1-2}} + (\alpha_2 - 1)\frac{V_2^2}{2g} - (\alpha_1 - 1)\frac{V_1^2}{2g} \qquad (154)$$

from which it is seen that the conventional head loss does not equal exact head loss unless $(\alpha_2 - 1)V_2^2/2g = (\alpha_1 - 1)V_1^2/2g$. However, constriction of the passage causes V_2 to be greater than V_1, while the flattening of the velocity profile causes $(\alpha_2 - 1)$ to be less than $(\alpha_1 - 1)$; thus there are some compensating features in equation 154 which tend to make exact head loss not very different from conventional head loss in this simple example. Although in other cases these conventional and exact head losses may differ considerably, it should be noted that this is no serious obstacle in most engineering problems; no matter how these losses may be defined or related to each other, there can be (for a given flowrate) only one value for the change of $(p/\gamma + z)$ between sections 1 and 2; if calculations are made with this fact in mind, reliable predictions can be made from conventional head losses even though these losses are never precisely equal to the exact ones.

Illustrative Problem

Assuming Fig. 94 to represent a parabolic velocity profile in a passage bounded by two infinite planes of spacing $2R$ and maximum velocity v_c, calculate q, α, and β.

Solution. Taking r as the distance from centerline of passage to any local velocity, v, and element of area dA, $dA = dr$. The equation of the parabola is $v = v_c(1 - r^2/R^2)$.

$$q = 2\int_0^R v_c\left(1 - \frac{r^2}{R^2}\right)dr = \tfrac{2}{3}(2R\,v_c) \qquad (56)$$

Since q also equals $2RV$, $V = 2v_c/3$.

$$\alpha = \frac{2\int_0^R v_c^3\left(1 - \frac{r^2}{R^2}\right)^3 dr}{\left(\frac{2v_c}{3}\right)^2 \frac{2}{3}(2R\,v_c)} = \frac{54}{35} = 1.543 \qquad (153)$$

$$\beta = \frac{2\int_0^R v_c^2\left(1 - \frac{r^2}{R^2}\right)^2 dr}{\left(\frac{2v_c}{3}\right) \frac{2}{3}(2R\,v_c)} = \frac{6}{5} = 1.200 \qquad (153)$$

The meaning of these figures is that the exact velocity head is more than 54 percent greater than $V^2/2g$, and the exact momentum flux 20 percent greater than $Q\rho V$. Differences of this magnitude warn the engineer that α and β should be considered when applying the energy and momentum equations to one-dimensional flow problems—unless their effects can be shown to have negligible consequence in the results desired.

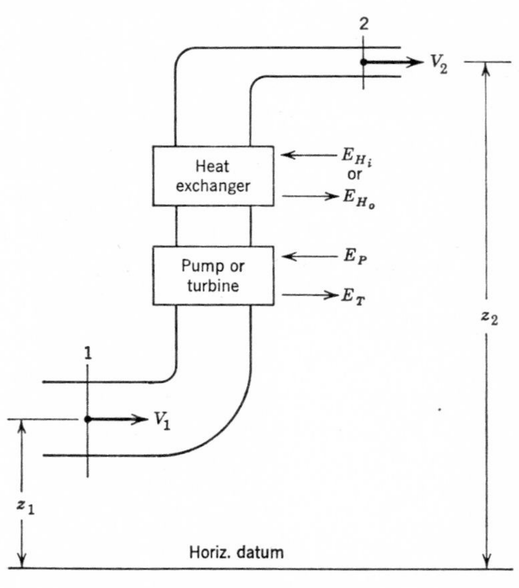

Fig. 97

56. The Energy Equation. The energy equation applied to the flow of real fluids is merely an accounting of the various energy changes within a portion of a flow system and presents few difficulties once these energies have been identified. Consider the reasonably (but not completely)[18] general situation of Fig. 97. Here the possibility of heat energy, E_{H_i} or E_{H_o}, being added to or extracted from the flow must be considered; in Fig. 97 this action is shown concentrated in the heat exchanger, but such a flow of heat energy may also occur through the walls of the streamtube, its direction dependent upon the relative temperatures at the surfaces of the tube wall. Such flows of heat energy are calculated in engineering problems by use of the principles of heat transfer, a large field of engineering science which cannot be explored in an elementary fluid mechanics text.

[18] Chemical, electrical, atomic energies, and the kinetic energy of turbulence are excluded from this analysis.

Writing the energy equation as a generalization of equation 76, it becomes [19]

$$I_1 + \frac{p_1}{\gamma_1} + \frac{V_1^2}{2g} + z_1 + E_{H_i} + E_P$$
$$= I_2 + \frac{p_2}{\gamma_2} + \frac{V_2^2}{2g} + z_2 + E_{H_o} + E_T \quad (155)$$

Without heat exchanger, pump, or turbine, this energy equation reduces to

$$I_1 + \frac{p_1}{\gamma_1} + \frac{V_1^2}{2g} + z_1 + E_H = I_2 + \frac{p_2}{\gamma_2} + \frac{V_2^2}{2g} + z_2 \quad (156)$$

in which E_H is the heat energy (ft-lb/lb) passing *into* [20] the fluid through the walls of the streamtube. For gases or vapors, equation 156 is usually written in terms of enthalpy (H) and with neglect of $(z_2 - z_1)$; it then becomes

$$H_1 + \frac{V_1^2}{2g} + E_H = H_2 + \frac{V_2^2}{2g} \quad (157)$$

For perfect gases where $H_1 - H_2 = c_p\,(T_1 - T_2)$, equation 157 frequently appears as

$$c_pT_1 + \frac{V_1^2}{2g} + E_H = c_pT_2 + \frac{V_2^2}{2g} \quad (158)$$

For the flow of liquids and gases in the numerous engineering situations where there is negligible change of fluid density, the equation is written

$$\frac{p_1}{\gamma} + \frac{V_1^2}{2g} + z_1 = \frac{p_2}{\gamma} + \frac{V_2^2}{2g} + z_2 + h_{L_{1-2}} \quad (159)$$

in which the head loss, $h_{L_{1-2}}$, represents the fall of the energy line between sections 1 and 2, and in terms of the other quantities is given (through comparison with equation 156) by

$$h_{L_{1-2}} = I_2 - I_1 - E_H \quad (160)$$

[19] Because of the use of mean velocity, pressure, density, and temperature, correction factors should be applied to the first three terms on each side of this equation to allow for the nonuniform distribution of these quantities across the flow. However, these correction factors are usually close to unity, and such refinement of the equation is generally not necessary for engineering use; these factors are omitted here.

[20] If heat energy flows *outward* through the walls of the streamtube, E_H will be negative or will appear on the opposite side of the equation.

In such problems the separate values of I_2, I_1, and E_H are usually not required, and the "packaging" of them into a single term, $h_{L_{1-2}}$, proves highly effective for engineering use. The equation offers proof that head loss is not a loss of total energy but rather a conversion of energy into heat, part of which leaves the fluid, the remainder serving to increase its internal energy. This is the practical case of incompressible flow as it appears in many engineering applications; here head loss is a permissible and useful concept because heat energy leaving the flow and energy converted into internal energy are seldom recoverable and are in effect lost from the useful total of pressure, velocity, and potential energies. For compressible flow this is not generally true, since the useful total of energies will include the internal energy.

Illustrative Problems

A flowrate of 240 lb/sec of air occurs in a streamtube similar to that of Fig. 97 but without pump or turbine. $A_1 = 4$ sqft, $A_2 = 2$ sqft, $z_1 = 30$ ft, $z_2 = 80$ ft, $V_1 = 600$ fps, $V_2 = 800$ fps, $p_1 = 20$ psia, $p_2 = 35$ psia. Calculate the mean temperatures of the air at sections 1 and 2 and the net heat energy added to the fluid between these sections.

Solution.

$$240 = 4 \times 600 \times \gamma_1 = 2 \times 800 \times \gamma_2; \quad \gamma_1 = 0.1, \quad \gamma_2 = 0.15 \text{ lb/cuft} \qquad (52)$$

$$T_1 = \frac{20 \times 144}{53.3 \times 0.1}, \quad T_2 = \frac{35 \times 144}{53.3 \times 0.15}; \quad T_1 = 540°\text{R}, \quad T_2 = 630°\text{R} \qquad (3)$$

$$H_2 - H_1 = 186.5(630 - 540) = 16{,}800 \text{ ft-lb/lb}$$

$$(600)^2/2g + 30 + E_H = 16{,}800 + (800)^2/2g + 80 \qquad (157)$$

$E_H = 22{,}210$ ft-lb/lb. Disregarding the z terms (30 and 80), $E_H = 22{,}160$ ft-lb/lb, a difference of about 0.2 percent. The total heat energy added to the fluid represents $240 \times 22{,}160/550 = 9330$ hp.

A flowrate of 50 cfs of water occurs in a streamtube, similar to that of Fig. 97, containing a pump (but no heat exchanger) which is delivering 400 hp to the flowing fluid. $A_1 = 4$ sqft, $A_2 = 2$ sqft, $z_1 = 30$ ft, $z_2 = 80$ ft, $p_1 = 20$ psi, $p_2 = 10$ psi. Calculate the head lost between sections 1 and 2.

Solution.

$$50 = 4V_1 = 2V_2; \quad V_1 = 12.5 \text{ fps}, \quad V_2 = 25.0 \text{ fps} \qquad (53)$$

$$E_P = 400 \times 550/62.4 \times 50 = 70.6 \text{ ft-lb/lb} \qquad (68)$$

Applying equation 155 with $h_L = I_2 - I_1 - E_{H_i} + E_{H_o}$, and $E_T = 0$,

$$\frac{20 \times 144}{62.4} + \frac{(12.5)^2}{2g} + 30 + 70.6$$

$$= \frac{10 \times 144}{62.4} + \frac{(25)^2}{2g} + 80 + h_L; \quad h_L = 36.4 \text{ ft}$$

57. Resistance Force and Energy Dissipation. Although the energy equations of Art. 56 are essential to engineering analysis, they contain no information about the basic resistance forces which cause

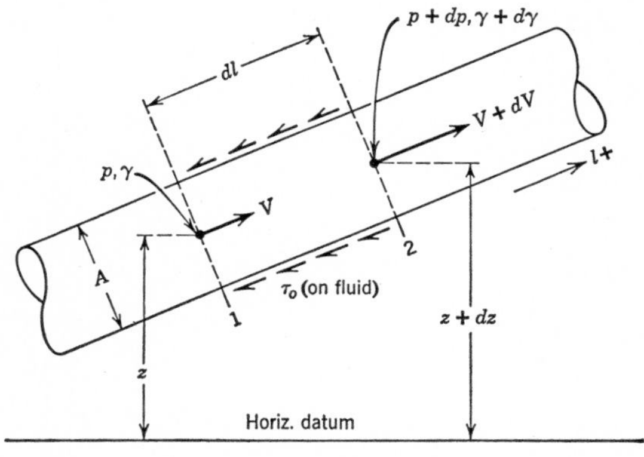

FIG. 98

energy dissipation in fluid flow. These forces may be isolated, identified, and related to the energy equations through a one-dimensional analysis of the (compressible or incompressible) flow in an element of a cylindrical passage (Fig. 98). The stress τ_o is the basic resistance (shear) stress to be investigated and will produce a force on the periphery of the streamtube opposing the direction of fluid motion. The other variables shown on the figure will be recognized from previous treatment (Art. 21). An impulse-momentum analysis applied along the direction of the streamtube to the control volume bounded by sections 1, 2, and the streamtube boundary yields

$$pA - (p + dp)A - \tau_o P\, dl - (\gamma + d\gamma/2)A\, dl(dz/dl)$$

$$= \beta_2(V + dV)^2 A(\rho + d\rho) - \beta_1 V^2 A\rho$$

in which P is the perimeter of the streamtube. Assuming $\beta_1 = \beta_2 = 1$, dividing the equation by $A\gamma$, neglecting $d\gamma/\gamma$, and neglecting terms containing products of differential quantities, the equation may be reduced to [21]

$$\frac{dp}{\gamma} + d\left(\frac{V^2}{2g}\right) + dz = -\frac{\tau_o\,dl}{\gamma R_h} \tag{161}$$

and compared with the differential form of (energy) equation 156,

$$d\left(\frac{p}{\gamma}\right) + d\left(\frac{V^2}{2g}\right) + dz = dE_H - dI$$

Since $d(p/\gamma) = dp/\gamma + p\,d(1/\gamma)$, comparison of these equations yields

$$\frac{\tau_o\,dl}{\gamma R_h} = dI - dE_H + p\,d\left(\frac{1}{\gamma}\right) \tag{162}$$

which gives a basic relationship between the resistance stress, τ_o, and various terms of the energy equation. Its integration may be indicated,

$$\frac{1}{R_h}\int_1^2 \frac{\tau_o\,dl}{\gamma} = I_2 - I_1 - E_H + \int_1^2 p\,d\left(\frac{1}{\gamma}\right)$$

but cannot be performed for compressible flow without information on the thermodynamic process, which will be affected by the amount of heat transfer E_H. For compressible flow it is to be noted that τ_o will also depend upon l, and if the former is considered constant the result will be the mean value $\bar{\tau}_o$, over the length $(l_2 - l_1)$. Another observation to be made here is that thermodynamics shows (for frictionless processes) that

$$dI - dE_H + p\,d\left(\frac{1}{\gamma}\right) = 0$$

and that this is also obtained from equation 162 when $\tau_o = 0$.

For established incompressible flow, τ_o is no longer a function of l, γ is constant, and $d(1/\gamma) = 0$. Integration then gives

$$\frac{\tau_o(l_2 - l_1)}{\gamma R_h} = I_2 - I_1 - E_H$$

However, the right-hand side of this equation has been defined (equation 160) as fall of energy line, or head loss $h_{L_{1-2}}$. Thus for incompressible flow

$$h_{L_{1-2}} = \frac{\tau_o\,(l_2 - l_1)}{\gamma R_h} \tag{163}$$

[21] The ratio A/P is known as the *hydraulic radius* R_h; further discussion of it will be found in Art. 73.

giving a simple relation between resistance stress τ_o and the head loss caused by the action of resistance.

It should be apparent that the foregoing analysis may be similarly applied to any streamtube of the flow; it is useful to do this for a streamtube of radius r and concentric with the axis of a cylindrical pipe (Fig. 99). For such a streamtube the frictional stress, τ, will be that exerted on the outermost fluid layer of the streamtube by the adjacent (more slowly moving) fluid. Without repeating the

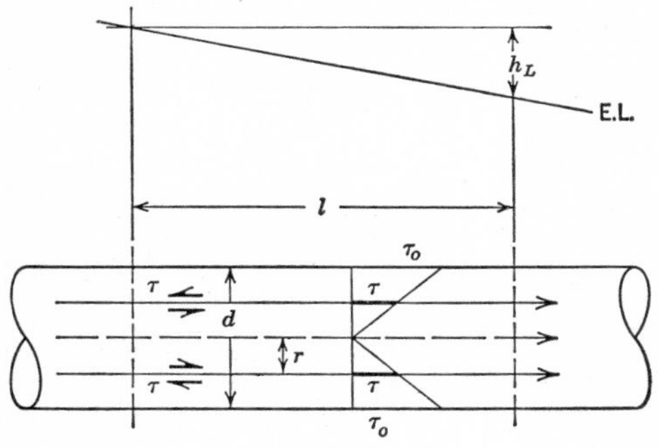

FIG. 99

foregoing development, the following substitutions may be made in equation 163; τ for τ_o, $r/2$ for R_h, l for $l_2 - l_1$; this yields

$$\tau = \left(\frac{\gamma h_L}{2l}\right) r \tag{164}$$

and shows that, in established pipe flow, the shear stress τ in the fluid varies linearly with distance from the centerline of the pipe. The relationships of equations 163 and 164 have been developed without regard to the flow regime; thus it follows that they are equally applicable to both laminar and turbulent flow in pipes. These equations will be used later as the first step in the analytical treatment of these problems.

ILLUSTRATIVE PROBLEMS

Water flows in a 3-ft by 2-ft rectangular conduit. The head lost in 200 ft of this conduit is determined (experimentally) to be 30 ft. Calculate the resistance stress exerted between fluid and conduit walls.

Solution.

$$\tau_o = 30 \times 62.4 \times 6/10 \times 200 = 5.62 \text{ psf} = 0.039 \text{ psi} \tag{163}$$

Since this flow is *not* axisymmetric, τ_o must be presumed to be the *mean* shear stress on the perimeter of the conduit.

If the results cited in the preceding problem are obtained for water flowing in a cylindrical pipe 2 ft in diameter, what shear stress is to be expected (*a*) between fluid and pipe wall, and (*b*) in the fluid at a point 8 in. from the wall?

Solution.

$$\tau_o = (30 \times 62.4/2 \times 200)1 = 4.69 \text{ psf} = 0.0325 \text{ psi} \qquad (164)$$

From the linear variation of τ with r,

$$\tau = \tfrac{4}{12}(0.0325) = 0.0108 \text{ psi}$$

58. Flow Establishment—Boundary Layers. Up to this point developments have been confined to *established* flow in prismatic passages where from section to section there is no change of velocity

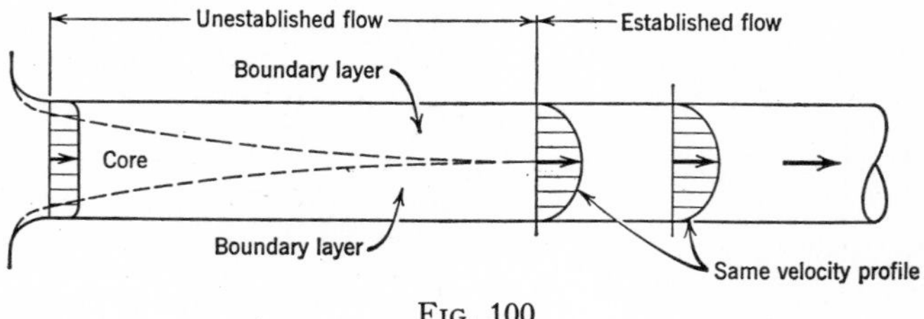

FIG. 100

profile (Fig. 100). However, the zone of *unestablished* flow is of paramount importance in many engineering problems; it will be described here for the simplest case—flow from a large reservoir into a cylindrical passage.

The unestablished flow zone may contain many different flow phenomena. With an unrounded entrance, separation (Art. 59) will dominate the flow picture—featured by localized eddy formation close to the entrance and followed by decay of the eddies in the final established flow; measurements show that this process extends over a distance of some 50 pipe diameters. However, with a rounded (streamlined) entrance this problem is of more engineering interest; here the nominal length of establishment is some 20 pipe diameters at high Reynolds numbers [22] and steadily diminishes with decreasing Reynolds number. For this case the unestablished flow zone will be dominated by the growth of *boundary layers* along the walls (accompanied by a diminishing *core* of fluid at the center of the passage);

[22] Calculated for the established flow.

established flow results from the merging of these boundary layers. Prandtl[23] first suggested the boundary layer concept in 1904 in explaining the resistance of streamlined bodies, flat plates parallel to the flow, etc.; the essential point is that the *frictional aspects of the flow are confined to the boundary layer* (where the flow is *rotational*) but outside the boundary layer (in this case in the core) the viscosity of the fluid is inoperative; i.e., the flow is frictionless and *irrotational*. The boundary layer idea has found wide application in numerous problems of fluid dynamics and has provided a powerful tool for the analysis of problems of fluid resistance; it has probably contributed more to progress in modern fluid mechanics than any other single idea. The engineer of today, although he need not know all the analytical and mathematical aspects of boundary layer theory, must thoroughly understand the basic concepts and phenomena if he is to make or follow modern approaches to fluid-flow problems.

The mechanism of boundary layer growth may be described as follows. As the fluid flows into the entrance, fluid particles at the boundary walls remain at rest, and a high velocity gradient (dv/dy) develops in the vicinity of the boundary. These high velocity gradients are associated with large frictional stresses in the boundary layer which "eat their way" into the flow by slowing down successive fluid elements. Thus the boundary layers steadily thicken until they meet and envelop the whole flow; downstream from this point the influence of wall friction is felt throughout the flowfield, the flow is *established* and everywhere *rotational*.

The flow in a boundary layer may be either laminar or turbulent, and Fig. 100 may be used to visualize both. If Reynolds number Vd/ν for the established flow is less than 2100 it may be safely inferred that the established laminar flow has resulted from the growth of laminar boundary layers, and if this Reynolds number is greater than 5000 the boundary layers are turbulent. However, in the latter case this conclusion is adequate only when the entrance is rough or irregular or where the reservoir fluid contains residual turbulence. For a well-shaped and smooth entrance the boundary layer may be laminar at the upstream end of the zone of unestablished flow, followed by a turbulent boundary layer in the downstream portion of this region. A laminar boundary layer is to be expected at the upstream end where the boundary layer is thin, because of the inhibiting effect of the smooth wall on the development of turbulence, and the dominance of viscous action. One of two Reynolds numbers is widely used to define the character of flow

[23] *Proc. Third Intern. Math. Congress*, Heidelberg, 1904.

in a boundary layer; they are

$$\mathbf{R}_x = Vx/\nu \quad \text{and} \quad \mathbf{R}_\delta = V\delta/\nu \tag{165}$$

and experiments on flat plates (Fig. 101) have shown their nominal critical values to be approximately 500,000 and 4000, respectively. For appropriate Reynolds numbers below these values, laminar boundary layers are to be expected, whereas Reynolds numbers above these values will define turbulent boundary layers. Comparison of flat plate boundary layers with those of the pipe entrance (Fig. 100) will reveal certain subtle but critical differences between them. The superficial similarity between Figs. 100 and 101 is noted

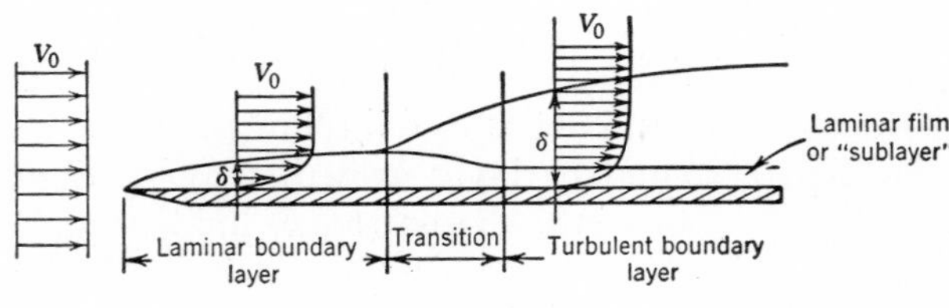

Fig. 101

immediately; however, for the pipe entrance: (*a*) the plate has been rolled into a cylinder so it is not flat, (*b*) the core velocity steadily increases downstream whereas the corresponding free stream velocity of Fig. 101 remains essentially constant, and (*c*) the pressure in the fluid diminishes [24] in a downstream direction whereas for the flat plate there is no such pressure variation. Because of these differences the critical Reynolds numbers for the boundary layers in the pipe entrance are somewhat greater than those of the flat plate, and critical values of 600,000 and 5000 may be taken as nominal [25] for approximate predictions of maximum length and thickness of the laminar boundary layer.

Although this is not the place for a detailed analysis of the pipe entrance boundary layer problem, certain qualitative facts should be noted (Fig. 102). Between the turbulence (of boundary layer or established flow) and solid boundary there exists the laminar film (or sublayer) cited in Art. 54. The velocity of the core flow increases over the region of establishment from a value slightly more than the mean velocity, Q/A, to the centerline velocity of the established

[24] This is known as a *favorable pressure gradient*, since it extends the length of the laminar boundary layer and thus reduces frictional effects.

[25] Much higher values have been obtained by using passages with porous or slotted walls, which serve to remove part of the boundary layer.

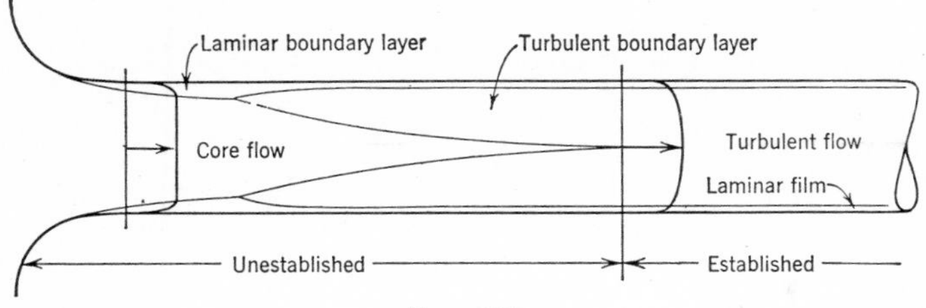

FIG. 102

flow; from this it follows that the values of α and β increase from slightly more than 1 at the upstream end of the pipe to their values associated with the established velocity profiles. The thickening of the laminar boundary layer will cause a decrease in velocity gradient and thus a decrease in wall shearing stress in a downstream direction. With the change of velocity profile in the turbulent boundary layer a rather sudden increase of wall shearing stress can be expected after the boundary layer has changed from laminar to turbulent, after which this shear stress will continue to decrease in a downstream direction. Since the core flow may be treated as frictionless, the fall of pressure may be predicted from the simple Bernoulli equation (without head loss) once the core velocities are known; this has been proved by experiment to be reliable and accurate in the upstream portion of the region of establishment (where the boundary layer is thin) but less accurate in the downstream portion.

59. Separation. Separation of moving fluid from boundary surfaces is another important difference between the flow of ideal and real fluids. The mathematical theory of the ideal fluid yields no information about the expectation of separation even in simple cases where intuition alone would predict separation with complete certainty. Examples are shown in Fig. 103 for a sharp projection on a wall, a flat plate normal to a rectilinear flow, and a diaphragm orifice in a pipe line. For the ideal fluid the flowfields will be found to be symmetrical upstream and downstream from such obstructions, the fluid rapidly accelerating toward the obstruction and decelerating in the same pattern downstream from it. However, the engineer would reason that the inertia of the moving fluid would prevent its following the sharp corners of such obstructions and that consequently separation of fluid from boundary surface is to be expected there, resulting in asymmetric flowfields featured by eddies and wakes downstream from the obstructions. Motion pictures of such eddies disclose that they are basically unsteady—forming, being swept away, and re-forming—thus absorbing energy from the flow and

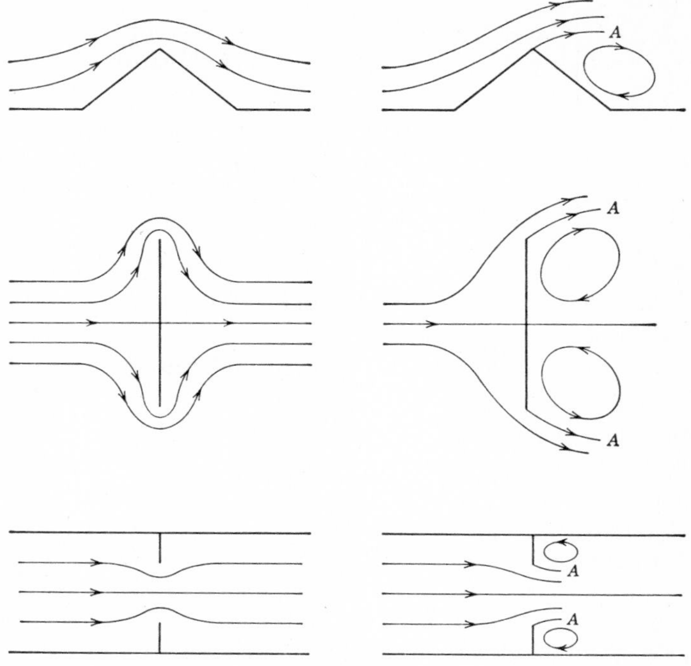

Fig. 103

dissipating it in heat as they decay in an extensive zone downstream from the obstruction; thus the sketches of Fig. 103 are to be taken as time-average flow pictures which are intended to convey the essentials but not the complete details of flow separation.

Surfaces of discontinuity (indicated by A on Fig. 103) divide the live stream from the adjacent and more sluggishly moving eddies. Across such surfaces there will be high velocity gradients and accompanying high shear stress, but no discontinuity of pressure. The tendency for surfaces of discontinuity to break up into smaller eddies may be seen from the simplified velocity profile of Fig. 104*a*. An observer moving at velocity V would see the relative velocity profile of Fig. 104*b* from which the tendency for eddy formation is immediately evident.

Under special circumstances the streamlines of a surface of discontinuity become *free streamlines*,[26] which are streamlines along which the pressure is constant. It is apparent that the surfaces of discontinuity of Fig. 103 do not quite satisfy this requirement, and thus their streamlines may be considered free streamlines only as a

[26] Some examples of free streamlines have been cited in Art. 28.

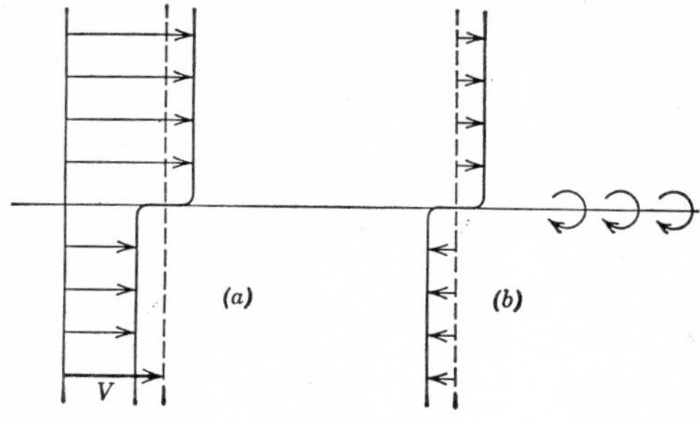

FIG. 104

crude approximation. Another example of interest is the cavitation zone behind a disk (Fig. 105) (or other sharp-cornered object) in a liquid flowfield; here the constant pressure imposed on the free streamline is the vapor pressure of the liquid. Bernoulli's equation applied along any free streamline shows that, if pressure is constant, $z + v^2/2g$ is also constant; if variations in z are negligible or zero, free streamlines are also lines of constant velocity. Modern free streamline theory is concerned with the prediction of form and positions of such lines from their unique properties and the hydrodynamical equations of flowfield theory.

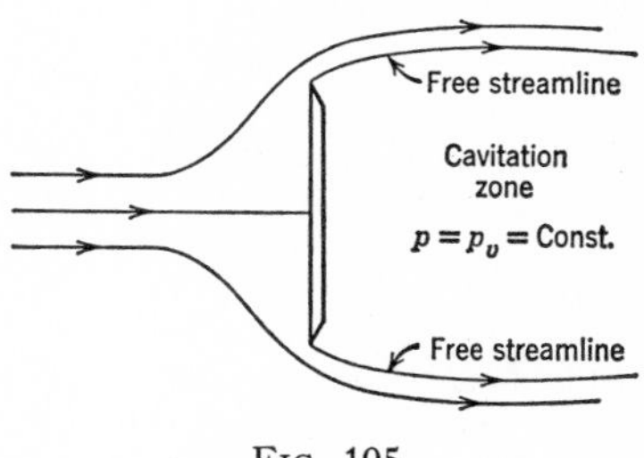

FIG. 105

Although the prediction of separation may be quite simple for sharp-cornered obstructions, this is a considerably more complex matter for gently curved (streamlined) objects or surfaces. Suppose that separation of flow from the streamlined strut of Fig. 106 does not occur. This is a very reasonable assumption for small ratios of thickness to length but much less reasonable for high values of this ratio. For no separation, the flowfield is virtually identical with that of the ideal fluid except for the growth of a tiny boundary layer between the strut and the remainder of the flowfield. The boundary layers coalesce at the trailing edge of the strut, producing a narrow

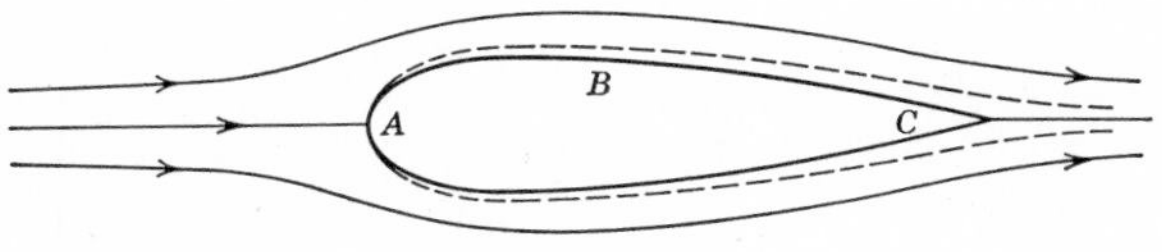

FIG. 106

wake of fine-grained eddies. Proceeding along the surface of the strut from (the stagnation point) A to B, the pressure falls, producing a favorable pressure gradient which "strengthens" the boundary layer. From B to C, however, the pressure rises, producing an adverse (unfavorable) pressure gradient which may "weaken" the boundary layer sufficiently to cause separation. The likelihood of separation is enhanced with increase of thickness-to-length ratio of the strut which from the increased divergence of the streamlines adjacent to BC will produce a larger unfavorable pressure gradient. This adverse pressure gradient penetrates the boundary layer and serves to produce a force opposing the motion of its fluid; if the gradient is large enough, the slowly moving fluid near the wall will be brought to rest and begin to accumulate, diverting the live flow outward from the surface and producing an eddy accompanied by separation of the flow from the body surface. After separation has occurred, the flowfield near the separation point will appear about as shown in Fig. 107. Obviously the analytical prediction of separation

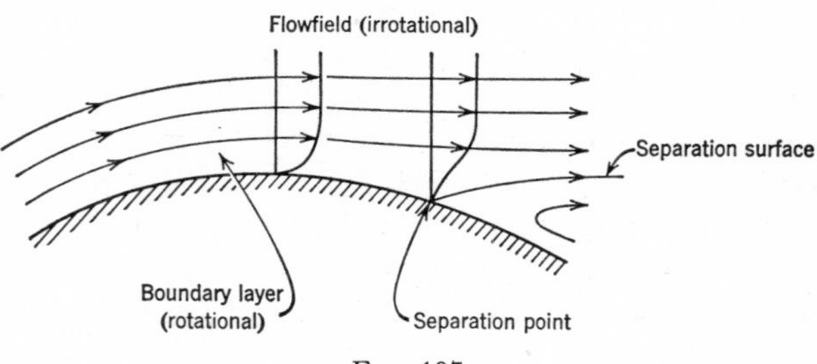

Fig. 107

point location is an exceedingly difficult problem requiring accurate quantitative information on the phenomena cited above; for this reason the prediction of separation and location of separation points on gently curved bodies or obstructions are usually obtained more reliably from experiment than from analysis.

Lest the reader get the impression that separation is of engineering importance only in external-flow problems, the classic internal-flow problem of the *diffuser* will be cited (Fig. 108). Here the engineering

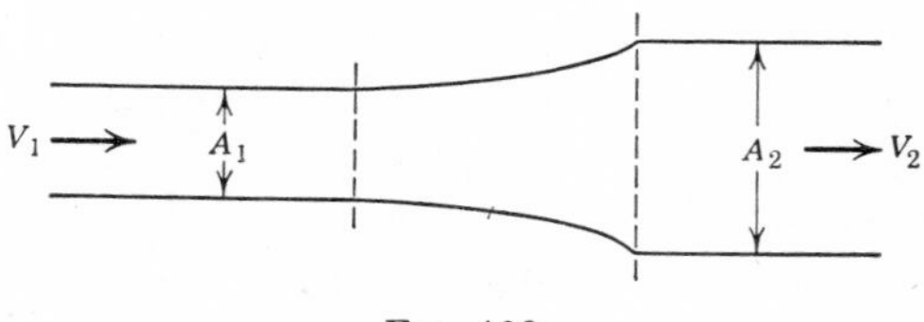

Fig. 108

objective is to provide an expanding passage of proper shape and minimum length which will yield minimum head loss or maximum pressure rise for reduction of mean velocity from V_1 to V_2. A short passage with sharp wall curvatures will prove inefficient because of flow separation and large energy dissipation; a longer one will be efficient but too space-consuming and will produce too much boundary resistance. The optimum design lies between these extremes but (as yet) cannot be determined wholly by analytical methods; again experiment must be used to supplement and confirm analytical attacks on problems where separation is involved.

The foregoing examples may now be generalized into the following simple axiom: *Acceleration of real fluids tends to be an efficient process, deceleration an inefficient one.* Accelerated motion, as it occurs through a convergent nozzle or along the surface of the front end of a submerged object, is accompanied by favorable pressure gradient which serves to stabilize the boundary layer and thus minimize energy dissipation. Decelerated motion is accompanied by adverse pressure gradient which tends to promote separation, instability, eddy formation, and large energy dissipation. This axiom may even be extended to explain the difference between the maximum efficiencies obtained in comparable complex machines such as hydraulic turbines and centrifugal pumps. In turbines the flow passages are predominantly convergent and the flow is accelerated; in the pump the opposite situation obtains. Hydraulic turbine efficiencies have been obtained up to 94 percent, but maximum centrifugal pump efficiencies are around 87 percent. The striking difference between these figures can be attributed primarily to the inherent efficiency and inefficiency of the acceleration and deceleration flow processes, respectively.

60. Secondary Flow. Another consequence of wall friction (in established or unestablished flow) is the creation of a flow within a flow—a *secondary flow* superposed on the main or *primary* flow. The secondary flow most easily described is the double spiral motion produced by a gradual bend in a closed passage. Consider the circular pipe bend of Fig. 109 and assume that its curves are gentle enough that separation is not to be expected. For an ideal fluid flowing under these conditions it has been shown (Art. 28) that a

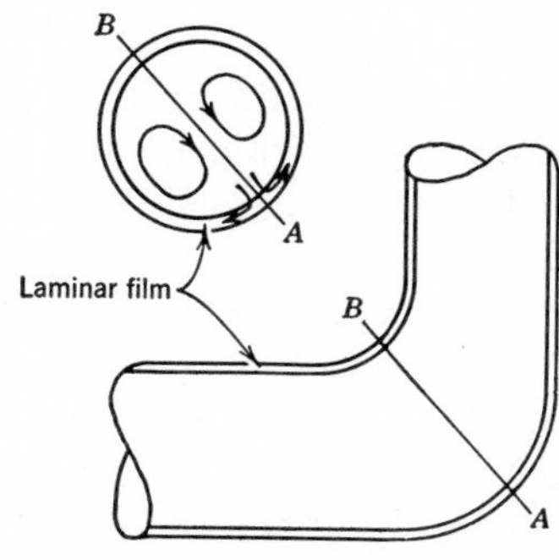

FIG. 109

pressure gradient due to the centrifugal forces of fluid particles as they move through the bend develops across the bend. Stability occurs in the ideal fluid when this pressure gradient brings about a balance between the centrifugal and centripetal forces on the fluid particles. In a real fluid this stability is disrupted by the velocity reduction toward the walls. The reduction of velocity at the outer part, A, of the bend reduces the centrifugal force of the particles moving near the wall, causing the pressure at the wall to be below that which would be maintained in an ideal fluid. However, the velocities of fluid particles toward the center of the bend are about the same as those of the ideal fluid, and the pressure gradient developed by their centrifugal forces is about the same. The "weakening" of the pressure gradient at the outer wall will cause a flow to be set up from the center of the pipe toward the wall which will develop into the twin eddy motion shown, and this secondary motion added to the main flow will cause a double spiral motion, the energy of which will be dissipated in heat as the motion is destroyed by viscous action. The energy of the secondary motion has been derived from the available useful fluid energy, and, as viscosity causes this energy to be dissipated, fluid energy is made unavailable in much the same way as in the eddies of a wake downstream from a solid object in the flow.

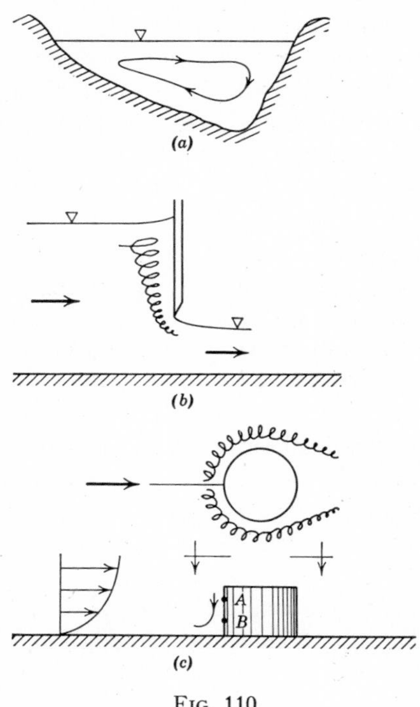

FIG. 110

Other but less easily definable secondary flows are shown in Fig. 110. Figure 110*a* shows a cross section through a river bend where the secondary motion serves to deposit material at the inside of the bend and to assist in scouring the outer side, thus producing the well-known meandering characteristic of natural streams. Figure 110*b* depicts the periodic vortex type of secondary flow to be found in the corner between a channel wall and sluice gate. In Fig. 110*c* is shown the horseshoe-shaped vortex which is produced by projections from a boundary surface. Its origin can be easily seen from the velocity profile along the wall which leads to the stagnation pressure at A being larger than that at B. This pressure difference maintains a downward secondary flow from A to B, thus inducing a vortex type of motion, the core of the vortex being swept downstream around the sides of the projection.

Because of the complex origin and geometry of secondary flows they have (so far) defied rigorous analysis. They are cited in this elementary text only so that the reader will become aware of them and make due allowance for their possible existence in analyzing new problems. A conservative attitude is to expect secondary flow phenomena wherever there are severe irregularities in boundary geometry.

REFERENCES

C. V. Drysdale (Ed.), *The Mechanical Properties of Fluids*, Blackie and Son, 1925.

L. Prandtl, "The Mechanics of Viscous Fluids," Part G of Vol. III of *Aerodynamic Theory* (W. F. Durand, Ed.), Julius Springer, Berlin, 1935.

B. A. Bakhmeteff, *The Mechanics of Turbulent Flow*, Princeton University Press, 1936.

S. Goldstein, *Modern Developments in Fluid Dynamics*, Vols. I and II, Oxford University Press, 1938.

H. Rouse, *Fluid Mechanics for Hydraulic Engineers*, McGraw-Hill Book Co., 1938.

H. F. P. Purday, *An Introduction to the Mechanics of Viscous Flow*, Dover Publications, Inc., 1949.

E. G. Richardson, *Dynamics of Real Fluids*, Edward Arnold and Co., 1950.

L. Prandtl, *Essentials of Fluid Dynamics*, Hafner Publishing Co., 1952.

G. K. Batchelor, *The Theory of Homogeneous Turbulence*, Cambridge University Press, 1953.

A. A. Townsend, *The Structure of Turbulent Shear Flow*, Cambridge University Press, 1956.

S.-I Pai, *Viscous Flow Theory—Laminar Flow*, D. Van Nostrand Co., 1956.

S.-I Pai, *Viscous Flow Theory—Turbulent Flow*, D. Van Nostrand Co., 1957.

J. O. Hinze, *Turbulence*, McGraw-Hill Book Co., 1959.

C. C. Lin (Ed.), *Laminar Flow and Transition to Turbulence*, Princeton University Press, 1959.

PROBLEMS

511. When 30 gpm of water flow in a 3-in. pipe line at 70°F, is the flow laminar or turbulent?

512. Glycerin flows in a 1-in. pipe at a mean velocity of 1 fps and temperature 80°F. Is the flow laminar or turbulent?

513. Carbon dioxide flows in a 2-in. pipe at a velocity of 5 fps, temperature 150°F, and pressure 55 psia. Is the flow laminar or turbulent?

514. What is the maximum flowrate of air which may occur at laminar condition in a 4-in. pipe at 30 psia and 100°F?

515. What is the smallest-diameter pipe line which may be used to carry 100 gpm of linseed oil at 80°F if the flow is to be laminar?

516. Derive an expression for pipe-line Reynolds number in terms of Q, d, and ν.

517. A fluid flows in a 3-in. pipe which discharges into a 6-in. line. What is the Reynolds number in the 6-in. pipe if that in the 3-in. pipe is 20,000?

518. If water (68°F) flows at constant depth (i.e., uniformly) in a wide open channel, below what flowrate q may the regime be expected to be laminar?

519. What is the maximum speed at which a spherical sand grain of diameter 0.01 in. may move through water (68°) and the flow regime be laminar?

520. In the laminar flow of an oil of viscosity 0.02 lb-sec/sqft the velocity at the center of a 12-in. pipe is 15 fps and the velocity distribution is parabolic. Calculate the shear stress at the pipe wall and within the fluid 3 in. from the pipe wall.

521. If the turbulent velocity profile in a pipe 2 ft in diameter may be approximated by $v = 10y^{\frac{1}{7}}$ (v fps, y ft) and the shearing stress in the fluid 6 in. from the pipe wall is 0.13 psf, calculate the eddy viscosity, mixing length, and turbulence constant at this point. Specific gravity of the fluid is 0.90.

522. The thickness of the laminar film (Fig. 93) is usually defined by the intersection of a laminar velocity profile (adjacent to the wall) with a turbulent velocity profile. If the former may be described by $v = c_1 y$ and the latter by $v = c_2 y^{\frac{1}{7}}$, derive an expression for the thickness of the film in terms of c_1 and c_2.

523. When oil (kinematic viscosity 0.001 sqft/sec, specific gravity 0.92) flows at a mean velocity of 5 fps through a 2-in. pipe line, the head lost in 100 ft of pipe is 18 ft. What will be the head loss when the velocity is increased to 10 fps?

524. If in the preceding problem the oil has kinematic viscosity 0.0001 sqft/sec, the head lost will be 37 ft if the pipe is smooth and 50 ft if the roughness is very great. What head losses will occur when the mean velocity is increased to 10 fps?

525. Calculate α, β, and the momentum flux for the velocity profiles of problem 251. Assume that the passage is 2 ft high, 1 ft wide normal to the paper, and that water is flowing.

526. Calculate α, β, and the momentum flux for the velocity profiles of problem 252. Assume that the passage is 2 ft in diameter and that water is flowing.

527. Calculate α and β for the velocity profile of problem 253.

528. Just downstream from the nozzle tip the velocity distribution is as shown. Calculate the flowrate past section 1, α, β, and the momentum flux. Assume water flowing.

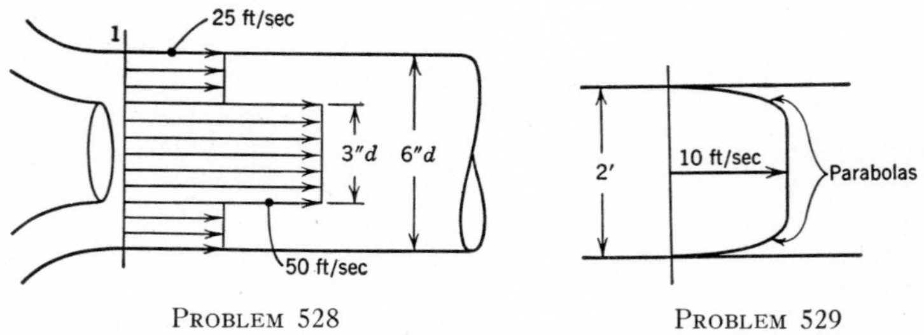

PROBLEM 528 PROBLEM 529

529. Calculate α and β for the flow in this two-dimensional passage if q is 16 cfs/ft.

530. A horizontal nozzle having a cylindrical tip of 3-in. diameter attached to a 6-in. water pipe discharges 1.77 cfs. In the pipe just upstream from the nozzle the pressure is 8.50 psi and α is 1.05. In the issuing jet α is 1.01. Calculate the conventional and exact head losses in the nozzle.

531. In a reach of cylindrical pipe in which air is flowing the mean velocity is observed to increase from 400 to 500 fps and the mean temperature to rise from 100°F to 120°F. How much heat is being added to each pound of air in this reach of pipe?

532. In a reach of cylindrical pipe wrapped with perfect insulation ($E_H = 0$) the pressure is observed to drop from 100 to 50 psia and the mean velocity to increase from 300 to 575 fps. Predict the temperatures in the pipe at each end of the reach if air is flowing in the pipe.

533. When water flows at a mean velocity of 10 fps in a 12-in. pipe, the head loss in 500 ft of pipe is 15 ft. Estimate the rise of mean temperature of the water if the pipe is wrapped with perfect insulation ($E_H = 0$). If the pipe is not insulated, how much heat must be extracted from the water to hold its temperature constant?

534. When fluid of specific weight 50 lb/cuft flows in a 6-in. pipe line, the frictional stress between fluid and pipe is 0.5 psf. Calculate the head lost per foot of pipe. If the flowrate is 2.0 cfs, how much power is lost per foot of pipe?

535. If the head lost in 100 ft of 3-in. pipe is 25 ft when a certain quantity of water flows therein, what is the total dragging force exerted by the water on this reach of pipe?

536. Air ($\gamma = 0.08$ lb/cuft) flows through a horizontal 1-ft by 2-ft rectangular duct at a rate of 200 lb/min. Find the mean shear stress at the wall of the duct if the pressure drop in a 1000 ft length is 0.07 psi. Compute the horsepower lost per foot of duct length.

537. If a zone of unestablished flow may be idealized to the extent shown and the centerline may be treated as a streamline in an ideal fluid, calculate the drag force exerted by the sidewalls (between sections 1 and 2) on the fluid if the flow is: (*a*) two-dimensional and 1 ft wide normal to the paper and (*b*) axisymmetric. The fluid flowing has specific gravity 0.90.

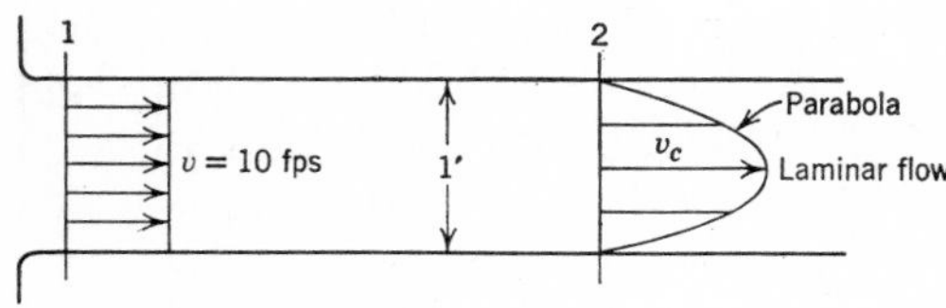

PROBLEM 537

chapter 8

Similitude and Dimensional Analysis

61. Similitude and Models. Near the latter part of the last century, models began to be used to study flow phenomena which could not be solved by analytical methods or by means of available experimental results. Over the era of modern engineering, the use of models and confidence in model studies have steadily increased: the aeronautical engineer obtains data from model tests in wind tunnels; the naval architect tests ship models in towing basins; the mechanical engineer tests models of turbines and pumps and predicts the performance of the full-scale machines from these tests; the civil engineer works with models of hydraulic structures and rivers to obtain more reliable solutions to his design problems. The justification for the use of models is an economic one—a model, being small, costs little compared to the *prototype* from which it is built, and its results may lead to savings of many times its cost; however, models are usually expensive when compared to the cost of analytical work and thus are not justified in problems where analytical methods will yield a reliable solution. In fields where calculations are made on the basis of simplified theory, a model adds a certainty to design which can never be obtained from calculations alone. Although the basic theory for the interpretation of model tests is quite simple, it is seldom possible to design and operate a model of a fluid phenomenon from theory alone; here the *art* of engineering must be practiced with experience, judgment, ingenuity, and patience if useful results are to be obtained, correctly interpreted, and prototype performance predicted therefrom.

Similitude of flow phenomena not only occurs between a prototype and its model but also may exist between various natural phenomena if certain laws of similarity are satisfied. Similarity thus becomes a means of correlating the apparently divergent results obtained from similar fluid phenomena and as such becomes a valuable tool of modern fluid mechanics; the application of the laws of similitude will be found to lead to more comprehensive solutions and, therefore, to a better understanding of fluid phenomena in general.

There are many types of similitude, all of which must be obtained if complete similarity is to exist between fluid phenomena.[1] The

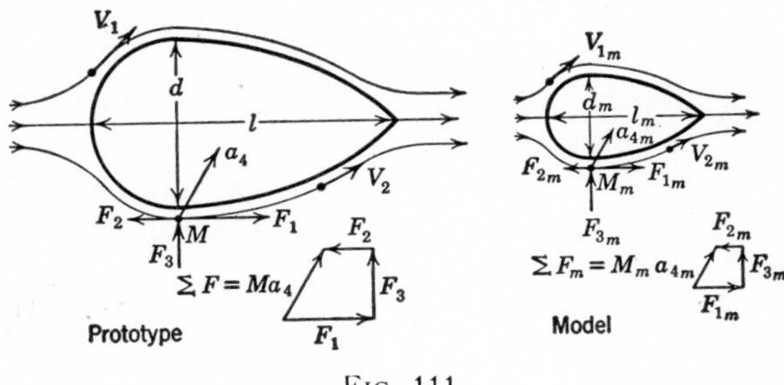

FIG. 111

first and simplest of these is the familiar geometrical similarity which states that the flowfields and boundary geometry of model and prototype have the same shape[2] and, therefore, that the ratios between corresponding lengths in model and prototype are the same. In the model and prototype of Fig. 111, for example,

$$\frac{d}{d_m} = \frac{l}{l_m}$$

Corollaries of geometric similarity are that corresponding areas vary with the squares of their linear dimensions,

$$\frac{A}{A_m} = \left(\frac{d}{d_m}\right)^2 = \left(\frac{l}{l_m}\right)^2$$

and that volumes vary with the cubes of their linear dimensions.

Consider now the flows past the geometrically similar objects of Fig. 111; if these are flowfields of the same shape, and if ratios of

[1] However, in some special cases effective and useful similarity is obtained without satisfying this condition.

[2] Departure from geometric similarity (resulting in a *distorted* model) is frequently used (for economic and physical reasons) in models of rivers, harbors, estuaries, etc.; however, most models are geometrically similar to their prototypes.

corresponding velocities and accelerations are the same throughout the flow,[3] the two flows are said to possess kinematic similarity.

In order to maintain geometric and kinematic similarity between flowfields, the forces acting on *corresponding* fluid masses must be related by ratios similar to those above; this similarity is known as *dynamic similarity*. With the forces shown acting on the corresponding fluid masses M and M_m of Fig. 111, the vector polygons may be drawn, and from the geometric similarity of these polygons and from Newton's second law (which, of course, is operative in both model and prototype)

$$\frac{F_1}{F_{1m}} = \frac{F_2}{F_{2m}} = \frac{F_3}{F_{3m}} = \frac{Ma_4}{M_m a_{4m}} \tag{166}$$

For dynamic similarity these force ratios must be maintained on all the corresponding fluid masses throughout the flowfields; thus it is evident that they can be governed only by relations between the dynamic and kinematic properties of the flows and by the physical properties of the fluids involved.

Referring again to Fig. 111, it is apparent that

$$\vec{F_1} + \vec{F_2} + \vec{F_3} = \overrightarrow{Ma_4}$$

and

$$\vec{F_{1m}} + \vec{F_{2m}} + \vec{F_{3m}} = \overrightarrow{M_m a_{4m}}$$

and from this it may be concluded that, if the ratios between three of the four corresponding terms in these equations are the same, the ratio between the corresponding fourth terms must be the same as that between the other three. Thus one of the ratios of equation 166 becomes unnecessary, and dynamic similarity is characterized by an equality of force ratios numbering one less that the forces.[4] When the first force ratio [5] is eliminated from equation 166, the equation may be rewritten in various ways. It is usually written in the form of simultaneous equations,

$$\frac{Ma_4}{F_2} = \frac{M_m a_{4m}}{F_{2m}} \tag{167}$$

$$\frac{Ma_4}{F_3} = \frac{M_m a_{4m}}{F_{3m}} \tag{168}$$

[3] For example, $V_1/V_{1m} = V_2/V_{2m}$ and $a_3/a_{3m} = a_4/a_{4m}$.

[4] The force, Ma, is the "inertia force."

[5] Any force ratio may be selected for elimination, depending upon the quantities which are desired in the equations.

The forces which *may* affect a flowfield are those of pressure, F_P; inertia, F_I; gravity, F_G; viscosity, F_V; elasticity, F_E; and surface tension, F_T. Since these forces are taken to be those on or of any fluid mass, they may be generalized by the following fundamental relationships:

$$F_P = (\Delta p)A = (\Delta p)l^2$$

$$F_I = Ma = \rho l^3(V^2/l) = \rho V^2 l^2$$

$$F_G = Mg = \rho l^3 g$$

$$F_V = \mu\,(dv/dy)A = \mu(V/l)l^2 = \mu V l$$

$$F_E = EA = El^2$$

$$F_T = \sigma l$$

To obtain dynamic similarity between two flowfields when all these forces act, all corresponding force ratios must be the same in model and prototype; thus dynamic similarity between two flowfields when all possible forces are acting may be expressed (after the pattern of equations 167 and 168) by the following five simultaneous equations:

$$\left(\frac{F_I}{F_P}\right)_p = \left(\frac{F_I}{F_P}\right)_m = \left(\frac{\rho V^2}{\Delta p}\right)_p = \left(\frac{\rho V^2}{\Delta p}\right)_m; \quad \mathbf{E}_p = \mathbf{E}_m \tag{169}$$

$$\left(\frac{F_I}{F_V}\right)_p = \left(\frac{F_I}{F_V}\right)_m = \left(\frac{Vl\rho}{\mu}\right)_p = \left(\frac{Vl\rho}{\mu}\right)_m; \quad \mathbf{R}_p = \mathbf{R}_m \tag{170}$$

$$\left(\frac{F_I}{F_G}\right)_p = \left(\frac{F_I}{F_G}\right)_m = \left(\frac{V^2}{lg}\right)_p = \left(\frac{V^2}{lg}\right)_m; \quad \mathbf{F}_p = \mathbf{F}_m \tag{171}$$

$$\left(\frac{F_I}{F_E}\right)_p = \left(\frac{F_I}{F_E}\right)_m = \left(\frac{\rho V^2}{E}\right)_p = \left(\frac{\rho V^2}{E}\right)_m; \quad \mathbf{M}_p = \mathbf{M}_m \tag{172}$$

$$\left(\frac{F_I}{F_T}\right)_p = \left(\frac{F_I}{F_T}\right)_m = \left(\frac{\rho l V^2}{\sigma}\right)_p = \left(\frac{\rho l V^2}{\sigma}\right)_m; \quad \mathbf{W}_p = \mathbf{W}_m \tag{173}$$

Defining the dimensionless numbers of dynamic similarity as follows:

$$\text{Euler number, } \mathbf{E} = V\sqrt{\rho/2\,\Delta p} \tag{174}$$

$$\text{Reynolds number, } \mathbf{R} = Vl/\nu \tag{175}$$

$$\text{Froude number, } \mathbf{F} = V/\sqrt{lg} \tag{176}$$

$$\text{Cauchy}^{6} \text{ number, } \mathbf{C} = \rho V^2/E \tag{177}$$

$$\text{Weber number, } \mathbf{W} = \rho l V^2/\sigma \tag{178}$$

[6] Since the sonic velocity $a = \sqrt{E/\rho}$ (Art. 4), the Mach number, **M**, and Cauchy number, **C**, are related by $\mathbf{C} = \mathbf{M}^2$.

it is apparent that the foregoing force-ratio equations may be written in terms of the dimensionless numbers as indicated above. Following the argument leading to equations 167 and 168, it will be noted that only four of these equations are independent; thus, if any four of them are simultaneously satisfied, dynamic similarity will be ensured if geometric similarity of boundaries and flowfield has also been provided.

Fortunately, in most engineering problems four simultaneous equations are not necessary, since some of the forces stated above (1) may not act, (2) may be of negligible magnitude, or (3) may oppose other forces in such a way that the effect of both is reduced. In each new problem of similitude a good understanding of fluid phenomena is necessary to determine how the problem may be satisfactorily simplified by the elimination of the irrelevant, negligible, or compensating forces. The reasoning involved in such analysis is best illustrated by citing certain simple and recurring engineering examples.

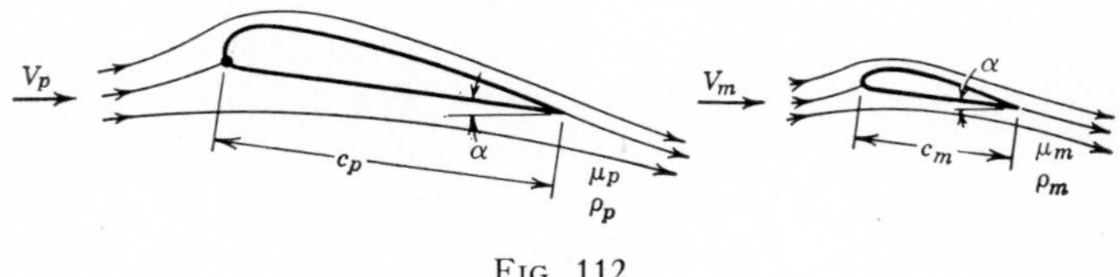

FIG. 112

In the classical low-speed submerged body problem typified by the conventional airfoil of Fig. 112 there are no surface tension phenomena, negligible compressibility (elastic) effects, and gravity does not affect the flowfield. Thus three of the four equations are not relevant to the problem, and dynamic similarity is obtained between model and prototype when

$$(Vl/\nu)_p = \mathbf{R}_p = \mathbf{R}_m = (Vl/\nu)_m \tag{170}$$

providing that model and prototype are geometrically similar and are similarly oriented to their oncoming flows. Since the equation places no restriction on the fluids of the model and prototype, the latter could move through air and the former be tested in water; if the Reynolds numbers of model and prototype could be made the same, dynamic similitude would result. If for practical reasons the same fluid is used in model and prototype, the product (Vl) must be the same in both; this means that the velocities around the model will be *larger* than the corresponding ones around the prototype. In aeronautical research the model is frequently tested with compressed air in a *variable density* (or *pressure*) *wind tunnel;* here $\nu_m < \nu_p$ and large velocities past the model are not required. Once equality of Reynolds numbers is obtained in model and prototype, it follows that the ratio of any corresponding forces (such as lift or drag) will be equal to the ratio of any other relevant corresponding forces. Thus (for drag force)

$$(D/\rho V^2 l^2)_p = (D/\rho V^2 l^2)_m \tag{179}$$

from which the drag of the prototype may be predicted directly from drag and velocity measurements in the model; no corrections for

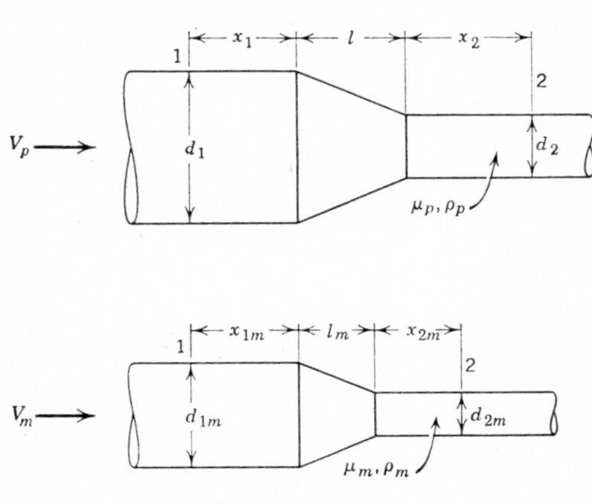

FIG. 113

"scale effect" are needed if the Reynolds numbers are the same in model and prototype.

Much of the foregoing reasoning may also be applied to the flow of incompressible fluids through closed passages. Consider, for example, the flows through prototype and model of the contraction of Fig. 113. For geometric similarity $(d_2/d_1)_p = (d_2/d_1)_m$, $(l/d_1)_p = (l/d_1)_m$, $(x_1/d_1)_p = (x_1/d_1)_m$, $(x_2/d_2)_p = (x_2/d_2)_m$; and the roughness pattern of the two passages must be similar in every detail. Surface tension and elastic effects are nonexistent and gravity does not affect the flowfields. Accordingly dynamic similarity results when

$$(Vd_1/\nu)_p = \mathbf{R}_p = \mathbf{R}_m = (Vd_1/\nu)_m \tag{170}$$

from which it follows that

$$\left(\frac{p_1 - p_2}{\rho V^2}\right)_p = \left(\frac{p_1 - p_2}{\rho V^2}\right)_m \tag{169}$$

allowing (for example) prediction of prototype pressure drop $(p_1 - p_2)$ from model measurements. Here again it is immaterial whether the fluids are the same, dynamic similarity being ensured by the equality[7] of the Reynolds numbers in model and prototype.

Another example of wide engineering interest is the modeling of the flowfield about an object (such as the ship of Fig. 114) moving on the surface of a liquid. Here geometric similarity is obtained by a carefully dimensioned model suitably weighted so that $(d/l)_p = (d/l)_m$. Compressibility of the liquid is of no consequence in such problems, and surface tension may also be ignored if the model is not

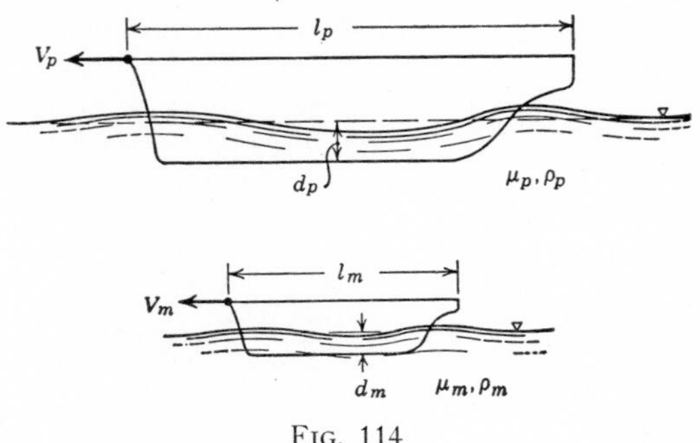

FIG. 114

too small. The drag of the ship will be associated with the energy dissipated in the surface wave pattern (a gravitational action) and in the frictional action of liquid on hull. *If frictional effects are assumed to be negligible,*[8] dynamic similitude is characterized by

$$(V/\sqrt{lg})_p = \mathbf{F}_p = \mathbf{F}_m = (V/\sqrt{lg})_m \qquad (171)$$

From this it follows (as for the submerged object of Fig. 112) that

$$(D/\rho V^2 l^2)_p = (D/\rho V^2 l^2)_m \qquad (179)$$

which means that, if the model is tested with $\mathbf{F}_p = \mathbf{F}_m$, and the drag measured, the drag of the prototype may be predicted for the corresponding speed; the latter is (from equation 171) $V_m\sqrt{l_p/l_m}$.

For ship hulls of good design the contribution of wave pattern and frictional action to the drag are of the same order, and neither can be ignored. Here the phenomena will be associated with both the

[7] Equality of the Reynolds numbers will not produce dynamic similarity in this example unless established flow exists upstream of the contraction; this has been assumed.

[8] As they could be for an unstreamlined object where most of the resistance is associated with the wave pattern.

Froude and the Reynolds numbers, requiring for dynamic similitude the two simultaneous equations

$$(V/\sqrt{gl})_p = \mathbf{F}_p = \mathbf{F}_m = (V/\sqrt{gl})_m \qquad (171)$$

$$(Vl/\nu)_p = \mathbf{R}_p = \mathbf{R}_m = (Vl/\nu)_m \qquad (170)$$

Taking g the same for model and prototype and solving these equations (by eliminating V) yields $\nu_p/\nu_m = (l_p/l_m)^{\frac{3}{2}}$, indicating that a relation between the viscosities of the liquids is required once the model scale is selected. This means (1) that a liquid of appropriate viscosity must be found for the model test, or (2) if the same liquid is used for model and prototype, the model must be as large as the prototype! Since liquids of appropriate viscosity may not exist and full-scale models are obviously impractical, the engineer is forced to choose between the two equations, since both cannot be satisfied simultaneously with the same liquid for model and prototype. This is done by operating the model so that $\mathbf{F}_p = \mathbf{F}_m$ (resulting in $\mathbf{R}_p \gg \mathbf{R}_m$) and then correcting[9] the test results by experimental data dependent on Reynolds number. William Froude originated this technique for ship model testing in England around 1870, but the same equations and principles apply to any flowfield controlled by the combined action of gravity and viscous action; models of rivers, harbors, hydraulic structures, and open-flow problems in general are good examples. However, such models are considerably more difficult to operate and interpret than ship models because of the less well-defined frictional resistance caused by variations of surface roughness and complex boundary geometry.

An example of similitude in compressible fluid flow is that of the projectile of Fig. 115. Here gravity and surface tension do not affect the flowfield, and similitude will result from the actions of resistance and elasticity (compressibility) characterized by

$$(Vl/\nu)_p = \mathbf{R}_p = \mathbf{R}_m = (Vl/\nu)_m \qquad (170)$$

$$(V/a)_p = \mathbf{M}_p = \mathbf{M}_m = (V/a)_m \qquad (172)$$

which may be solved to yield $l_p/l_m = (\nu_p/\nu_m)(a_m/a_p)$, showing that a relation must exist between model scale and the viscosities and sonic speeds in the gases used in model and prototype if dynamic similarity is to be complete. In this situation (unlike the analogous

[9] This is known as correcting for *scale effect* and is a correction necessitated by incomplete similitude; there would be no scale effect if equations 170 and 171 could both be satisfied.

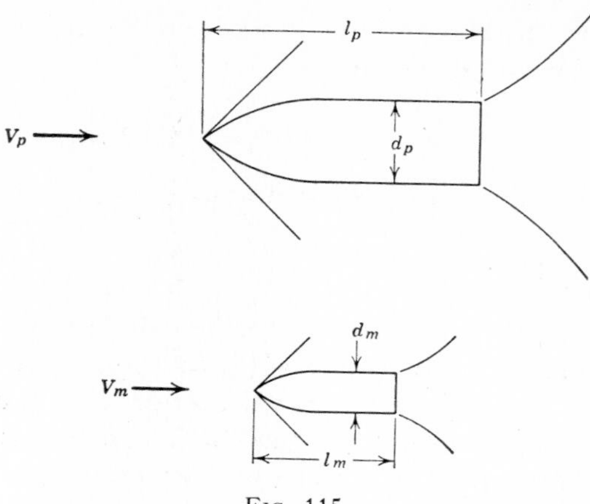

FIG. 115

one for the ship model), gases are available which will allow the equation to be satisfied.

When prototype cavitation is to be modeled, the foregoing equations of similitude are inadequate, since they do not include vapor pressure, the attainment of which is the unique feature of the cavitation.[10] Consider the torpedo prototype and model of Fig. 116, the

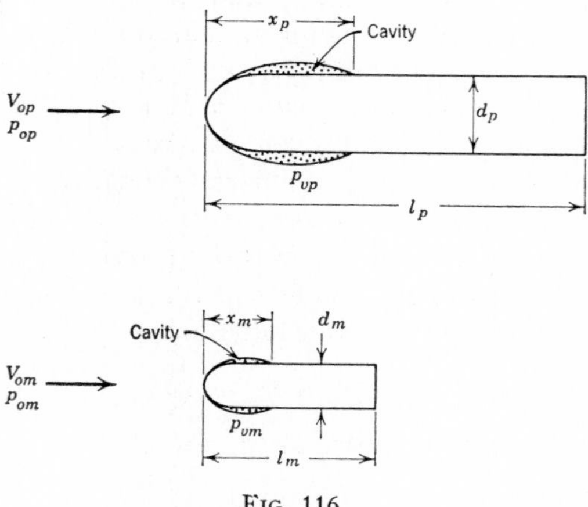

FIG. 116

model fixed in a water tunnel where p_o and V_o can be controlled. Geometric similarity and the same orientation to the oncoming flow are assumed, and complete dynamic similitude is desired. Here gravity will have small effect on the flowfield, compressibility of the fluid

[10] See Appendix VII.

will be insignificant, and surface tension may be neglected.[11] Accordingly, dynamic similitude is obtained when

$$(V_o l/\nu)_p = \mathbf{R}_p = \mathbf{R}_m = (V_o l/\nu)_m \tag{170}$$

$$\left(\frac{p_o - p_v}{\rho V_o^2}\right)_p = \left(\frac{p_o - p_v}{\rho V_o^2}\right)_m \tag{180}$$

and when these equations are satisfied the cavity shapes in model and prototype will be the same, i.e., $(x/l)_p = (x/l)_m$. The second of the two simultaneous equations represents an equality of special Euler numbers (known as the *cavitation number*) containing the (absolute) pressures, p_o and p_v, peculiar to cavitation. If these equations can be simultaneously satisfied by the use of appropriate liquid or adjustment of tunnel velocities and pressures or both, then (as for the submerged body without cavitation)

$$(D/\rho V_o^2 l^2)_p = (D/\rho V_o^2 l^2)_m \tag{179}$$

and

$$(\Delta p/\rho V_o^2)_p = (\Delta p/\rho V_o^2)_m \tag{169}$$

in which Δp in model and prototype represents the pressure changes between any two *corresponding* points in the flowfields of model or prototype. Here (as for the ship model) it is virtually impossible to satisfy both of the simultaneous equations with essentially the same liquid in model and prototype. Again the engineer must select and satisfy the more important of the two equations and correct the test results to allow for the unsatisfied equation. In this problem it is clear that the existence and extent of the cavitation zone dominate the flowfield and frictional aspects are of comparatively small importance; accordingly the tests would be carried out with the cavitation numbers the same in model and prototype and empirical adjustments made for the frictional phenomena.

It is a rare occurrence in engineering problems when pure theory alone leads to a complete answer. This is particularly true in the construction, operation, and interpretation of models where compromises of the type above are almost invariably necessary; there is no substitute for experience and judgment in effecting such compromises. The foregoing treatment of similitude is intended as an introduction to the theory and to the numerous practical applications cited in the literature at the end of the chapter.

[11] Surface tension may have critical influence in such problems if the cavitation is incipient; the gas content of the liquid is also important. Both of these points are ignored here to simplify the problem.

Illustrative Problems

Water (32°F) flows in a 3-in. horizontal pipe line at a mean velocity of 10 fps. The pressure drop in 30 ft of this pipe is 2.0 psi. With what velocity must benzene (68°F) flow in a geometrically similar 1-in. pipe line for the flows to be dynamically similar, and what pressure drop is to be expected in 10 ft of the 1-in. pipe?

Solution.

$$\frac{10 \times (3/12) \times 1.935}{0.0000375} = \frac{V \times (1/12) \times 1.935 \times 0.88}{0.0000137}; \quad V = 12.45 \text{ fps} \tag{170}$$

$$2.0/1.00 \times (10)^2 = \Delta p/0.88 \times (12.45)^2; \quad \Delta p = 2.73 \text{ psi} \tag{169}$$

A surface vessel of 400-ft length is to be tested by a model 10 ft long. If the vessel travels at 30 fps, at what speed must the model be towed for dynamic similitude between model and prototype? If the drag of the model is 2 lb, what prototype drag is to be expected?

Solution. Determine the model speed from equality of the Froude numbers.

$$(30)^2/400 \times 32.2 = V^2/10 \times 32.2; \quad V = 4.73 \text{ fps} \tag{171}$$

Assuming the same liquid for model and prototype and neglecting frictional effects,

$$2/(4.73)^2 \times (10)^2 = D/(30)^2 \times (400)^2; \quad D = 128{,}000 \text{ lb} \tag{179}$$

With no information on hull form, frictional effects cannot be included in this problem; however, if they were included the predicted drag of the prototype would be somewhat *less* than 128,000 lb. See the illustrative problem of Art. 117.

A short smooth overflow structure passes a flowrate of 20,000 cfs. What flowrate should be used with a 1:15 model of this structure to obtain dynamic similitude if friction may be neglected?

Solution. Because of the predominance of gravitational action, similitude will result when equation 171 is satisfied. Writing V as Q/A or Q/l^2, the equality of Froude numbers may be expressed as

$$(Q/l^2\sqrt{gl})_p = (Q/l^2\sqrt{gl})_m \tag{171}$$

whence

$$Q_m = 20{,}000 \times (1/15)^{\frac{5}{2}} = 23 \text{ cfs}$$

A 1:10 model of a projectile traveling at a velocity of 3000 fps at 15,000 altitude in the ICAO Standard Atmosphere (Appendix IV) is to be tested by firing it in a tank of carbon dioxide at 68°F. Calculate the required velocity of the model and the pressure required in the tank for dynamic similitude between model and prototype.

Solution. For dynamic similarity $\mathbf{R}_p = \mathbf{R}_m$ and $\mathbf{M}_p = \mathbf{M}_m$. Using the second equation, $V_m = 3000\ (a_m/a_p)$. Calculating a_m and a_p from equation 11, $a_m =$ 870 fps and $a_p = 1057$ fps, yielding $V_m = 2470$ fps. Now, using the first equation (with numerical values from Fig. 6, Table II, and Appendix IV),

$$3000 \times 10 \times 0.001495/0.033 = 2470 \times 1 \times \rho_m/0.031; \quad \rho_m = 0.017 \text{ slug/cuft} \tag{170}$$

Finally,

$$32.2 \times 0.017 = p_m \times 144/34.9 \times 528; \quad p_m = 70 \text{ psia} \tag{3}$$

62. Dimensional Analysis. Another useful tool of modern fluid mechanics, and one closely related to the principle of similitude, is the field of mathematics known as *dimensional analysis*—the mathematics of the dimensions of quantities. Although it can be successfully argued that similitude and dimensional analysis are virtually identical since they imply the same things and frequently lead to the same results, their methods are sufficiently different to justify developing them as separate topics. After developing a certain facility with both, the student will become familiar with their interrelation and will learn to think in terms of both topics when attacking new problems.

The methods of dimensional analysis are built upon Fourier's *principle of dimensional homogeneity* (1822), which states that an equation expressing a physical relationship between quantities must be dimensionally homogeneous; i.e., the dimensions of each side of the equation must be the same. This principle has already been utilized in Chapter 1 in obtaining the dimensions of density and kinematic viscosity, and it has been recommended as a valuable means of checking engineering calculations. Further investigation of the principle will reveal that it affords a means of ascertaining the forms of physical equations from knowledge of relevant variables and their dimensions. Although dimensional manipulations cannot be expected to produce analytical solutions to physical problems, dimensional analysis proves a powerful tool in formulating problems which defy analytical solution and must be solved experimentally. Here dimensional analysis comes into its own by pointing the way toward a maximum of information from a minimum of experiment. It accomplishes this by the formation of dimensionless groups, some of which are identical with the force ratios developed with the principle of similitude.

Before examining the methods of dimensional analysis, recall that there are two different systems by which the dimensions of physical quantities may be expressed. These systems are those of force-length-time and mass-length-time. The force-length-time system, generally preferred by engineers, becomes the familiar *foot-pound-second* system when expressed in English dimensions; the mass-length-time system in English dimensions becomes the *foot-slug-second* system widely used in engineering mechanics. The foot-slug-second system is generally preferred in dimensional analysis, and, since the student is familiar with the foot-pound-second system, the use of the foot-slug-second system will help to develop versatility in the use of dimensions.

A summary of the fundamental quantities of fluid mechanics and their dimensions in the various systems is given in Appendix I, the conventional system of capital letters being followed to indicate the dimensions of quantities. In problems of mechanics, the basic relation between the force-length-time and mass-length-time systems of dimensions is given by the Newtonian law, force or weight = mass × acceleration; and, therefore, dimensionally,

$$F = ML/T^2$$

through which the dimensions of any quantity may be converted from one system to the other.

To illustrate the mathematical steps in a simple dimensional problem, consider the familiar equation of fluid statics,

$$p = \gamma h$$

but assume that the dimensions of γ and h are known and those of p unknown. The dimensions of p can be only some combination of M, L, and T, and this combination may be discovered by writing the equation dimensionally as

$$(\text{Dimensions of } p) = (\text{Dimensions of } \gamma) \times (\text{Dimensions of } h)$$

or

$$M^a L^b T^c = \left(\frac{M}{L^2 T^2}\right) \times (L)$$

in which a, b, and c are unknowns. The principle of dimensional homogeneity being applied, the exponent of *each* of the fundamental dimensions is the same on each side of the equation, giving

$$a = 1, \quad b = -2 + 1 = -1, \quad c = -2$$

whence

$$(\text{Dimensions of } p) = ML^{-1}T^{-2} = \frac{M}{LT^2}$$

It is obvious, of course, that this result might have been obtained more directly by cancellation of L on the right-hand side of the equation, for this has been, and will continue to be, the usual method of obtaining the unknown dimensions of a quantity. It is of utmost importance, however, to note the mathematical steps which lie unrevealed in this hasty cancellation, if the techniques of dimensional analysis are to be understood.

The foregoing methods may now be used in quite another and more important way. To illustrate by another familiar example, suppose that it is known that the power, P, which can be derived from a hydraulic turbine is dependent upon the rate of flow through the machine, Q, the specific weight of the fluid flowing, γ, and the unit mechanical energy, E, which is given up by every pound of fluid as it passes through the machine. Suppose that the relation between these four variables is unknown but it is known that these are the only variables involved in the problem.[12] With this meager knowledge the following mathematical statement may be made:

$$P = f(Q, \gamma, E)$$

From the principle of dimensional homogeneity it is apparent that the quantities involved cannot be added or subtracted since their dimensions are different. This principle limits the equation to a combination of products of powers of the quantities involved, which may be expressed in the general form

$$P = CQ^a\gamma^bE^c$$

in which C is a dimensionless constant which may exist in the equation but cannot, of course, be obtained by dimensional methods. Writing the equation dimensionally,

$$\frac{ML^2}{T^3} = \left(\frac{L^3}{T}\right)^a\left(\frac{M}{L^2T^2}\right)^b(L)^c$$

the following equations in the exponents of the dimensions are obtained:

$$M: \quad 1 = b$$

$$L: \quad 2 = 3a - 2b + c$$

$$T: \quad -3 = -a - 2b$$

whence

$$a = 1, \quad b = 1, \quad c = 1$$

[12] Note that experience and analytical ability in determining the relevant variables are necessary before the methods of dimensional analysis can be successfully applied.

and resubstitution of these values in the equation above for P gives

$$P = CQ\gamma E$$

The form of the equation (confirmed by equation 68) has, therefore, been derived, without physical analysis, solely from consideration of the dimensions of the quantities which were known to enter the problem. The magnitude of C may be obtained either (1) from a physical analysis of the problem or (2) from experimental measurements of P, Q, γ, and E.

From the foregoing problem it appears that in dimensional analysis (of problems in mechanics) only three equations can be written since there are only three fundamental dimensions: M, L, and T. This fact limits the completeness with which a problem with more than three unknowns may be solved, but it does not limit the utility of dimensional analysis in obtaining the form of the terms of the equation. This point may be fruitfully illustrated by re-examining the ship model problem previously treated (Art. 61) by the principle of similitude. Consider the ship of Fig. 114 having a certain *shape* [13] and draft, d. The drag of the ship will depend [14] upon the size of the ship (characterized by its length), the viscosity and density of the fluid, the velocity of the ship, and acceleration due to gravity (which dominates the surface wave pattern). Thus, with no further knowledge than that of the *independent* variables involved in the problem, an equation may be written:

$$D = f(l, \rho, \mu, V, g)$$

which for dimensional reasons must have the form [15]

$$D = Cl^a \rho^b \mu^c V^d g^e$$

and the equation of the dimensions of the terms is

$$\frac{ML}{T^2} = (L)^a \left(\frac{M}{L^3}\right)^b \left(\frac{M}{LT}\right)^c \left(\frac{L}{T}\right)^d \left(\frac{L}{T^2}\right)^e$$

resulting in the three equations of the exponents of M, L, and T:

[13] *Shape* is emphasized here so that it cannot be confused with *size*. Size of the ship is *not* fixed by *algebraic* dimensions such as l, d, etc. With shape constant and size variable, a series of geometrically similar objects is implied.

[14] Selection of the relevant variables is the crucial first step in any problem of dimensional analysis, and obviously this cannot be done without some experience and "feel" for the problem; the mathematical processes of dimensional analysis cannot be expected to overcome the selection of irrelevant variables!

[15] More rigorously, this should be expressed as a series.

$$M: \quad 1 = b + c$$

$$L: \quad 1 = a - 3b - c + d + e$$

$$T: \quad -2 = -c - d - 2e$$

which, when solved [16] in terms of c and e, yield

$$b = 1 - c, \quad d = 2 - c - 2e, \quad a = 2 + e - c$$

Substituting these values in the second equation of this analysis,

$$D = Cl^{2+e-c} \rho^{1-c} \mu^c V^{2-c-2e} g^e$$

and collecting terms gives

$$D = C(V^2/lg)^{-e} (Vl\rho/\mu)^{-c} \rho l^2 V^2$$

But

$$\mathbf{F} = V/\sqrt{lg} \quad \text{and} \quad \mathbf{R} = Vl\rho/\mu$$

allowing the equation to be written in the more general form [17]

$$D = f'(\mathbf{F}, \mathbf{R}) \rho l^2 V^2$$

or

$$D/\rho l^2 V^2 = f'(\mathbf{F}, \mathbf{R})$$

showing without experiment, but from dimensional analysis alone, that $D/\rho l^2 V^2$ depends only upon the Froude and Reynolds numbers. This sort of result is the main objective of dimensional analysis in engineering problems and may always be obtained by application of the technique (orginated by Lord Rayleigh) used here. Although it gives no clue to the functional relationship between $D/\rho l^2 V^2$, $\mathbf{F}$, and $\mathbf{R}$, it has arranged the numerous original variables into a relation between a smaller number of dimensionless groups of variables and has thus indicated how test results should be processed for concise presentation.

The principles of similitude have shown that the flowfield about a prototype surface ship is dynamically similar to that of its model if (with geometric similarity) $\mathbf{F}_p = \mathbf{F}_m$ and $\mathbf{R}_p = \mathbf{R}_m$—and that satisfaction of these conditions leads to $(D/\rho l^2 V^2)_p = (D/\rho l^2 V^2)_m$. The

[16] Nine other equally valid solutions are possible in terms of: a, b; a, c,; a, d; a, e; b, c; b, d; b, e; c, d; d, e. However, many of these are inconvenient for comparison with the results of similitude.

[17] C, e, and c of the preceding equation are not necessarily constants. If a more rigorous development (see footnote 15) had been followed, a series of similar terms would have appeared on the right-hand side of the equation and the general form of the function would have been more rigorously proved.

result of dimensional analysis is the same since it has demonstrated the existence of a unique functional relationship between $D/\rho l^2 V^2$, **F**, and **R** for a ship of one shape. Thus (whatever this relationship) for the same **F** and same **R** in model and prototype, $(D/\rho l^2 V^2)_p = (D/\rho l^2 V^2)_m$.

The Rayleigh method of dimensional analysis was improved upon by Buckingham[18] with a broad generalization known as the Π-theorem. Buckingham showed (the details cannot be repeated here) that, if n variables (such as D, l, ρ, μ, V, g of the preceding example) were functions of each other, k equations (three for the preceding example) of their exponents could be written. Dimensional analysis would assemble the variables into $(n - k)$ dimensionless groups ($D/\rho l^2 V^2$, **F**, and **R** of the preceding example) which were functionally related. Buckingham designated these dimensionless groups by the Greek (capital) letter Π. The Π-theorem offers considerable advantage over the Rayleigh approach in that it shows in advance of the analysis how many groups are to be expected and allows the engineer more flexibility in formulating them. Attacking the foregoing problem once again will illustrate the method. Transferring all terms to the left-hand side of the equation,

$$f(D, l, \rho, \mu, V, g) = 0$$

the number of Π's to be expected is $n - k = 3$, so the relation between dimensionless groups will be expressed as

$$f'(\Pi_1, \Pi_2, \Pi_3) = 0$$

Taking D as the *dependent* variable (dependent on the other five, which are to be adjusted independently), and assuming that it is desired to combine D with ρ, V, and l, Π_1 will be a dimensionless combination of D, ρ, V, l. Π_2 and Π_3 must exclude D but between them contain the independent variables, so Π_2 may be taken as the dimensionless combination of V, l, ρ, and μ and Π_3 that of V, l, ρ, g. Thus it remains only to determine the detailed form of these dimensionless groups. Using Π_1 as an example to illustrate the method,

$$\Pi_1 = D^a \rho^b V^c l^d$$

Writing the equation dimensionally,

$$M^o L^o T^o = \left(\frac{ML}{T^2}\right)^a \left(\frac{M}{L^3}\right)^b \left(\frac{L}{T}\right)^c (L)^d$$

[18] See reference at end of this chapter.

The following equations in the exponents of the dimensions are obtained.

$$M: \quad 0 = a + b$$

$$L: \quad 0 = a - 3b + c + d$$

$$T: \quad 0 = 2a + c$$

Solving these equations in terms of a, b, c, or d (a will be used here),

$$b = -a, \quad c = -2a, \quad d = -2a$$

and substituting them in the equation above for Π_1,

$$\Pi_1 = (D/\rho l^2 V^2)^a$$

Since a dimensionless group raised to a power is of no more significance than the group itself, the exponent may be taken as any convenient number other than zero, and Π_1 as $D/\rho V^2 l^2$. Similarly Π_2 is obtained as $Vl\rho/\mu$ and Π_3 as $V/\sqrt{lg}$, giving

$$f'(D/\rho l^2 V^2, \mathbf{R}, \mathbf{F}) = 0$$

as before. Use of the Π-theorem in dimensional analysis is thus seen to be highly efficient in formulating dimensionless groups which may be easily interpreted in terms of those of geometric, kinematic, and dynamic similitude. An amateur using the Rayleigh method might have had to make ten [19] solutions of the problem to accomplish the same result.

REFERENCES

Lord Rayleigh, "Stability of Flow of Fluids," and "Investigations in Capillarity," *Phil. Mag.*, Vol. 34, No. 59, 1892.

E. Buckingham, "Model Experiments and the Forms of Empirical Equations," *Trans. A.S.M.E.*, Vol. 37, p. 263, 1915.

P. W. Bridgman, *Dimensional Analysis*, Yale University Press, 1922.

A. H. Gibson, "The Principle of Dynamic Similarity with Special Reference to Model Experiments," *Engineering*, Vol. 117, pp. 325, 357, 391, 422, 1924.

J. R. Freeman, *Hydraulic Laboratory Practice* (J. R. Freeman, Ed.), p. 775, A.S.M.E., 1929.

A. C. Chick, "Dimensional Analysis," *Hydraulic Laboratory Practice* (J. R. Freeman, Ed.), p. 782, A.S.M.E., 1929.

K. C. Reynolds, "Notes on the Laws of Hydraulic Similitude as Applied to Experiments with Models," *Hydraulic Laboratory Practice* (J. R. Freeman, Ed.), p. 759, A.S.M.E., 1929.

A. C. Chick, "Dimensional Analysis," *Hydraulic Laboratory Practice* (J. R. Freeman, Ed.), p. 796, A.S.M.E., 1929.

O. G. Tietjens, "The Use of Models in Aerodynamics and Hydrodynamics," *Trans. A.S.M.E.*, Vol. 54, p. 225, 1932.

[19] See footnote 16.

"Hydraulic Models," *A.S.C.E. Manual of Engineering Practice*, No. 25, A.S.C.E., 1942.

J. Allen, *Scale Models in Hydraulic Engineering*, Longmans, Green and Co., 1947.

G. Murphy, *Similitude in Engineering*, The Ronald Press Co., 1950.

J. E. Warnock, "Hydraulic Similitude," Chapter II of *Engineering Hydraulics* (H. Rouse, Ed.), John Wiley and Sons, 1950.

H. L. Langhaar, *Dimensional Analysis and Theory of Models*, John Wiley and Sons, 1951.

H. E. Huntley, *Dimensional Analysis*, Rinehart and Co., 1951.

G. H. Hickox, "Hydraulic Models," Chapter 24 of *Handbook of Applied Hydraulics* (C. V. Davis, Ed.), 2nd Edition, McGraw-Hill Book Co., 1952.

W. J. Duncan, *Physical Similarity and Dimensional Analysis*, Edward Arnold and Co., 1953.

C. M. Focken, *Dimensional Methods and Their Applications*, Edward Arnold and Co., 1953.

L. I. Sedov, *Similarity and Dimensional Methods in Mechanics*, Academic Press, 1959.

D. C. Ipsen, *Units, Dimensions, and Dimensionless Numbers*, McGraw-Hill Book Co., 1960.

PROBLEMS

538. An airplane wing of 10-ft chord length moves through still air at 59°F and 14.7 psia at a speed of 200 mph. A 1:20 scale model of this wing is placed in a wind tunnel, and dynamic similarity between model and prototype is desired. (*a*) What velocity is necessary in a tunnel where the air has the same pressure and temperature as that in flight? (*b*) What velocity is necessary in a variable-density wind tunnel where pressure is 200 psia and temperature is 59°F? (*c*) At what speed must the model move through water (60°F) for dynamic similarity?

539. In a water (68°F) tunnel test the velocity approaching the model is 80 fps. The pressure head difference between this point and a point on the object is 200 ft of water. Calculate the corresponding pressure difference (psi) on a 12:1 prototype in an air stream (14.7 psia and 59°F) if model and prototype are tested under conditions of dynamic similarity.

540. A flat plate 5 ft long and 1 ft wide is towed at 10 fps in a towing basin containing water at 68°F, and the drag force is observed to be 3 lb. Calculate the dimensions of a similar plate which will yield dynamically similar conditions in an air stream (14.7 psia and 59°F) having a velocity of 60 fps. What drag force may be expected on this plate?

541. It is desired to obtain dynamic similarity between 2 cfs of water at 50°F flowing in a 6-in. pipe and linseed oil flowing at a velocity of 30 fps at 90°F. What size of pipe is necessary for the linseed oil?

542. When castor oil flows at 60°F in a 2-in. horizontal pipe line at 10 fps, a pressure drop of 480 psi occurs in 200 ft of pipe. Calculate the pressure drop in the corresponding length of 1-in. pipe when linseed oil at 100°F flows therein at the same Reynolds number.

543. A flowrate of 7.86 cfs of water (68°F) discharges from a 12-in. pipe through a 6-in. nozzle into the atmosphere. The axial force component exerted by water on nozzle is 683 lb. If frictional effects may be ignored, what corresponding force will be exerted on a 4:1 prototype of nozzle and pipe discharging 40 cfs of air (14.7 psia and 59°F) to the atmosphere? If frictional effects are included the axial force component is 800 lb. What flowrate of air is then required for dynamic similarity? What is the corresponding force on the nozzle discharging air?

544. The pressure drop in a certain length of 12-in. horizontal water line is 10 psi when the mean velocity is 15 fps and the water temperature 68°F. If a 1:6 model of this pipe line using air as the working fluid is to produce a pressure drop of 0.8 psi in the corresponding length when the mean velocity is 100 fps, calculate the air pressure and temperature required for dynamic similarity between model and prototype.

545. A model of this disk rotating in a casing is to be constructed with air as the working fluid. What should be the corresponding dimensions and speed of the model for the torques of model and prototype to be the same? The air is at 59°F and 5.0 psia.

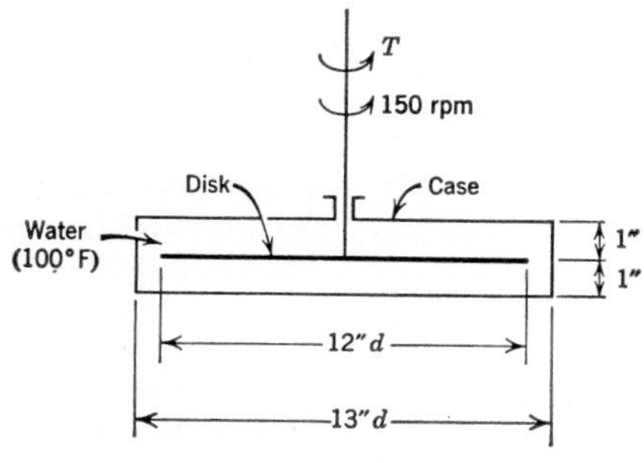

PROBLEM 545

546. A ship 200 ft long is to be tested by a 1:50 scale model. If the ship is to travel at 30 mph, at what speed must the model be towed to obtain dynamic similarity (neglecting friction) with its prototype?

547. A ship model 3 ft long (with negligible skin friction) is tested in a towing basin at a speed of 2 fps. To what ship velocity does this correspond if the ship is 200 ft long? A force of 1 lb is required to tow the model; what propulsive force does this represent in the prototype?

548. A seaplane is to take off at 80 mph. If the maximum speed available for testing its model is 15 fps, what is the largest model scale which can be used?

549. A ship 400 ft long moves through fresh water at 60°F at 20 mph. A 1:100 model of this ship is to be tested in a towing basin containing a liquid of specific gravity 0.92. What viscosity must this liquid have for both Reynolds' and Froude's laws to be satisfied? At what velocity must the model be towed? What propulsive force on the ship corresponds to a towing force of 2 lb in the model?

550. A perfect fluid discharges from a orifice under a static head. For this orifice and a geometrically similar model of the same, what are the ratios between velocities, flowrates, and jet horsepowers in model and prototype in terms of the model scale?

551. An overflow structure 1600 ft long is designed to pass a flood flow of 120,000 cfs. A 1:20 model of the *cross section* of the structure is built in a laboratory channel 1 ft wide. Calculate the required laboratory flowrate if the actions of viscosity and surface tension may be neglected. When the model is tested at this flowrate the pressure at a point on the model is observed to be 2-in. of mercury vacuum; how should this be interpreted for the prototype?

552. A hydraulic jump from 2 ft to 5 ft is to be modeled in a laboratory channel at a scale of 1:10. What (two-dimensional) flowrate should be used in the laboratory channel? What are the Froude numbers upstream and downstream from the jump in model and prototype?

553. If a 1:2000 tidal model is operated to satisfy Froude's law, what length of time in the model represents a day in the prototype?

554. An open cylindrical tank of 4-ft diameter contains water to a depth of 4 ft. The tank is rotated at 100 rpm about its axis (which is vertical). A half-size model of this tank is to be made with mercury as the fluid. At what speed must the model be rotated for similarity? What is the ratio of the pressures at corresponding points in the liquids? What model law is operative in this situation?

555. The flowrate from a 1.2-in. diameter orifice under a 12-ft static head is 0.133 cfs of water at 68°F. A 1:12 model of this setup is to be operated at conditions completely dynamically similar to those of the prototype. The liquid used in the model has a specific gravity of 1.40 and surface-tension effects may be ignored. What flowrate is required in the model? What viscosity should the liquid have?

556. When a sphere of 0.01-in. diameter and specific gravity 5.54 is dropped in water at 80°F it will attain a constant velocity of 0.233 fps. What specific gravity must a 0.1-in. sphere have so that when it is dropped in linseed oil (80°F) the two flows will be dynamically similar when the terminal velocity is attained?

557. The flow about a 6-in. artillery projectile which travels at 2000 fps through still air at 90°F and 14.7 psia is to be modeled in a high-speed wind tunnel with a 1:6 model. If the wind tunnel air has a temperature of 0°F, and pressure of 10 psia, what velocity is required? If the drag force on the model is 8 lb, what is the drag force on the prototype if skin friction may be neglected?

558. A cavitation zone is expected on an overflow structure when the flowrate is 5000 cfs, atmospheric pressure 14.7 psia, and water temperature 40°F. The cavitation is to be reproduced on a 1:20 model of the structure operating in a vacuum tank with water at 120°F. Disregarding frictional and surface-tension effects, determine the flowrate and pressure (psia) to be used in the tank for dynamic similarity.

559.[20] A 1:15 scale model of a hydraulic turbine of 15-ft diameter is found to develop 3.21 hp, with a flowrate of 10.0 cfs of water, under a head of 3.75 ft while rotating at 500 rpm. Calculate the corresponding horsepower, flowrate, head, and speed of the prototype, assuming the same efficiency in model and prototype.

560.[20] A centrifugal pump has a 10-in. diameter impeller which rotates at 1750 rpm and delivers a flowrate of 2.5 cfs of water at a head of 30 ft. Calculate the corresponding quantities for a half-scale model of the pump if the efficiencies and fluid of model and prototype are the same.

561.[20] A 1-ft diameter model of a ship propeller rotates at 500 rpm. The velocity well upstream from the model is 5 fps. The thrust produced is observed to be 30 lb. If the prototype (of 6-ft diameter) drives a ship through still water at a speed of 10 fps, what must its thrust and rotational speed be for its operation to be dynamically similar to that of its model? Assume the same efficiency and same (fresh) water in model and prototype.

562. A model is to be built of a flow phenomenon which is dominated by the

[20] In such problems kinematic similarity is obtained by the same ratio of flow velocity to peripheral velocity in model and prototype; with geometric similarity this is equivalent to similarity of the velocity triangles of Art. 51.

action of gravity and surface tension. Derive an expression for the model scale in terms of the physical properties of the fluid.

563. A liquid rises a certain distance in a capillary tube. If this phenomenon is to be modeled, derive an expression which must be equal in model and prototype. Compare this with equation 15.

564. For small models and (small) prototypes of surface ships and overflow structures, the actions of gravity, viscosity, and surface tension may be of equal importance. For dynamic similarity between model and prototype, what relation must exist between viscosity, surface tension, and model scale?

565. Prove by dimensional analysis that centrifugal force $= CMV^2/r$.

566. Prove by dimensional analysis that $G = CA\gamma V$.

567. Assume that the velocity acquired by a body falling from rest (without resistance) depends upon weight of body, acceleration due to gravity, and distance of fall. Prove by dimensional analysis that $V = C\sqrt{gh}$ and is thus independent of the weight of the body.

568. If $\mathbf{C} = f(V, \rho, E)$, prove the expression for Cauchy number by dimensional analysis.

569. If $\mathbf{R} = f(V, l, \rho, \mu)$, prove the expression for Reynolds number by dimensional analysis.

570. If $\mathbf{W} = f(V, l, \rho, \sigma)$, prove the expression for Weber number by dimensional analysis.

571. Derive a dimensionless number containing length, linear acceleration, density, and viscosity.

572. Derive a dimensionless number which includes linear acceleration, surface tension, viscosity, and density.

573. A physical problem is characterized by a relation between length, velocity, density, viscosity, and surface tension. Derive all possible dimensionless groups significant to this problem.

574. If the velocity of deep water waves depends only on wave length and acceleration due to gravity, derive an expression for wave velocity.

575. Derive an expression for the velocity of very small ripples on the surface of a liquid if this velocity depends only upon ripple length and density and surface tension of the liquid.

576. Derive an expression for the axial thrust exerted by a propeller if it depends only on forward speed, angular speed, size, and viscosity and density of the fluid.

577. Derive an expression for drag force on a smooth submerged object moving through incompressible fluid if this force depends only on speed and size of object and viscosity and density of the fluid.

578. Derive an expression for the head lost in an established incompressible flow in a smooth pipe if this loss of head depends only upon diameter and length of pipe, density, viscosity, and mean velocity of the fluid, and acceleration due to gravity.

579. Derive an expression for the drag force on a smooth object moving through compressible fluid if this force depends only upon speed and size of object, and viscosity, density, and modulus of elasticity of the fluid.

580. Derive an expression for the velocity of a jet of viscous liquid issuing from an orifice under static head if this velocity depends only upon head, orifice size, acceleration due to gravity, and viscosity and density of the fluid.

581. Derive an expression for the flowrate over an overflow structure if this flowrate depends only upon size of structure, head on the structure, acceleration due to gravity, and viscosity, density, and surface tension of the liquid flowing.

582. Derive an expression for terminal velocity of smooth solid spheres falling through incompressible fluids if this velocity depends only upon size and density of sphere, acceleration due to gravity, and density and viscosity of the fluid.

583. A circular disk of diameter d and of negligible thickness is rotated at a constant angular speed, ω, in a cylindrical casing filled with a liquid of viscosity μ and density ρ. The casing has an internal diameter D, and there is a clearance y between the surfaces of disk and casing. Derive an expression for the torque required to maintain this speed if it depends only upon the foregoing variables.

584. Two cylinders are concentric, the outer one fixed and the inner one movable. A viscous incompressible fluid fills the gap between them. Derive an expression for the torque required to maintain constant-speed rotation of the inner cylinder if this torque depends only upon the diameters and lengths of the cylinders, and the viscosity and density of the fluid, and the angular speed of the inner cylinder.

585. Derive an expression for the frictional torque exerted on the journal of a bearing if this torque depends only upon the diameters of journal and bearing, their axial lengths (these are the same), viscosity of the lubricant, angular speed of the journal, and the transverse load (force) on the bearing.

586. Derive an expression for the power of hydraulic machines if this power depends only upon the angular speed of the rotating element, size of the machine, flowrate, and the density and viscosity of the fluid flowing.

chapter 9

Fluid Flow in Pipes

The problems of fluid flow in pipe lines—the prediction of flow-rate through pipes of given characteristics, the calculation of energy conversions therein, etc.—are widely distributed through engineering practice; they afford an opportunity of applying many of the foregoing principles to (essentially one-dimensional) fluid flows of a comparatively simple and controlled nature. The subject of pipe flow embraces only those problems in which pipes flow completely full; pipes which flow partially full, such as sewer lines and culverts, are treated as open channels and are discussed in the next chapter.

The solution of practical pipe flow problems results from application of the energy principle, equation of continuity, and the principles and equations of fluid resistance. Resistance to flow in pipes is offered not only by long reaches of pipe but also by pipe fittings, such as bends and valves, which dissipate energy by producing relatively large-scale turbulences.

Problems of compressible fluid flow in pipes require the application of special techniques and an extensive knowledge of thermodynamics and heat-transfer processes; the advanced nature of such problems precludes their general treatment in an elementary textbook, and discussion of this topic will therefore be restricted to isothermal flow, a practical and relatively simple application.

63. Fundamental Equations—Incompressible Flow. The Bernoulli equation for incompressible fluid motion in pipes is, from Art. 55,

$$\frac{p_1}{\gamma} + \alpha_1 \frac{V_1^2}{2g} + z_1 = \frac{p_2}{\gamma} + \alpha_2 \frac{V_2^2}{2g} + z_2 + h_{L_{1-2}} \qquad (181)$$

However, in most problems of pipe flow the α terms may be omitted for several reasons: (1) Most engineering pipe flow problems involve turbulent flow in which α is only slightly more than unity. (2) In laminar flow where α is large, velocity heads are usually negligible when compared to the other Bernoulli terms. (3) The velocity heads in most pipe flows are usually so small compared to the other terms that inclusion of α has little effect on the final result. (4) The effect of α tends to cancel since it appears on both sides of the equation. (5) Engineering answers are not usually required to an accuracy which would justify the inclusion of α in the equation. Application of equation 181 to practical problems thus depends (the Bernoulli terms having been presented in Chapter 4) primarily on an understanding of the factors which affect the head loss, h_l, and the methods available for calculating this quantity.

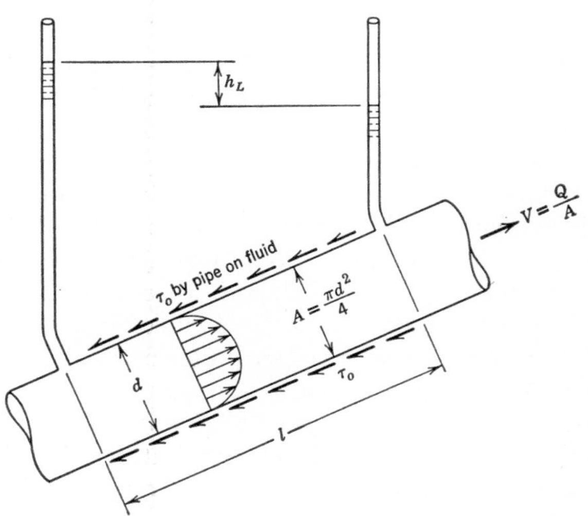

FIG. 117

Early experiments (circa 1850) on the flow of water in long, straight, cylindrical pipes (Fig. 117) indicated that head loss varied (approximately) directly with velocity head and pipe length, and inversely with pipe diameter. Using a coefficient of proportionality, f, called the *friction factor*, Darcy, Weisbach, and others proposed equations of the form

$$h_L = f \frac{l}{d} \frac{V^2}{2g} \tag{182}$$

Observations indicated that the friction factor depended primarily on pipe roughness but also upon velocity and pipe diameter; more recently it was observed that the friction factor also depended upon the viscosity of the fluid flowing. This equation, usually called the "Darcy equation," is still the basic equation for head loss caused by established pipe friction (not pipe fittings) in long, straight, uniform pipes.

Equations 182 and 164 may now be combined to give a basic relation between frictional stress, τ_o, and friction factor,[1] f; this is

$$\tau_o = f\rho V^2/8 \tag{183}$$

In this fundamental equation relating wall shear to friction factor, density, and mean velocity it is apparent that, with f dimensionless, $\sqrt{\tau_o/\rho}$ must have the dimensions of velocity; this is known as the *friction velocity*, $v_\star$, which (from equation 183) is given by

$$v_\star = \sqrt{\tau_o/\rho} = V\sqrt{f/8} \tag{184}$$

However, the physical meaning of the friction velocity is not revealed by this algebraic definition; since it is a velocity which embodies only wall shear and fluid density, it is defined by the same equation whatever the flow regime (laminar or turbulent) or whatever the boundary texture (rough or smooth). For this reason it is a useful generalization which will find wide application in further developments.

Illustrative Problem

Water flows in a 6-in. diameter pipe line at a mean velocity of 15 fps. The head lost in 100 ft of this pipe is measured experimentally and found to be 17.5 ft. Calculate the friction velocity.

Solution.

$$17.5 = f\frac{100}{0.5}\frac{(15)^2}{2g}; \quad f = 0.025 \tag{182}$$

$$v_\star = 15\sqrt{0.025/8} = 0.56 \text{ fps} \tag{183}$$

64. Dimensional Analysis of the Pipe Friction Problem. An introduction to the pipe friction problem may be obtained by application of the methods of similitude and dimensional analysis. This will allow the problem to be treated in a general way and will indicate the variables on which the friction factor depends; although such

[1] Another method of defining the friction factor is that used by Fanning, in which, following the pattern of the standard drag formula (see Art. 116), equation 183 is written $\tau_o = f'\rho V^2/2$ and equation 182 becomes $h_L = 4f'lV^2/2gd$. The hydraulic radius, R_h, is then substituted for $d/4$, yielding $h_L = f'lV^2/2gR_h$.

an analysis cannot alone produce definite mathematical relations, it will indicate the significant combinations of variables and justify the now generally accepted methods of presenting the experimental results of pipe friction tests.

The pipe of Fig. 118 has diameter d and contains fluid of density ρ and viscosity μ flowing with mean velocity V; the roughness projections have a mean height e and may be assumed to be of the same shape and distribution pattern. It appears that the frictional

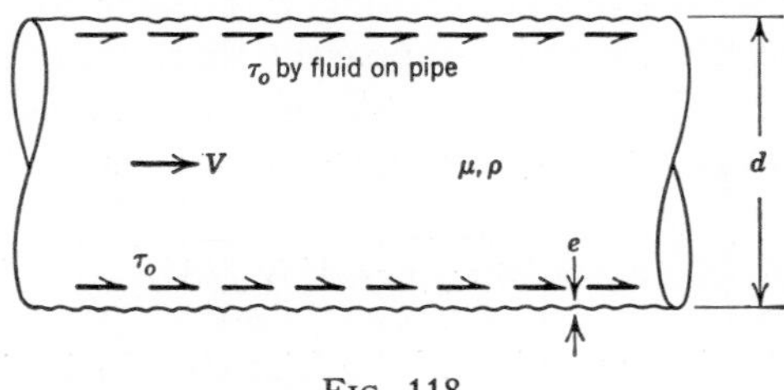

FIG. 118

stress τ_o will depend on all these independent variables but on no others; thus

$$\tau_o = F(V, d, \rho, \mu, e)$$

By the methods of dimensional analysis it may be easily shown [2] that

$$\tau_o = F'(Vd\rho/\mu, e/d)\rho V^2$$

and, by comparison with equation 183,

$$f = F''(\mathbf{R}, e/d) \tag{185}$$

Thus without prior knowledge of experimental results it has been shown that the friction factor depends only upon the Reynolds number of the flow and upon the ratio e/d, termed the *relative roughness* [3] of the pipe; thus dimensional analysis has yielded a single general result which is true for all fluids flowing in all circular pipes.

The physical significance of equation 185 may be stated briefly: the friction factors of pipes will be the same if their Reynolds numbers, roughness patterns, and relative roughnesses are the same. When this is interpreted by the principle of similitude its basic meaning is: the friction factors of pipes are the same if their flow pictures in every detail are geometrically and dynamically similar.

[2] See Appendix IX for details.

[3] "Relative roughness," since e/d expresses the size of the roughness protuberances relative to the diameter of the pipe.

65. Results of Pipe Friction Experiments. The relationships of equation 185 indicate a convenient means of presenting experimental data on friction factor. This was used by Stanton[4] (1914) and consists of a logarithmic plot of friction factor against Reynolds number with surface roughness the parameter as in Fig. 119. From such a plot, complete data on friction factor may be obtained for laminar and turbulent flow of any fluid in smooth or rough pipes, the only difficulty being the definition of a reliable index of pipe roughness.

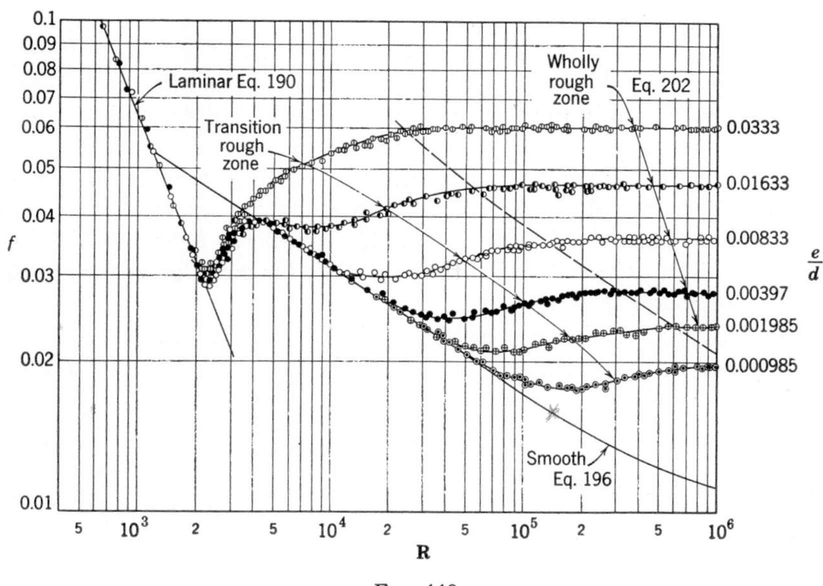

Fig. 119

The results of systematic tests by Nikuradse[5] on turbulent flow in smooth and rough pipes demonstrate perfectly the relationship between f, **R**, and relative roughness. In these tests geometric similarity of the roughness pattern was obtained by fixing a coating of uniform sand grains to the pipe wall, thus giving an easily measurable index for the relative roughness, e being the diameter of the sand grain. Although the sand grain roughness of Nikuradse is quite different from that of commercial pipes, the former is an easily measured definite quantity which provides a reliable basis for quan-

[4] It is now generally known as a *Stanton diagram.*

[5] J. Nikuradse, "Strömungsgesetze in rauhen Rohren," *VDI-Forschungsheft,* 361, 1933. Translation available in *N.A.C.A., Tech. Mem.* 1292.

titative measurement of roughness effects; Nikuradse's results are generally accepted today as the basic standard for this measurement. The test results when plotted logarithmically in Fig. 119 illustrate the following important fundamentals:[6]

1. The physical difference between the laminar and turbulent flow regimes is indicated by the change in the relationship of f to $\mathbf{R}$ near the critical Reynolds number of 2100.

2. The laminar regime is characterized by a single curve given by the equation $f = 64/\mathbf{R}$ for all surface roughnesses and thus shows that head loss in laminar flow is independent of surface roughness.

3. In turbulent flow a curve of f vs. $\mathbf{R}$ exists for every relative roughness, e/d, and from the horizontal aspect of the curves it may be concluded that for rough pipes the roughness is more important than the Reynolds number in determining the magnitude of the friction factor.

4. At high Reynolds numbers the friction factors of rough pipes become constant, dependent wholly upon the roughness of the pipe, and thus independent of the Reynolds number. From the Darcy equation it may be concluded that $h_L \propto V^2$ for completely turbulent flow over rough surfaces.

5. Although the lowest curve was obtained from tests on hydraulically smooth pipes, many of Nikuradse's rough pipe test results coincide with it for $5000 < \mathbf{R} < 50{,}000$. Here the roughness is submerged in the laminar film (Art. 54) and can have no effect upon friction factor and head loss, which depend upon viscosity effects alone. Blasius[7] has shown that this curve (for $3000 < \mathbf{R} < 100{,}000$) may be closely approximated by a line whose equation is

$$f = 0.316/\mathbf{R}^{0.25} \tag{186}$$

When this is substituted in (Darcy's) equation 182, it will be noted that $h_L \propto V^{1.75}$ for turbulent flow in smooth pipes with $\mathbf{R} < 10^5$.

6. The series of curves for the rough pipes diverge from the smooth pipe curve as the Reynolds number increases. In other words, pipes which are smooth at low values of $\mathbf{R}$ become rough at high values of $\mathbf{R}$. This may be explained by the thickness of the laminar film decreasing (Art. 54) as the Reynolds number increases, thus exposing smaller roughness protuberances to the turbulent region and causing the pipe to exhibit the properties of a rough pipe.

[6] Figure 119 is a summary of Nikuradse's test results. Curves drawn through the points have the equations indicated; the forms of these will be justified later.

[7] H. Blasius, *Forschungsarbeiten auf dem Gebiete des Ingenieurwesens*, 131,1913.

ILLUSTRATIVE PROBLEM

Water at 100°F flows in a 3-in. pipe at Reynolds number 80,000. If the pipe contains a uniform sand roughness of grain size 0.006-in. diameter, how much head loss is to be expected in 1000 ft of the pipe? How much head loss would be expected if this pipe were smooth?

Solution. From Fig. 119, with $\mathbf{R} = 80{,}000$ and $e/d = 0.006/3 = 0.002$,

$$f \cong 0.021$$

From Appendix II,

$$\nu = 0.729 \times 10^{-5} \text{sqft/sec}$$

$$80{,}000 = V \times 0.25/0.729 \times 10^{-5}, \quad V = 2.33 \text{ fps} \tag{175}$$

$$h_L \cong 0.021 \times \frac{1000}{0.25} \times \frac{(2.33)^2}{2g} = 7.1 \text{ ft} \tag{182}$$

For a smooth pipe, since $3000 < \mathbf{R} < 100{,}000$,

$$f = 0.316/(80{,}000)^{0.25} = 0.0188 \tag{186}$$

which is confirmed by the "smooth" line of Fig. 119. Thus, by proportion,

$$h_L \cong \left(\frac{0.0188}{0.021}\right) 7.1 = 6.35 \text{ ft}$$

66. Analysis of Experimental Results. Unfortunately the excellent results of Nikuradse cannot be applied directly to engineering problems since the roughness patterns of commercial pipe are entirely different, much more variable, and much less definable than the artificial roughnesses used by Nikuradse. However, Colebrook [8] has shown how these results may be applied toward a quantitative measure of commercial pipe roughness. Subtracting 2 log d/e from both sides of Nikuradse's equations (196 and 202) for smooth and wholly rough pipes, these become

Smooth: $$\frac{1}{\sqrt{f}} - 2 \log \frac{d}{e} = -0.80 + 2 \log \mathbf{R} \left(\frac{e}{d}\right) \sqrt{f}$$

Wholly rough: $$\frac{1}{\sqrt{f}} - 2 \log \frac{d}{e} = 1.14$$

[8] C. F. Colebrook, "Turbulent Flow in Pipes, with Particular Reference to the Transition Region between the Smooth and Rough Pipe Laws," *Jour. Inst. Civil Engrs., London*, p. 133, February 1939.

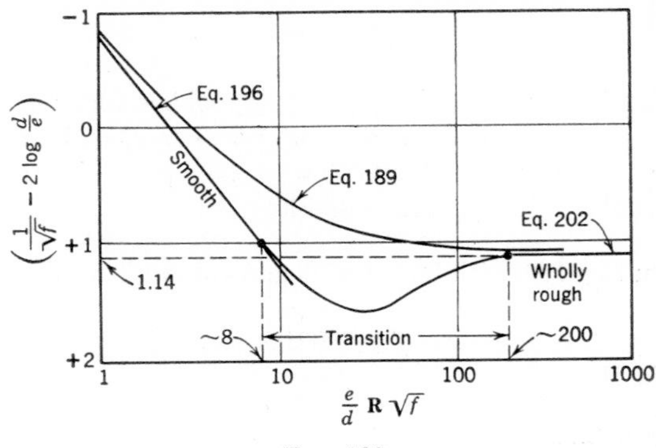

FIG. 120

These equations may now be plotted as in Fig. 120 using $(1/\sqrt{f} - 2 \log d/e)$ as ordinate and $\mathbf{R}(e/d)\sqrt{f}$ as abscissa. The result is two single straight lines characterizing the smooth and wholly rough surfaces, these lines being connected by a curve through the transition region. The family of curves of Fig. 119 has thus been reduced to the single curve of Fig. 120. Now Colebrook found that any pipe of commercial roughness when tested to a high enough Reynolds number gave a friction factor which no longer varied with Reynolds number. This allowed comparison with equation 202 and, from measurement of the friction factor, permitted an *equivalent* sand grain size, e, to be computed, giving the values indicated.

	e (in.)
Uncoated cast iron	0.01
Galvanized iron	0.006
Asphalted cast iron	0.005
Wrought iron	0.0017

Obtaining e in this manner and plotting test results on many pipes, Colebrook found results from all pipes to be closely clustered about a single line having the equation

$$\frac{1}{\sqrt{f}} - 2 \log \frac{d}{e} = 1.14 - 2 \log \left[1 + \frac{9.28}{\mathbf{R}(e/d)\sqrt{f}}\right] \tag{187}$$

This equation, although conveniently summarizing the data on pipes of commercial roughness, is hardly suitable for engineering use and has been plotted by Moody[9] in the form of the Stanton diagram of Fig. 121 (now known as the *Moody diagram*), which may be used in

[9] L. F. Moody, "Friction Factors for Pipe Flow," *Trans. A.S.M.E.*, Vol. 66, p. 671, 1944.

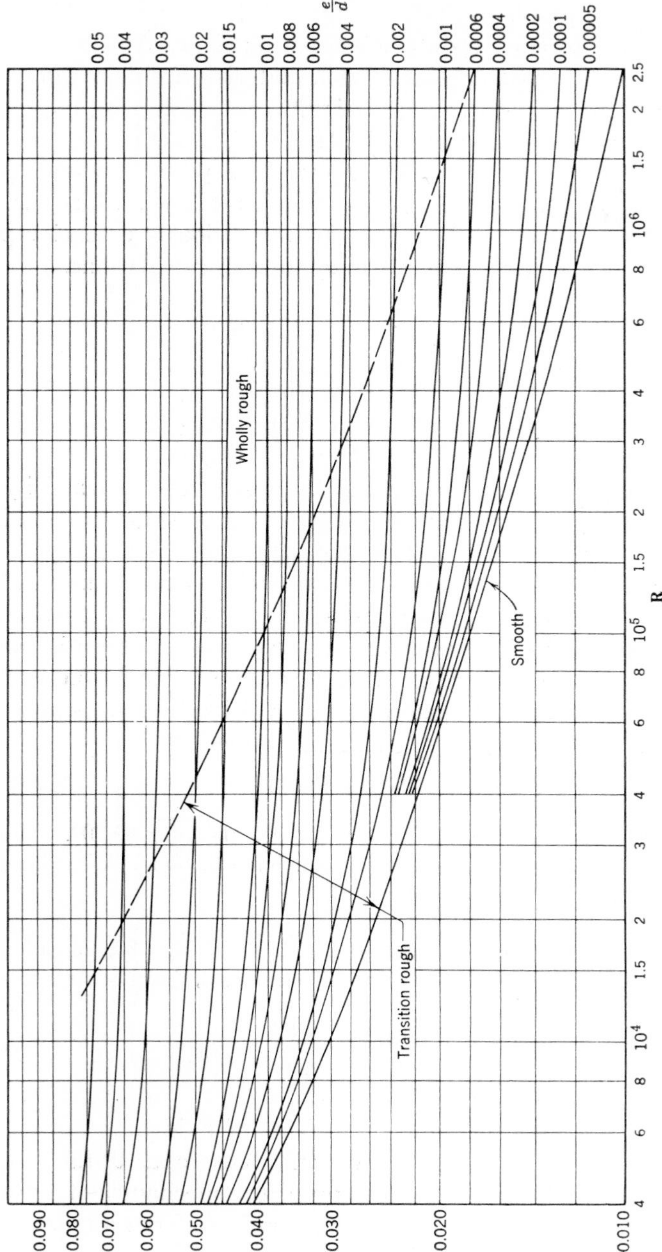

FIG. 121. Relation of friction factor, Reynolds number, and roughness for commercial pipes.

the solution of problems. Moody extended the work of Colebrook by including information on commercial pipes of other materials as listed below. The roughnesses of such materials are more difficult to classify because of variations in design, workmanship, and age—and are seen to vary between wide limits. This is one of the practical considerations of pipe flow calculations which poses a difficult problem for the designer in predicting friction factors which will be realized after a pipe line is constructed. Obviously extensive practical experience is needed for accuracy in such calculations.

	e (in.)
Wood stave	0.007 to 0.036
Concrete	0.012 to 0.12
Riveted steel	0.035 to 0.35

The accuracy of pipe friction calculations is also lessened by the somewhat unpredictable change in the roughness and friction factor due to the accumulation of dirt and corrosion on the pipe walls. This accumulation not only increases surface roughness but also reduces the effective pipe diameter and may lead to an extremely large increase in the friction factor after the pipe has been in service over a long period. Pipe-line designers have different methods to allow for these effects, but no attempt will be made to summarize them here.

Illustrative Problem

Water at 100°F flows in a 3-in. pipe at Reynolds number 80,000. This is a commercial pipe of equivalent sand grain roughness 0.006 in. What head loss is to be expected in 1000 ft of this pipe?

Solution. From the Moody diagram of Fig. 121 with $\mathbf{R} = 80{,}000$ and $e/d = 0.006/3 = 0.002$, $f \cong 0.0255$. From the illustrative problem of Art. 65, $V = 2.33$ fps. Therefore

$$h_L \cong 0.0255 \frac{1000}{0.25} \times \frac{(2.33)^2}{2g} = 8.6 \text{ ft} \tag{182}$$

67. Laminar Flow. Although the facts of fluid flow can usually be established by experiment, an analytical approach to the problem is also necessary to an understanding of the mechanics of the flow. For the mechanics of a real fluid this consists of the application of basic physical laws which have themselves been established by experiment; the use of "pure theory" alone is seldom possible in this field.

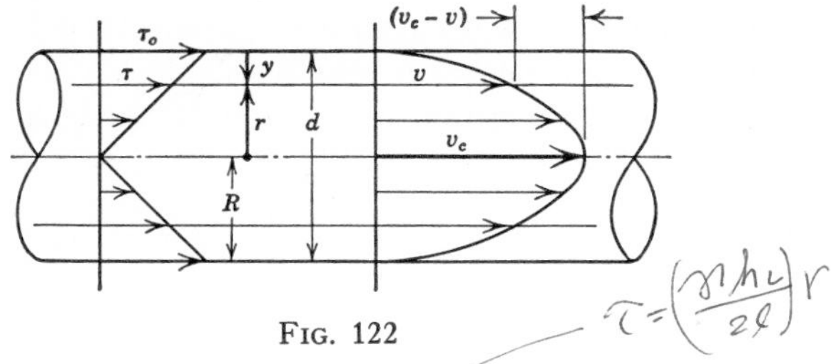

FIG. 122

Analysis of laminar flow in a pipe line (Fig. 122) may be begun with the following established facts: (1) symmetrical distribution of shear stress and velocity, (2) maximum velocity at the center of the pipe and no velocity at the wall, (3) linear shear stress distribution in the fluid given by equation 164, (4) shear stress in the fluid also given by equation 12. When these expressions [10] are equated,

$$\tau = \left(\frac{\gamma h_L}{2l}\right) r = \mu \frac{dv}{dy} = -\mu \frac{dv}{dr}$$

The equation of the velocity profile may be obtained by noting that, when $v = v_c$, $r = 0$. Separating variables and integrating gives

$$v = v_c - \left(\frac{\gamma h_L}{4\mu l}\right) r^2 = v_c - Kr^2 \tag{188}$$

which shows the velocity profile to be a parabolic curve. Since $v = 0$ when $r = R$, K is found to be v_c/R^2.

The ratio of mean velocity to centerline velocity has already been shown [11] to be 0.50 for a parabolic velocity distribution in a cylindrical pipe. This fact may now be used in equation 188 to derive a useful expression for head loss. Substituting $v = 0$, $r = d/2$, and $v_c = 2V$, there results

$$h_L = \frac{32\mu l V}{\gamma d^2} \tag{189}$$

which shows that, *in laminar flow, head loss varies with the first power of the velocity*. By setting equation 189 equal to (Darcy) equation 182 an expression for the friction factor for laminar pipe flow is obtained; it is

$$f = \frac{64\mu}{Vd\rho} = \frac{64}{\mathbf{R}} \tag{190}$$

[10] With y increasing toward, and r increasing away from, the pipe centerline, $dy = -dr$.

[11] See illustrative problem, Art. 18.

confirmation of which is seen by the proximity of the line of this equation to the experimental points of Fig. 119.

When $4Q/\pi d^2$ is substituted for V in equation 189,

$$Q = \frac{\pi d^4 \gamma h_L}{128 \mu l} \tag{191}$$

results. This equation shows that in laminar flow the flowrate, Q, which will occur in a circular pipe varies directly with the head loss and with the fourth power of the diameter but inversely with the length of pipe and viscosity of the fluid flowing. These facts of laminar flow were established experimentally, independently, and almost simultaneously by Hagen (1839) and Poiseuille (1840), and thus the law of laminar flow expressed by the equations above is termed the *Hagen-Poiseuille law.*

The experimental verification (by Hagen, Poiseuille, and later investigators) of the derivations above serves to confirm the assumptions (1) that there is no velocity adjacent to a solid boundary (i.e., no "slip" between fluid and pipe wall), and (2) that in laminar flow the shear stress is given by $\tau = \mu \, dv/dy$ which were taken for granted in the foregoing derivations.

Illustrative Problem

One hundred gallons of oil (s.g. = 0.90, and μ = 0.0012 lb-sec/sqft) flow per minute through a pipe line of 3-in. diameter. Calculate the center velocity, the lost head in 1000 ft of this pipe, and the shear stress and velocity at a point 1 in. from the centerline.

Solution.

$$V = \frac{100}{60 \times 7.48} \frac{1}{(\pi/4)(\frac{3}{12})^2} = 4.53 \text{ fps} \tag{53}$$

$$\mathbf{R} = \frac{4.53 \times (\frac{3}{12}) \times 0.90 \times 1.935}{0.0012} = 1645 \tag{175}$$

Since $\mathbf{R} < 2100$, laminar flow exists, and

$$h_L = \frac{64}{1645} \times \frac{1000}{3/12} \times \frac{(4.53)^2}{2g} = 49.6 \text{ ft of oil} \tag{182}$$

$$v = 9.06 \left[1 - \left(\frac{1}{1.5}\right)^2\right] = 5.04 \text{ fps} \tag{188}$$

$$\tau = \left(\frac{62.4 \times 0.90 \times 49.6}{2 \times 1000}\right) \frac{1}{12} = 0.116 \text{ psf} \tag{164}$$

68. Turbulent Flow—Smooth and Rough Pipes. In laminar flow the perfection of the expression for shearing stress allowed the derivation of equations for velocity profile and friction factor, which were accurately confirmed by experiment. Since this perfection is lacking in the basic equation (148) for turbulent shear, similar confirmation by experiment cannot be expected; the imperfections of this equation are reflected in the variation of the "constant" κ. A comprehensive equation for shear stress in turbulent flow has not yet been developed; although the Prandtl-Kármán equation represents great progress toward this end, it is not perfect in the sense that equation 12 is perfect for laminar flow.

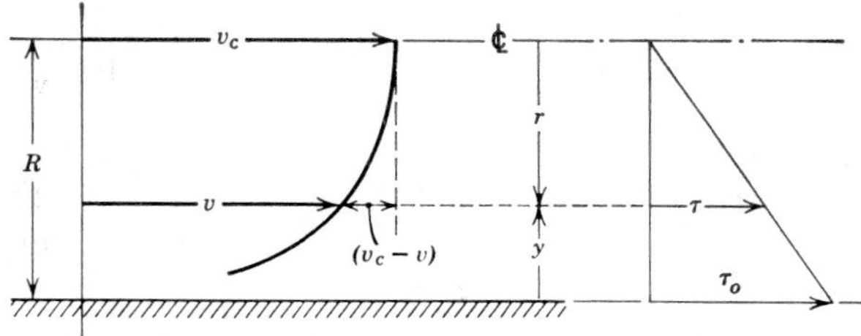

FIG. 123

An analytical treatment of turbulent flow in pipes begins (see Fig. 123) by equating the expression for linear variation of shear stress (equation 164) with that for shear stress caused by turbulence:

$$\tau = \tau_o\left(1 - \frac{y}{R}\right) = \rho\kappa^2 \frac{(dv/dy)^4}{(d^2v/dy^2)^2}$$

and integrating twice to obtain an equation for velocity profile. The result is [12]

$$\frac{v_c - v}{v_\star} = -\frac{1}{\kappa}\left[\sqrt{1 - \frac{y}{R}} + \ln\left(1 - \sqrt{1 - \frac{y}{R}}\right)\right] \qquad (192)$$

from which the value of κ may be deduced from experimental observations: measurements of velocity profile by pitot tube yield values of v, v_c, and y; measurements of head loss allow computation of τ_o from $\tau_o = \gamma R h_L / 2l$ and $v_\star$ from $v_\star = \sqrt{\tau_o/\rho}$. Nikuradse's systematic and comprehensive measurements of velocity profiles in smooth pipes ($5 \times 10^3 < \mathbf{R} < 3 \times 10^6$) and pipes of uniform sand grain roughness in the wholly rough zone showed that all velocity profiles

[12] See Appendix X for details.

could be characterized by the single equation

$$\frac{v_c - v}{v_\star} = -2.5 \ln \frac{y}{R} \tag{193}$$

Formal mathematical comparison of equations 192 and 193 yields the following values of κ:

y/R	0.1	0.2	0.3	0.4	0.5	0.6	0.7	0.8	0.9
κ	0.352	0.336	0.325	0.313	0.301	0.288	0.276	0.261	0.242

which seem to show κ to be anything but constant through the flow. However, this apparent variation of κ is due to the convenient assumption (see Appendix X) that, for $y \to 0$, $dv/dy \to \infty$. Had dv/dy been assumed to approach values less than infinity as y approaches zero, higher and more constant values of κ would have resulted; the widely accepted value of κ is 0.40.

The useful ratio V/v_c may be derived for turbulent flow by applying the methods of Art. 18 to the *velocity defects* $(v_c - V)$ and $(v_c - v)$:

$$(v_c - V)\pi R^2 = \int_0^R (v_c - v) 2\pi r \, dr$$

Substituting $-2.5 v_\star \ln (y/R)$ for $(v_c - v)$ (from equation 193) and $V\sqrt{f/8}$ for $v_\star$, the final result,[12] which conforms well with experimental observations, is

$$\boxed{\frac{V}{v_c} = \frac{1}{1 + 4.07\sqrt{f/8}}} \tag{194}$$

Equations 192 and 193 are general equations for velocity distribution in circular pipes. Since they have been developed from the turbulence structure, and without regard to the texture of the pipe wall, it is evident that they are equally valid for smooth and rough pipes. Figure 124 shows a comparison (for the same flowrate and

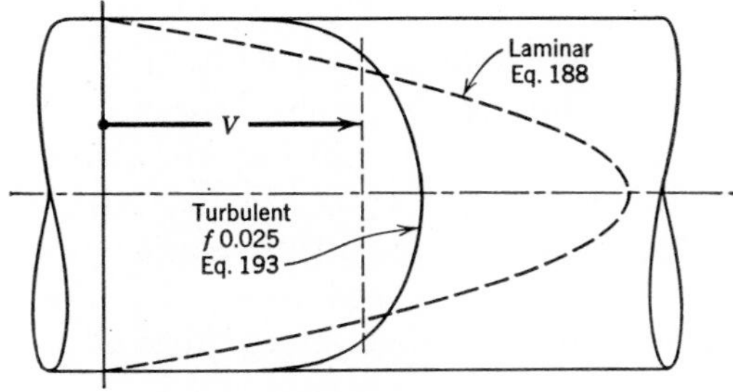

FIG. 124

mean velocity) of typical velocity profiles for laminar and turbulent pipe flow.

Illustrative Problem

Water (68°F) flows in a long pipe line made of 12-in. diameter clean cast iron. If the center velocity is 10 fps, what is the flowrate?

Solution. A trial-and-error solution is required since V/v_c depends on f, which in turn depends on $\mathbf{R}$, which contains V. Assuming $V/v_c = 0.80$, $V = 8.0$ fps, $\mathbf{R} = 737{,}000$, and f (from Fig. 121) $= 0.019$. When this value of f is substituted in equation 194, $V/v_c = 0.835$ results. Repeating the process with $V =$ 8.35 fps, f (within the accuracy of Fig. 121) is again 0.019; thus all equations are satisfied, $V = 8.35$ fps, and $Q = \pi \times 8.35/4 = 6.55$ cfs.

69. Turbulent Flow—Smooth Pipes. Since pipe friction has been seen to be a viscosity-inertia phenomenon and thus characterized by a Reynolds number (Art. 61) and, in smooth pipes, to be associated with viscous action in a laminar film, the general equation (193) for turbulent velocity profile may be rearranged in terms of a Reynolds number as follows; written in terms of common logarithms, equation 193 becomes

$$\frac{v}{v_\star} = \frac{v_c}{v_\star} + 5.75 \log \frac{y}{R} = \frac{v_c}{v_\star} + 5.75 \log \frac{\nu}{v_\star R} + 5.75 \log \frac{v_\star y}{\nu}$$

Nikuradse's tests on smooth pipes showed that the sum of the first two terms on the right-hand side of the equation was a constant, 5.50; thus

$$\frac{v}{v_\star} = 5.50 + 5.75 \log \frac{v_\star y}{\nu} \tag{195}$$

is a general equation of the velocity distribution for turbulent flow in smooth pipes. This equation is especially useful in extending the applicability of Nikuradse's work to other flows over smooth surfaces since it expresses velocity profile in terms of wall shear ($v_\star = \sqrt{\tau_o/\rho}$) and without reference to centerline velocity; thus it may be used to describe the velocity distributions of turbulent flow in a two-dimensional smooth passage or in a boundary layer on a smooth flat plate. Also, from this equation, the relation between friction factor and Reynolds number for turbulent flow in smooth pipes may be derived. Substitute v_c for v, $d/2$ for y, $V\sqrt{f/8}$ for $v_\star$, $V(1 + 4.07\sqrt{f/8})$ from equation 194 for v_c; this yields the correct *form* of the equation but necessitates some adjustment of coefficients to conform with experimental results. The modified equation [13] is

$$\frac{1}{\sqrt{f}} = -0.80 + 2.0 \log \mathbf{R}\sqrt{f} \tag{196}$$

and it may be used to obtain friction factors for smooth pipes for Reynolds numbers above [14] 5000.

The laminar film (Art. 54) which covers smooth boundary surfaces in turbulent flow may be defined in general terms with the aid of the dimensionless velocity distribution of equation 195 shown in Fig. 125. Here distance from the surface y appears in the Reynolds

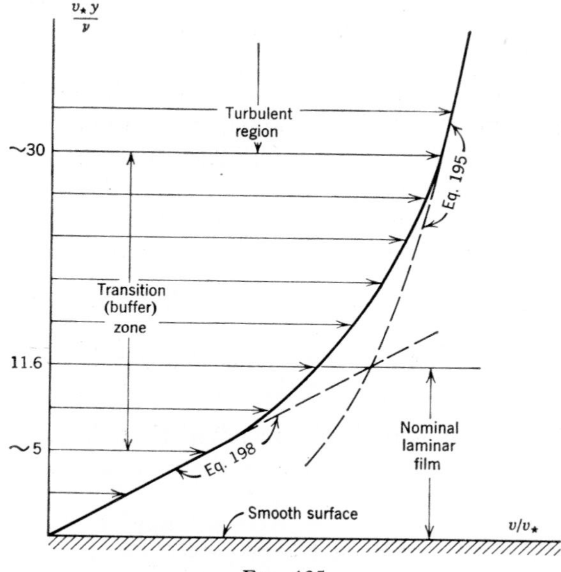

Fig. 125

number $v_\star y/\nu$ and velocity v in the ratio $v/v_\star$. In a thin laminar film the velocity distribution may be assumed linear (which implies that the shearing stress is constant through the film) and the equation of the velocity profile obtained directly from the equation for laminar shearing stress,

$$\tau_o = \tau = \mu \frac{dv}{dy} = \mu \frac{v}{y} \tag{197}$$

[13] The indicated substitutions will give 1.03 and 2.03 instead of 0.80 and 2.0, respectively.

[14] Although Nikuradse's tests did not extend beyond Reynolds numbers of 3×10^6, there is little doubt that the equation may be used for larger Reynolds numbers, at least to engineering accuracy.

Dividing this equation by ρ, substituting $v_\star^2$ for τ_o/ρ, ν for μ/ρ, and rearranging, the equation for the dimensionless velocity profile in the laminar film becomes

$$\frac{v}{v_\star} = \frac{v_\star y}{\nu} \tag{198}$$

The intersection of this curve with the turbulent velocity distribution is obtained by solving equations 195 and 198 simultaneously; this intersection defines the arbitrary [15] boundary of the film and the film thickness δ. The solution yields

$$\frac{v_\star \delta}{\nu} = 11.6 \tag{199}$$

This may be rearranged for practical use by dividing by d, substituting $V\sqrt{f/8}$ for $v_\star$ and **R** for Vd/ν, which results in

$$\frac{\delta}{d} = \frac{32.8}{\mathbf{R}\sqrt{f}} \tag{200}$$

From this equation the decrease of film thickness with increasing Reynolds number is readily seen; this principle when applied to a particular pipe flow shows that film thickness decreases with increasing velocity and may result in the same boundary surface being *smooth* at a low velocity and *rough* at a higher one (Arts. 54 and 65).

Illustrative Problem

Ninety gallons per minute of water at 68°F flow in a 3-in. diameter smooth brass pipe line. Calculate the head lost in 3000 ft of this pipe. Calculate the wall shearing stress, the center velocity, the shearing stress and velocity 1 in. from the pipe centerline, and the thickness of the laminar film.

Solution.

$$V = \frac{90}{60 \times 7.48} \times \frac{1}{(\pi/4)(\frac{3}{12})^2} = 4.08 \text{ fps} \tag{53}$$

$$\mathbf{R} = \frac{4.08 \times (\frac{3}{12}) \times 1.935}{0.000021} = 93{,}800 \tag{175}$$

[15] There is no precise boundary of the film, and no actual intersection of the two velocity distribution curves; a gradual transition from laminar to turbulent action occurs through a "buffer zone" extending from $5 < v_\star y/\nu < 30$. The velocity profile of Fig. 125 is taken to be a universal one for the turbulent flow of fluids over smooth surfaces, providing there is no transfer of heat between fluid and surface.

From the Moody diagram of Fig. 121,

$$f = 0.018$$

$$h_L = 0.018 \frac{3000}{0.25} \frac{(4.08)^2}{2g} = 56.0 \text{ ft of water} \quad (182)$$

$$\tau_o = \frac{0.018 \times 1.935 \times (4.08)^2}{8} = 0.073 \text{ psf} \quad (183)$$

From the linear variation of shear stress with radial distance,

$$\tau_1 = \frac{1}{1.5}(0.073) = 0.049 \text{ psf}$$

$$v_c = 4.08(1 + 4.07\sqrt{0.018/8}) = 4.88 \text{ fps} \quad (194)$$

$$v_\star = \sqrt{0.073/1.935} = 0.194 \text{ fps} \quad (184)$$

$$v_1 = 0.194\left(5.50 + 5.75 \log \frac{0.194 \times 0.5/12}{0.00001085}\right) = 4.27 \text{ fps} \quad (195)$$

or

$$\frac{4.88 - v_1}{0.194} = 5.75 \log \frac{1.5}{0.5}; \quad v_1 = 4.35 \text{ fps} \quad (193)$$

$$\delta = 3 \times \frac{32.8}{93{,}800\sqrt{0.018}} = 0.0078 \text{ in.} \quad (200)$$

70. Definition of Smoothness and Roughness. Some interesting and useful quantitative information may be derived by rewriting equation 200 as

$$\frac{\delta}{d}\mathbf{R}\sqrt{f} = 32.8$$

and comparing it with the point on Fig. 120 at the end of the *smooth* zone for Nikuradse's uniform sand grain roughness; this point is seen to be characterized by

$$\frac{e}{d}\mathbf{R}\sqrt{f} \cong 8$$

leading directly to the conclusion that here $e/\delta \cong \frac{1}{4}$. This means that, for a uniform roughness, the size of the roughness may be as great as one-fourth of the thickness of the laminar film before the pipe begins to behave as a rough pipe; or, in other words, such roughnesses are so completely submerged in the laminar film that the pipe is virtually a smooth one, and resistance and head loss are entirely unaffected by roughness up to this size. The general conclusion to be drawn is that absolute size is not a measure of the effect of surface roughness on fluid flow; the effect is dependent upon the size of the roughness *relative to the thickness of the laminar film.*

Application of the foregoing reasoning to commercial pipes cannot lead to the same quantitative conclusion since the lines of Fig. 121 are asymptotic to the smooth pipe line and there is, therefore, no precise end to the smooth surface zone. However, from this fact may be drawn the conclusions that for roughness of irregular size the larger roughnesses become effective first as the Reynolds number is increased (and the laminar film becomes thinner) and that this process goes on until all the roughness sizes become effective in determining the friction factor. This condition is reached at about the same point for both uniform sand grain roughness and commercial roughness and is characterized by

$$\frac{e}{d}\mathbf{R}\sqrt{f} \cong 200$$

on Fig. 120. Comparing this fact with equation 200, it may be concluded that rough surfaces will be *wholly rough*, and their friction factors no longer vary with Reynolds number, when e/δ is greater than approximately 6. Comparison [16] of laminar film thickness and size of roughness has thus led to the establishment of a lower limit above which a rough surface will behave as a *wholly rough* one.

71. Turbulent Flow—Wholly Rough Pipes. Pipe friction in rough pipes at high Reynolds numbers will be governed primarily by the size and pattern of the roughness, since disruption of the laminar film will render viscous action negligible. Thus the velocity distribution and friction factor can be expected to depend upon parameters containing the magnitude of the roughness rather than a Reynolds number. Rearranging equation 193 on this basis, and writing it with common logarithms,

$$\frac{v}{v_\star} = \frac{v_c}{v_\star} + 5.75 \log \frac{y}{R} = \frac{v_c}{v} + 5.75 \log \frac{e}{R} + 5.75 \log \frac{y}{e}$$

Nikuradse's tests showed the sum of the first two terms on the right-hand side of the equation to be a constant, 8.48; and thus

$$\frac{v}{v_\star} = 8.48 + 5.75 \log \frac{y}{e} \tag{201}$$

[16] It should be noted in passing that the comparison of 32.8 with 200 is somewhat artificial (and therefore approximate), since the first figure was derived on the basis of a smooth pipe velocity distribution and the latter for rough surfaces. The comparison implies that the laminar film forms over the individual projections of roughness—a somewhat dubious assumption which has not yet been proved.

for turbulent flow in *wholly rough pipes* at high Reynolds number. This equation (like the analogous one for smooth pipes) may be used as a generalized equation for the velocity profile to be expected over surfaces of large roughness at high Reynolds number. Also, from this equation, a relation between friction factor and relative roughness may be derived for turbulent flow in pipes of uniform sand grain roughness in the *wholly rough* zone. Substitute v_c for v, $d/2$ for y, $V\sqrt{f/8}$ for $v_\star$, and $V(1 + 4.07\sqrt{f/8})$ from equation 194 for v_c. The equation (after some adjustment [17] of coefficients to conform with experimental results) is

$$\frac{1}{\sqrt{f}} = 1.14 + 2.0 \log \frac{d}{e} \tag{202}$$

which characterizes the *horizontal portions* of the curves of Fig. 119, 120, and 121, where frictional effects are produced by roughness alone without viscous action.

Illustrative Problem

The mean velocity in a 12-in. pipe line is 10 fps. The relative roughness of the pipe is 0.002, and kinematic viscosity of the water 0.00001 sqft/sec. Determine the friction factor, centerline velocity, velocity 2 in. from the pipe wall, and the head lost in 1000 ft of this pipe.

Solution.

$$\mathbf{R} = 10 \times 1/10^{-5} = 10^6 \tag{175}$$

Entering the Moody diagram of Fig. 121 with $\mathbf{R} = 10^6$ and $e/d = 0.002$, f is found to be 0.0235; or, having found f to be in the wholly rough zone,

$$\frac{1}{\sqrt{f}} = 1.14 + 2.0 \log \frac{1}{0.002}\,; \quad f = 0.0234 \tag{202}$$

$$\frac{10}{v_c} = \frac{1}{1 + 4.07\sqrt{0.0235/8}}\,; \quad v_c = 12.25 \text{ fps} \tag{194}$$

$$\tau_o = 0.0235 \times 1.935 \times (10)^2/8 = 0.570 \text{ psf} \tag{183}$$

$$\frac{12.25 - v_2}{0.570} = 5.75 \log \frac{6}{2}\,; \quad v_2 = 10.7 \text{ fps} \tag{193}$$

or

$$\frac{v_2}{0.570} = 8.48 + 5.75 \log \frac{\frac{2}{12}}{0.002}\,; \quad v_2 = 11.1 \text{ fps} \tag{201}$$

[17] The indicated substitutions will give 0.95 and 2.03 instead of 1.14 and 2.0, respectively.

The 4 percent difference between the above velocities might be expected in view of the imperfections of the theory and the adjustment of coefficients at various points in the developments.

$$h_L = 0.0235 \frac{1000}{1} \frac{(10)^2}{2g} = 36.5 \text{ ft of water} \tag{182}$$

72. Seventh-Root Law of Turbulent Flow. Before the development of the foregoing generalizations by Prandtl, von Kármán, and Nikuradse, a pioneering effort was made by Blasius to relate velocity profile, wall shear, and friction factor for turbulent flow in smooth

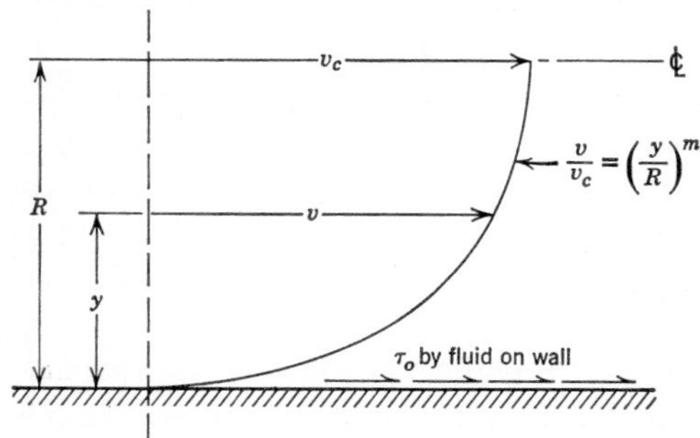

FIG. 126

pipes. Although Blasius' work has been superseded by these generalizations, it is still of some importance in engineering (in spite of its limited scope and empiricism) because of a mathematical simplicity which allows easy visualization and leads directly to useful (but approximate) results.

Substituting the Blasius expression for friction factor (equation 186), which is limited to smooth pipes with $3 \times 10^3 < \mathbf{R} < 10^5$, into equation 183

$$\tau_o = \frac{0.316}{(2RV\rho/\mu)^{0.25}} \frac{\rho V^2}{8} = 0.0332 \mu^{\frac{1}{4}} R^{-\frac{1}{4}} V^{\frac{7}{4}} \rho^{\frac{3}{4}} \tag{203}$$

Blasius then assumed that the turbulent velocity profile could be approximated by (see Fig. 126)

$$\frac{v}{v_c} = \left(\frac{y}{R}\right)^m \tag{204}$$

For this equation the mean velocity V may be related to the center velocity v_c by applying equation 56:

$$V\pi R^2 = \int_R^0 v_c \left(\frac{y}{R}\right)^m 2\pi(R - y)(-dy)$$

which gives

$$\frac{V}{v_c} = \frac{2}{(m+1)(m+2)} \tag{205}$$

whence, by substitution into equation 204,

$$V = \frac{2}{(m+1)(m+2)} v \left(\frac{R}{y}\right)^m$$

Substituting this expression for V into equation 203,

$$\tau_o = 0.0332 \left[\frac{2}{(m+1)(m+2)}\right]^{\frac{7}{4}} \mu^{\frac{1}{4}} R^{-\frac{1}{4}+\frac{7m}{4}} v^{\frac{7}{4}} y^{-\frac{7m}{4}} \rho^{\frac{3}{4}}$$

However (Blasius reasoned), wall shear τ_o could depend only upon the form of the velocity profile and the physical properties of the fluid but could not be affected by pipe size R; thus the exponent of R must be zero, and from this the exponent m must be equal to $\frac{1}{7}$. Validation of this depended upon experimental measurements of turbulent velocity profiles which were found to agree quite well with the hypothesis above; thus the so-called *seventh-root law* for turbulent velocity distribution has been widely accepted. It is written

$$\frac{v}{v_c} = \left(\frac{y}{R}\right)^{\frac{1}{7}} \tag{206}$$

A useful corollary of this law is an equation for wall shear τ_o in terms of v_c and R. Using $m = \frac{1}{7}$ in equation 205 yields $V/v_c = 49/60$, and, substituting $49v_c/60$ into equation 203 for V and rearranging,

$$\tau_o = 0.0464 \left(\frac{\mu}{v_c \rho R}\right)^{\frac{1}{4}} \frac{\rho v_c^2}{2} \tag{207}$$

which will be used later in the approximate analysis of turbulent boundary layers.

The extension of Blasius' methods to smooth pipes with $\mathbf{R} > 10^5$ may be done on a piecemeal basis by approximating the smooth pipe line of Fig. 119 or 121 with lengths of straight lines having the equation $f = \text{constant } \mathbf{R}^{-n}$. The general relation between m and n may be shown to be $m = n/(2 - n)$. Since $n < \frac{1}{4}$ when $\mathbf{R} > 10^5$, $m < \frac{1}{7}$ for this region, and the turbulent velocity profile is a flatter curve—which confirms the expected trend with increasing Reynolds number.

Illustrative Problem

For the illustrative problem of Art. 69, calculate the friction factor, wall shear, center velocity, and velocity 1 in. from the pipe centerline, using the seventh-root law.

Solution. From the illustrative problem cited, $V = 4.08$ fps and $\mathbf{R} = 93{,}800$.

$$f = 0.316/(93{,}800)^{\frac{1}{4}} = 0.018 \tag{186}$$

$$\frac{4.08}{v_c} = \frac{2}{(\frac{8}{7})(\frac{15}{7})}; \; v_c = 5.00 \text{ fps} \tag{205}$$

$$\frac{v_1}{5.00} = \left(\frac{1}{3}\right)^{\frac{1}{7}}; \quad v_1 = 4.27 \text{ fps} \tag{206}$$

$$\tau_o = 0.018 \times 1.935 \times (4.08)^2/8 = 0.073 \text{ ps} \tag{183}$$

Compare results with the illustrative problem of Art. 69.

73. Pipe Friction in Noncircular Pipes—The Hydraulic Radius. Although the majority of pipes used in engineering practice are of circular cross section, occasions arise when calculations must be carried out on head loss in rectangular passages and other conduits of noncircular form. The foregoing equations for circular pipes may be adapted to these special problems through use of the *hydraulic radius* concept.

The hydraulic radius, R_h, is defined as the area, A, of flow cross section divided by its "wetted perimeter," P (see Art. 57). In a circular pipe of diameter d,

$$R_h = d/4 \quad \text{or} \quad d = 4R_h \tag{208}$$

This value may be substituted into the Darcy equation for head loss and into the expression for the Reynolds number with the following results:

$$h_L = \frac{f}{4}\frac{l}{R_h}\frac{V^2}{2g} \tag{209}$$

and

$$\mathbf{R} = \frac{V(4R_h)\rho}{\mu} \tag{210}$$

from which the head loss in many conduits of noncircular cross section may be calculated with the aid of the Moody diagram of Fig. 121.

The calculation of lost head in noncircular conduits thus involves the calculation of the hydraulic radius of the flow cross section and the use of the friction factor obtained for an *equivalent* circular pipe having a diameter d equal to $4R_h$. In view of the complexities of laminar films, turbulence, roughness, shear stress, etc., it seems surprising at first that a circular pipe equivalent to a noncircular conduit may be obtained so easily, and it would, therefore, be expected that the method might be subject to certain limitations.

The method gives satisfactory results when the problem is one of turbulent flow but, if it is used for laminar flow, large errors are introduced.

The foregoing facts may be justified analytically by examining further the structure of equation 209 in which h_L varies with $1/R_h$. From the definition of the hydraulic radius, its reciprocal is the *wetted perimeter per unit of flow cross section* and is, therefore, an index of the extent of the boundary surface in contact with the flowing fluid. The hydraulic radius may be safely used in the equation above when resistance to flow and head loss are primarily dependent upon the extent of the boundary surface, as for turbulent flow in which pipe friction phenomena are confined to a thin region adjacent to the boundary surface and thus vary with the size of this surface. However, severe deviation from a circular flow cross section will prevent the hydraulic radius from accounting for changes in head loss, even for cases of turbulent flow. Tests on turbulent flow through annular passages,[18] for example, show a large increase in friction factor with increase of the ratio of core diameter to pipe diameter.

In laminar flow, friction phenomena result from the action of viscosity throughout the whole body of the flow, are independent of roughness, and are not primarily associated with the region close to the boundary walls. In view of these facts the hydraulic radius technique cannot be expected to give reliable conversions from circular to noncircular passages in laminar flow. This expectation is borne out both by experiment and by analytical solutions of laminar flow in noncircular passages.

Illustrative Problem

Calculate the loss of head and the pressure drop when air at 14.7 psia and 60°F flows through 2000 ft of an 18-in. by 12-in. smooth rectangular duct with a mean velocity of 10 fps.

Solution.

$$R_h = \frac{18 \times 12}{2 \times 18 + 2 \times 12} = 3.6 \text{ in.} = 0.30 \text{ ft}$$

$$\mathbf{R} = \frac{10 \times 4 \times 0.3 \times 0.0763/32.2}{0.000000375} = 75{,}900 \tag{210}$$

[18] See W. M. Owen, "Experimental Study of Water Flow in Annular Pipes," *Trans. A.S.C.E.*, Vol. 117, 1952; and J. E. Walker, G. A. Whan, R. R. Rothfus, "Fluid Friction in Noncircular Ducts," *Jl.A.I.Ch.E.*, Vol. 3, 1957.

From the Moody diagram of Fig. 121,

$$f \cong 0.019$$

$$h_L \cong 0.019 \times \frac{2000}{4 \times 0.30} \times \frac{(10)^2}{2g} \cong 49.2 \text{ ft of air} \tag{209}$$

$$\Delta p \cong 0.0763 \times 49.2 \cong 3.75 \text{ psf} \cong 0.026 \text{ psi}$$

74. Pipe Friction Calculations by Empirical Methods. Many empirical formulas are used for pipe friction calculations in engineering practice. These formulas have been derived mostly from tests of flow of water under turbulent conditions and usually cannot be employed reliably for the flow of other fluids. The Hazen-Williams formula [19]

$$V = 1.318\, C_{hw} R_h^{\,0.63} S^{0.54} \tag{211}$$

is typical of such formulas and is probably the one most widely used in American practice. In this formula V is the mean velocity (fps), R_h the hydraulic radius (ft), S the head lost per unit length of pipe (ft/ft), and C_{hw} a roughness (or smoothness) coefficient, typical values of which are given in Table III.

TABLE III

Hazen-Williams Coefficient, C_{hw}

Pipes extremely straight and smooth	140
Pipes very smooth	130
Smooth wood, smooth masonry	120
New riveted steel, vitrified clay	110
Old cast iron, ordinary brick	100
Old riveted steel	95
Old iron in bad condition	60–80

In view of the shape of the formula and the rather indefinite descriptions to be interpreted for selection of coefficients, it is difficult to judge its validity or meaning without wide experience with its application. It is not evident whether C_{hw} is a measure of absolute or relative roughness, whether there is any effect of Reynolds number in the formula, whether it applies only to the wholly rough zone of boundary roughness, etc. Such questions can be answered fairly conclusively if the equation is rewritten so that the more basic yard-

[19] G. S. Williams and A. Hazen, *Hydraulic Tables*, 3rd ed., John Wiley and Sons, 1933.

stick of Fig. 121 may be applied. Substituting $d/4$ for R_h, h_L/l for S, solving for h_L, and multiplying by $2g/2g$, there results

$$h_L = \frac{194}{C_{hw}{}^{1.85}} \frac{1}{(Vd)^{0.15}} d^{0.015} \frac{l}{d} \frac{V^2}{2g}$$

If $d^{0.015}$ is dropped, because of the small exponent, and the bracketed quantity multiplied by $(\nu/\nu)^{0.15}$, a Reynolds number appears and the equation may be rewritten.

$$h_L = \frac{194}{\nu^{0.15} C_{hw}{}^{1.85}} \frac{1}{\mathbf{R}^{0.15}} \frac{l}{d} \frac{V^2}{2g}$$

Comparing this with the Darcy equation (182) and taking a nominal value of ν for water of, say, 0.00001 sqft/sec,

$$f = \frac{1090}{C_{hw}{}^{1.85} \mathbf{R}^{0.15}}$$

which may be plotted and compared with Fig. 121. The appearance of the curves is indicated on Fig. 127 which compares roughly (but

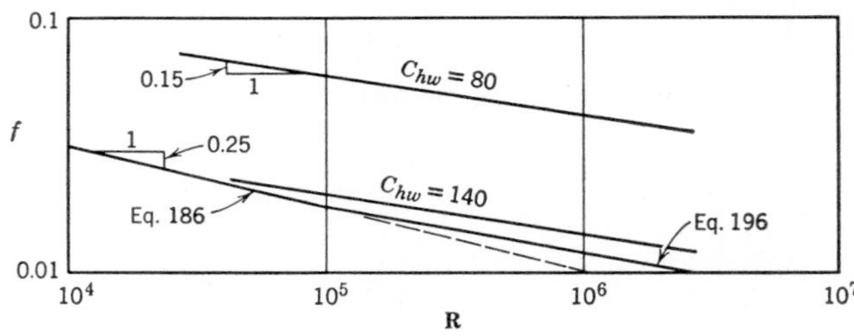

FIG. 127

favorably) with the curves for the rough pipes of Fig. 121. In view of this favorable comparison, it may be concluded that C_{hw} is an index of relative roughness rather than the absolute roughness which might be assumed from Table III.

The advantages and disadvantages of the Hazen-Williams method are evident from the formula and foregoing discussion. Among its advantages are: (1) the coefficient is a rough measure of relative roughness, (2) the effect of Reynolds number is included in the formula, and (3) the effect of roughness and the other variables upon the velocity are given directly. Its disadvantages are: (1) its empirical nature, and (2) the impossibility of applying it to all fluids under all conditions. Although the formula appears cumbersome with its fractional exponents, this disadvantage is overcome in engineering practice by the use of tables and diagrams in its solution.

ILLUSTRATIVE PROBLEM

If 90 gpm of water flow through a smooth 3-in. pipe line, calculate the loss of head in 3000 ft of this pipe.

Solution.

$$V = \frac{90}{60 \times 7.48} \times \frac{1}{(\pi/4)(\frac{3}{12})^2} = 4.08 \text{ fps} \tag{53}$$

$$R_h = (\tfrac{3}{12})/4 = 0.0625 \text{ ft} \tag{208}$$

From Table III,

$$C_{hw} = 140$$

$$4.08 = 1.318 \times 140 \times (0.0625)^{0.63} S^{0.54} \tag{211}$$

$$S = 0.0218 = h_L/3000; \quad h_L = 65.3 \text{ ft of water}$$

Compare results with the illustrative problem of Art. 69.

75. Pipe Friction for Compressible Fluids. The calculation of pressure drop due to friction when gases and vapors flow in uninsulated pipe lines is in general a rather complex thermodynamic problem and therefore cannot be treated exhaustively here. However, the isothermal (constant-temperature) flow of gas in a pipe line is a simple problem which has wide application in practice and will serve as a practical illustration of a pipe flow situation with a fluid of varying density. To secure isothermal flow in a pipe the heat transferred out of the fluid through the pipe walls and the energy converted into heat by the friction process must be adjusted so that the fluid temperature remains constant. Such an adjustment is approximated naturally in uninsulated pipes where velocities are low (well below the sonic) and where temperatures inside and outside the pipe are of the same order; frequently the flow of gases in long pipe lines may be treated isothermally.

The weight flowrate, G, through the line (see Fig. 128) being constant and given by $G = A\gamma V$, the Reynolds number may be calculated as

$$\mathbf{R} = \frac{V d\rho}{\mu} = \frac{G}{A\gamma}\frac{d\rho}{\mu} = \frac{Gd}{\mu g A} \tag{212}$$

and is seen to be a constant of the flow inasmuch as μ does not vary if there is no temperature change. Hence the friction factor is a

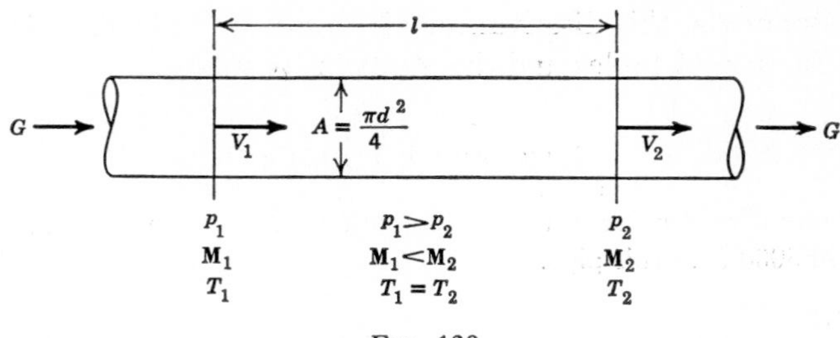

FIG. 128

constant of the problem even though the velocity of the gas will increase and its density decrease in a downstream direction.

Applying equation 161 (Art. 57), omitting the z terms,

$$\frac{dp}{\gamma} + \frac{V\,dV}{g} + \frac{\tau_o}{\gamma R_h}\,dl = 0$$

Substituting $f\rho V^2/8$ for τ_o and $d/4$ for R_h, and multiplying through by $2g/V^2$,

$$\frac{2g}{V^2}\frac{dp}{\gamma} + 2\frac{dV}{V} + \frac{f}{d}\,dl = 0$$

Substituting $G^2/A^2\gamma^2$ for V^2 and p/RT for γ,

$$\frac{2gA^2}{G^2RT}\int_1^2 p\,dp + 2\int_1^2 \frac{dV}{V} + \frac{f}{d}\int_1^2 dl = 0$$

and integrating, there results

$$p_1^2 - p_2^2 = \frac{G^2RT}{gA^2}\left[2\ln\frac{V_2}{V_1} + f\frac{l}{d}\right] \tag{213}$$

This equation can be rearranged by use of the continuity equation $A_1\gamma_1 V_1 = A_2\gamma_2 V_2$ in which $A_1 = A_2$, $\gamma_1 = p_1/RT$, and $\gamma_2 = p_2/RT$. Thus, by substitution, $p_1V_1 = p_2V_2$ and $V_2/V_1 = p_1/p_2$. Hence equation 213 may be written

$$p_1^2 - p_2^2 = \frac{G^2RT}{gA^2}\left[2\ln\frac{p_1}{p_2} + f\frac{l}{d}\right] \tag{214}$$

Generally the solution of this equation for p_2 must be accomplished by trial, but frequently $2\ln p_1/p_2$ is so small (in comparison with fl/d) that it may be neglected, thus allowing a direct solution.

An important limitation to the applicability of this equation occurs for large pressure drops ($p_2 \ll p_1$), which may be seen by further rearrangement. For brevity let $G^2RT/gA^2 = B$ and rearrange the equation to read

$$(p_2^2 - 2B \ln p_2) = (p_1^2 - 2B \ln p_1) - fl/d$$

For a fixed value of p_1, an increase of l will be expected to produce a reduction of p_2 and an increase of pressure drop ($p_1 - p_2$), all of which may be anticipated from the form of the equation. However, deeper investigation of the equation will reveal that there is a point beyond which a *reduction of l is required for further reduction of* p_2; the pressure cannot drop below this point and thus the equation is applicable only between the pressure p_1 and the limiting value of p_2.

Equation 214 may also be written (in conformance with modern practice in gas dynamics) in terms of pressure ratio and Mach number, **M**. Divide the equation by p_1^2 and multiply the right-hand side by k/k and substitute $A\gamma_1 V_1$ for G; whereupon $G^2RTk/kp_1^2gA^2$ will be found to be equal to $k\mathbf{M}_1^2$. Since the sonic velocity, a, is constant for isothermal flow, $V_2/V_1 = \mathbf{M}_2/\mathbf{M}_1$, and, since $V_2/V_1 = p_1/p_2$ (see above), $p_2/p_1 = \mathbf{M}_1/\mathbf{M}_2$ and equation 214 becomes

$$\frac{\mathbf{M}_1^2}{\mathbf{M}_2^2} = 1 - k\mathbf{M}_1^2 \left[2\, ln \frac{\mathbf{M}_2}{\mathbf{M}_1} + f\frac{l}{d}\right] \tag{215}$$

Considering the flow process in terms of Mach number, $\mathbf{M}_1 \ll 1$ and $\mathbf{M}_2 > \mathbf{M}_1$. From the above-mentioned limitation of the equation the limiting value of $\mathbf{M}_2$ may be found by differentiating equation 215 with respect to l and setting $dp_2/dl = 0$; the result is $\mathbf{M}_2 = \sqrt{1/k}$. In other words the equation is applicable only to that portion of the subsonic flow regime where $\mathbf{M}_1 < \mathbf{M}_2 \leqq \sqrt{1/k}$. Such limitations are of far-reaching importance in engineering problems of gas flow; for nonisothermal flows there are analogous limitations, but these are considerably more difficult of computation and cannot be included here.

Illustrative Problem

If 40 lb/min of air flow isothermally through a smooth 3-in. pipe line at a temperature of 100°F, and the pressure at a point in this line is 50 psia, calculate the pressure in the line 2000 ft downstream from this point.

Solution.

$$\mathbf{R} = \frac{(\frac{40}{60})(\frac{3}{12})}{4 \times 10^{-7} \times 32.2 \times 0.0491} = 262{,}000 \tag{212}$$

From the Moody diagram of Fig. 121,

$$f \cong 0.015$$

$$\gamma_1 = 50 \times 144/53.3 \times 560 = 0.241 \text{ lb/cuft} \quad (3)$$

$$V_1 = (\tfrac{40}{60})/0.0491 \times 0.241 = 56.3 \text{ fps} \quad (52)$$

$$a = \sqrt{1.4 \times 32.2 \times 53.3 \times 560} = 1160 \text{ fps} \quad (11)$$

$$\mathbf{M}_1 = 56.3/1160 = 0.0485$$

The limiting value of $\mathbf{M}_2$ is $\sqrt{1/1.4} = 0.845$. Calculating the limiting value of l for applicability of equation 214 from equation 215,

$$\left(\frac{0.0485}{0.845}\right)^2 = 1 - 1.4(0.0485)^2\left[2 \ln \frac{0.845}{0.0485} + 0.015\frac{l}{0.25}\right]; \quad l = 7000 \text{ ft}$$

Since 2000 < 7000, equation 214 is applicable. Substituting values therein,

$$(144)^2[(50)^2 - p_2{}^2] = \frac{(\frac{40}{60})^2 \times 53.3 \times 560}{32.2 \times (0.0491)^2}\left[2 \ln \frac{50}{p_2} + 0.015\frac{2000}{0.25}\right]$$

Solving for p_2 by trial, $p_2 = 38.9$ psia.

76. Minor Losses in Pipe Lines. Into the category of minor losses in pipe lines fall those losses incurred by change of section, bends, elbows, valves, and fittings of all types. Although in long pipe lines these are distinctly "minor" losses and can often be neglected without serious error, in shorter pipe lines an accurate knowledge of their effects must be known for correct engineering calculations.

The general aspects of minor losses in pipe lines may be obtained from a study of the flow phenomena about an abrupt obstruction placed in a pipe line (Fig. 129), which creates flow conditions typical of those which dissipate energy and cause minor losses. Minor losses usually result from rather abrupt changes (in magnitude or direction) of velocity; in general, increase of velocity (acceleration) is associated with small head loss but decrease of velocity (deceleration) causes large head loss because of the production of large scale turbulence (Art. 59). In Fig. 129 useful energy is extracted in the creation of eddies as the fluid decelerates between sections 2 and 3, and this energy is dissipated in heat as the eddies decay between sections 3 and 4. Minor losses in pipe flow are, therefore, accomplished in the pipe downstream from the source of the eddies, and the pipe friction processes in this length of pipe are hopelessly complicated by the superposition of large-scale turbulence upon the normal turbulence pattern. To make minor loss calculations possible it is necessary to assume separate action of the normal turbulence and large-scale turbulence, although in reality a complex combina-

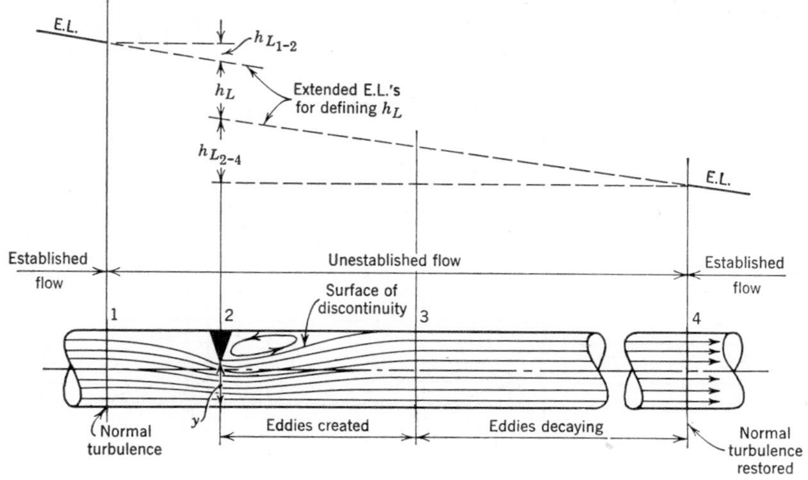

FIG. 129

tion of the two processes exists. Assuming the processes independent allows calculation of the losses due to established pipe friction, $h_{L_{1-2}}$ and $h_{L_{2-4}}$, and also permits the loss h_L, due to the obstruction alone, to be assumed concentrated at section 2. This is a great convenience for engineering calculations since the total lost head in a pipe line may be obtained by simple addition of established pipe friction and minor losses without detailed consideration of the above-mentioned complications.

Early experiments with water (at high Reynolds number) indicated that minor losses vary approximately with the square of velocity and led to the proposal of the basic equation

$$h_L = K_L \frac{V^2}{2g} \tag{216}$$

in which K_L, the *loss coefficient*, is, for a given flow geometry, practically constant at high Reynolds number; the loss coefficient tends to increase with increasing roughness and decreasing Reynolds number,[20] but these variations are usually of minor importance in turbulent flow. The magnitude of the loss coefficient is determined primarily by the flow geometry, i.e., by the shape of the obstruction or pipe fitting.

When an *abrupt enlargement* of section (Fig. 130) occurs in a pipe line, a rapid deceleration takes place, accompanied by characteristic large-scale turbulence, which may persist in the larger pipe

[20] Note that this is the same trend followed by the friction factor, f.

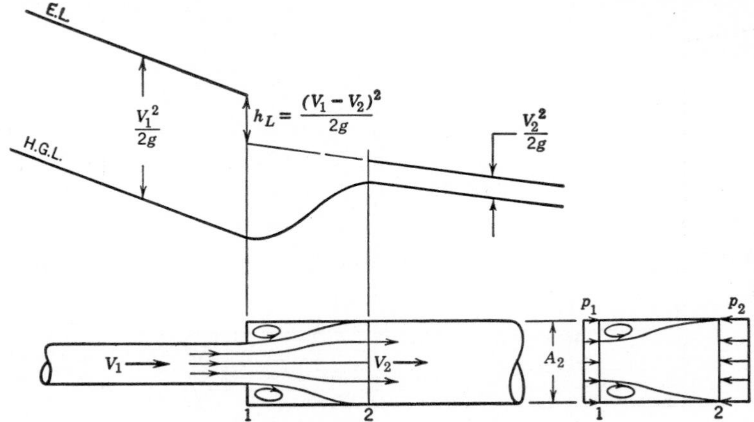

FIG. 130. Abrupt enlargement.

for a distance of 50 diameters or more before the normal turbulence pattern of established flow is restored. Simultaneous application of the continuity, Bernoulli, and momentum principles to this problem has shown (Art. 43) that (with certain simplifying assumptions)

$$h_L = K_L \frac{(V_1 - V_2)^2}{2g} \tag{217}$$

in which $K_L \cong 1$. Experimental determinations of K_L confirm this value within a few percent, making it quite adequate for engineering use.

A special case of an abrupt enlargement exists when a (relatively small) pipe discharges into a (relatively large) tank or reservoir (Fig. 131). Here the velocity downstream from the enlargement may

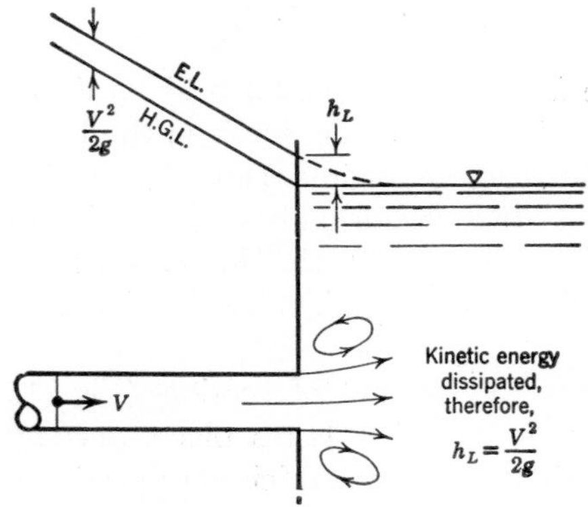

FIG. 131. Pipe exit.

be taken to be zero, and when the lost head (called the exit loss) is calculated from equation 217 it is found to be the velocity head in the pipe. This agrees closely with the result of the nonmathematical analysis indicated in Fig. 131.

The loss of head due to *gradual enlargement* is, of course, dependent upon the shape of the enlargement. Tests have been carried out by Gibson on the losses in conical enlargements, and the results are expressed by equation 217, in which K_L is primarily dependent upon the cone angle but is also a function of the area ratio, as shown in Fig. 132. Because of the large surface of the conical enlargement

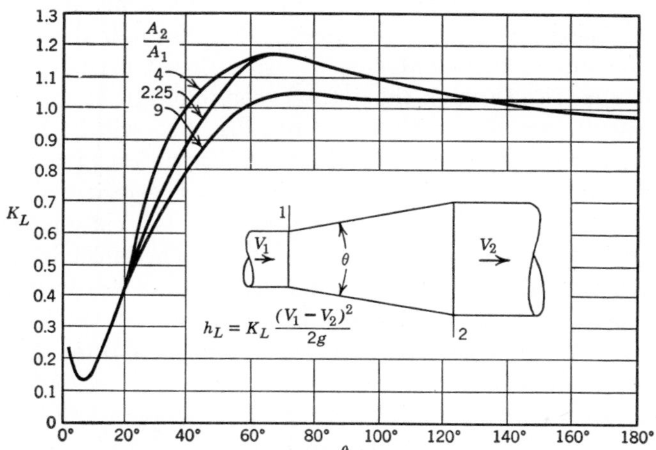

FIG. 132. Loss coefficients for conical enlargements.[21]

which contacts the fluid, the coefficient K_L embodies the effects of wall friction as well as those of large-scale turbulence. In an enlargement of small central angle, K_L will result almost wholly from surface friction; but, as the angle increases and the enlargement becomes more abrupt, not only is the surface area reduced but also separation occurs, producing large eddies, and here the energy dissipated in the eddies determines the magnitude of K_L. From the plot it may be observed that (1) there is an optimum cone angle of about 7° where the combination of the effects of surface friction and eddying turbulence is a minimum; (2) it is better to use a sudden enlargement than one of cone angle around 60°, since K_L is smaller for the former.

Gradual enlargements in passages (termed *diffusers*) of various forms are widely used in engineering practice for *pressure recovery*,

[21] A. H. Gibson, *Hydraulics and Its Applications*, 4th Edition, p. 93, D. Van Nostrand Co., 1930.

i.e., pressure rise in the direction of flow. No attempt will be made here to review the extensive literature [22] on diffusers, but it should be realized that Gibson's results cannot give a reliable solution to this problem. His tests were made, as are all such tests for minor losses, with long straight lengths of pipe upstream and downstream from the enlargement (see Fig. 129). The designer, however, is frequently more interested in the pressure rise through the diffuser and substitution of a short nozzle for the upstream pipe length. The tests of Gibson and others have shown that the pressure will continue to rise for a few pipe diameters downstream from section 2, owing primarily to readjustment of velocity distribution from a rather pointed one caused by deceleration through the diffuser to the flatter one of turbulent flow.[23] From this it may be concluded that pressure rise through the diffuser, computed from application of the Bernoulli equation with data from Fig. 132, will be larger than that which will actually be realized. Substitution of a nozzle for the upstream pipe length will alter the inlet velocity distribution from the standard one of turbulent flow to a practically uniform one with a thin boundary layer (see Fig. 100). The effect is to reduce the losses by stabilizing the flow and delaying separation; not only will smaller loss coefficients result, but also the cone angle for minimum losses will become larger, thus allowing a shorter diffuser for the same area ratio.

Illustrative Problem

A 12-in. horizontal water line enlarges to a 24-in. line through a 20° conical enlargement. When 10 cfs flow through this line the pressure in the smaller pipe is 20 psi. Calculate the pressure in the larger pipe, neglecting pipe friction.

Solution.

$$V_{12} = 10/(\pi/4) = 12.7 \text{ fps}, \quad V_{24} = 12.7/4 = 3.18 \text{ fps} \qquad (53)$$

From the plot of Fig. 132,

$$K_L = 0.43$$

$$\frac{20 \times 144}{62.4} + \frac{(12.7)^2}{2g} = \frac{p_{24}}{\gamma} + \frac{(3.18)^2}{2g} + 0.43\frac{(12.7 - 3.18)^2}{2g} \qquad (159)$$

$$p_{24}/\gamma = 47.9 \text{ ft}, \quad p_{24} = 20.7 \text{ psi}$$

[22] See, for example, E. G. Reid, "Performance Characteristics of Plane-Wall Two-Dimensional Diffusers," *N.A.C.A., Tech. Note* 2888, 1953; S. J. Kline, D. E. Abbott, and R. W. Fox, "Optimum Design of Straight-Walled Diffusers," *Trans. A.S.M.E.*, Vol. 81, 1959; and J. M. Robertson and H. R. Fraser, "Separation Prediction in Conical Diffusers," *Trans. A.S.M.E.* (*Series D*), Vol. 82, No. 1, 1960.

[23] Compare this with the opposite situation discussed in Art. 55.

This is the pressure to be expected a few feet downstream from the end of the enlargement.

Flow through an *abrupt contraction* is shown in Fig. 133 and is featured by the formation of a vena contracta (Art. 28) and subsequent re-expansion of the live stream of flowing fluid.

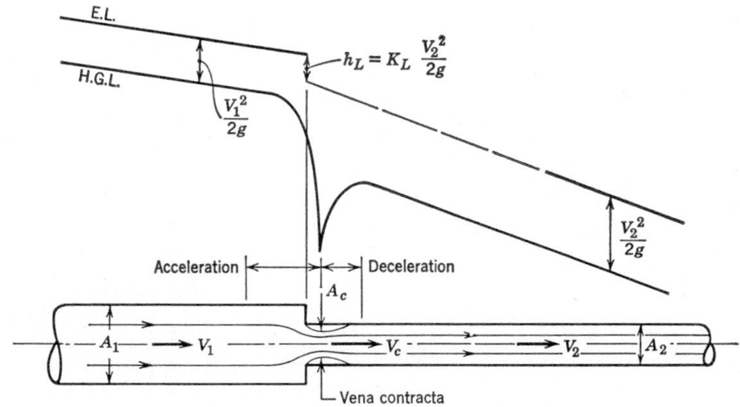

FIG. 133. Abrupt contraction.

Experimental measurements of K_L are somewhat conflicting in magnitude although they exhibit a well-established trend from 0.5 for $A_2/A_1 = 0$ to 0 for $A_2/A_1 = 1$. In view of this it is more enlightening and entirely adequate for engineering practice to use an analytical approach to the problem based on generally accepted experimental information. Early measurements by Weisbach [24] established accurately the magnitude of the jet contraction; these results, expressed as a coefficient of contraction C_c (ratio of A_c to A_2), were found (for flows at high Reynolds number) to be dependent upon A_2/A_1, as given in Table IV.

Taking the head loss caused by the abrupt contraction as the sum of the head losses in the acceleration and deceleration regions,

TABLE IV

A_2/A_1	0	0.1	0.2	0.3	0.4	0.5	0.6	0.7	0.8	0.9	1.0
C_c	0.617	0.624	0.632	0.643	0.659	0.681	0.712	0.755	0.813	0.892	1.00
$(1/C_c - 1)^2$	0.385	0.365	0.340	0.308	0.266	0.219	0.164	0.106	0.053	0.015	0
K_a/C_c^2	0.115	0.093	0.074	0.056	0.041	0.029	0.018	0.010	0.005	0.001	0
K_L (approx.)	0.50	0.46	0.41	0.36	0.30	0.24	0.18	0.12	0.06	0.02	0

[24] J. Weisbach, *Die experimental Hydraulik*, 1855,

and treating these as gradual contraction and abrupt enlargement, respectively,

$$h_L = K_L \frac{V_2^2}{2g} = K_a \frac{V_c^2}{2g} + \frac{(V_c - V_2)^2}{2g}$$

From continuity considerations, $A_2 V_2 = A_c V_c$ or $V_c = V_2/C_c$; substituting V_c and canceling $V_2^2/2g$

$$K_L = \frac{K_a}{C_c^2} + \left(\frac{1}{C_c} - 1\right)^2$$

from which K_L can be estimated after a reasonable assumption is made for K_a/C_c^2. With K_L for $A_2/A_1 = 0$ established at 0.50, K_a/C_c^2 must be close to 0.115 since $C_c = 0.617$ and $[(1/C_c) - 1]^2 = 0.385$. Evidently, K_a/C_c^2 must be zero when $A_2/A_1 = 1$, inasmuch as there will be no contraction at all for this limiting case. Assuming for convenience that K_a/C_c^2 diminishes linearly (from 0.115 to 0) with the diameter ratio (parabolically with the area ratio), the tabulated values will be obtained, and they may be totaled to give reasonable loss coefficients for the abrupt contraction. In passing, it should be observed that although such methods are not rigorous they give results of adequate precision for practical use and are frequently the only methods available for simple analysis of such problems. The relative sizes of the acceleration and deceleration portions of the loss coefficient should also be noted; although the increase of velocity from section 1 to vena contracta is always considerably more than the decrease from vena contracta to section 2, the loss produced by the latter is always the larger; this is another example of the characteristic efficiency associated with acceleration and the inefficiency associated with deceleration in fluid flow.

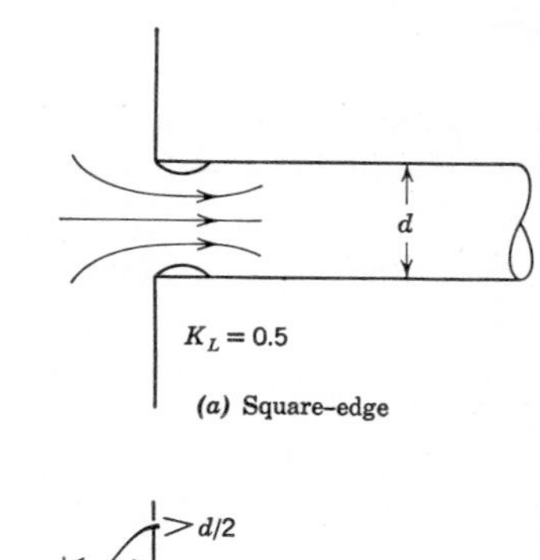

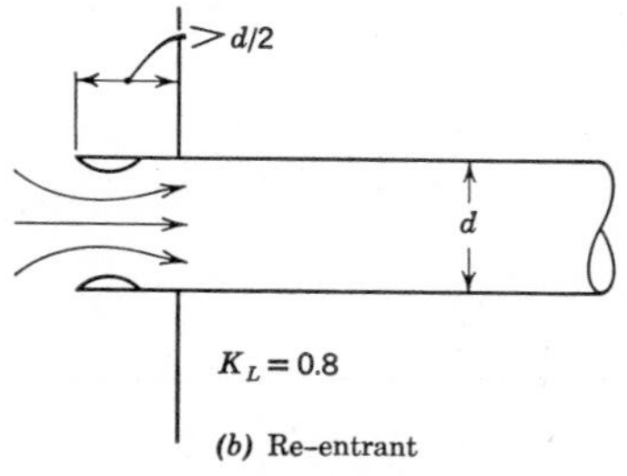

FIG. 134. Pipe entrances.

A square-edged pipe entrance (Fig. 134*a*) from a large body of fluid is the limiting case of the abrupt contraction, with $A_2/A_1 = 0$ (A_1 is virtually infinite here). The head loss is expressed by equation 216 in which K_L is close to 0.5 for highly turbulent flow, as mentioned above.

The entrance of Fig. 134*b* is known as a re-entrant one. If the pipe wall is very thin and if the plane of the opening is more than one pipe diameter upstream from the reservoir wall, the loss coefficient will be close to 0.8, this high value resulting mainly from the small vena contracta and consequent large deceleration loss. For thick-walled pipes the vena contracta can be expected to be larger and the loss coefficient less than 0.8; Harris [25] has shown that pipes of wall thicknesses greater than $0.05d$ (if square-edged) will give a loss coefficient equal to that of the square-edged entrance.

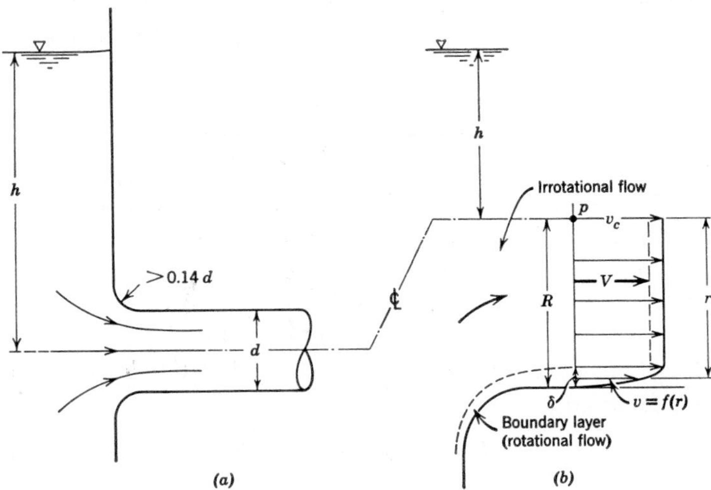

FIG. 135. Bell-mouth entrance.

If the edges of a pipe entrance are rounded to produce a streamlined *bell-mouth* (Fig. 135*a*) the loss coefficient can be materially reduced. Hamilton [26] has shown that any radius of rounding greater than $0.14d$ will prevent the formation of a vena contracta and thus eliminate the head loss due to flow deceleration. The nominal value of K_L for such an entrance is about 0.1, but its exact magnitude will depend upon the detailed geometry of the entrance and structure of the boundary layer (see Fig. 100 of Art. 58); recent experiments at Stanford University with Reynolds numbers around 100,000 have given small negative values in the case of a long laminar boundary layer, and above 0.1 when the boundary layer was turbulent.

[25] C. W. Harris, "The Influence of Pipe Thickness on Reentrant Intake Losses," *Univ. Wash. Eng. Expt. Sta., Bull.* 48, 1928.

[26] J. B. Hamilton, "The Suppression of Intake Losses by Various Degrees of Rounding," *Univ. Wash. Eng. Expt. Sta., Bull.* 51, 1929.

The growth of boundary layers in the bell-mouth of a rounded pipe entrance is the frictional mechanism by which energy is dissipated, which in turn produces a value of K_L greater than zero. Some insight into the relation between boundary layer thickness and K_L may be gained by an approximate analysis of the problem (see Fig. 135*b*). From the continuity principle,

$$V\pi R^2 = v_c\pi(R - \delta)^2 + \int_{R-\delta}^{R} 2\pi v r\, dr$$

in which $v = F(r)$. Assuming that the relation between v and r is a parabolic one,[27] the integration may be completed to yield a relation between V and v_c; the result is

$$\frac{v_c}{V} = \frac{1}{1 - (\frac{2}{3})(\delta/R) + (\frac{1}{6})(\delta/R)^2} \tag{218}$$

For the irrotational flow there are no head losses, so the Bernoulli equation written along the central streamline is

$$H = \frac{p}{\gamma} + \frac{v_c^{\,2}}{2g}$$

For the whole flow, assumed one-dimensional at the downstream end of the bell-mouth,

$$H = \frac{p}{\gamma} + \frac{V^2}{2g} + K_L\frac{V^2}{2g}$$

Comparing these equations and substituting the velocity ratio of equation 218, a relation between K_L and δ is obtained:

$$K_L = \left(\frac{v_c}{V}\right)^2 - 1 = \left[\frac{1}{1 - (\frac{2}{3})(\delta/R) + (\frac{1}{6})(\delta/R)^2}\right]^2 - 1 \tag{219}$$

For the limiting case where there is no boundary layer and the whole flow is irrotational (i.e., frictionless), head loss and K_L would be expected to be zero; this is confirmed by the equation. The equation also shows that K_L will increase with boundary layer thickness; this also is to be expected since, from the nature of boundary layer growth (Art. 58), increase of boundary layer thickness implies larger viscous action. The reader is reminded that equation 219 is *not a general relationship* between loss coefficient and boundary layer thickness for all flow geometries and boundary layer regimes; it is intended only as an example to prove the existence of a relation-

[27] This assumption is quite accurate if the boundary layer is laminar.

ship. However, it may be used for estimating the order of boundary layer thickness from loss coefficient, or vice versa. Nominal values of K_L and δ/R for short smooth bell-mouths or nozzles are 0.04, and 0.03, respectively. It is of interest to note that K_a for the square-edged pipe entrance (Table IV) is also close to 0.04, and this implies a velocity profile at the vena contracta similar to that of Fig. 135*b*.

The head loss caused by short well-streamlined gradual contractions (Fig. 136) is so small that it may usually be neglected in engineering problems. However, an appreciable fall of the hydraulic grade line over such contractions is to be expected; in the pipe downstream from the contraction the hydraulic grade line will be found to slope more steeply than that for established flow because of the change of velocity distribution and boundary layer growth, which cause an increase of α (see Fig. 100 of Art. 58). These effects are not yet fully understood but are known to be small and may be ignored in many problems. A nominal value of K_L (for use in equation 216 with $V = V_2$) for short well-streamlined contractions is 0.04; by careful design this figure may be lowered to 0.02, but for long contractions values much larger than 0.04 are to be expected because of extensive wall friction.

FIG. 136. Gradual contraction.

Losses of head in *smooth pipe bends* are caused by the combined effects of separation, wall friction, and the twin-eddy secondary flow described in Art. 60; for bends of large radius of curvature, the last two effects will predominate, whereas, for small radius of curvature, separation and the secondary flow will be the more significant. The loss of head in a bend is expressed by equation 216 in which the head loss is the drop of the extended energy lines (see Fig. 129) between the entrance and exit of the bend. Reliable and extensive information on this subject will be found in the recent work of Itō,[28] a small but typical portion of which is shown in Fig. 137. Head loss coefficients for smooth pipe bends provide another example of the dependence of K_L on shape of passage (determined by θ and R/d)

[28] H. Itō, "Pressure Losses in Smooth Pipe Bends," *Trans. A.S.M.E. (Series D)*, Vol. 82, No. 1, 1960.

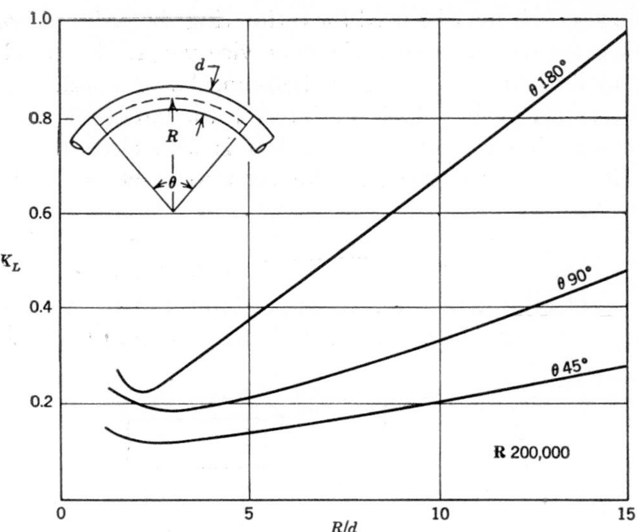

FIG. 137. Itō's loss coefficients for smooth bends (**R** = 200,000).

and Reynolds number; the research of Hofmann[29] provides information on the (expected) dependence of loss coefficient on relative roughness as well. To the engineer the most significant feature of head loss in bends is the minimum value of K_L occurring at certain values of R/d which allows selection of bend shapes for maximum efficiency in pipe-line design.

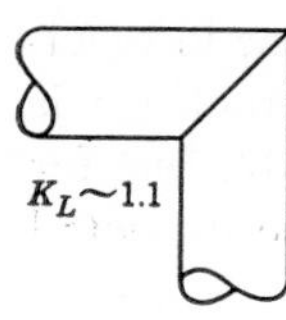

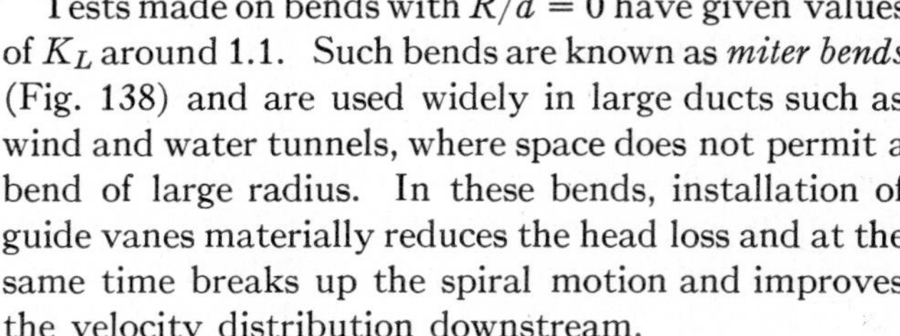

Tests made on bends with $R/d = 0$ have given values of K_L around 1.1. Such bends are known as *miter bends* (Fig. 138) and are used widely in large ducts such as wind and water tunnels, where space does not permit a bend of large radius. In these bends, installation of guide vanes materially reduces the head loss and at the same time breaks up the spiral motion and improves the velocity distribution downstream.

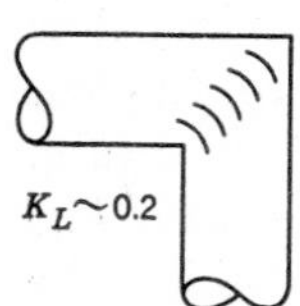

FIG. 138. Miter bends.

The losses of head caused by commercial pipe fittings occur because of their rough and irregular shapes which produce excessive large-scale turbulence. The shapes of commercial pipe fittings are determined more by structural properties, ease in handling, and production methods than by head loss considerations, and it is, therefore,

[29] A. Hofmann, "Loss in 90° Pipe Bends of Constant Circular Cross Section," *Trans. Hydr. Inst. of Munich Techn. Univ., Bull.* 3, A.S.M.E., 1935.

not feasible or economically justifiable to build pipe fitting having completely streamlined interiors in order to minimize head loss. The loss of head in commercial pipe fittings is usually expressed by equation 216 with V the mean velocity in the pipe and K_L a constant (at high Reynolds numbers), the magnitude of which depends upon the shape of the fitting. Values of K_L for various common fittings, compiled by the Crane Co.,[30] are presented in Table V.

TABLE V

LOSS COEFFICIENT, K_L, FOR COMMERCIAL PIPE FITTINGS

Globe valve, wide open	10.0
Angle valve, wide open	5.0
Gate valve, wide open	0.19
Three-fourths open	1.15
One-half open	5.6
One-fourth open	24.0
Return bend	2.2
90° elbow	0.90
45° elbow	0.42

It is generally recognized that when fittings are placed in close proximity the total head loss caused by them is less than their numerical sum obtained by the foregoing methods. Systematic tests have not been made on this subject because a simple numerical sum of losses gives a result in excess of the actual losses and thus produces an error on the conservative side when design calculations of pressures and flowrates are to be made.

77. Pipe-Line Problems—Single Pipes. All steady-flow pipe-line problems may be solved by application of the Bernoulli and continuity equations, and the most effective method of doing this is the construction of energy and hydraulic grade lines. From such lines the variations of pressure, velocity, and unit energy can be clearly seen for the whole problem; thus the construction of these lines becomes equivalent to writing numerous equations, but the lines lend a clarity to the solution of the problem which equations alone never can.

In engineering offices, tables, charts, nomograms, etc., are employed where numerous pipe-flow problems are to be solved. Although all

[30] *Engineering Data on Flow of Fluids in Pipes and Heat Transmission*, The Crane Co., 1935. More accurate and comprehensive information will be found in a later issue of this publication entitled *Flow of Fluids through Valves, Fittings, and Pipe*, Techn. Paper 410, The Crane Co., 1957.

these methods are different they have their foundations in the Bernoulli principle, usually with certain approximations; no attempt will be made here to cover these many methods—the following discussion will be confined to the application of the Bernoulli and continuity principles and the use of certain approximations.

Engineering pipe-flow problems usually consist of (1) calculation of head loss and pressure variation from flowrate and pipe-line characteristics, (2) calculation of flowrate from pipe-line characteristics and the head which produces flow, and (3) calculation of required pipe diameter to pass a given flowrate between two regions of known pressure difference. The first of these problems can be solved directly, but solution by trial is required for the other two.[31] Trial-and-error solutions are necessitated by the fact that the friction factor, f, and loss coefficients, K_L, depend upon the Reynolds number, which in turn depends upon flowrate and pipe diameter, the unknowns of problems 2 and 3, respectively. However, many engineering pipe-line problems involve flow in rough pipes at high Reynolds numbers. Here trial solutions are seldom required (1) because of the tendency of f and K_L toward constancy in this region, (2) because of the inevitable error [32] in selecting f from Fig. 121, and (3) because engineering answers are usually not needed to a precision which warrants trial-and-error solution in the light of the foregoing facts. Construction of energy and hydraulic grade lines for some typical pipe-line problems will indicate further approximations which may frequently be used in the solution of engineering problems.

Consider the calculation of flowrate in a pipe line laid between two tanks or reservoirs having a difference of surface elevation H (Fig. 139). The energy line must start in one reservoir surface and end in the other; using a gradual drop to represent head loss due to pipe friction, h_{L_f}, and abrupt drops to represent entrance and exit losses, h_{L_e} and h_{L_x}, the energy line is constructed as shown. It is apparent from the energy line that

$$h_{L_e} + h_{L_f} + h_{L_x} = H$$

which is the Bernoulli equation written between the reservoir surfaces. When the appropriate expressions for the head losses are substituted (Arts. 63 and 76),

$$\left(0.5 + f\frac{l}{d} + 1\right)\frac{V^2}{2g} = H$$

[31] Unless special plots are devised for circumventing this.

[32] Due to the inexactness of definition of the roughness.

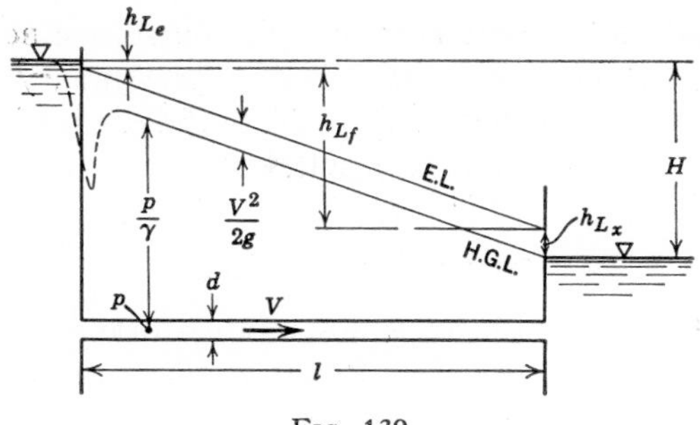

FIG. 139

If turbulent flow is assumed and an approximate value for f of 0.03 is selected (Fig. 121), the quantity in parentheses becomes 4.5, 31.5, and 301.5 for l/d of 100, 1000, and 10,000, respectively. The quantities 0.5 and 1.0, which result from inclusion of minor losses, have a decreasing effect on the equation with increasing l/d; if these were omitted entirely, errors of about 18, 2, and 0.3 percent, respectively, would be produced in the velocity and flowrate. Evidently, the effect of minor losses in pipe lines of reasonable length is so small that they may often be neglected entirely and calculations appreciably simplified. Another convenient approximation accompanies the above; increasing l/d also decreases $V^2/2g$ and thus brings the energy line and the hydraulic grade line closer together; since $V^2/2g$ is of the order of the minor losses, it is consistent to neglect this also, thus making energy and hydraulic grade lines coincident (except near the entrance) and necessitating the construction of only one line. When the single line is drawn for this pipe-line problem (Fig. 140), the equation

$$h_L = f\frac{l}{d}\frac{V^2}{2g} = H$$

may be written, and the velocity and flowrate may be obtained by trial-and-error procedure. The foregoing approximations are con-

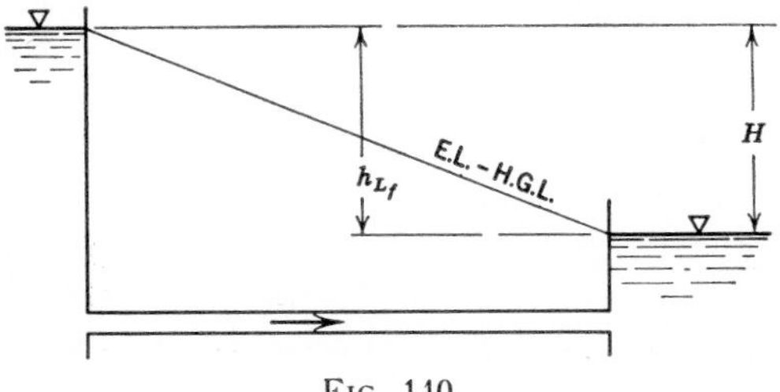

FIG. 140

venient in engineering problems but, of course, cannot be applied blindly and without some experience; preliminary calculations similar to those above will usually indicate the effect of such approximations on the accuracy of the result.

ILLUSTRATIVE PROBLEMS

A clean cast iron pipe line of 12-in. diameter and 1000 ft long connects two reservoirs having surface elevations 200 and 250. Calculate the flowrate through this line, assuming water at 50°F and a square-edged entrance.

Solution.

$$\mathbf{R} = V \times 1/0.0000141 = 70{,}800\,V$$

Applying the Bernoulli equation (159) with appropriate values of loss coefficients,

$$50 = \left(0.5 + f\frac{1000}{1} + 1\right)\frac{V^2}{2g}$$

Assuming $f = 0.030$ gives $V = 10.1$ fps and $\mathbf{R} = 715{,}000$. When these values of f and $\mathbf{R}$ are plotted on Fig. 121, a point is obtained considerably above the (interpolated) curve for $e/d = 0.000833$. Assuming $f = 0.019$ and repeating the above calculation yields $\mathbf{R} = 890{,}000$, which gives (within the accuracy of the plot) a point on the (interpolated) curve. The velocity, V, is then 12.6 fps, and the flowrate 9.9 cfs.

A smooth brass pipe line 200 ft long is to carry a flowrate of 0.1 cfs between two water tanks whose difference of surface elevation is 5 ft. If a square-edged entrance and water at 50°F are assumed, what diameter of pipe is required?

Solution.

$$V = 0.1/(\pi/4)d^2 = 0.1272/d^2 \qquad (53)$$

$$\mathbf{R} = \frac{0.1272}{d^2}\,\frac{d}{0.0000141} = \frac{9020}{d} \qquad (175)$$

Applying the Bernoulli equation (159) with appropriate values of loss coefficients,

$$5.0 = \left(0.5 + f\frac{200}{d} + 1\right)\frac{V^2}{2g}$$

Assuming a value of d allows computation of $\mathbf{R}$ and f. The solution is obtained when these values give a point on the smooth pipe curve of Fig. 121. This occurs (within the accuracy of the plot) when $d = 0.187$ ft, $V = 3.63$ fps, $\mathbf{R} = 48{,}300$, and $f = 0.0214$. A pipe diameter of 2.24 in. is thus theoretically required. Commercial brass pipe of 2.5-in. diameter should be specified.

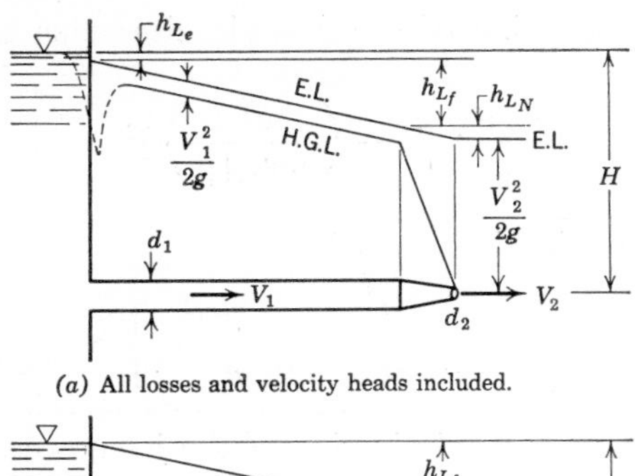

(*a*) All losses and velocity heads included.

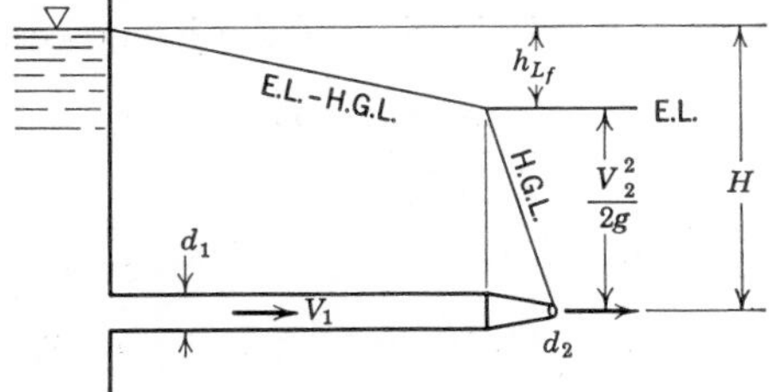

(*b*) Minor losses and pipe line velocity head neglected.

FIG. 141

Energy and hydraulic grade lines are shown in Fig. 141 for a pipe line terminating in a nozzle. Here, if l/d is large and d_2/d_1 small, minor losses and pipe-line velocity head may be neglected. However, the velocity head at the tip of the nozzle cannot be neglected and therefore the energy and hydraulic grade lines, which are coincident over the pipe for the approximate solution, separate as shown over the nozzle. The flowrate may be predicted from the simultaneous solution of the continuity and Bernoulli equations:

$$Q = A_1 V_1 = A_2 V_2 \tag{53}$$

$$H = \frac{V_2^2}{2g} + f\frac{l}{d}\frac{V_1^2}{2g} \tag{159}$$

If the jet from the nozzle were to drive an impulse turbine (Art. 50), the jet power (which is the input to the machine) would be of more interest to the engineer than the flowrate; specifically he would be most concerned with providing pipe and nozzle of such relative proportions that this power could be maximized. The general expression for jet power is (from equation 68)

$$P = Q\gamma V_2^2/2g \tag{220}$$

However, by substitution of the first simultaneous equation in the second,

$$\frac{V_2^2}{2g} = H - \left(\frac{fl}{2gd_1A_1^2}\right)Q^2$$

whence by substitution in equation 220

$$P = Q\gamma\left[H - \frac{flQ^2}{2gd_1A_1^2}\right]$$

Taking $dP/dQ = 0$ for maximization of the jet power,

$$\frac{flQ^2}{2gd_1A_1^2} = \frac{H}{3}$$

which shows that maximum jet power may be expected when [33]

$$flV_1^2/2gd_1 = H/3 \quad \text{and} \quad V_2^2/2g = 2H/3$$

from which pipe and nozzle may be sized for any available flowrate.

When a pipe line runs above its hydraulic grade line, negative pressure in the pipe is indicated (Art. 25). Sketching pipe and hydraulic grade line to scale indicates regions of negative pressures and critical points which may place limitations on the flowrate. Theoretically the absolute pressure in a pipe line may fall to the vapor pressure of the liquid, at which point cavitation (Appendix VII) sets in; however, this extreme condition is to be avoided in pipe lines, and much trouble can be expected before such low pressures are attained. Most engineering liquids contain dissolved gases which will come out of solution well before the cavitation point is reached; since such gases go back into solution very slowly, they move with the liquid as large bubbles, collect in the high points of the line, reduce the flow cross section, and tend to disrupt the flow. In practice, large negative pressures in pipes should be avoided if possible by improvements of design; where such negative pressures cannot be avoided they should be prevented from exceeding about two-thirds of the difference between barometric and vapor pressures.

Illustrative Problem

How many horsepower must be applied to pump 2.5 cfs of water (50°F) from the lower to the upper reservoir? Neglect minor losses and velocity

[33] In hydro power practice, $V_2^2/2g$ will be found to be considerably larger than $(\frac{2}{3})H$, depending upon the economic value of the water. The derived result could be used for design only in a situation where the liquid was of no value.

heads. Assume clean cast iron pipe. What is the maximum dependable flow which can be pumped through this system?

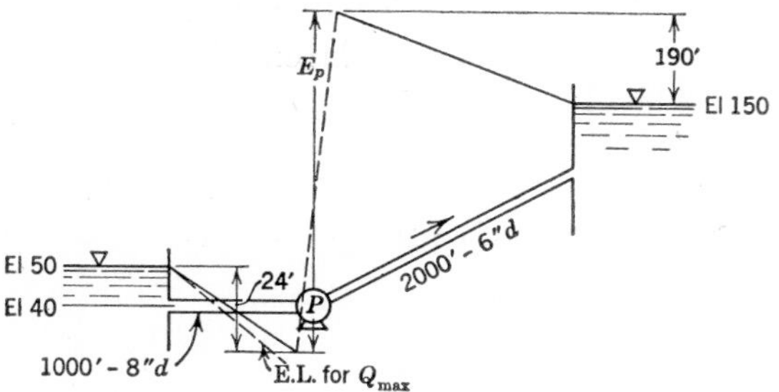

Solution.

$$V_8 = 2.5/0.349 = 7.16 \text{ fps}, \quad V_6 = 2.5/0.1964 = 12.72 \text{ fps} \qquad (53)$$

$$\mathbf{R}_8 = 7.16 \times 0.667/0.0000141 = 338{,}000; \text{ from Fig. 121, } f_8 \cong 0.020$$

$$\mathbf{R}_6 = 12.72 \times 0.5/0.0000141 = 450{,}000; \text{ from Fig. 121, } f_6 \cong 0.019$$

$$h_{L_8} \cong 0.020 \frac{1000}{0.667} \frac{(7.16)^2}{2g} = 24 \text{ ft} \qquad (182)$$

$$h_{L_6} \cong 0.019 \frac{2000}{0.5} \frac{(12.72)^2}{2g} = 190 \text{ ft} \qquad (182)$$

From the E.L:

$$E_P = 100 + 24 + 190 = 314 \text{ ft-lb/lb}$$

$$\text{Horsepower of pump (100\% efficiency)} = \frac{2.5 \times 62.4 \times 314}{550} = 89 \text{ hp} \qquad (68)$$

From the energy line the point of maximum negative pressure is at the suction flange of the pump. Allowing this to be a maximum of 20 ft of water places the energy line 20 ft below the suction flange and fixes the head loss in the suction pipe at 30 ft of water. The maximum reliable flow may then be obtained ($f = 0.020$ is approximate) from

$$30 \cong 0.020 \frac{1000}{0.667} \frac{V_8^2}{2g}; \quad V_8 = 8 \text{ fps} \qquad (182)$$

$$Q_{max} = 8 \times 0.349 = 2.8 \text{ cfs} \qquad (53)$$

78. Pipe-Line Problems—Multiple Pipes. Some of the more complex problems of pipe-line design involve the flow of fluids in pipes which intersect. The principles involved in problems of this type may be obtained by a study of pipes which (1) divide and rejoin, and (2) lead from regions of known pressure and elevation and meet at a common point. In such problems, velocity heads, minor losses, and variation of f with $\mathbf{R}$ are usually neglected, and calculations are made on the basis of coincident energy and hydraulic grade lines.

In pipe-line practice, *looping* or laying a pipe line B parallel to an existing pipe line A and connected with it (Fig. 142) is a standard method of increasing the capacity of the line. Here an analogy between fluid flow and the flow of current through a parallel electric circuit is noted if head loss is compared with drop in electric potential and flowrate with electric current. Evidently, the distribution of flow in the branches must be such that the same head loss occurs in each branch; if this were not so there would be more than one energy line for the pipe upstream and downstream from the junctions—an obvious impossibility.

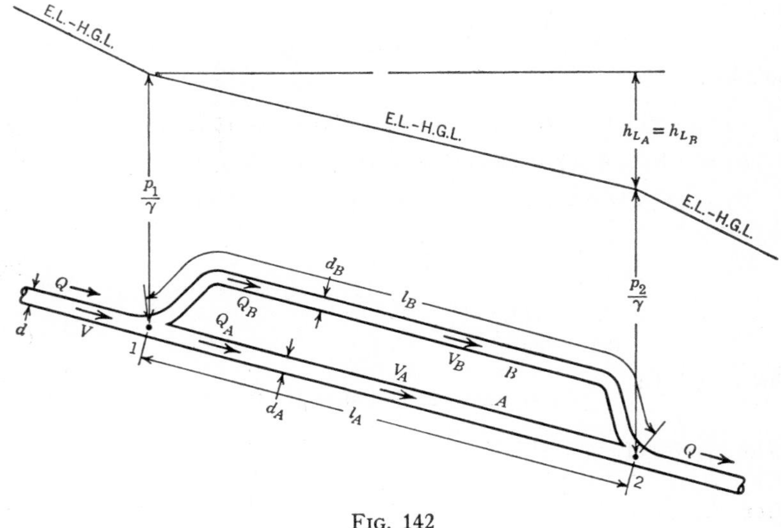

FIG. 142

Application of the continuity principle shows that the flowrate in the main line is equal to the sum of the flowrates in the branches. Thus the following simultaneous equations may be written:

$$h_{L_A} = h_{L_B}$$

$$Q = Q_A + Q_B$$

Head losses may be expressed in terms of flowrate either through the Darcy equation (182),

$$h_L = f\frac{l}{d}\frac{V^2}{2g} = \frac{fl}{2gd}\frac{16Q^2}{\pi^2 d^4} = \left(\frac{16fl}{2\pi^2 g d^5}\right)Q^2 \tag{221}$$

or from the Hazen-Williams formula (Art. 74),

$$h_L = \left(\frac{4.7l}{C_{hw}^{1.85}d^{4.86}}\right)Q^{1.85} \tag{222}$$

both of which may be generalized by writing them

$$h_L = KQ^n \tag{223}$$

Substituting this in the first of the simultaneous equations above,

$$K_A Q_A^n = K_B Q_B^n$$

$$Q = Q_A + Q_B$$

Solution of these simultaneous equations will allow prediction of the division of a flowrate Q into flowrates Q_A and Q_B when the pipe characteristics are known. Application of these principles will also allow prediction of the increased flowrate obtainable by looping an existing pipe line.

Illustrative Problem

A 12-in. pipe line 5000 ft long is laid between two reservoirs having a difference of surface elevation of 70 ft. The maximum flowrate obtainable through this line (with all valves wide open) is 5.0 cfs. When this pipe is looped with a 2000-ft pipe of the same size and material laid parallel and connected to it, what increase of maximum flowrate may be expected?

Solution. For the single pipe, using the Hazen-Williams expression,

$$70 = K(5.0)^{1.85}; \quad K = 3.55$$

Since $K \propto l$, K for each branch of the looped section will be $(2000/5000)3.55 = 1.42$, and for the unlooped section $(3000/5000)3.55 = 2.13$.

For the looped pipe line,

$$70 = 2.13Q^{1.85} + 1.42Q_A^{1.85}$$

or

$$70 = 2.13Q^{1.85} + 1.42Q_B^{1.85}$$

in which Q_A and Q_B are the flowrates in the parallel branches. Solving these equations by eliminating Q shows that $Q_A = Q_B$ (which is to be expected from symmetry in this problem). Since, from continuity, $Q = Q_A + Q_B$, $Q_A = Q/2$. Substituting this in the first equation yields $Q = 6.0$ cfs. Thus the gain of capacity by looping the pipe is 1.0 cfs or 20 percent.

Another engineering example of a multiple pipe system is the classic *three-reservoir problem* of Fig. 143 in which pipes lead from (three or more) reservoirs to a common point; this problem may be

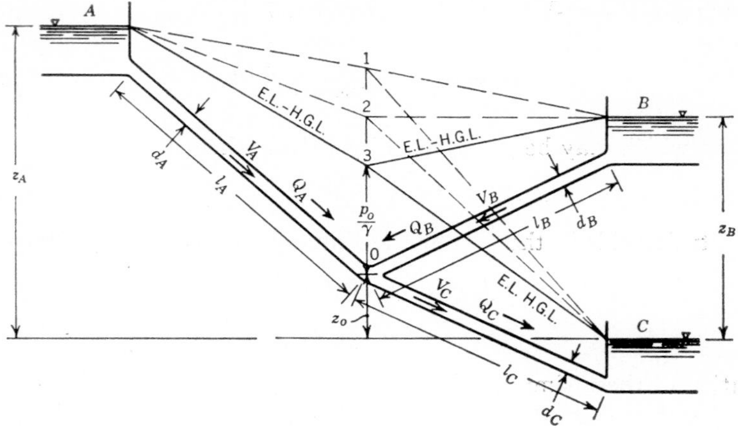

FIG. 143

solved advantageously by the use of the energy line. Here flow may take place (1) from reservoir A into reservoirs B and C, or (2) from reservoir A to C without inflow or outflow from reservoir B, or (3) from reservoirs A and B into reservoir C. For situation (1), writing head losses in the form of KQ^n,

$$z_A - K_A Q_A^n = K_C Q_C^n$$

$$z_A - K_A Q_A^n = z_B + K_B Q_B^n$$

$$Q_A = Q_B + Q_C$$

For situation (3),

$$z_A - K_A Q_A^n = K_C Q_C^n$$

$$z_A - K_A Q_A^n = z_B - K_B Q_B^n$$

$$Q_A = Q_C - Q_B$$

For situation (2), $Q_B = 0$ and the sets of equations above become identical. In view of the physical flow picture only one of these sets of equations can be satisfied; this set may be discovered by a preliminary calculation using situation (2) in which Q_A and Q_C may be computed from the first two equations. If $Q_A > Q_C$, it can be seen from the continuity equation that the first set of equations should be used; if $Q_A < Q_C$, the second set will yield a solution. Having identified the set of equations valid for the problem, these may then be solved (by trial) to yield the flowrates Q_A, Q_B, and Q_C from which the pressures at all points in the lines may be predicted.

Multiple pipe systems reach an extreme of complexity in the problems of distribution of flow in pipe networks such as those of a city water distribution system. Space does not permit comprehensive treatment of this professional engineering problem here, but the flow in a single loop of such a network may be analyzed fruitfully from foregoing principles to obtain a basic understanding.

The solution of any network problem must satisfy the continuity and Bernoulli principles throughout the network. The continuity principle states that the total flowrate toward and away from any

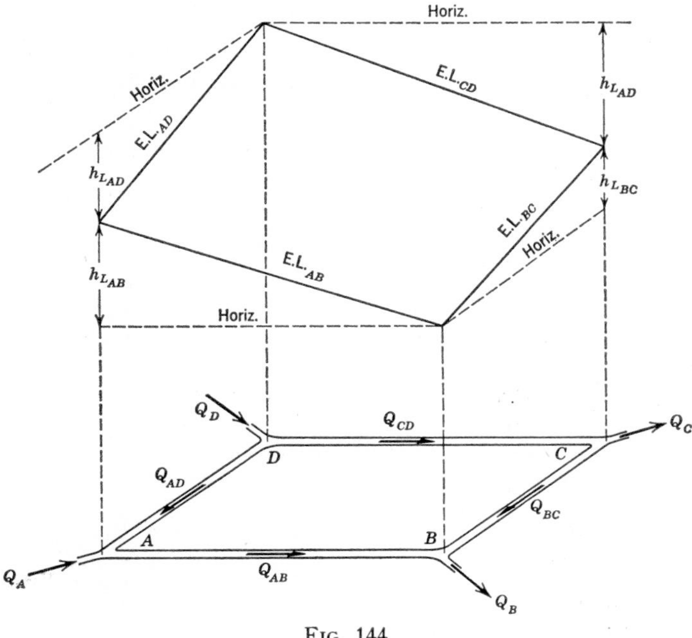

FIG. 144

pipe junction must be exactly equal; the Bernoulli principle requires that at any junction there be only one position of the energy line; in addition, the direction of flow must always be in the direction of falling energy line. Applying these principles to the single loop of Fig. 144, for which Q_A, Q_B, Q_C, and Q_D are known and Q_{AB}, Q_{BC}, Q_{CD}, and Q_{AD} are to be found,

$$Q_A + Q_{AD} = Q_{AB}$$

$$Q_{AB} + Q_{BC} = Q_B$$

$$Q_{CD} = Q_C + Q_{BC}$$

$$Q_D = Q_{AD} + Q_{CD}$$

$$(KQ^n)_{AD} + (KQ^n)_{AB} = (KQ^n)_{CD} + (KQ^n)_{BC}$$

Since any one of the first four equations may be derived from the other three, the foregoing set represents *four independent equations* which may be solved (by trial) for the four unknowns. However, the writing of the foregoing equations *has been predicated on assumed directions of flow* in the pipes which may or may not be correct; if they are not, no solution will be found, other directions must be assumed, and other equations written and solved. In view of these difficulties the reader is justified in concluding that there is little hope of a trial-and-error solution of a network containing numerous contiguous and interacting loops in any reasonable time. To overcome this difficulty Cross[34] developed a controlled trial-and-error technique which yields satisfactorily accurate solutions from but a few trials. Since about 1950, analog and digital computers[35] have been increasingly used in the solution of such problems.

REFERENCES

T. E. Stanton and J. R. Pannell, "Similarity of Motion in Relation to the Surface Friction of Fluids," *Trans. Roy. Soc. London* (A), Vol. 214, 1914.

R. J. S. Pigott, "The Flow of Fluid in Closed Conduits," *Mech. Eng.*, Vol. 55, No. 8, 1933.

C. F. Colebrook and C. M. White, "The Reduction of Carrying Capacity of Pipes with Age," *Jour. Inst. Civil Engrs., London*, Vol. 7, p. 99, 1937.

H. Rouse, "Modern Conceptions of the Mechanics of Fluid Turbulence," *Trans. A.S.C.E.*, Vol. 102, 1937.

C. E. S. Bardsley, "Historical Sketch of the Flow of Fluids through Pipes and Suggested Solutions of Pipe Flow Problems," *Univ. Okla. Eng. Expt. Sta., Publ.* 44, 1940.

J. Hinds, "Comparison of Formulas for Pipe Flow," *Jour. Am. Water Works Assoc.*, November 1946.

C. H. McClain, *Fluid Flow in Pipes*, The Industrial Press, 1947.

C. W. Harris, "An Engineering Concept of Flow in Pipes," *Trans. A.S.C.E.*, Vol. 115, 1950.

R. J. S. Pigott, "Pressure Losses in Tubing, Pipe, and Fittings," *Trans. A.S.M.E.*, Vol. 72, 1950.

V. L. Streeter, "Steady Flow in Pipes and Conduits," Chapter VI of *Engineering Hydraulics* (H. Rouse, Ed.), John Wiley and Sons, 1950.

C. T. B. Donkin, *Elementary Practical Hydraulics of Flow in Pipe Lines*, Oxford University Press, 1959.

[34] H. Cross, "Analysis of Flow in Networks of Conduits and Conductors," *Univ. Illinois Engr. Expt. Sta., Bull.* 286, 1936.

[35] See, for example, M. S. McIlroy, "Direct Reading Electric Analyzer for Pipe Line Networks," *Journ. A.W.W.A.*, Vol. 42, 1950; L. N. Hoag and G. Weinberg, "Analysis of Pipeline Networks by Electronic Digital Computers," *Journ. A.W.W.A.*, Vol. 49, 1957; and C. C. Lomax, Jr., "Network Flow Distribution Using McIlroy Analyzer or I.B.M. 650 Digital Computer," *Wash. State Inst. Tech., Bull.* 247, 1959.

PROBLEMS

587. When 10 cfs of water flow through a 6-in. constriction in a 12-in. horizontal pipe line, the pressure at a point in the pipe is 50 psi, and the head lost between this point and the constriction is 10 ft. Calculate the pressure in the constriction.

588. A 2-in. nozzle terminates a vertical 6-in. pipe line in which water flows downward. At a point on the pipe line a pressure gage reads 40 psi. If this point is 12 ft above the nozzle tip and the head lost between point and tip is 5 ft, calculate the flowrate.

589. A 12-in. pipe leaves a reservoir of surface elevation 300 at elevation 250 and drops to elevation 150, where it terminates in a 3-in. nozzle. If the head lost through line and nozzle is 30 ft, calculate the flowrate.

590. A vertical 6-in. pipe leaves a water tank of surface elevation 80. Between the tank and elevation 40, on the line, 8 ft of head are lost when 2 cfs flow through the line. If an open piezometer tube is attached to the pipe at elevation 40, what will be the elevation of the water surface in this tube?

591. A pump of what horsepower is required to pump 20 cfs of water from a reservoir of surface elevation 100 to one of surface elevation 250, if in the pump and pipe line 40 feet of head are lost?

592. Through a hydraulic turbine flow 100 cfs of water. On the 42-in. inlet pipe at elevation 145, a pressure gage reads 50 psi. On the 60-in. discharge pipe at elevation 130 a vacuum gage reads 6 in. of mercury. If the total head lost through pipes and turbine between elevations 145 and 130 is 30 ft, what horsepower may be expected from the machine?

593. In a 9-in. pipe line 5 cfs of water are pumped from a reservoir of surface elevation 100 over a hill of elevation 165. A pump of what horsepower is required to maintain a pressure of 50 psi on the hilltop if the head lost between reservoir and hilltop is 20 ft?

594. Water flows through a section of 12-in. pipe line 1000 ft long running from elevation 300 to elevation 250. A pressure gage at elevation 300 reads 40 psi, and one at elevation 250 reads 50 psi. Calculate head loss, direction of flow, and shear stress at the pipe wall and 3 in. from the pipe wall. If the flowrate is 5.0 cfs, calculate the friction factor and friction velocity.

595. When a liquid flows in a horizontal 6-in. pipe, the shear stress at the walls is 0.015 psi. Calculate the pressure drop in 100 ft of this pipe line. What is the shearing stress in the liquid 1 in. from the pipe centerline?

596. If the friction factor for a 10-in. water line is 0.030, and the shearing stress in the water 2 in. from the pipe centerline is 0.002 psi, what is the flowrate?

597. When 10 cfs of water flow in a 12-in. pipe line, 85 hp are lost in friction in 1000 ft of pipe. Calculate head lost, friction factor, friction velocity, and shear stress at the pipe wall.

598. In a 12-in. pipe, 15 cfs of water flow upward. At a point on the line at elevation 100, a pressure gage reads 130 psi. Calculate the pressure at elevation 150, 2000 ft up the line, assuming the friction factor to be 0.02.

599. A 2-in. water line possesses two pressure connections 50 ft apart (along the pipe) having a difference of elevation of 10 ft. When water flows upward through the line at a velocity of 10 fps, a differential manometer attached to the pressure connections and containing mercury and water shows a reading of 10 in. Calculate the friction factor of the pipe.

600. If 180 gpm of water at 70°F flow in a 6-in. pipe line having roughness protuberances of average height 0.030 in., and if similar roughness having height 0.015 in. exists in a 3-in. pipe, what flowrate of linseed oil (100°F) must take place therein for the friction factors of the two pipes to be the same?

601. If two cylindrical horizontal pipes are geometrically similar and the flows in them dynamically similar, how will the pressure drops in corresponding lengths vary with density and mean velocity of the fluid?

602. Calculate the loss of head in 1000 ft of 3-in. brass pipe when water at 80°F flows therein at a mean velocity of 10 fps.

603. In a 2-in. pipe line, 25 gpm of glycerin flow at 70°F. Calculate the loss of head in 200 ft of this pipe.

604. If 100 lb/min of air flow in a 3-in. galvanized iron pipe line at 175 psia and 80°F, calculate the pressure drop in 300 ft of this pipe. Assume the air to be of constant density.

605. If 12 cfs of water flow in a 12-in. riveted steel pipe at 70°F, calculate the smallest loss of head to be expected in 500 ft of this pipe.

606. A 3-in. brass pipe line 100 ft long carries 100 gpm of linseed oil. Calculate the head loss when the oil is at (*a*) 80°F, (*b*) 110°F.

607. In a laboratory test 490 lb/min of water at 60°F flow through a section of 2-in. pipe 30 ft long. A differential manometer connected to the ends of this section shows a reading of 19 in. If the fluid in the bottom of the manometer has a specific gravity of 3.20, calculate the friction factor and the Reynolds number.

608. Carbon dioxide flows in a horizontal 4-in. wrought iron pipe line at a velocity of 10 fps. At a point in the line a pressure gage reads 100 psi and the temperature is 100°F. What pressure is lost as the result of friction in 100 ft of this pipe? Barometric pressure is 14.7 psia. Assume the fluid of constant density.

609. When water at 68°F flows through 20 ft of 2-in. smooth brass pipe, the head lost is 1.00 ft. Calculate the flowrate.

610. When glycerin (80°F) flows through a 100-ft length of 3-in. pipe, the head loss is 120 ft. Calculate the flowrate.

611. A pump of what horsepower is required to pump 50 gpm of linseed oil from a tank of surface elevation 40 to one of elevation 60 through 1500 ft of 3-in. pipe, if the oil is at (*a*) 80°F, (*b*) 100°F?

612. If the head lost in 500 ft of 3-in. smooth pipe is 70 ft when the flowrate is 0.3 cfs, is the flow laminar or turbulent?

613. When the flowrate in a certain smooth pipe is 5 cfs, the friction factor is 0.06. What friction factor can be expected for a flowrate of 25 cfs of the same fluid in the same pipe?

614. A fluid of kinematic viscosity 0.005 sqft/sec flows in a certain length of smooth pipe of 1-ft diameter with a mean velocity of 10 fps and head loss of 15 ft. Predict the head loss in this length of pipe when the velocity is increased to 20 fps.

615. When 5 cfs of water (68°F) flow in a smooth 6-in. pipe line the head lost in a certain length is 15 ft. What head loss can be expected in the same length for a flowrate of 10 cfs?

616. Warm oil (s.g. 0.92) flows in a 2-in. smooth brass pipe line at a mean velocity of 8 fps and Reynolds number 7500. Calculate the wall shear stress. As the oil cools, its viscosity increases; what higher viscosity will produce the same shear stress? Neglect variation in specific gravity. The flowrate does not change.

617. The same fluid flows through 1000 ft of 3-in. and 1000 ft of 4-in. smooth pipe. The two flows are adjusted so that their Reynolds numbers are the same. What is the ratio between their head losses?

618. When 2.0 cfs of liquid (s.g. 1.27, viscosity 0.00025 lb-sec/sqft) flow in 500 ft of 6-in. pipe, the head lost is 34.0 ft. Is this pipe rough or smooth?

619. Fluid of specific gravity 0.92 and viscosity 0.002 lb-sec/sqft flows in a 2-in. smooth brass pipe line. If **R** = 2100, calculate the head lost in 100 ft of pipe if the flow is (*a*) laminar, (*b*) turbulent.

620. The head lost in 500 ft of 12-in. pipe having sand grain roughness projections 0.10 in. high is 40 ft for a flowrate of 10 cfs. What head loss can be expected when the flowrate is 20 cfs?

621. A liquid of specific gravity 0.85 flows in a 4-in. diameter brass pipe. The flowrate is 0.15 cfs. If the pressure drop over a 200-ft length of horizontal pipe is 0.25 psi, determine the viscosity of the liquid.

622. Derive expressions for wall shearing stress, wall velocity gradient, and friction velocity in terms of V, d, ρ, and μ for laminar flow in a pipe.

623. When a horizontal laminar flow occurs between two parallel plates of infinite extent 1 ft apart, the velocity at the midpoint between the plates is 9 fps. Calculate (*a*) the flowrate through a cross section 3 ft wide, (*b*) the velocity gradient at the surface of the plate, (*c*) the wall shearing stress if the fluid has viscosity 0.03 lb-sec/sqft, (*d*) the pressure drop in each 100 ft along the flow.

624. Glycerin (50°F) flows in a 2-in. pipe line. The center velocity is 8 fps. Calculate the flowrate and the head loss in 10 ft of pipe.

625. In a laminar flow of 0.25 cfs in a 3-in. pipe line the shearing stress at the pipe wall is known to be 1.0 psf. Calculate the viscosity of the fluid.

626. Oil of viscosity 0.01 lb-sec/sqft and specific gravity 0.90 flows with a mean velocity of 5 fps in a 12-in. pipe line. Calculate shearing stress and velocity 3 in. from the pipe centerline.

627. A flowrate of 0.25 gpm of oil of specific gravity 0.92 exists in this pipe line. Is this flow laminar? What is the viscosity of the oil? For the same flow in the opposite direction, what manometer reading is to be expected?

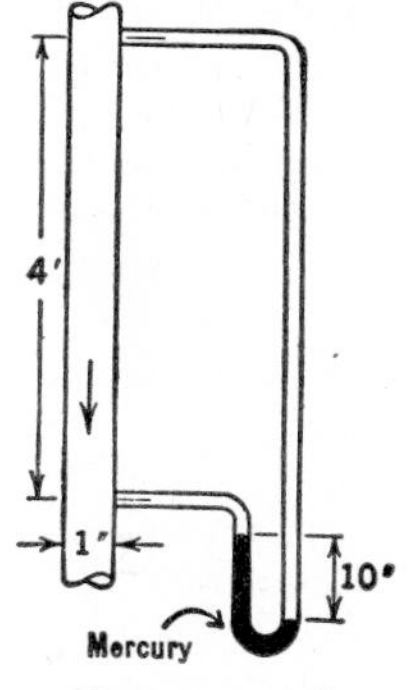

PROBLEM 627

628. In a laminar flow in a 12-in. pipe the shear stress at the wall is 1.0 psf and the fluid viscosity 0.002 lb-sec/sqft. Calculate the velocity gradient 1 in. from the centerline.

629. A fluid of specific gravity 0.90 flows at a Reynolds number of 1500 in a 12-in. pipe line. The velocity 2 in. from the wall is 10 fps. Calculate the flowrate and the velocity gradient at the wall.

630. In a turbulent flow in a 12-in. pipe the centerline velocity is 20.0 fps, and that 2-in. from the pipe wall 17.2 fps. Calculate the friction factor and flowrate.

631. Water flows in a 4-in. brass pipe line at 60°F. If the center velocity is 3 fps, what is the flowrate?

632. Carbon dioxide flows in a 3-in. wrought iron pipe at 50 psia and 50°F. If the center velocity is 2 fps, calculate the weight flowrate.

633. If the velocity past a smooth surface in turbulent flow depends only upon the distance from the surface, viscosity and density of fluid, and wall shear, show by dimensional analysis that $v/v_\star = F(v_\star y/\nu)$. See equation 195.

634. Solve problem 623 with turbulent flow, smooth plates, and fluid density and viscosity 1.94 slugs/cuft and 0.000030 lb-sec/sqft, respectively.

635. Solve problem 624 assuming water (50°F) flowing in a smooth pipe.

636. Solve problem 625 assuming turbulent flow and a smooth pipe with fluid density 1.94 slugs/cuft and wall shear 0.10 psf.

637. Solve problem 626 assuming a viscosity of 0.00010 lb-sec/sqft and the pipe a smooth one.

638. Solve problem 628 assuming a wall shear of 0.10 psf, viscosity 0.00002 lb-sec/sqft, and density 1.94 slugs/cuft. Assume turbulent flow.

639. Solve problem 629 for a Reynolds number of 150,000 and a smooth pipe.

640. Ten cubic feet per second of liquid (s.g. 1.25) flow in a smooth brass pipe of 12-in. diameter at Reynolds number 10,000. Predict the velocity where $y = \delta$.

641. Solve problem 623 for turbulent flow, rough plates with $e = 0.02$ in. and fluid density and viscosity 1.94 slugs/cuft and 0.000030 lb-sec/sqft, respectively.

642. Solve problem 624 assuming water (50°F) flowing in a rough pipe with $e =$ 0.02 in.

643. Solve problem 626 assuming turbulent flow and a rough pipe having $e =$ 0.02 in. with fluid viscosity 0.000010 lb-sec/sqft.

644. Solve problem 628 assuming a wall shear of 0.20 psf, rough surface with $e =$ 0.03 in. and fluid viscosity and density 0.000020 lb-sec/sqft and 1.94 slugs/cuft, respectively. Assume turbulent flow.

645. Solve problem 629 for a Reynolds number of 1.5×10^6 and rough surface having $e = 0.02$ in.

646. Water flows in a smooth pipe line at a Reynolds number of 10^6. After many years of use it is observed that half the original flowrate produces the same head loss as for the original flow. Estimate the size of the relative roughness of the deteriorated pipe.

647. Show that the seventh-root law velocity profile gives velocity gradients of $v_c/7R$ at the center of the pipe and infinity at the wall.

648. Fluid flows in a 6-in. smooth pipe at a Reynolds number of 25,000. Compare values of V/v_c computed from equation 194 and from the seventh-root law.

649. If the f–$\mathbf{R}$ relationship for $10^5 < \mathbf{R} < 10^6$ may be approximated by a straight line (on log-log plot) between $\mathbf{R} = 10^5, f = 0.0180$ and $\mathbf{R} = 10^6, f = 0.0115$, what value of m should be used in equation 204 to define the velocity profile?

650. Calculate a value of $v_\star\delta/\nu$ which defines the laminar film thickness when the seventh-root law is used for the turbulent velocity profile.

651. Air at 14.7 psia and 60°F flows in a horizontal triangular smooth duct, having 8-in. sides, at a mean velocity of 12 fps. Calculate the pressure drop per foot of duct. Assume the air of constant density.

652. Ten cubic feet of water flow per second in a smooth 9-in. square duct at 50°F. Calculate the head lost in 100 ft of this duct.

653. A concrete conduit of cross-sectional area 10 sqft and wetted perimeter 12 ft carries water at 50°F at a mean velocity of 8 fps. Calculate the smallest head loss to be expected in 200 ft of this conduit.

654. A semicircular concrete conduit of 5-ft diameter carries water at 70°F at a velocity of 10 fps. Calculate the smallest loss of head to be expected per foot of conduit.

655. What relative roughness is equivalent to a Hazen-Williams coefficient of 140 for $10^5 < \mathbf{R} < 10^6$?

656. A new 12-in. riveted steel pipe line carries a flowrate of 2.5 cfs of water. Calculate the head loss for 1000 ft of this pipe using the Hazen-Williams method. Calculate f and an approximate value of e. Compare the latter with the values given in Art. 66.

657. An 18-in. new riveted steel pipe line 1000 ft long runs from elevation 150 to elevation 200. If the pressure at elevation 150 is 100 psi and at elevation 200 is 72 psi, what flowrate can be expected through the line?

658. Smooth masonry pipe of what diameter is necessary to carry 50 cfs between two reservoirs of surface elevations 250 and 100 if the pipe line is to be 2 miles long?

659. What Hazen-Williams coefficient will yield the same head loss as the Darcy equation for a 2-in. smooth pipe with flow at Reynolds number 10^5? Compare the result with the values of Table III.

660. In the early hydraulic literature there are many empirical head loss formulas of the form $h_L/l = CV^x/d^y$. Show from the shape of the curves on the Moody diagram that $x \leqq 2$, $y \geqq 1$, and $x + y = 3$.

661. Air flows isothermally in a 3-in. pipe line at a mean velocity of 10 fps, pressure of 40 psia, and a temperature of 50°F. If pressure is lost by friction, calculate the mean velocity where the pressure is 30 psia.

662. Through a horizontal section of 6-in. cast iron pipe 1000 ft long in which the temperature is 60°F, 200 lb/min of air flow isothermally. If the pressure at the upstream end of this length is 30 psia, calculate the pressure at the downstream end and the mean velocities at these two points.

663. Carbon dioxide flows isothermally in a 2-in. horizontal wrought iron pipe, and at a certain point the mean velocity, pressure, and temperature are 60 fps, 60 psia, and 80°F, respectively. Calculate the pressure and mean velocity 500 ft downstream from this point.

664. The Mach number, V/a, at a point in an isothermal (100°F) flow of air in a 4-in. pipe is 0.20, and the pressure at this point is 50.0 psia. If the friction factor of the pipe is 0.020, what pressure and Mach number may be expected 200 ft downstream from this point?

665. If 5 cfs of water flow through a 6-in. horizontal pipe which enlarges abruptly to 12-in. diameter, and if the pressure in the smaller pipe is 20 psi, calculate the pressure in the 12-in. pipe, neglecting pipe friction.

666. The fluid flowing has specific gravity 0.90; $V_3 = 20$ fps; $\mathbf{R} = 10^5$. Calculate the gage reading.

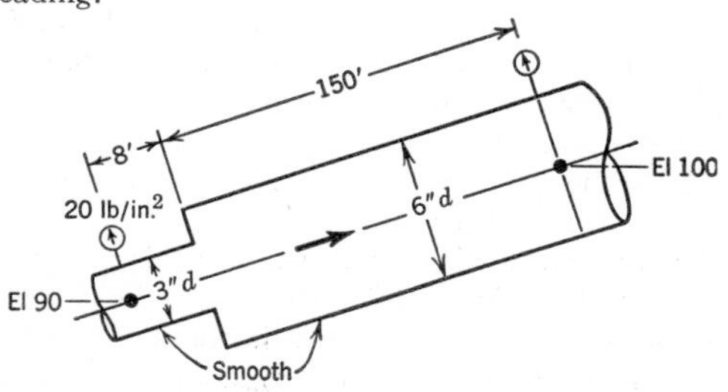

PROBLEM 666

667. Estimate the flowrate of benzene through this abrupt enlargement if the upper gage reads 20 psi and the lower one 23 psi.

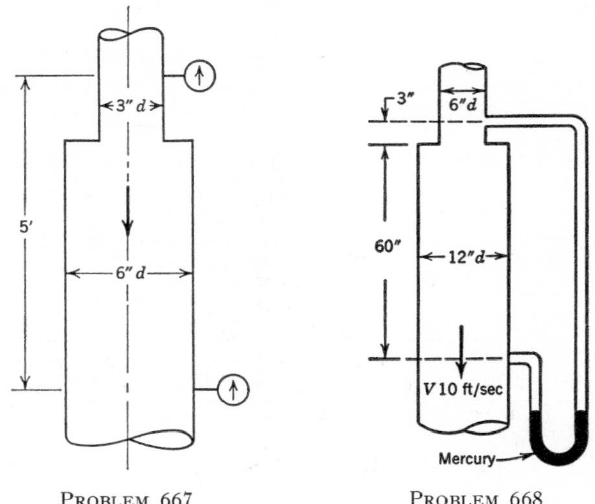

PROBLEM 667 PROBLEM 668

668. Water is flowing. Calculate direction and approximate magnitude of the manometer reading.

669. Solve problem 668 assuming conical enlargements of 70° and 7°.

670. Calculate the magnitude and direction of the manometer reading. Water is flowing.

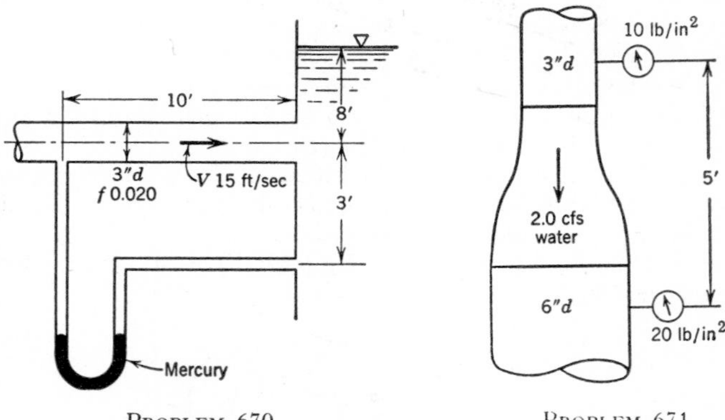

PROBLEM 670 PROBLEM 671

671. Calculate the approximate loss coefficient for this gradual enlargement.

672. Experimental determination of minor losses and loss coefficients are made from measurements of the hydraulic grade lines in zones of established flow. Calculate the head loss and loss coefficient for this gradual enlargement from the data given.

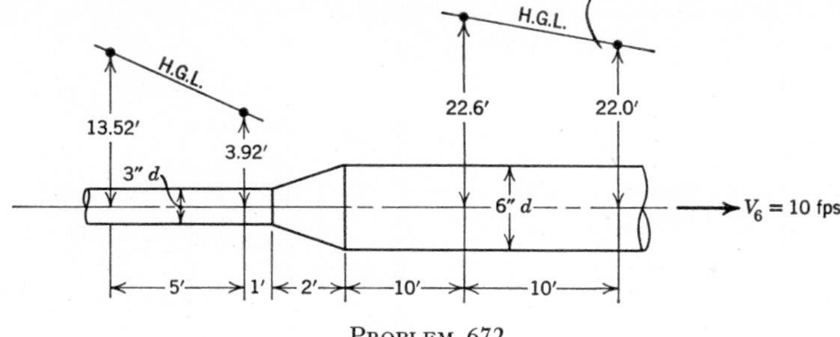

Problem 672

673. The mean velocity of water in a 6-in. horizontal pipe is 3 fps. Calculate the loss of head through an abrupt contraction to 2-in. diameter. If the pressure in the 6-in. pipe is 50 psi, what is the pressure in the 2-in. pipe, neglecting pipe friction?

674. A 6-in. horizontal water line contracts abruptly to 3-in. diameter. A pressure gage 6 in. upstream from the contraction reads 5 psi when the mean velocity in the 6-in. pipe is 5 fps. What will pressure gages read 24 in. downstream and just downstream from the contraction if the diameter of the vena contracta is 2.4 in? Neglect pipe friction.

675. Water is flowing. Calculate the gage reading when V_{12} is 8 fps.

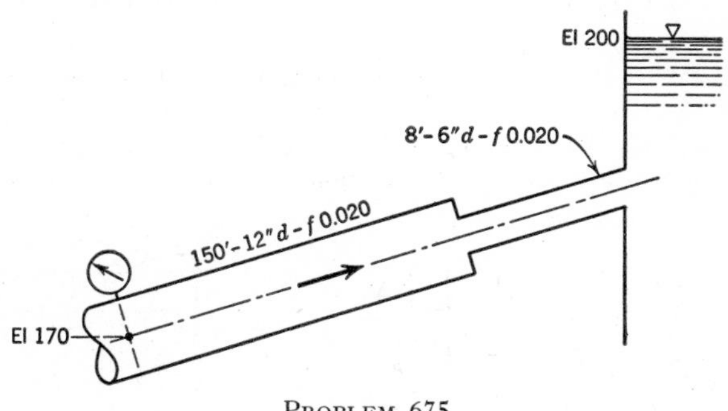

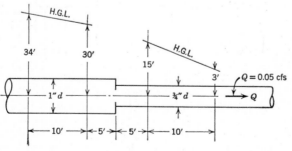

Problem 675

676. Calculate the head loss and loss coefficient caused by this restricted contraction.

Problem 676

677. Estimate as accurately as possible the loss coefficient for this pipe-line contraction.

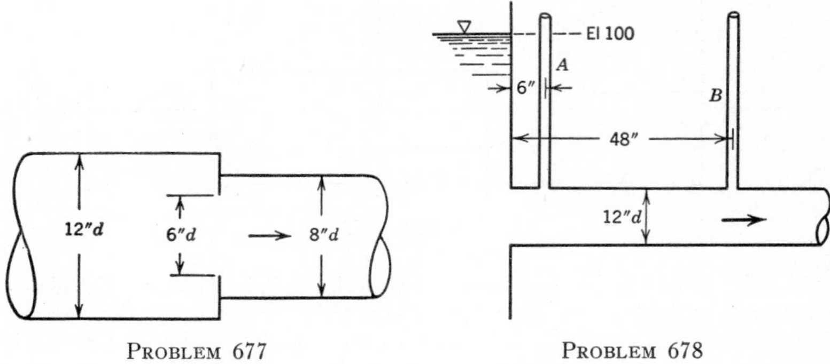

PROBLEM 677 PROBLEM 678

678. Calculate the approximate water surface elevations in tubes A and B if the flowrate is 10 cfs. Assume tube A to be opposite the vena contracta.

679. Water at 60°F flows through this smooth pipe line at Reynolds number 200,000. Predict the elevations (relative to the pipe centerline) of the hydraulic grade line at sections A (vena contracta) and B.

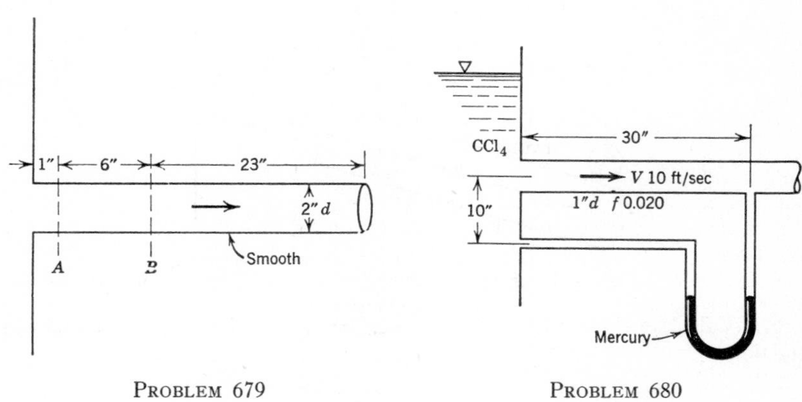

PROBLEM 679 PROBLEM 680

680. Calculate magnitude and direction of manometer reading.

681. A sharp-edged orifice discharges a flowrate Q under a certain static head. What (percent) increase or decrease of flowrate will occur for the same head when a pipe of the same diameter as the orifice is attached to the downstream side of the orifice, if the pipe is 6 diameters in length? Assume a friction factor of 0.020 and $C_c = 0.617$.

682. Determine the head loss and loss coefficient for this restricted pipe entrance.

683. Calculate the approximate loss coefficient for this short conical pipe entrance. The piezometer opening is opposite the vena contracta.

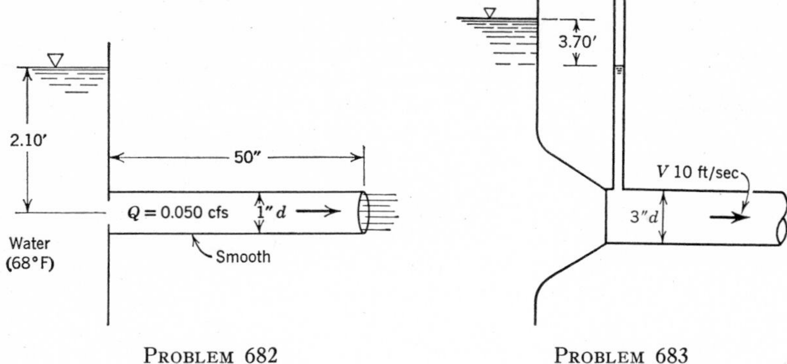

PROBLEM 682 PROBLEM 683

684. Calculate the flowrate and loss coefficient for this (re-entrant) pipe entrance if the diameter of the vena contracta is 1.50 in.

685. The 6-in. suction pipe for a pump extends 10 ft vertically below the free surface of water in a tank. If the mean velocity in the pipe is 20 fps, what is the pressure in the pipe at the level of the liquid surface if the entrance is (*a*) rounded, (*b*) re-entrant? Assume the friction factor of the pipe to be 0.020.

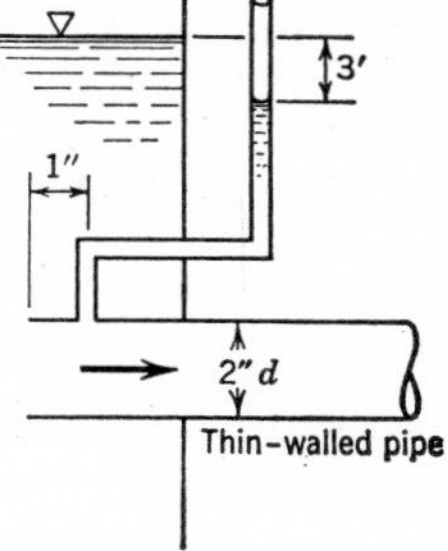

PROBLEM 684

686. Water is drawn out of a tank through a 2-in. diameter thin-walled pipe having a re-entrant entrance. Calculate the approximate diameter of the vena contracta.

687. A horizontal pipe line of 6-in. diameter leaves a tank (square-edged entrance) 50 ft below its water surface and enters another tank 20 ft below its water surface. If the flowrate in the line is 4.0 cfs, what will gages read on the pipe line a short distance (say 2 ft) from the tanks? Neglect pipe friction.

688. If the boundary layer in this nozzle is laminar, what is its approximate thickness at the nozzle exit?

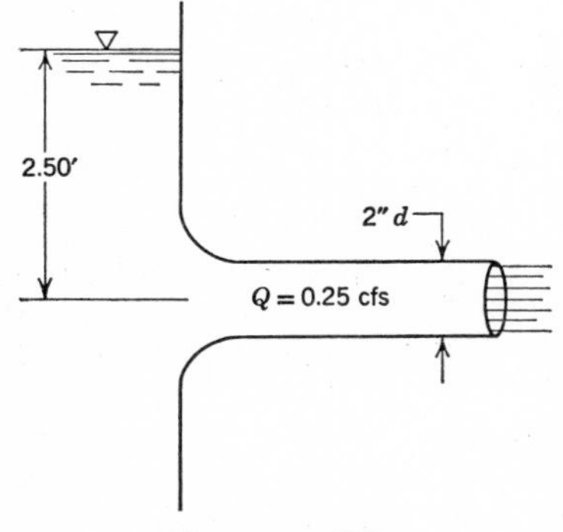

PROBLEM 688

689. If the length of the unestablished flow zone downstream from a rounded pipe entrance (see Fig. 100 of Art. 58) is 50 pipe diameters when the Reynolds number (Vd/ν) is 1800, what is the loss coefficient of the entrance if the total head may be assumed to remain constant along the central streamline?

690. Solve the preceding problem for a turbulent flow in a smooth pipe at Reynolds number 180,000 if the length of the unestablished flow zone is 25 diameters. Assume that the seventh-root law (Art. 72) is applicable.

691. A 90° smooth bend in a 6-in. pipe line has a radius of 5 ft. If the mean velocity through the bend is 10 fps and the Reynolds number 200,000, what head loss is caused by the bend? If the bend were unrolled and established flow assumed to exist in the length of pipe, what percent of the total head loss could be considered due to wall friction?

692. A 90° elbow is installed in a 2-in. pipe line having a friction factor of 0.03. The head lost at the elbow is equivalent to that lost in how many feet of the pipe? Repeat the calculation for a 1-in. pipe.

693. A 2-in. pipe line 5 ft long leaves a tank of water and discharges into the atmosphere at a point 12 ft below the water surface. In the line close to the tank is a gate valve. What flowrate can be expected when the valve is (*a*) one-half open, (*b*) one-fourth open? Assume a square-edged entrance and a friction factor of 0.020.

694. Calculate the total tension in the bolts. Neglect entrance loss.

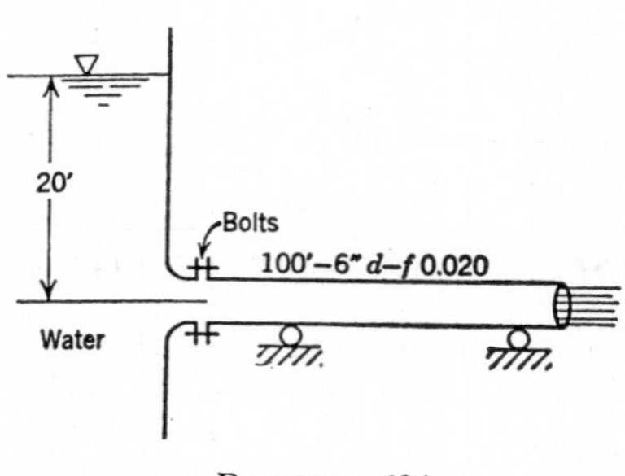

PROBLEM 694

695. Water flows at 50°F from a reservoir through a 1-in. pipe line 2000 ft long which discharges into the atmosphere at a point 1 ft below the reservoir surface. Calculate the flowrate, assuming it to be laminar and neglecting minor losses and velocity head in the pipe line. Check the assumption of laminar flow.

696. Glycerin flows through a 2-in. horizontal pipe line leading from a tank and discharging into the atmosphere. If the pipe line leaves the tank 20 ft below the liquid surface and is 100 ft long, calculate the flowrate when the glycerin has a temperature of (*a*) 50°F, (*b*) 70°F. Neglect minor losses and velocity head.

697. A horizontal 2-in. brass pipe line leaves (square-edged entrance) a water tank 10 ft below its free surface. At 50 ft from the tank, it enlarges abruptly to a 4-in. pipe which runs 100 ft horizontally to another tank, entering it 2 ft below its surface. Calculate the flowrate through the line (water temperature 68°F), including all head losses.

698. Water flows from a tank through 200 ft of horizontal 2-in. brass pipe and discharges into the atmosphere. If the water surface in the tank is 4 ft above the pipe, calculate the flowrate, considering losses due to pipe friction only, when the water temperature is (*a*) 50°F, (*b*) 100°F.

699. A smooth 12-in. pipe line leaves a reservoir of surface elevation 500 at elevation 460. A pressure gage is located on this line at elevation 400 and 1000 ft from the reservoir (measured along the line). Calculate the gage reading when 10 cfs of water (68°F) flow in the line, using (*a*) the Hazen-Williams formula, (*b*) the Moody diagram. Neglect minor losses.

700. Nine cubic feet per second of liquid (68°F) are to be carried between two tanks having a difference of surface elevation of 30 ft. If the pipe line is smooth and 300 ft long, what pipe size is required if the liquid is (*a*) glycerin, (*b*) water? Neglect minor losses.

701. A horizontal 2-in. pipe line leaves a water tank 20 ft below the water surface. If this line has a square-edged entrance and discharges into the atmosphere, calculate the flowrate, neglecting and considering the entrance loss, if the pipe length is (*a*) 15 ft, (*b*) 150 ft. Assume a friction factor of 0.025.

702. Calculate the flowrate from this water tank if the 6-in. pipe line has a friction factor of 0.020 and is 50 ft long. Is cavitation to be expected in the pipe entrance? The water in the tank is 5 ft deep.

703. A 12-in. horizontal pipe 1000 ft long leaves a reservoir of surface elevation 200 at elevation 180. This line connects (abrupt contraction) to a 6-in. pipe 1000 ft long running to elevation 100, where it enters a reservoir of surface elevation 130. Assuming friction factors of 0.02, calculate the flowrate through the line.

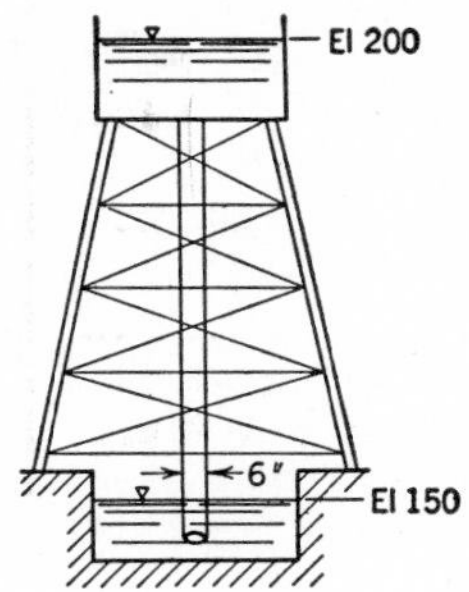

PROBLEM 702

704. What is the maximum flow which may be theoretically obtained in problem 703 when the 6-in. and 12-in. pipes are interchanged?

705. A long 12-in. pipe line laid between two reservoirs carries a flowrate of 5 cfs of water. A parallel pipe of the same friction factor is laid beside this one. Calculate the approximate diameter of the second pipe if it is to carry 10 cfs.

706. A 6-in. horizontal smooth pipe 1000 ft long takes oil from a large tank and discharges it into the atmosphere. At the midpoint of the pipe the pressure is 10.0 psi. If the specific gravity and viscosity of the oil are 0.88 and 0.0005 lb-sec/sqft, respectively, calculate (*a*) the flowrate and (*b*) the pressure in the tank on the same level as the pipe.

707. There is a leak in a horizontal 12-in. pipe line having a friction factor of 0.025. Upstream from the leak two gages 2000 ft apart on the line show a difference of 20 psi. Dowstream from the leak two gages 2000 ft apart show a difference of 18 psi. How much water is being lost from the pipe per second?

708. The pipe is filled and the plug then removed. Estimate the steady flowrate.

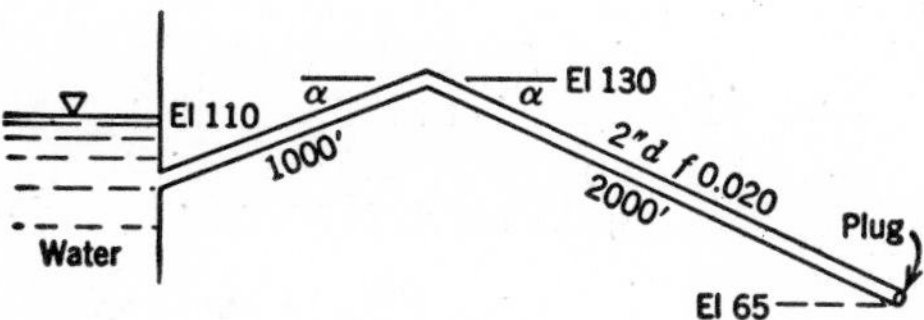

PROBLEM 708

709. An irrigation siphon has the dimensions shown. Estimate the flowrate to be expected under a head of 1 ft. Assume a re-entrant entrance, a friction factor of 0.020, and bend loss coefficients of 0.20.

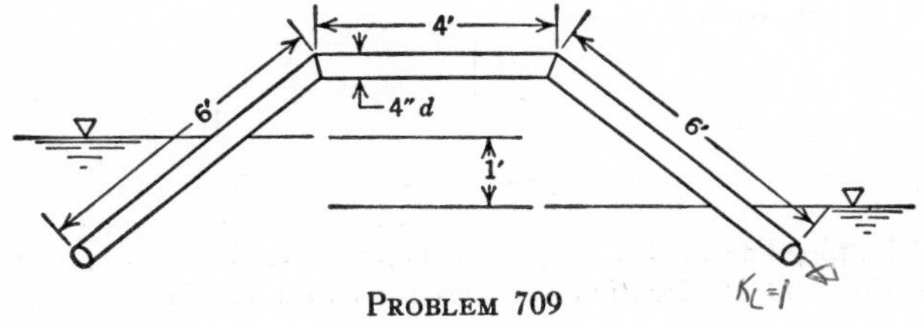

PROBLEM 709

710. Calculate the flowrate and the gage reading, neglecting minor losses and velocity heads.

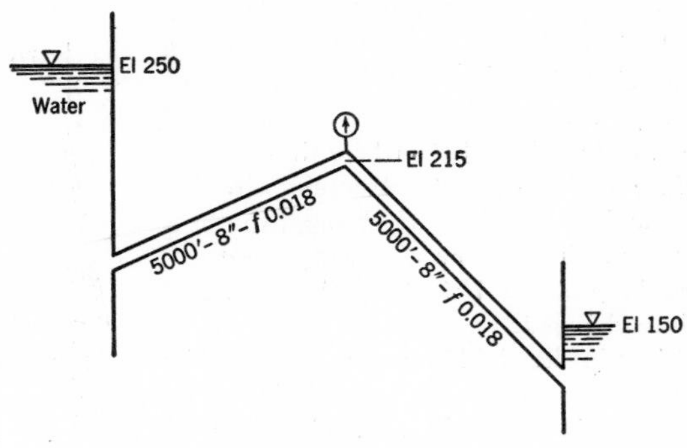

PROBLEM 710

711. At least 2.8 cfs of oil (ν = 0.00018 sqft/sec) are to flow between two reservoirs having a 50-ft difference of free surface elevation. Twelve-inch rough steel pipe of equivalent sand grain size 0.12-in. and 10-in. pipe of ultimate smoothness are available at the same cost. Which pipe should be used? Provide calculations to justify the choice.

712. A 15-in. pipe line having equivalent sand grain roughness of 0.25 in. carries 12 cfs of water (68°F). If a smooth liner is installed in the pipe, thereby reducing the diameter to 14 in., what (percent) reduction of head loss can be expected in the latter pipe for the same flowrate?

713. A 12-in. pipe line 1500 ft long leaves (square-edged entrance) a reservoir of surface elevation 500 at elevation 460 and runs to elevation 390, where it discharges into the atmosphere. Calculate the flowrate and sketch the energy and hydraulic grade lines (assuming that f = 0.022) (*a*) for the conditions above, (*b*) when a 3-in. nozzle is attached to the end of the line, assuming the lost head caused by the nozzle to be 5 ft. How many horsepower are available in the jet?

714. The flowrate is 0.80 cfs and the nozzle is frictionless. Calculate elevations A and B as accurately as possible.

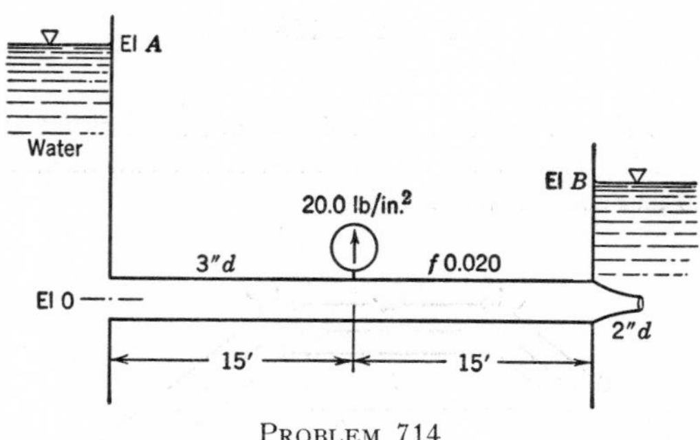

PROBLEM 714

715. A 24-in. pipe line 3000 ft long leaves (square-edged entrance) a reservoir of surface elevation 500 at elevation 450 and runs to a turbine at elevation 200.

Water flows from the turbine through a 36-in. vertical pipe ("draft tube") 20 ft long to tail water of surface elevation 185. When 30 cfs flow through pipe and turbine, what horsepower is developed? Take $f = 0.020$, include exit loss; neglect other minor losses and those within the turbine. How many horsepower may be saved by replacing the above draft tube with a 7° conical diffuser of the same length?

716. At the rate of 50 gpm, linseed oil is to be pumped through 1000 ft of 2-in. brass pipe line between two tanks having a 10-ft difference of surface elevation. Neglecting minor losses, what pump horsepower is required if the oil temperature is (*a*) 80°F, (*b*) 120°F?

717. A pump close to a reservoir of surface elevation 100 pumps water through a 6-in. pipe line 1500 ft long and discharges it at elevation 200 through a 2-in. nozzle. Calculate the pump horsepower necessary to maintain a pressure of 50 psi behind the nozzle, and sketch accurately the energy line, taking $f = 0.020$.

718. The horizontal 8-in. suction pipe of a pump is 500 ft long and is connected to a reservoir, of surface elevation 300, 10 ft below the water surface. From the pump, the 6-in. discharge pipe runs 2000 ft to a reservoir of surface elevation 420, which it enters 30 ft below the water surface. Taking f to be 0.020 for both pipes, calculate the pump horsepower required to pump 3.0 cfs from the lower reservoir. What is the maximum dependable flowrate which may be pumped through this system (*a*) with the 8-in. suction pipe, (*b*) with a 6-in. suction pipe?

719. A 12-in. pipe line 2 miles long runs on an even grade between reservoirs of surface elevations 500 and 400, entering the reservoirs 30 ft below their surfaces. The flowrate through the line is inadequate, and a pump is installed at elevation 420 to increase the capacity of the line. Assuming f to be 0.020, what pump horsepower is required to pump 6.0 cfs downhill through the line? Sketch accurately the energy line before and after the pump is installed. What is the maximum dependable flowrate which may be obtained through the line?

720. Calculate the smallest reliable flowrate which can be pumped through this pipe line. Assume atmospheric pressure 14.7 psia.

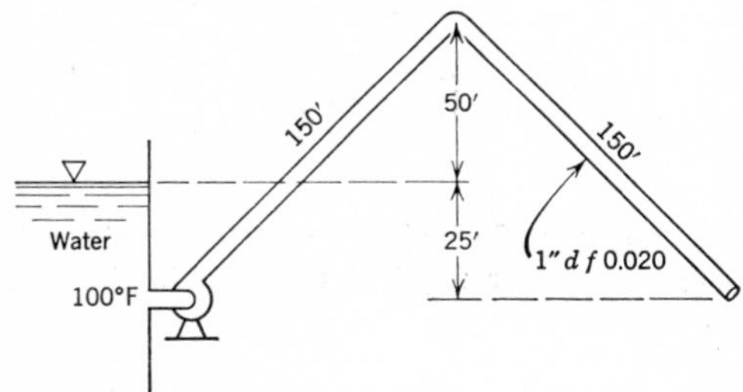

PROBLEM 720

721. The suction side of a pump is connected directly to a tank of water at elevation 100. The water surface in the tank is at elevation 130. The discharge pipe from the pump is horizontal, 1000 ft long, of 6-in. diameter, and has a friction factor of 0.025. The pipe runs from west to east and terminates in a 3-in. frictionless nozzle. The nozzle stream is to discharge into a tank whose western upper edge is horizontal and is located at elevation 50 and 100 ft east of the tip of the nozzle. What is the minimum horsepower that the pump may supply for the stream to pass into the tank?

722. The pump is required to maintain the flowrate which would have occurred without any friction. What horsepower pump is needed? Neglect minor losses.

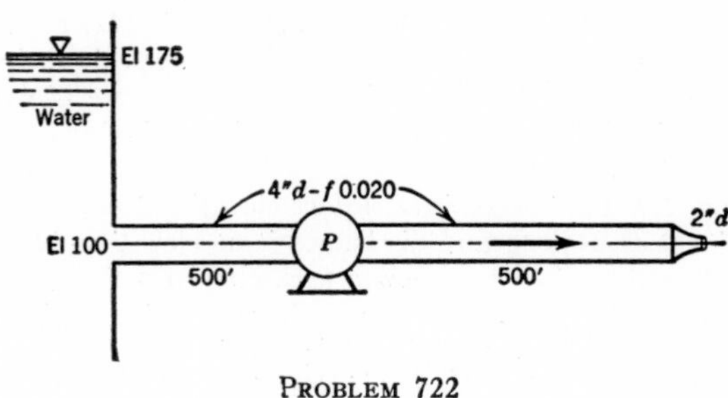

PROBLEM 722

723. If the turbine extracts 530 hp from the flow, what flowrate must be passing through the system? What is the maximum power obtainable from the turbine?

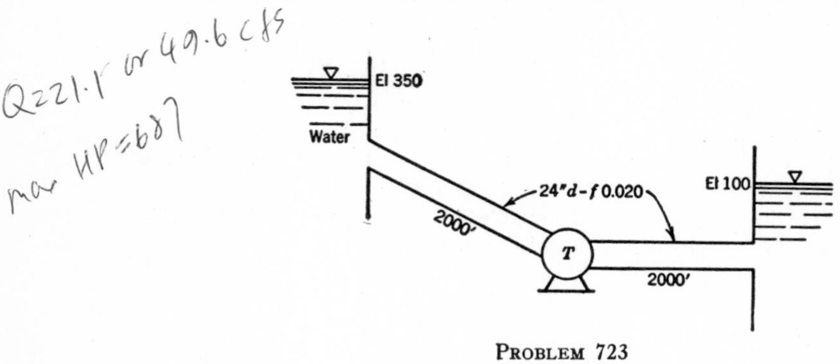

PROBLEM 723

724. If there were no pump, 5 cfs of water would flow through this pipe system. Calculate the pump horsepower required to maintain the same flowrate in the opposite direction.

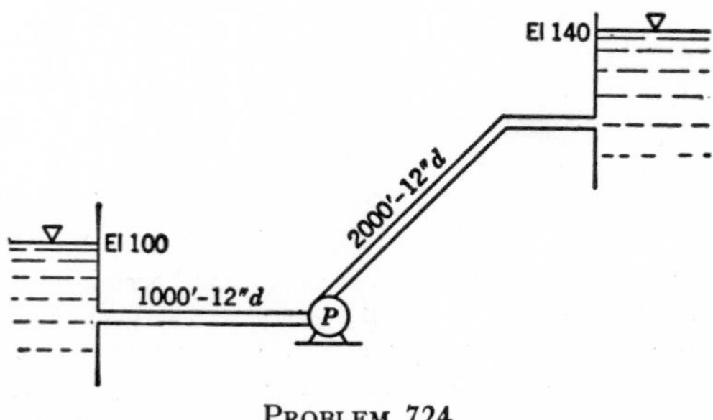

PROBLEM 724

725. The flowrate is 4.5 cfs without the pump. Calculate the approximate pump horsepower required to maintain a flowrate of 6.0 cfs.

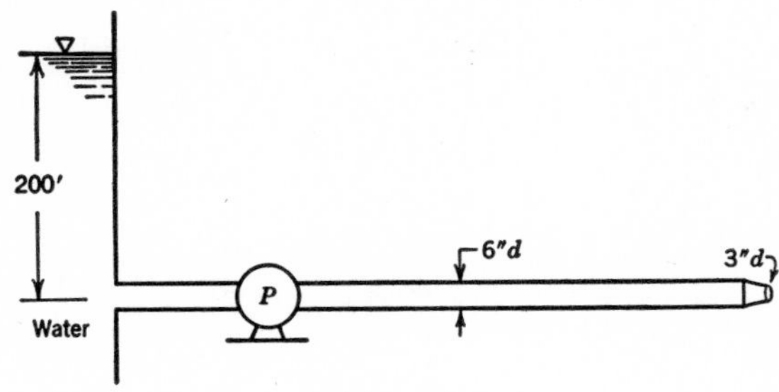

PROBLEM 725

726. The pitot tube is on the centerline of the pipe. The flowrate is 2 cfs of water. Calculate the horsepower transferred from pump to fluid.

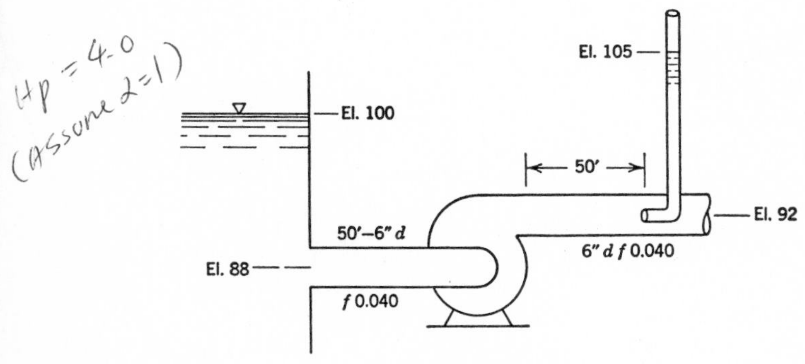

PROBLEM 726

727. Five 1-in. diameter brass pipes will conduct 0.085 cfs of water (68°F) between two reservoirs whose respective surface levels are constant. Find the diameter of a single brass pipe which will carry the same flowrate between these reservoirs.

728. A 24-in. pipe line branches into a 12-in. and an 18-in. pipe, each of which is 1 mile long, and they rejoin to form a 24-in. pipe. If 30 cfs flow in the main pipe, how will the flow divide? Assume that $f = 0.018$ for both branches.

729. A 24-in. pipe line carrying 30 cfs divides into 6-in., 8-in., and 12-in. branches, all of which are the same length and enter the same reservoir below its surface. Assuming that $f = 0.020$ for all pipes, how will the flow divide?

730. A straight 12-in. pipe line 3 miles long is laid between two reservoirs of surface elevations 500 and 350 entering these reservoirs 30 ft beneath their free surfaces. To increase the capacity of the line a 12-in. line 1.5 miles long is laid from the original line's midpoint to the lower reservoir. What increase in flowrate is gained by installing the new line? Assume that $f = 0.020$ for all pipes.

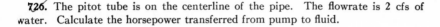

731. If the pump supplies 300 hp to the water, what flowrate will occur in the pipes? Assume a friction factor of 0.020 for both pipes.

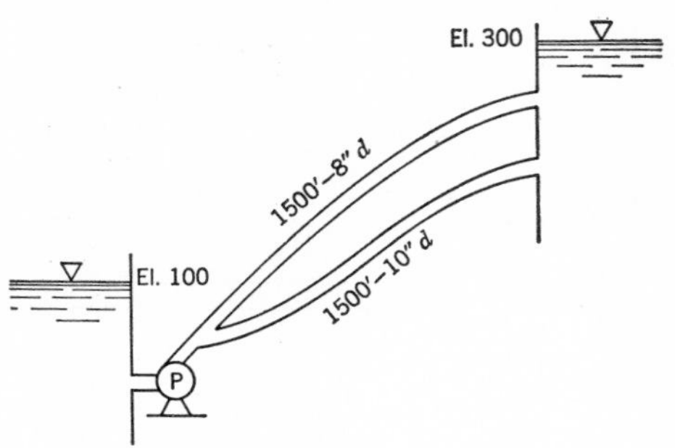

PROBLEM 731

732. A 36-in. pipe divides into three 18-in. pipes at elevation 400. The 18-in. pipes run to reservoirs which have surface elevations 300, 200, and 100, these pipes having respective lengths of 2, 3, and 4 miles. When 50 cfs flow in the 36-in. line, how will the flow divide? Assume that $f = 0.017$ for all pipes.

733. Reservoirs A, B, and C have surface elevations 500, 400, and 300, respectively. A 12-in. pipe 1 mile long leaves reservoir A and runs to point O at elevation 450. Here the pipe divides and an 8-in. pipe 1 mile long runs from O to B and a 6-in. pipe 1.5 miles long runs from O to C. Assuming that $f = 0.020$, calculate the flowrates in the lines.

734. Three pipes join at a common point at elevation 350. One, a 12-in. line 2000 ft long, goes to a reservoir of surface elevation 400; another, a 6-in. line 3000 ft long, goes to a reservoir of surface elevation 500; the third (6 in.) runs 1000 ft to elevation 250, where it discharges into the atmosphere. Assuming that $f = 0.020$, calculate the flowrate in each line. Calculate these flowrates when a 2-in. nozzle is attached to end of the third pipe.

735. A 12-in. pipe line 2000 ft long leaves a reservoir of surface elevation 500 and runs to elevation 400 where it divides into two 6-in. lines each 1000 ft long, both of which discharge into the atmosphere, one at elevation 450, the other at elevation 350. Calculate the flowrates in the three pipes if all friction factors are 0.025.

736. The pump is to deliver 4 cfs to the outlet at elevation 550 and 8 cfs to the upper reservoir. Calculate pump horsepower and the required diameter of the 1000-ft pipe.

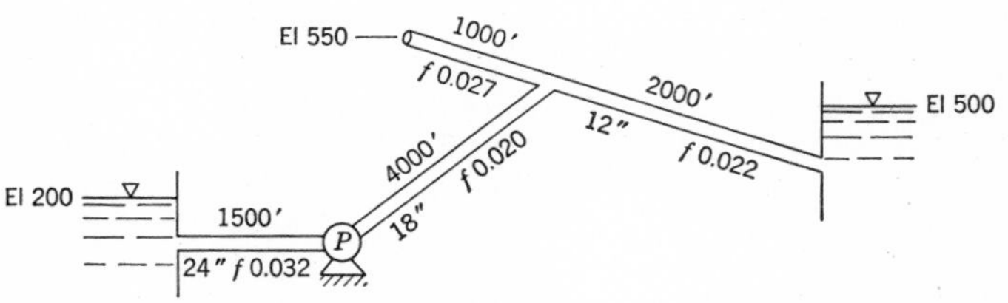

PROBLEM 736

737. Water is flowing. For $Q_8 = 4.0$ cfs, calculate Q_6, Q_{12}, and pump horsepower.

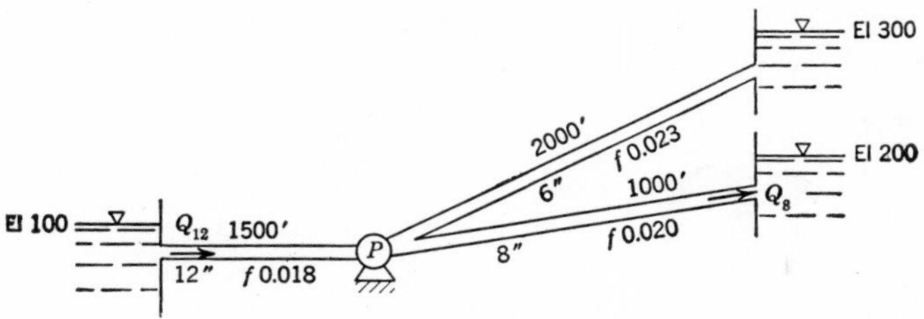

PROBLEM 737

738. Calculate the flowrates in the pipes of this loop (*a*) if all friction factors are 0.020 and (*b*) if all Hazen-Williams coefficients are 140.

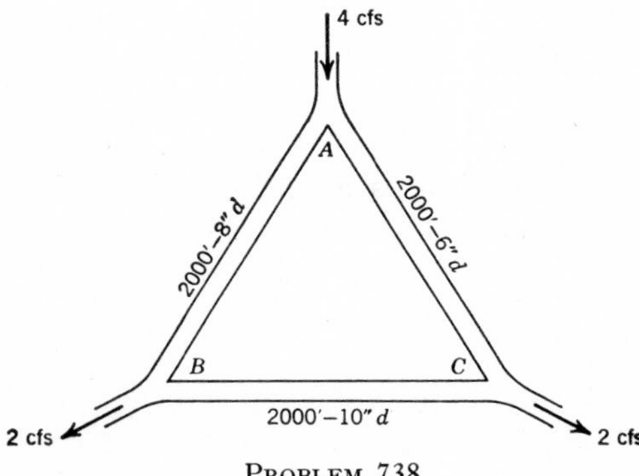

PROBLEM 738

chapter 10
Liquid Flow in Open Channels

Open-channel flow embraces a variety of problems which arise when water flows in natural water courses, regular canals, irrigation ditches, sewer lines, flumes, etc.—a province of paramount importance to the civil engineer. Although open-channel problems practically always involve the flow of water, and although the experimental results used in these problems were obtained by hydraulic tests, modern fluid mechanics indicates the extent to which these results may be applied to the flow of other liquids in open channels.

79. Fundamentals. In the problems of pipe flow, as may be seen from the hydraulic grade line, the pressures in the pipe may vary along the pipe and depend upon energy losses and the conditions imposed upon the ends of the line. Open flow, however, is characterized by the fact that pressure conditions are determined by the constant pressure, usually atmospheric, existing on the entire surface of the flowing liquid. Usually pressure variations within an open-channel flow may be determined by the principles of hydrostatics since the streamlines are ordinarily straight, parallel, and approximately horizontal (compare Fig. 45, Art. 24); when these conditions are satisfied, the hydraulic grade lines for all the streamtubes will be the same and will coincide with the liquid surface. There are, of course, many exceptions to the foregoing situation, and, in flow fields where streamlines are convergent, divergent, curved, or steeply sloping, each streamline will in general have its own hydraulic grade line and these will not lie in the liquid surface.

A typical reach of open-channel flow is shown in Fig. 145.[1] The individual streamlines of the flow are very slightly convergent but may safely be assumed to be parallel, and the hydraulic grade line coincides with the liquid surface. The energy line is located one velocity head [2] above the hydraulic grade line, and the head loss is defined, as usual, as the drop in the energy line. Thus the energy line-hydraulic grade line approach to one-dimensional flow problems can be expected to find wide application in many problems of open flow.

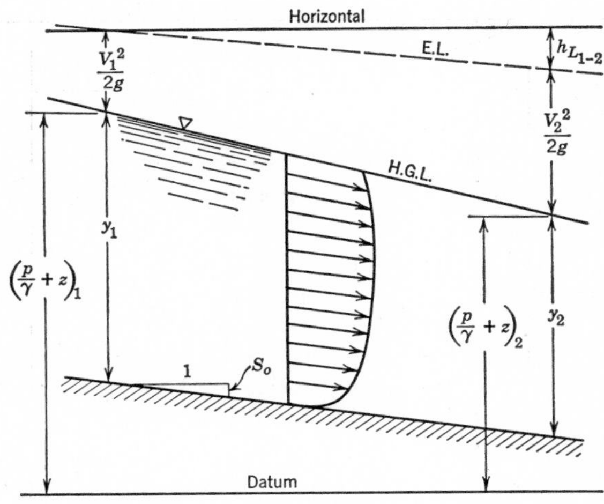

FIG. 145

Open-channel flow may be laminar or turbulent, steady or unsteady, *uniform* or *varied*, *subcritical* or *supercritical*.[3] The complexity of unsteady open-flow problems forbids their treatment in an elementary text but the other categories will be examined herein; the emphasis, however, will be directed toward steady turbulent flow, which is the problem generally encountered in practice. The definitions of subcritical and supercritical flow will be presented subsequently, but the significance, causes, and limits of uniform

[1] The channel slopes in many of the illustrations in this chapter have been exaggerated to emphasize their existence. In open-channel practice, slopes are very seldom encountered which are greater than 1 ft (vertical) in 100 ft (horizontal), or 0.01.

[2] Strictly, this distance should be $\alpha V^2/2g$, but, in view of the relatively flat velocity profile, α will in most cases be only slightly greater than unity.

[3] For an excellent discussion of the first and last pairs of categories see J. M. Robertson and H. Rouse, "The Four Regimes of Open Channel Flow," *Civil Eng.*, Vol. 11, No. 3, 1941.

and varied flow must first be examined. The meaning of these terms and also the fundamentals of open-channel flow may be seen from a comparison of ideal-fluid flow and real-fluid flow in identical prismatic channels leading from reservoirs of the same surface elevation (Fig. 146). No resistance will be encountered by the ideal fluid as it flows down the channel, and because of this lack of resistance it will continually accelerate under the influence of gravity. Thus the mean velocity continually increases, and with this increase of velocity

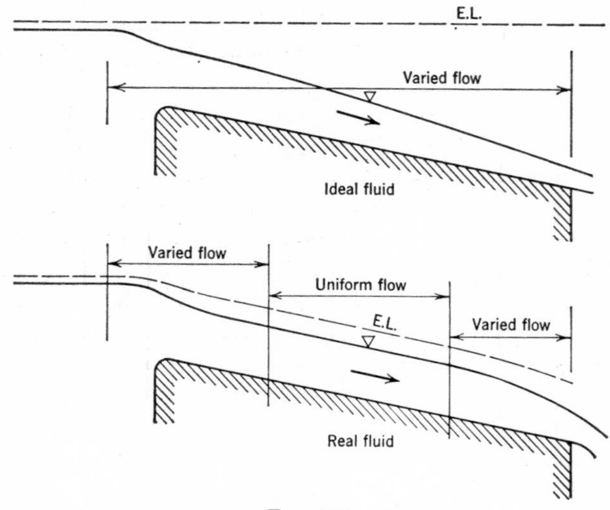

FIG. 146

a reduction in flow cross section is required by the continuity principle. Reduction in flow cross section is characterized by a decrease in depth of flow; since the depth of flow continually *varies* (from section to section, not with time) this type of fluid motion is termed *varied flow*.[4]

When real fluid flows in the same channel, there are resistance forces due to fluid viscosity and channel roughness. Analysis of the resistance forces originating from these same properties in pipes has shown that such forces depend upon the velocity of flow (equation 183). Thus, in the upper end of the channel where motion is slow, resistance forces are small, but the components of gravity forces in the direction of motion are about the same as for the ideal fluid. The resulting unbalanced forces in the direction of motion bring about acceleration and varied flow in the upper reaches of the chan-

[4] The term *nonuniform flow* is also used, but *varied flow* is increasingly preferred.

nel. However, with an increase of velocity, the forces of resistance increase until they finally balance those caused by gravity. Upon the occurrence of this force balance, constant-velocity motion is attained; it is characterized by no change of flow cross section and thus no change in the depth of flow; this is *uniform flow*. Toward the lower end of the channel, pressure and gravity forces again exceed resistance forces and varied flow again results.

Obviously, an inequality between the above-mentioned forces is more probable than a balance of these forces, and, hence, varied flow occurs in practice to a far greater extent than uniform flow. In short channels, for example, uniform-flow conditions may never be attained because of the long reach of channel necessary for the establishment of uniform flow; nevertheless, solution of the uniform-flow problem forms the basis for open-channel flow calculations.

80. Uniform Flow—The Chezy Equation. The fundamental equation for uniform open-channel flow may be derived readily by equating (since there is no change of momentum) the equal and opposite force components of gravity and resistance and applying some of the fundamental notions of fluid mechanics encountered in the analysis of pipe flow. Consider the uniform flow of a liquid between sections 1 and 2 of the open channel of Fig. 147. The forces

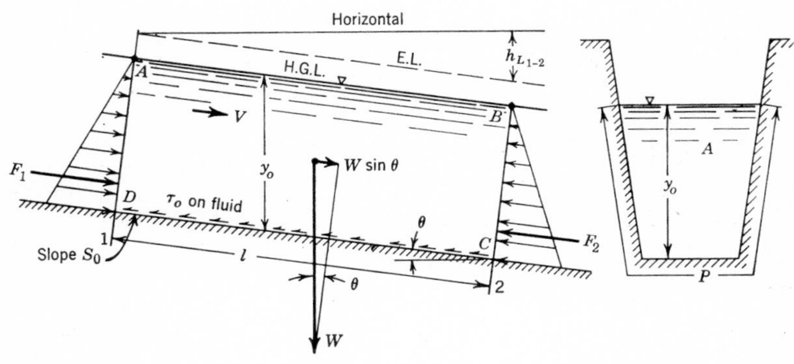

FIG. 147

acting upon the control volume of liquid $ABCD$ are (1) the forces of static pressure, F_1 and F_2, acting on the ends of the body; (2) the weight, W, which has a component $W \sin \theta$ in the direction of motion; (3) the pressure forces exerted by the bottom and sides of the channel;[5] and (4) the force of resistance exerted by the bottom

[5] These are not shown since they do not enter the following equations.

and sides of the channel $Pl\tau_o$. A summation of forces along the direction of motion gives

$$F_1 + W \sin\theta - F_2 - Pl\tau_o = 0$$

since there is no change of momentum between sections 1 and 2. Obviously, $F_1 = F_2$, $W = A\gamma l$, and $\sin\theta = h_L/l$. The slope, S_o, of both channel bed and liquid surface is $\tan\theta$, and for the small slopes encountered in open-channel practice the approximation $\tan\theta = \sin\theta$ may be made. A/P is recognized (Arts. 57 and 73) as the hydraulic radius, R_h. When these substitutions are made in the equation above an expression for the *mean shear stress* results; it is

$$\tau_o = \gamma R_h S_o \qquad (224)$$

In pipe flow, τ_o was shown (Art. 63) to be given by $f\rho V^2/8$ in which f, the friction factor, is dependent upon surface roughness and Reynolds number, but more dependent upon magnitude of roughness for highly turbulent flow. The mechanism of real-fluid motion is similar in pipes and open channels, and if it is assumed [6] that the hydraulic radius concept will account adequately for the differences in the cross-sectional shapes of circular pipes and open channels, these expressions for τ_o may be equated. Solving for V and replacing γ/ρ by g,

$$V = \sqrt{8g/f}\,\sqrt{R_h S_o}$$

or, if $C = \sqrt{8g/f}$,

$$V = C\sqrt{R_h S_o} \qquad (225)$$

This is called the *Chezy equation* after the French hydraulician who established this relationship experimentally in 1775. By applying the continuity principle, the equation may be placed in terms of flowrate as

$$Q = CA\sqrt{R_h S_o} \qquad (226)$$

the fundamental equation of uniform flow in open channels.

81. The Chezy Coefficient. From the time of Chezy to the present, many experiments in field and laboratory have been performed to determine the magnitude of the Chezy coefficient, C, and its dependence upon other variables. The simplest relation and the one most widely used in the United States may be derived from the

[6] Experience has shown this to be a valid assumption for regular prismatic channels (canals, flumes, etc.) with turbulent flow. It cannot be expected to apply to laminar flow (Art. 82) or to channels of irregular or distorted cross section (such as natural streams at flood stage).

work of Manning,[7] an analysis of experimental data obtained from his own experiments and from those of others. His results may be summarized by the empirical relation

$$C = 1.49\, R_h^{\frac{1}{6}}/n \tag{227}$$

which may be combined with the Chezy equation to give the so-called Chezy-Manning equation,

$$Q = (1.49/n) A R_h^{\frac{2}{3}} S_o^{\frac{1}{2}} \tag{228}$$

The Manning n is obtained from a descriptive statement of the channel roughness, some typical values [8] of which are given in Table VI. From the table it is evident that "there is no substitute for

TABLE VI

VALUES OF THE ROUGHNESS COEFFICIENT, n, $\text{ft}^{\frac{1}{6}}$.

Smooth cement, planed timber	0.010
Rough timber, canvas	0.012
Good ashlar masonry or brickwork	0.013
Vitrified clay	0.015
Rubble masonry	0.017
Firm gravel	0.020
Canals and rivers in good condition	0.025
Canals and rivers in bad condition	0.035

experience" in the interpretation and selection of values for n. Fortunately, however, the inevitable errors made in the selection of n have a reduced effect on engineering calculations since the Chezy equation is used primarily for determining the depth of uniform flow from flowrate, slope, and roughness; this may be seen in the following numerical example.

ILLUSTRATIVE PROBLEM

A rectangular channel lined with rubble masonry is 20 ft wide and laid on a slope of 0.0001. Calculate the depth of uniform flow in this channel when the flowrate is 400 cfs.

Solution. Using y_o as the depth of uniform flow,

$$A = 20y_o, \quad P = 20 + 2y_o, \quad R_h = 20y_o/(20 + 2y_o)$$

[7] *Trans. Inst. Civil Engrs. Ireland*, Vol. 20, p. 161, 1890.

[8] More comprehensive tables are available in many handbooks; see, for example, H. W. King, *Handbook of Hydraulics*, 4th Edition, McGraw-Hill Book Co., 1954.

From Table VI,

$$n = 0.017 \text{ ft}^{\frac{1}{6}}$$

$$400 = \left(\frac{1.49}{0.017}\right)(20y_o)\left(\frac{20y_o}{20 + 2y_o}\right)^{\frac{2}{3}}(0.0001)^{\frac{1}{2}} \tag{228}$$

whence

$$y_o\left(\frac{20y_o}{20 + 2y_o}\right)^{\frac{2}{3}} = 22.9$$

Solving by trial, $y_o = 8.34$ ft.

If a value of n of 0.018 $\text{ft}^{\frac{1}{6}}$ (6 percent greater than 0.017) had been selected, the depth would have been 8.7 ft (4 percent greater than 8.34 ft).

The somewhat unsatisfactory definition of open-channel roughness is being steadily improved as more knowledge is gained through analysis and systematic experiments similar to those of Nikuradse and Colebrook in the field of pipe flow. Application of these results to open-channel flow through use of the hydraulic radius concept leads to a more scientific interpretation of Chezy's C and Manning's n.

Through the relationship $C = \sqrt{8g/f}$ Nikuradse's plot (Fig. 119) of friction factor may be replotted to show the relation between C, Reynolds number, and relative roughness. Such a plot would be a distorted inversion of Fig. 119 and would have the general appearance of Fig. 148, $4R_h$ having been substituted for d. On this plot can be seen the

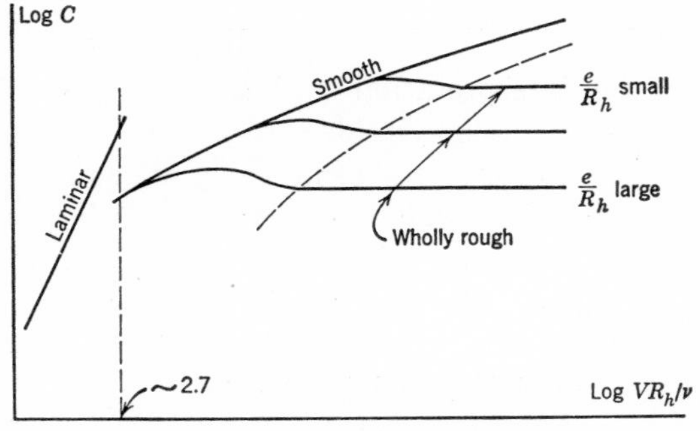

Fig. 148

general relation of C to both Reynolds number and roughness over the whole range of laminar and turbulent flow.

From this plot it may be seen that, in the *wholly rough zone*, the Chezy C is a direct index of *relative* roughness; outside the wholly rough zone this is only approximately true, and for surfaces of ultimate smoothness it is entirely untrue since there C is a function of Reynolds number only. Nevertheless it is useful to consider C to be primarily a measure of relative roughness since in most engineering problems it falls within or close to the wholly rough zone.

If the Manning n is now interpreted to be a measure of roughness only, it follows that: (1) it can apply only to the wholly rough region of the plot, which is featured by horizontal lines for constant relative roughness; and (2) its dimensions must be some function of length only. From Manning's equation,

$$C = \sqrt{8g/f} = 1.49R_h^{\frac{1}{6}}/n \quad \text{or} \quad 1/\sqrt{f} = (1.49/\sqrt{8g})(R_h^{\frac{1}{6}}/n)$$

From Nikuradse's equation (202) for the wholly rough region,

$$1/\sqrt{f} = 1.14 + 2 \log 4R_h/e$$

whence

$$(1.49/\sqrt{8g})(R^{\frac{1}{6}}/n) = 1.14 + 2 \log 4R_h/e \tag{229}$$

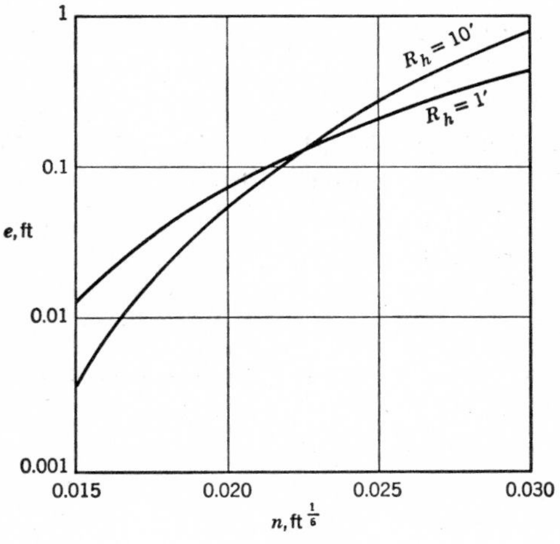

FIG. 149

This equation shows relative roughness $e/4R_h$ to be directly related to $n/R_h^{\frac{1}{6}}$ and the latter fraction to be a measure of *relative* roughness; this also suggests [9] that n may be assumed to have dimensions of $(\text{ft})^{\frac{1}{6}}$. The form of equation 229 also suggests that the effect of R_h tends to cancel, yielding a (rough) relationship between the absolute roughness (e) and n, and thus showing n to be primarily a measure of absolute roughness. This is documented on Fig. 149 for a wide range of R_h, which also shows n to be a relatively insensitive measure

[9] See V. T. Chow, *Open-Channel Hydraulics*, McGraw-Hill Book Co., 1959, p. 98, for a complete discussion of the dimensional aspects of n.

of e, a ten-fold increase of e producing (approximately) a 30 percent increase of n. Although these relationships are not exact, the foregoing general conclusions may be safely drawn and the equations may be used effectively for interpreting n in terms of tangible quantities. For example, in the construction of models of open channels and rivers, where the relative roughness of model and prototype are to be equal, the equations are useful in estimating the value of n required for the model.

Illustrative Problem

Estimate the size of the equivalent sand grain roughness in the preceding illustrative problem.

Solution.

$$R_h = \frac{20 \times 8.34}{20 + 2 \times 8.34} = 4.54 \text{ ft}$$

Entering Fig. 149 with $n = 0.017 \text{ ft}^{\frac{1}{6}}$, $e \cong 0.02 \text{ ft} \cong 0.25$ in., which seems of the right order of magnitude for rubble masonry.

82. Uniform Laminar Flow.[10] Laminar flow ($Vy_o/\nu \gtrsim 500$) in open channels was of little practical interest until recent times when it has been used effectively in the analysis of drainage from streets, airport runways, parking areas, etc. Here the flow is in thin sheets of virtually infinite width (i.e., without sidewalls) with resistance only at the bottom of the sheet; thus the flow is essentially a two-dimensional one. A definition sketch for uniform laminar flow is shown in Fig. 150. From the pipe flow analysis of Art. 67, a parabolic velocity profile and linear shear stress profile are to be expected. The hydraulic radius for such a flow may be deduced by considering a section of width b normal to the flow; the flow cross section is thus a rectangle of depth y_o and width b. However, the wetted perimeter (i.e., the boundary offering resistance to the flow) is only the channel bottom, of width b; therefore $R_h = A/P = by_o/b = y_o$. Inserting this in equation 224,

$$\tau_o = \gamma y_o S_o$$

However, in laminar flow, τ_o is also given by $\mu\,(dv/dy)_o$ in which $(dv/dy)_o$ is the velocity gradient at the channel bottom. From the properties of the parabolic velocity profile, $(dv/dy)_o$ may be shown

[10] See W. M. Owen, "Laminar to Turbulent Flow in a Wide Open Channel," *Trans. A.S.C.E.*, Vol. 119, 1954, and L. G. Straub, E. Silberman, and H. C. Nelson, "Open Channel Flow at Small Reynolds Number," *Trans. A.S.C.E.*, Vol. 123, 1957.

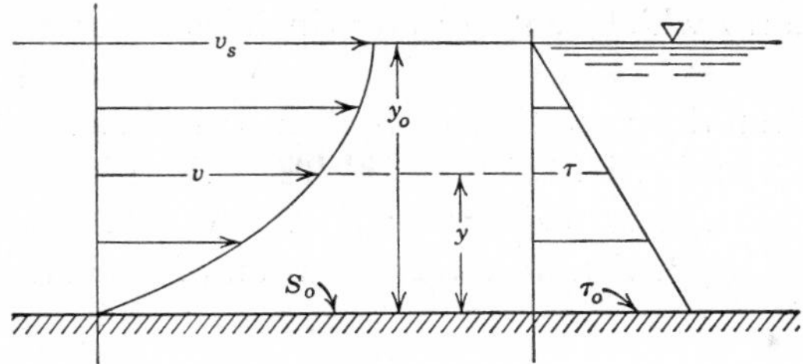

FIG. 150

to equal $2v_s/y_o$ and the mean velocity V to be $2v_s/3$. Combining these, equating the two expressions for τ_o, and substituting ρg for γ, and ν for μ/ρ, there results

$$V = \frac{gy_o{}^2S_o}{3\nu} \tag{230}$$

which relates mean velocity, slope, and depth when the flow is laminar. As in pipes, the limit of laminar flow is defined by an experimentally determined critical Reynolds number, in this case having a value of around 500 if the Reynolds number is defined as Vy_o/ν. From this a specific relationship between Chezy coefficient and Reynolds number may be derived for this laminar flow. Substitute V from equation 230 into the definition for Reynolds number; the result is

$$\mathbf{R} = gy_o{}^3S_o/3\nu^2 \tag{231}$$

Rearranging equation 230 to the Chezy form,

$$V = \frac{g\sqrt{S_o}\,y_o{}^{\frac{3}{2}}}{3\nu}\sqrt{y_oS_o} \tag{232}$$

and comparing equations 231 and 232,

$$C = \sqrt{g\mathbf{R}/3} \tag{233}$$

Although this relationship is not necessary for engineering use, it offers further justification of the straight line shown on the logarithmic plot of Fig. 148.

83. Hydraulic Radius Considerations. With the hydraulic radius R_h playing a prominent role in the equations of open-channel flow, and with depth variation a basic characteristic of such flows, the variation of hydraulic radius with depth (and other variables) becomes an important consideration. Clearly this is a problem of section geometry which requires no principles of mechanics for solution.

Consider first the variation of hydraulic radius with depth in a rectangular channel (Fig. 151*a*) of width B. Here $R_h = By/(B + 2y)$ and it is immediately evident that: for $y = 0$, $R_h = 0$ and, for $y \to \infty$, $R_h \to B/2$; therefore the variation of R_h with y must be as

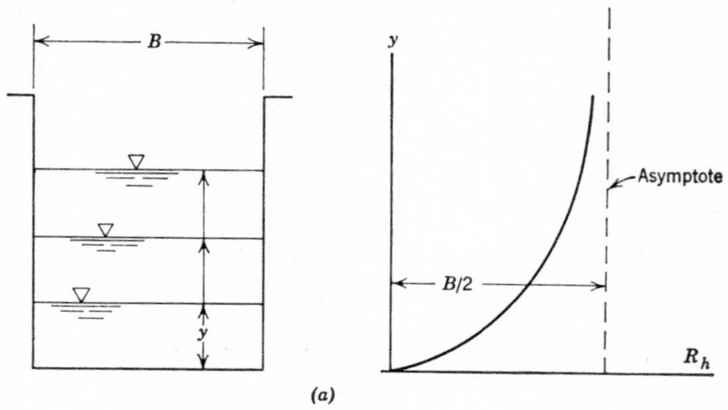

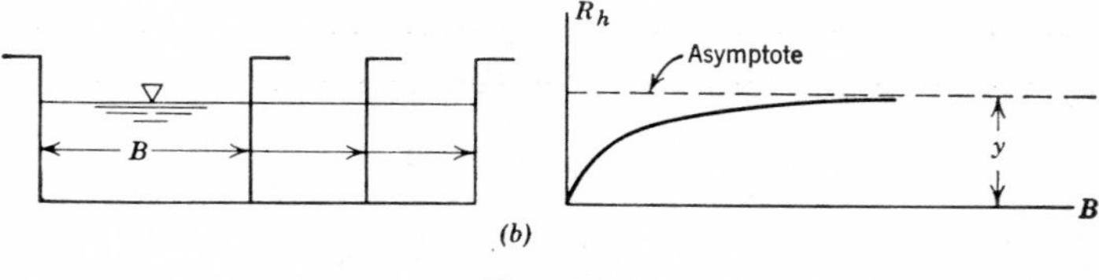

FIG. 151

shown. From this comes a useful engineering approximation: for *narrow deep sections* $R_h \cong B/2$; since any (nonrectangular) section when deep and narrow approaches a rectangle this approximation may be used for any deep and narrow section—for which the hydraulic radius may be taken to be one half of the mean width.

Another (and more useful) engineering approximation may be discovered by examining the variation of hydraulic radius with channel width (Fig. 151*b*) for a constant depth y. With $R_h = By/(B + 2y)$ it is noted that: for $B = 0$, $R_h = 0$ and, for $B \to \infty$, $R_h \to y$; thus the variation of R_h with B is as shown. From this it may be concluded that for wide shallow rectangular sections $R_h \cong y$; for nonrectangular sections the approximation is also valid if the section is wide and shallow—here the hydraulic radius approximates the mean depth.

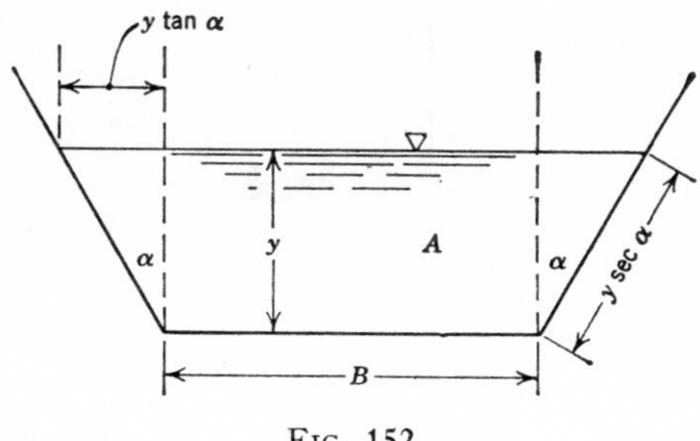

FIG. 152

A simple, useful, and typical nonrectangular section is the trapezoidal one of Fig. 152 for which $R_h = A/P = (By + y^2 \tan \alpha)/(B + 2y \sec \alpha)$. The derivative of R_h in respect to y will allow investigation of the form of the relationship between R_h and y. Performing this operation yields

$$\frac{dR_h}{dy} = \frac{1 + 2(y/B)\tan\alpha + 2(y/B)^2 \tan\alpha \sec\alpha}{1 + 4(y/B)\sec\alpha + 4(y/B)^2 \sec^2\alpha} \tag{234}$$

For $0 < \alpha < 90°$, $4 \sec \alpha > 2 \tan \alpha$ and therefore $4 \sec^2 \alpha > 2 \tan \alpha \sec \alpha$; from this it may be concluded that (for all values of y) the denominator of the fraction is larger than the numerator, $dR_h/dy < 1$, and also that dR_h/dy diminishes with increasing y. Thus the form of the variation of R_h with y is as shown on Fig. 153; it superficially

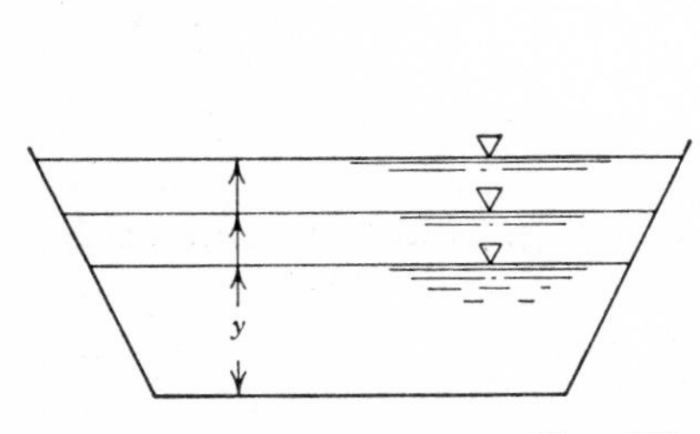

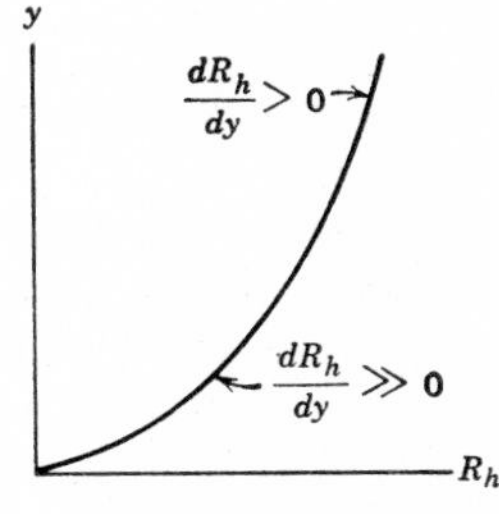

FIG. 153

resembles the curve of Fig. 151*a* but has no vertical asymptote since dR_h/dy does not approach zero as y approaches infinity. These are also general properties of most [11] (nontrapezoidal) channel sections with divergent side walls.

[11] Channel sections with diverging walls can be devised of such shape that there will be no variation of R_h with y, or even a decrease of R_h with increase of y; however, such sections are characterized by very small side slopes and great widths, and so they are of little practical engineering interest.

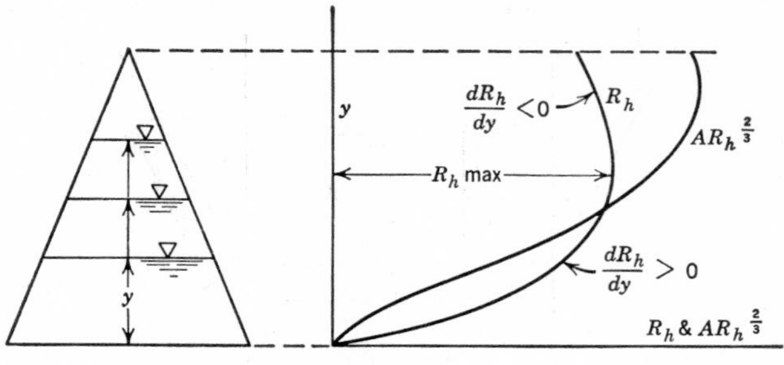

FIG. 154

For a trapezoidal channel with convergent sidewalls, Fig. 154 and equation 234 may be used with $-90° < \alpha < 0$. In this range $\tan \alpha < 0$, so the possibility of dR_h/dy (equation 234) becoming equal to or less than zero is evident. The variation of R_h with y for such a channel is shown in Fig. 154. The foregoing form of relation between R_h and y will be found valid for any section with convergent sidewalls; the circle (Fig. 155) is a simple example of this, and it is of wide use in engineering practice for underground open channels such as sewer lines, storm drains, and culverts. Here, for the full circle or semicircle, $R_h = d/4$ and, for $y = 0$, $R_h = 0$; thus a continuous variation of R_h with y may be expected to feature a maximum value as shown. Of more engineering significance in such problems is the variation of $AR_h^{\frac{2}{3}}$ with y since this combination of A and R_h appears in the Chezy-Manning equation 228. With $AR_h^{\frac{2}{3}}$ plotted against y (see Figs. 154 and 155) another maximum point is noted; from this it may be concluded that for given S_o and n there is a point at which the (uniform) flowrate is also maximum. Engineers usually disregard this in channel design, but it sometimes helps to

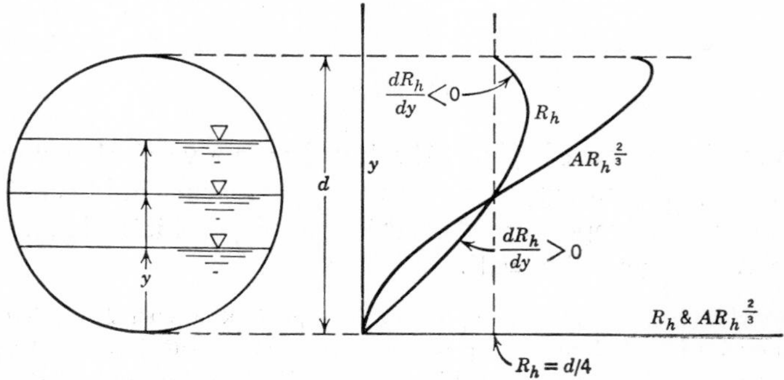

FIG. 155

explain phenomena which occur when channels of convergent walls flow nearly full.

Another important engineering problem of section geometry is the reduction of boundary resistance by minimization of the wetted perimeter for a given area of flow cross section. With A fixed, $R_h = A/P$, and, with P to be minimized, it is apparent that this may be considered a problem of maximization of the hydraulic radius. The desirability of this is evident from the above, but it is to be noted that reduction of wetted perimeter also tends to reduce the cost of lining material, grading, and construction. Hence a channel section of maximum hydraulic radius not only results in optimum hydraulic design but also tends toward a section of minimum cost. For this reason a section of maximum hydraulic radius is known as the *most efficient*, or *best, hydraulic cross section.* Since there is little hope of a generalized solution for all possible section shapes, the trapezoidal one (Fig. 152) will be again studied because of its wide use and good approximation to other sections. As before, $A = By + y^2 \tan \alpha$ and $P = B + 2y \sec \alpha$, but here [12] A and α are constants and the relation between B and y for maximum R_h is desired. Eliminating B, R_h may be written as a function of y:

$$R_h = \frac{A}{P} = \frac{A}{A/y - y \tan \alpha + 2y \sec \alpha}$$

Differentiating R_h in respect to y and equating to zero gives

$$A = y^2(2 \sec \alpha - \tan \alpha)$$

which, when substituted into the foregoing expression for R_h, yields

$$R_{h_{\max}} = y/2 \tag{235}$$

Thus, for best design conditions, a trapezoidal open channel should be proportioned so that its hydraulic radius is close to one-half of the depth of flow; this may also be used as a rough guide for the design of other channel sections which approach trapezoidal shape.

Since a rectangle is a special form of trapezoid, the foregoing results may be applied directly to the rectangular channel. Here again $R_{h_{\max}} = y/2$ and, for $\alpha = 0°$, $A = 2y^2$. Since A also equals By, it is evident that $B = 2y$ and that the best proportions for a rectangular channel exist when the depth of flow is one-half the width of the channel.

[12] The slope of the sidewalls, which determines the angle α, is limited in an earth canal by the angle of repose of the soil. If the canal is appropriately lined, α may have any value.

Illustrative Problem

What are the best dimensions for a rectangular channel which is to carry a uniform flow of 400 cfs if the channel is lined with rubble masonry and is laid on a slope of 0.0001?

Solution. From Table VI,

$$n = 0.017 \text{ ft}^{\frac{1}{6}}$$

$$A = 2y_o^2, \quad R_h = y_o/2 \tag{235}$$

$$400 = \left(\frac{1.49}{0.017}\right) 2y_o^2 \left(\frac{y_o}{2}\right)^{\frac{2}{3}} (0.0001)^{\frac{1}{2}} \tag{228}$$

whence

$$y_o^{\frac{8}{3}} = 364, \quad y_o = 9.15 \text{ ft}, \quad B = 2y_o = 18.30 \text{ ft}$$

A more general method of solution (necessary for nonrectangular sections) is:

$$400 = \left(\frac{1.49}{0.017}\right) By_o \left(\frac{By_o}{B + 2y_o}\right)^{\frac{2}{3}} (0.0001)^{\frac{1}{2}} \tag{228}$$

$$\frac{y_o}{2} = \frac{By_o}{B + 2y_o} \tag{235}$$

and, solving these equations simultaneously, $y_o = 9.15$ ft, $B = 18.30$ ft.

For an extended or distorted channel cross section such as that of Fig. 156, which represents a simplification of a river section in

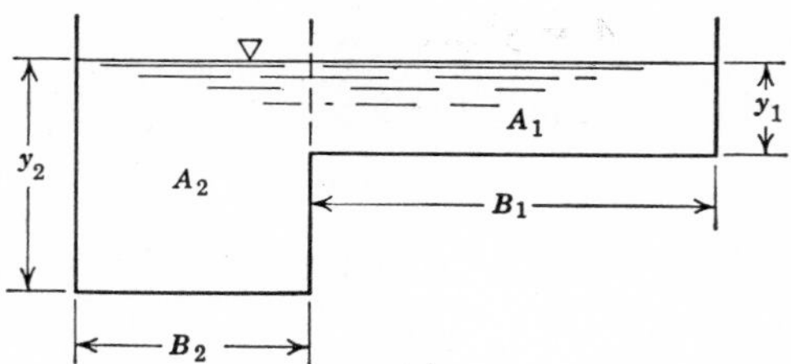

Fig. 156

flood, routine calculation of hydraulic radius from A/P will lead to large errors; such a calculation would imply that the effect of boundary resistance is uniformly distributed through the flow cross section, which is clearly not the case. Furthermore, accurate estimation of the effective value of n is virtually impossible since n for the narrow deep portion of the section may be very different from that of the wide shallow one. A logical (but not necessarily precise) means of treating such problems is to consider them composed of parallel channels separated by the vertical dashed line. The total flowrate is then expressed as $Q_1 + Q_2$, and the separate Q's by the Chezy-Manning equation,

$$Q = (1.49/n_1)A_1 R_{h_1}^{\frac{2}{3}} S_o^{\frac{1}{2}} + (1.49/n_2)A_2 R_{h_2}^{\frac{2}{3}} S_o^{\frac{1}{2}}$$

in which $A_1 = B_1 y_{o_1}$, $A_2 = B_2 y_{o_2}$, $P_1 = B_1 + y_{o_1}$, and $P_2 = B_2 + 2y_{o_2} - y_{o_1}$.

84. Specific Energy, Critical Depth, and Critical Slope—Wide Rectangular Channels. Many problems of open-channel flow are solved by a special application of the energy principle using the channel bottom as the datum plane. The concept and use of *specific energy*, the distance between channel bottom and energy line, were introduced by Bakhmeteff in 1912 and have proved fruitful in the explanation and analysis of new and old problems of open-channel flow. Today a knowledge of the fundamentals of specific energy is absolutely necessary in coping with the advanced problems of open flow; these fundamentals and a few of their applications are developed in the following paragraphs.

Considering the relative positions of channel bottom, liquid surface, and energy line in the typical reach of open channel shown in Fig. 157, it is not possible to predict the character of the change of specific energy between sections 1 and 2. Although the energy line must fall in the direction of the flow, *the specific energy may increase or decrease*, depending upon other factors to be investigated.

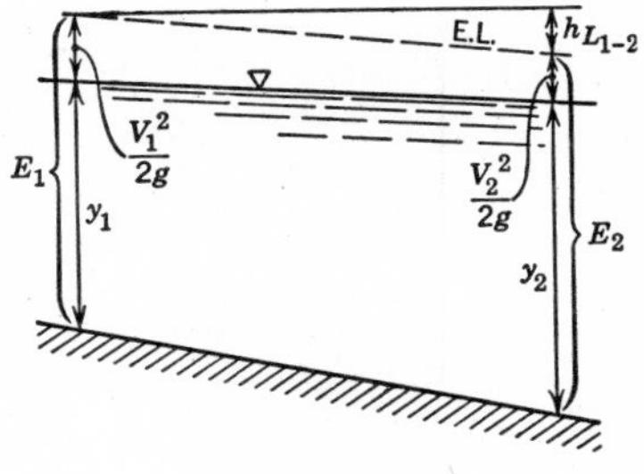

FIG. 157

With the specific energy defined as the vertical distance between channel bottom and energy line,

$$E = y + \frac{V^2}{2g} \tag{236}$$

or in terms of flowrate, which in steady flow is the same through each cross section,

$$E = y + \frac{1}{2g}\left(\frac{Q}{A}\right)^2 \tag{237}$$

It is convenient at this point to deal with flow in a wide channel of rectangular cross section to simplify the equations for better illustration of fundamentals; the principles may be applied, of course, to channels of other shapes (Art. 85), but the resulting equations are considerably more unwieldy. The assumption of a wide rectangular channel allows the use of the two-dimensional approximation and the use of the two-dimensional flowrate, q. Substituting q/y for V in equation 236 gives

$$E = y + \frac{1}{2g}\left(\frac{q}{y}\right)^2 \quad \text{or} \quad q = \sqrt{2g(y^2E - y^3)} \tag{238}$$

This equation gives clear and simple relationships between specific energy, flowrate, and depth. Although a three-dimensional plot of these variables may be visualized, a better understanding of the equation may more easily be acquired by (1) holding q constant and studying the relation between E and y and (2) holding E constant and examining the relation between q and y. Plotting these relations yields, respectively, the *specific energy diagram* and *q-curve* of Fig. 158.

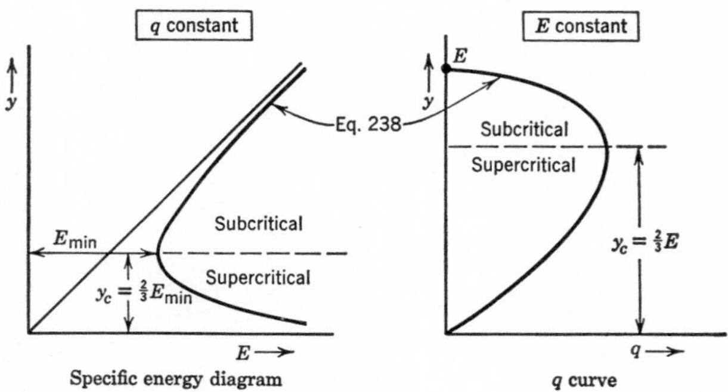

FIG. 158

Since these curves are merely different plots of the same equation, it may be expected (proof will be offered later) that the points of minimum E and maximum q are entirely equivalent. The depth associated with these points is known as the critical depth, y_c, and it is a boundary line between zones of open-channel flow which are very different in physical character. Flows at depths greater than critical are known as *subcritical* [13] flows; flows at depths less than critical are known as *supercritical* flows.

From the specific energy diagram the question raised by Fig. 157 can now be answered; an increase in depth is seen to cause an increase in specific energy in subcritical flow but a decrease in specific energy

[13] The terms *subcritical* and *supercritical* (analogous to the *subsonic* and *supersonic* of compressible-fluid motion) refer to the velocity; depths greater than critical give velocities less than that occurring at critical depth, and vice versa.

in supercritical flow; in other words, specific energy may be gained or lost in Fig. 157, depending on whether the depths of flow are greater or less than the critical depth. Furthermore, the form of the diagram shows that for a given specific energy two depths (and thus two flow situations) are, in general, possible, one of subcritical flow and one of supercritical flow; these two depths are known as the *alternate depths*.

The importance of the calculation of critical depth as a means of identifying the type of flow is apparent from the foregoing, and equations for this may be obtained from the specific energy diagram where $dE/dy = 0$, and from the q-curve where $dq/dy = 0$. Performing the differentiations of equations 238 yields (from the first equation)

$$q = \sqrt{g y_c^3} \quad \text{or} \quad y_c = \sqrt[3]{q^2/g} \tag{239}$$

and (from the second equation)

$$y_c = 2E_{\min}/3 \quad \text{or} \quad E_{\min} = 3y_c/2 \tag{240}$$

Substituting the latter relation into equation 238 for q gives $q = \sqrt{g y_c^3}$ as before, thus proving that the points of minimum E and maximum q have the same properties. Of great significance is the fact that (equation 239) *critical depth is dependent upon flowrate only;* the many other variables of open flow are not relevant to the computation of this important parameter. Equation 239 also suggests the use of critical flow as a means of flowrate measurement; if critical flow can be created or identified in a channel, the depth may be measured and flowrate determined.

Critical velocity, or velocity at the critical depth, may be obtained from $q = V_c y_c$ and $q = \sqrt{g y_c^3}$, whence

$$V_c = \sqrt{g y_c}$$

This equation is similar to another equation of open flow (see Appendix III),

$$V = \sqrt{g y}$$

which gives the velocity of propagation of a small wave on the surface of a liquid of depth y. The similarity of these equations offers a means of identifying subcritical and supercritical flows in the field.

In subcritical and supercritical flows, velocities are, respectively, less than and more than $\sqrt{gy}$; thus in subcritical flow small surface waves will progress upstream, but in supercritical flow such waves will be swept downstream (Fig. 159) to some angle β with the direction of flow. For such waves to remain stationary their component of propagation velocity (upstream) must be exactly equal and opposite to the flow velocity, V, whence $\sin \beta = \sqrt{gy}/V$. From this relationship the engineer in the field may estimate[14] velocity (and flowrate) from simple measurements of wave angle, water depth (and channel width).

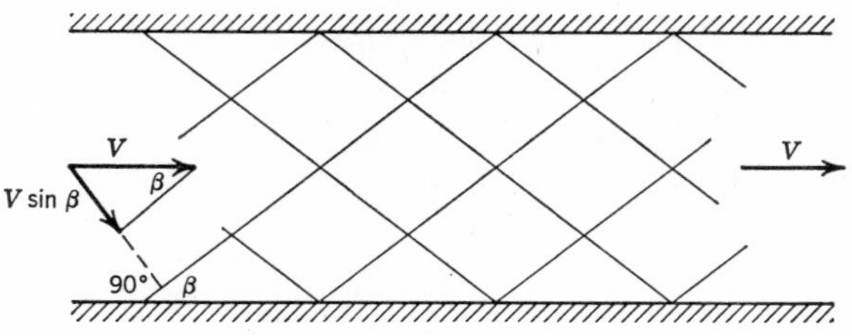

FIG. 159

Since wave phenomena are characterized by the interaction of inertia and gravity effects, it is to be expected that critical depth considerations may be written in terms of Froude number (Art. 61). Defining Froude number by $\mathbf{F} = V/\sqrt{gy}$ and applying the foregoing principles: for subcritical flow $\mathbf{F} < 1$, for critical flow $\mathbf{F} = 1$, for supercritical flow $\mathbf{F} > 1$, and $\sin \beta = 1/\mathbf{F}$. The similarity between wave phenomena in open flow and compressible flow (Arts. 44–47) is to be noted here. This analogy has found increasing use by aerodynamicists and nozzle designers in visualizing (hydraulically) the wave patterns in the flow of compressible fluids.

Uniform critical flow (i.e., uniform flow at critical depth) will occur in long open channels if they are laid on the *critical slope*, S_c. For a rectangular channel *of great width* a simple expression for S_c may be derived by equating the flowrates of equations 239 and 226:

$$q = \sqrt{gy_c^3} = Cy_c\sqrt{y_c S_c}$$

whence

$$S_c = g/C^2$$

However, from equation 227, $C = 1.49y^{\frac{1}{6}}/n$ so that S_c may also be expressed

$$S_c = gn^2/2.2y^{\frac{1}{3}} \tag{241}$$

[14] Great accuracy cannot be expected since surface velocity is not equal to mean velocity, nor are these velocities uniformly distributed through the flow cross section. Also it is re-emphasized that the application is to small waves only; larger waves produced by gross flow disturbances will stand at larger angles than arcsin $\sqrt{gy}/V$. See Art. 45.

showing critical slope to be a function of depth.[15] For a wide rectangular channel the form of this function is as shown on Fig. 160, featuring a relatively small change of S_c with y over a large range of depth. Although the form of this function is different for other cross sections, this feature is retained for practically all sections except those which are narrow and deep; it justifies the useful rough approximation that over a relatively large range of depths (and flowrates) S_c may be assumed constant. Uniform critical flow is, of course, a rare borderline situation between subcritical and supercritical flows. On the specific energy diagram it is noted that, in the vicinity of the critical depth, the depth may change considerably with little variation of specific energy; physically this means that, since many depths may occur for practically the same specific energy, flow near the critical depth will possess a certain instability (which manifests itself by undulations in the liquid surface). Uniform flow near the critical depth has been observed in both field and laboratory to possess these characteristics, and because of this the designer seeks to prevent flow situations close to the critical and uniform.

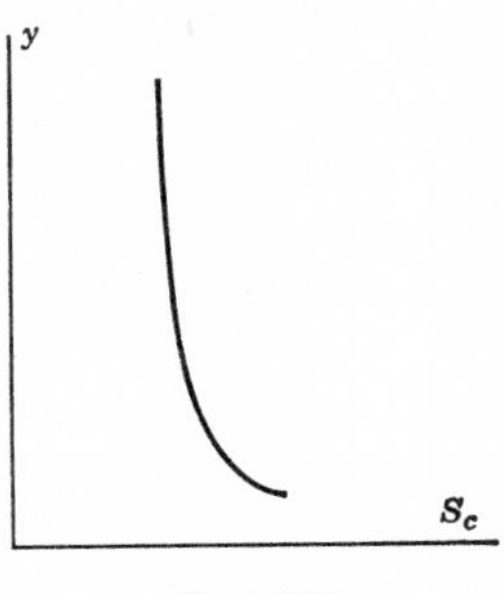

FIG. 160

Slopes greater than and less than the critical slope, S_c, are known, respectively, as *steep* and *mild* slopes. Evidently, channels of steep slope (if long enough) will produce supercritical uniform flows, and channels of mild slope (if long enough) will produce subcritical uniform flows. Whether or not uniformity is actually realized, the depth of uniform flow may always be computed from flowrate, slope, channel shape, and roughness through the Chezy-Manning equation. Depths computed in this manner are called *normal* or *neutral* depths; although such depths may not occur in short channels, they are, nevertheless, useful parameters of the flow and essential to an understanding of problems of varied flow.

ILLUSTRATIVE PROBLEM

For a flowrate of 500 cfs in a rectangular channel 40 ft wide, the water depth is 4 ft. Is this flow subcritical or supercritical? If $n = 0.017 \text{ ft}^{\frac{1}{6}}$, what is the critical slope of this channel for this flowrate? What channel slope should be provided to produce uniform flow at 4-ft depth?

[15] Although this equation shows a general relation between y and S_c, the latter may be computed from it only for $y = y_c$, which (through equation 239) implies a certain flowrate.

Solution.

$$q = 500/40 = 12.5 \text{ cfs/ft} \tag{55}$$

$$y_c = \sqrt[3]{(12.5)^2/32.2} = 1.69 \text{ ft} \tag{239}$$

Since $4 > 1.69$, this flow is subcritical:

$$S_c = 32.2(0.017)^2/2.2(1.69)^{\frac{1}{3}} \cong 0.0035 \tag{241}$$

The foregoing calculation is approximate since it implies that $y = R_h = 1.69$ ft; actually $y = 1.69$ ft and $R_h = 1.56$ ft. For uniform flow at 4-ft depth, $R_h = 160/48 = 3.33$ ft.

$$500 = (1.49/0.017)160(3.33)^{\frac{2}{3}}S_o^{\frac{1}{2}}; \quad S_o = 0.000255 \tag{228}$$

S_o is a *mild* slope since it is less than S_c.

85. Specific Energy, Critical Depth, and Critical Slope—Nonrectangular Channels. The application of the principles of Art. 84 to channels of nonrectangular cross section (Fig. 161) leads to more generalized and complicated mathematical expressions, owing to the more difficult geometrical aspects of such problems. The specific energy equation is written as before,

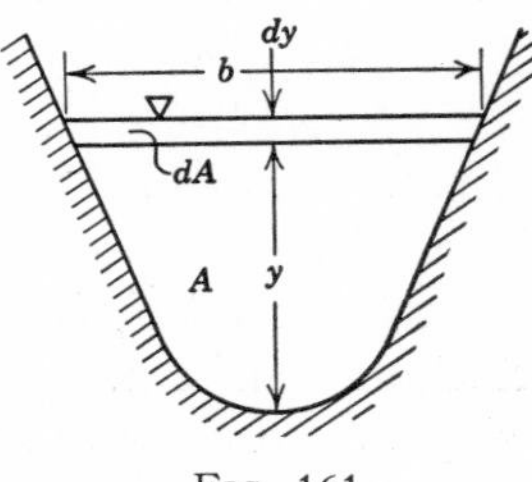

Fig. 161

$$E = y + \frac{1}{2g}\left(\frac{Q}{A}\right)^2 \tag{237}$$

Here A is some function of y, depending upon the form of the channel cross section, and the equation becomes

$$E = y + \frac{1}{2g}\left(\frac{Q}{f(y)}\right)^2 \tag{242}$$

which is analogous to equation 238 and leads to a specific energy diagram and a Q-curve (Fig. 162) having the same superficial appearance as those of Fig. 158 but for which there are other critical depth relationships. These may be worked out for the generalized channel cross section by differentiating equation 242 in respect to y and setting the result equal to zero:

$$\frac{dE}{dy} = 1 + \frac{Q^2}{2g}\left(-\frac{2}{A^3}\frac{dA}{dy}\right) = 0$$

From Fig. 161, $dA = b\,dy$ in which b is the channel width *at the liquid surface; dA/dy* is thus equal to b, and substitution gives

$$\frac{Q^2}{g} = \frac{A^3}{b} \quad \text{or} \quad \frac{Q^2 b}{gA^3} = 1 \tag{243}$$

as the equation which allows calculation of critical depth in nonrectangular channels. Substituting V for Q/A in equation 243 and defining a Froude number by

$$\mathbf{F} = \sqrt{\frac{Q^2 b}{gA^3}} = \frac{V}{\sqrt{g(A/b)}} \tag{244}$$

it is evident that (as for the wide rectangular channel): for subcritical flow $\mathbf{F} < 1$, for critical flow $\mathbf{F} = 1$, and for supercritical flow $\mathbf{F} > 1$.

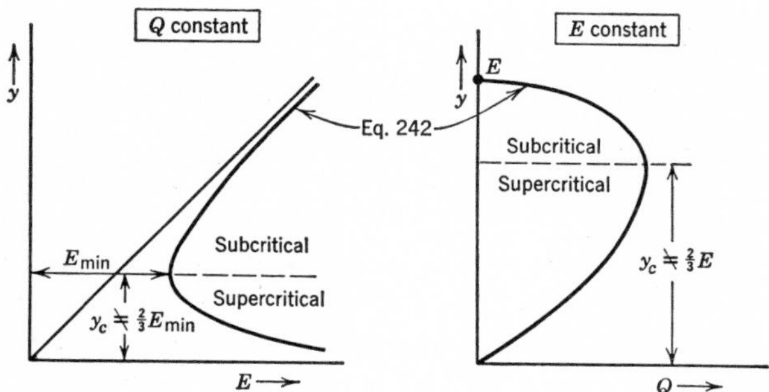

FIG. 162

Critical slope for a nonrectangular channel may be derived by following the same procedure as for the wide rectangular one:

$$Q = \sqrt{gA^3/b} = (1.49/n)AR_h^{\frac{2}{3}}S_c^{\frac{1}{2}}$$

whence

$$S_c = \frac{gn^2}{2.2}\left(\frac{A}{bR_h^{\frac{4}{3}}}\right) \tag{245}$$

In this expression A, b, and R_h are all functions of depth and dependent upon the section shape. Critical slope is thus a function of section shape also, and the form of the variation of S_c with y will usually be similar to that of Fig. 160.

ILLUSTRATIVE PROBLEM

A flow of 1000 cfs occurs in a trapezoidal canal having base width 10 ft, side slopes 1 (vert.) on 2 (horiz), and $n = 0.017$ ft$^{\frac{1}{6}}$. Calculate the critical depth and critical slope.

Solution.

$$A = 10y + 2y^2, \quad b = 10 + 4y, \quad P = 10 + 2\sqrt{5}\, y$$

$$R_h = A/P = (10y + 2y^2)/(10 + 2\sqrt{5}\, y)$$

$$\frac{(1000)^2}{32.2} = \frac{(10y_c + 2y_c^2)^3}{10 + 4y_c} \qquad (243)$$

When this is solved (by trial), $y_c = 4.91$ ft. Thus

$$A_c = 97.3 \text{ sqft}, \quad b_c = 29.65 \text{ ft}, \quad R_{h_c} = 97.3/32 = 3.04 \text{ ft}$$

$$S_c = \frac{32.2(0.017)^2}{2.2}\left(\frac{97.3}{29.65(3.04)^{\frac{4}{3}}}\right) = 0.00315 \qquad (245)$$

86. Occurrence of Critical Depth. The analysis of open-flow problems usually begins with the prediction of points in the channel

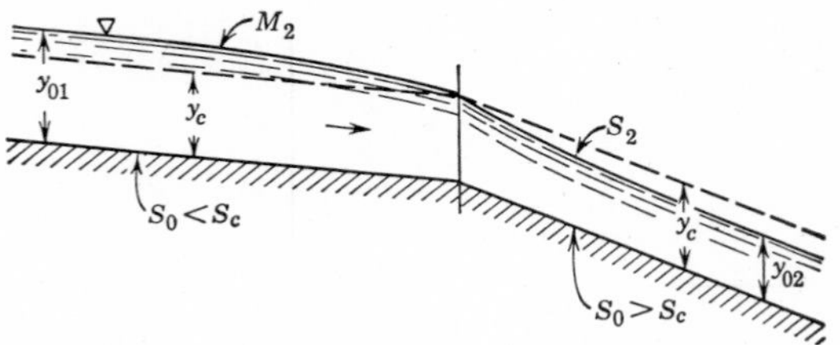

FIG. 163

at which, or near which, the critical depth will occur; such points, when they feature a change from subcritical to supercritical flow, are known as *controls* since their existence governs, or controls, the liquid depths in the reach of channel upstream from these points. Their prediction is a powerful tool in "blocking in" the variety of flow phenomena to be expected before making detailed calculations.

The most obvious place where critical depth can be expected is in the situation pictured in Fig. 163, where a long channel of mild slope ($S_o < S_c$) is connected to a long channel of steep slope ($S_o > S_c$). Far up the former channel uniform subcritical flow at normal depth, y_{o_1}, will occur, and far down the latter a uniform supercritical flow at a smaller normal depth, y_{o_2}, can be expected. These two uniform flows will be connected by a reach of varied [16] flow in which at some point the depth must pass through the critical. Experience shows that this point is close to (actually slightly upstream from) the break in slope and may be assumed there for most purposes. It is to be noted, however, that in the immediate vicinity of the change in bottom slope the streamlines are both curved and convergent, and

[16] See Art. 87 for explanation of the symbols M_2 and S_2.

the problem cannot be treated as one-dimensional flow with hydrostatic pressure distribution; the fact that the critical depth will not be found precisely at the point of slope change prevents the use of such points in the field for obtaining the exact flowrate by depth measurement and application of equations 239 or 243; however, the possibility of this method as an approximate means of flowrate measurement should not be overlooked.

When a long channel of steep slope discharges into one of mild slope (Fig. 164), normal depths will occur upstream and downstream

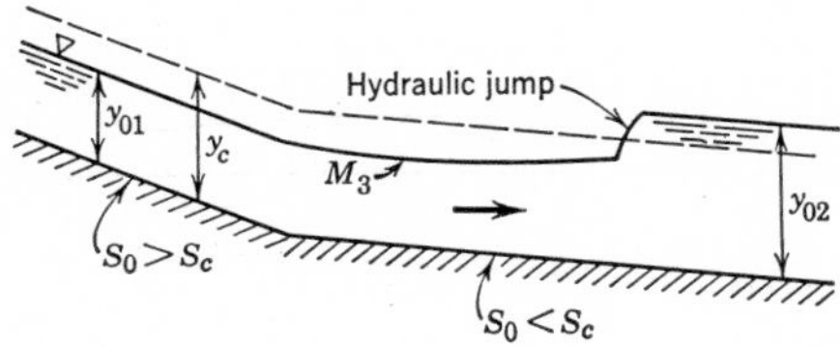

Fig. 164

from the point of slope change, but usually the critical depth will not be found near this point. Under these conditions a *hydraulic jump* (Arts. 44 and 88) will form whose location will be dictated (through varied[17] flow calculations) by the details of slopes, roughness, channel shapes, etc., but the critical depth will be found within the hydraulic jump. (See Art. 88.)

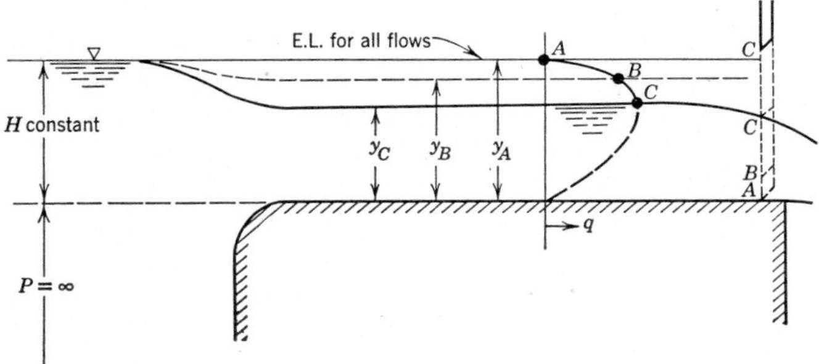

Fig. 165

The occurrence of critical depth on overflow structures may be proved by examining the flow over the top of a high frictionless broad-crested weir (Fig. 165) equipped with a movable sluice gate at the downstream end and discharging from a large reservoir of constant surface elevation. With gate closed (position A), the

[17] See Art. 87 for explanation of symbol M_3.

depth of liquid on the crest will be y_A, and the flowrate will obviously be zero, giving point A on the q-curve. With the gate raised to position B, a flowrate q_B will occur, with a decrease in depth from y_A to y_B. This process will continue until the gate is lifted clear of the flow (C) and can therefore no longer affect it. With the energy line fixed in position at the reservoir surface level and, therefore, giving constant specific energy, it follows that points A, B, and C have outlined the upper portion of the q-curve, that the flow occurring without gates is maximum, and that the depth on the crest is, therefore, the critical depth. For flow over weirs, a relation between

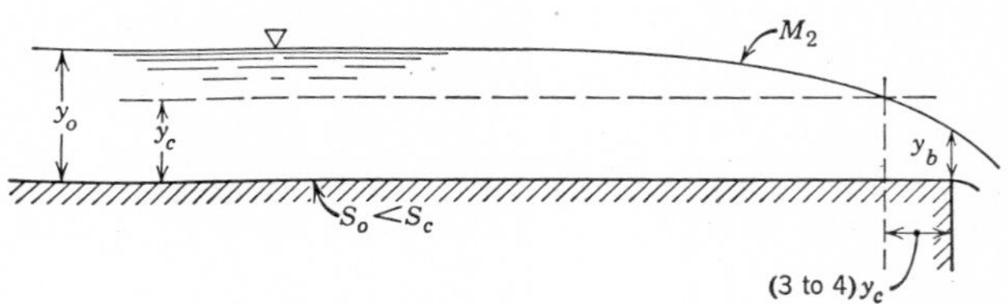

FIG. 166

head and flowrate is desired; this may be obtained by substituting (for a very high weir) $y_c = 2H/3$ in equation 239, which yields

$$q = \sqrt{g y_c^3} = \sqrt{g(2H/3)^3} = 0.577 \times \tfrac{2}{3}\sqrt{2g}\, H^{\frac{3}{2}} \qquad (246)$$

The last form of this equation is readily compared with the standard weir equation (Art. 105); here 0.577 is the *weir coefficient.* Because of neglect of friction in the analysis this coefficient of 0.577 is higher than those obtained in experiments; tests on high, broad-crested weirs give coefficients between 0.50 and 0.57 (depending upon the details of weir shape).

The reasoning of the foregoing paragraph may be extended to a *free outfall* (Fig. 166)[18] from a long channel of mild slope, to conclude that the depth must pass through the critical in the vicinity of the brink. Rouse [19] has found that for such rectangular channels the critical depth occurs a short distance (3 to $4y_c$) upstream from the brink and that the brink depth (y_b) is 71.5 percent of the critical depth.[20] Using this figure and equation 239, Rouse proposes the

[18] See Art. 87 for explanation of symbol M_2.

[19] H. Rouse, "Discharge Characteristics of the Free Overfall," *Civil Engr.*, Vol. 6, No. 4, p. 257, 1936.

[20] Experiments by the author on free outfalls from *circular conduits* of mild slope showed brink depth to be about 75 percent of critical depth when the brink depth is less than 60 percent of the diameter of the conduit.

free outfall as a simple device for metering the flowrate, which requires only measurement of the brink depth. Free outfalls also provide an opportunity for recognition of the limitations of specific energy theory. Upstream from the point of critical depth, the streamlines are essentially straight and parallel and the pressure distribution hydrostatic; between this point and the brink neither of these conditions obtains, so the one-dimensional theory is invalid in this region. For example, it *cannot* be concluded that the flow at the brink is supercritical merely because the brink depth is less than the critical depth.

Critical depth may be obtained at points in an open channel where the channel bottom is raised by the construction of a low hump or the channel is constricted by moving in the sidewalls. Such contractions (usually containing a rise in the channel bottom) are known generally as Venturi flumes, and specific designs [21] of these flumes are widely used in the measurement of irrigation water. Preliminary analysis of such contractions (which feature accelerated flow) may be made assuming one-dimensional flow and neglecting head losses, but final designs require more refined information either from other designs or from laboratory experiments. The advantages of such flumes are their ability to pass sediment-laden water without deposition and the small net change of water level required between entrance and exit channels.

Illustrative Problem

Uniform flow at a depth of 5 ft occurs in a long rectangular channel of 10-ft width, having a Manning n of 0.015 $\text{ft}^{\frac{1}{6}}$ and laid on a slope of 0.001. Cal-

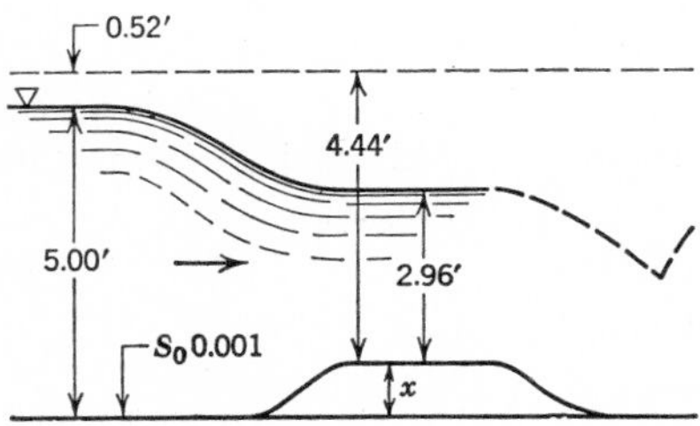

culate (*a*) the minimum height of hump which can be built in the floor of this channel to produce critical depth, and (*b*) the maximum width of contraction which can produce critical depth.

[21] See, for example, *Water Measurement Manual*, U.S. Dept. of the Interior, Bureau of Reclamation, 1953.

Solution.

$$Q = (1.49/0.015)50(50/20)^{\frac{2}{3}}(0.001)^{\frac{1}{2}} = 289 \text{ cfs} \quad (228)$$

$$E = 5.00 + (1/2g)(289/50)^2 = 5.52 \text{ ft} \quad (236)$$

(*a*)

$$y_c = \sqrt[3]{(28.9)^2/32.2} = 2.96 \text{ ft} \quad (239)$$

$$E_{\min} = 3 \times 2.96/2 = 4.44 \text{ ft} \quad (240)$$

A hump height smaller than x will lower the water surface over the hump but cannot produce critical depth; a hump height larger than this will produce critical depth, the hump then being a broad-crested weir. The latter condition would, however, raise the energy line in the vicinity of the hump and (this being a subcritical flow) increase the depth upstream from this point. Therefore the minimum height of hump to produce critical depth is that which will do so without raising the energy line. From the sketch, this is seen to be 5.52 − 4.44, or 1.08 ft.

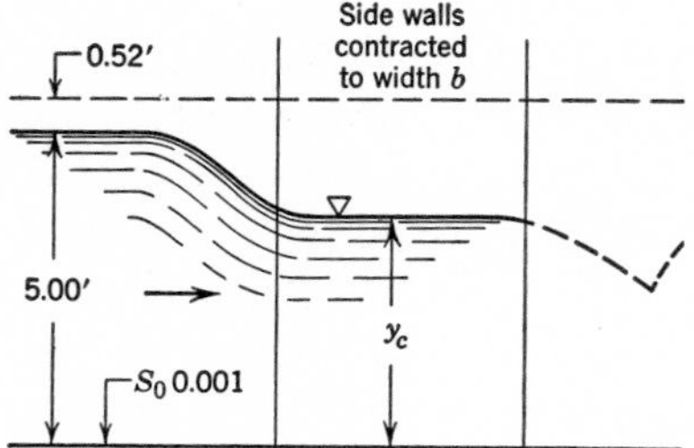

(*b*) The width b at the contraction being unknown but critical depth being specified,

$$y_c = \sqrt[3]{\frac{(289/b)^2}{32.2}} \quad (239)$$

From part (*a*) it may be seen that contraction widths smaller than that required to just produce critical depth will do so by raising the energy line and the water depth upstream from the contraction. Accordingly, the condition sought is that which will produce critical depth in the contraction without raising the energy line. The contraction being of rectangular cross section,

$$y_c = 2 \times 5.52/3 = 3.68 \text{ ft} \quad (240)$$

Equating this to the preceding expression and solving for b, $b = 7.2$ ft.

87. Varied Flow. For design of open channels and analyses of their performance, the engineer must be able to predict forms and positions of water-surface profiles of varied (nonuniform) flow and to acquire some facility in their calculation. The first objective may be attained by development and study of the differential equation of varied flow, and the second objective either by performing the integration or by the use of step-by-step calculations.

The differential equation of varied flow may be derived by considering the differential element of Fig. 167 in which the bottom slope S_o is small and over the horizontal distance dx, the depth is assumed to increase by dy, and the velocity head by $d(V^2/2g)$. Taking the slope of the energy line to be S and the slope of channel bottom S_o, the drops of these lines (in the distance dx) will be $S\,dx$

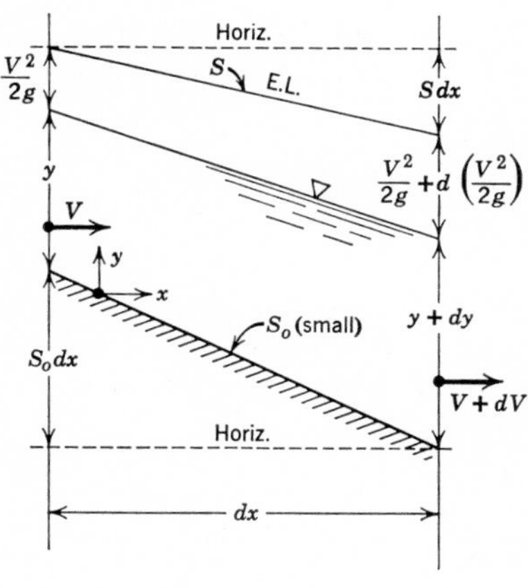

FIG. 167

and $S_o\,dx$, respectively, From the sketch may be written the equation

$$S_o\,dx + y + \frac{V^2}{2g} = y + dy + \frac{V^2}{2g} + d\left(\frac{V^2}{2g}\right) + S\,dx \qquad (247)$$

which, by appropriate cancellations and division by dx, may be reduced to

$$\frac{dy}{dx} + \frac{d}{dx}\left(\frac{V^2}{2g}\right) = S_o - S$$

Multiplying the second term by dy/dy and solving the equation for dy/dx,

$$\frac{dy}{dx} = \frac{S_o - S}{1 + \dfrac{d}{dy}\left(\dfrac{V^2}{2g}\right)}$$

The derivative in the denominator of the fraction must be recognized before the equation can be fully explored; this has been already calculated in the derivation of equation 243 and shown to be $-Q^2b/gA^3$ in which b is the top width of the flow cross section A.

Since $Q^2b/gA^2 = \mathbf{F}^2$ (see Art. 85), the differential equation of varied flow may be written alternatively as

$$\frac{dy}{dx} = \frac{S_o - S}{1 - Q^2b/gA^3} = \frac{S_o - S}{1 - \mathbf{F}^2} \tag{248}$$

from which the forms of all possible water-surface profiles of varied flow may be deduced.

From the first step (equation 247) in the development of this equation two limitations are evident: (1) Since hydrostatic pressure distribution has been assumed in the use of specific energy, application is limited to flows with streamlines essentially straight and parallel, and of small slope S_o. (2) Since the depth y is measured from the channel bottom, the slope of the water surface dy/dx is *relative to this channel bottom*, and thus does not have the same meaning as the conventional dy/dx of analytic geometry; its implications are shown on Fig. 168, which is basic to the prediction of surface profiles from analysis of equation 248.

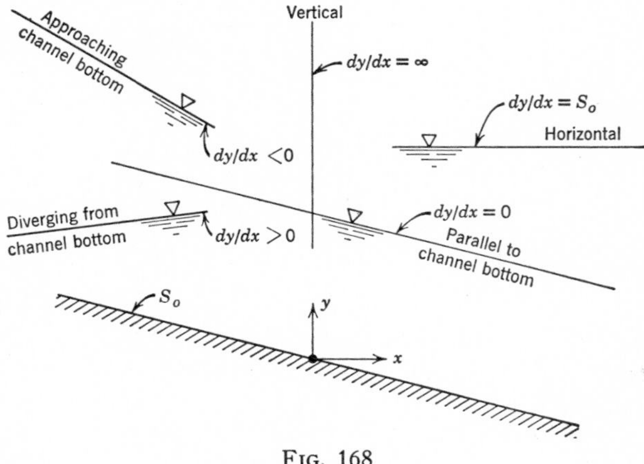

Fig. 168

Surface profiles are categorized in terms of bottom slope S_o as follows: (*a*) $S_o > S_c$ (S for *steep*), (*b*) $S_o = S_c$ (C for *critical*), (*c*) $S_o < S_c$ (M for *mild*), (*d*) $S_o = 0$ (H for *horizontal*), and (*e*) $S_o < 0$ (A for *adverse*). The form of the profile also depends on the water depth relative to both normal and critical depths; the relative positions of the latter are shown in column 2 of Table VII and show that twelve possible profiles are to be expected. Here the only point not previously discussed involves the values of y_o for $S_o = 0$ and $S_o < 0$; these are readily confirmed by solving the Chezy-Manning equation for y_o.

TABLE VII

SURFACE PROFILES OF VARIED FLOW

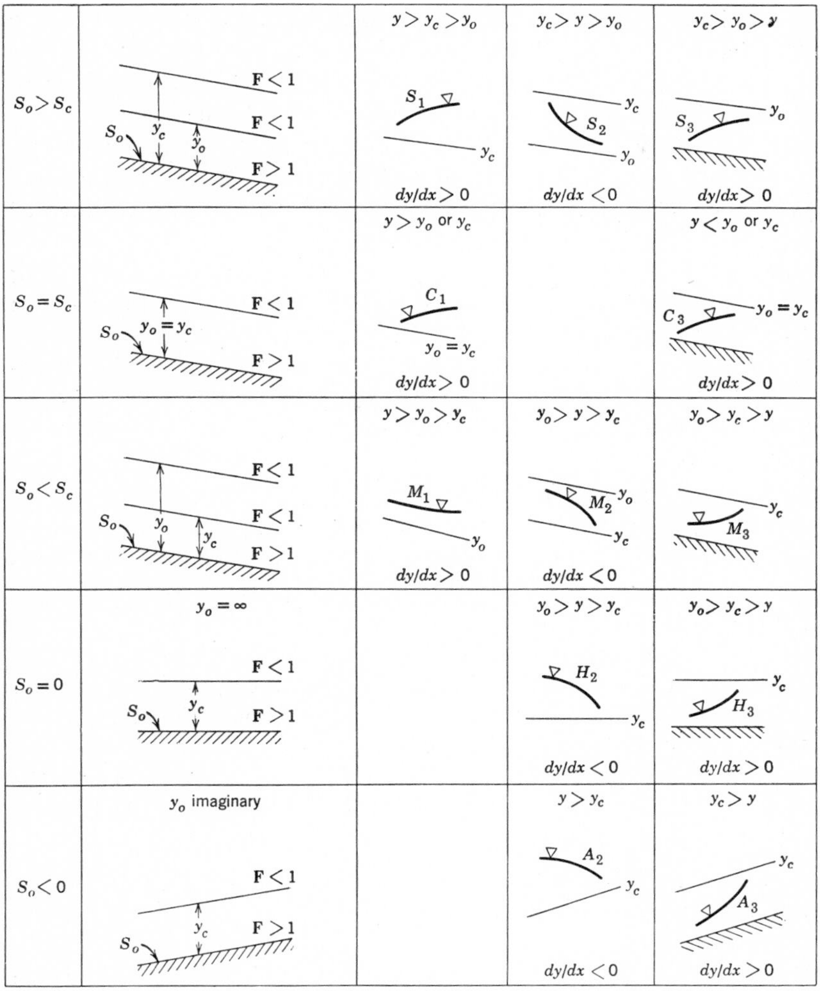

$S_o > S_c$	F < 1; F < 1; F > 1; S_o; y_c; y_o	$y > y_c > y_o$; S_1; y_c; $dy/dx > 0$	$y_c > y > y_o$; S_2; y_c; y_o; $dy/dx < 0$	$y_c > y_o > y$; S_3; y_o; $dy/dx > 0$
$S_o = S_c$	F < 1; F > 1; S_o; $y_o = y_c$	$y > y_o$ or y_c; C_1; $y_o = y_c$; $dy/dx > 0$		$y < y_o$ or y_c; C_3; $y_o = y_c$; $dy/dx > 0$
$S_o < S_c$	F < 1; F < 1; F > 1; S_o; y_o; y_c	$y > y_o > y_c$; M_1; y_o; $dy/dx > 0$	$y_o > y > y_c$; M_2; y_o; y_c; $dy/dx < 0$	$y_o > y_c > y$; M_3; y_c
$S_o = 0$	$y_o = \infty$; F < 1; F > 1; S_o; y_c		$y_o > y > y_c$; H_2; y_c; $dy/dx < 0$	$y_o > y_c > y$; H_3; y_c; $dy/dx > 0$
$S_o < 0$	y_o imaginary; F < 1; F > 1; S_o; y_c		$y > y_c$; A_2; y_c; $dy/dx < 0$	$y_c > y$; A_3; y_c; $dy/dx > 0$

Analysis of equation 248 will show $(S_o - S)$ to depend only upon the relative sizes of depth and normal depth—and $(1 - \mathbf{F}^2)$ to depend solely on the relative magnitudes of depth and critical depth; these statements may be validated in the following manner. From the Chezy-Manning equation it is easily seen that (for a given Q and n) S will decrease with increasing $AR_h^{\frac{2}{3}}$; since the latter term has been shown (Art. 83) to be a direct (not linear) function of depth for most open channels, S will also decrease with increasing depth. This is to be expected since increased depth implies larger flow cross sections and smaller velocities, with consequent reduction of energy dissipation. Since, when $y = y_o$, $S = S_o$ (uniform flow) the above variations may be applied to conclude that: for $y > y_o$, $S < S_o$, and $(S_o - S) > 0$; for $y < y_o$, $S > S_o$ and $(S_o - S) < 0$. Analysis of the quantity $(1 - \mathbf{F}^2)$ is much easier since it has been shown (Art. 85) that, for $y > y_c$, $\mathbf{F} < 1$ and, for $y < y_c$, $\mathbf{F} > 1$; hence, for $y > y_c$, $(1 - \mathbf{F}^2) > 0$ and, for $y < y_c$, $(1 - \mathbf{F}^2) < 0$. With this information on $(S_o - S)$ and $(1 - \mathbf{F}^2)$ the signs of dy/dx may be determined as indicated in columns 3, 4, and 5 of Table VII.

The final forms of the profiles (excepting those for $S_o = S_c$) may be established (from equation 248) by observing that (1) for $y \rightarrow \infty$: $S \rightarrow 0$, $\mathbf{F} \rightarrow 0$, and $dy/dx \rightarrow S_o$; (2) for $y \rightarrow y_o$: $S \rightarrow S_o$, and $dy/dx \rightarrow 0$; (3) for $y \rightarrow y_c$: $\mathbf{F} \rightarrow 1$ and $dy/dx \rightarrow \infty$. The exception occurs in the case of $S_o = S_c$ since here, as $y \rightarrow (y_o = y_c)$, dy/dx cannot approach both zero and infinity. Of the foregoing conclusions: (1) expresses the well-known fact that for large depths (and accompanying small velocities) liquid surfaces approach the horizontal; (2) shows that all surface profiles approach uniform flow asymptotically; (3) shows (if taken literally) that the free surface is vertical when passing through the critical depth! Although the last conclusion is invalid because the original development assumed essentially straight and parallel streamlines, it is nevertheless useful as a guide in the final formulation of the surface profiles, which, to satisfy the conditions above, must be as shown in Table VII. Examples of some of these surface profiles are shown in Figs. 163, 164, 166, 172, and 173.

Investigation of the interesting (but relatively unimportant) borderline situation where $S_c = S_o$ may be undertaken using a wide rectangular section as an approximation to most other sections.[22] Here equation 248 may be rearranged to

$$\frac{dy}{dx} = S_c \left(\frac{1 - S/S_c}{1 - \mathbf{F}^2} \right) \tag{249}$$

For a section of unit width $A = R_h = y$, and from the Chezy-Manning equation,

$$q = (1.49/n) y^{\frac{5}{3}} S^{\frac{1}{2}} = (1.49/n) y_c^{\frac{5}{3}} S_c^{\frac{1}{2}}$$

whence

$$S/S_c = (y_c/y)^{\frac{10}{3}} \tag{250}$$

[22] For narrow and deep sections this is obviously a poor approximation.

However, with $\mathbf{F} = V/\sqrt{gy}$ and $q^2 = gy_c^3$ (equation 239),

$$\mathbf{F}^2 = \frac{V^2}{gy} = \frac{q^2}{gy^3} = \frac{gy_c^3}{gy^3} = \left(\frac{y_c}{y}\right)^3 \tag{251}$$

By substitution of relations 250 and 251 into equation 249, it becomes

$$\frac{dy}{dx} = S_c\left(\frac{1-(y_c/y)^{\frac{10}{3}}}{1-(y_c/y)^3}\right) \tag{252}$$

which shows that: (1) for $y > y_c$ or $y < y_c$, $dy/dx > S_c$, and (2) for $y \rightarrow \infty$, $dy/dx \rightarrow S_c$, which justifies the form of the C_1 and C_3 profiles shown in Table VII.

Lack of space precludes review of the numerous attempts (over the last century) to integrate the differential equation of varied flow to provide the engineer with y as a function of x so that water-surface profiles may be plotted for known channel cross section, flowrate, slope, and Manning n. However, the recent and comprehensive method developed by Chow[23] has proved so successful that it bids fair to become the standard technique of solving this problem.

To handle isolated or occasional problems of varied flow the engineer may resort to a simple step-by-step method of computation after having sketched in the appropriate surface profiles from Table VII.

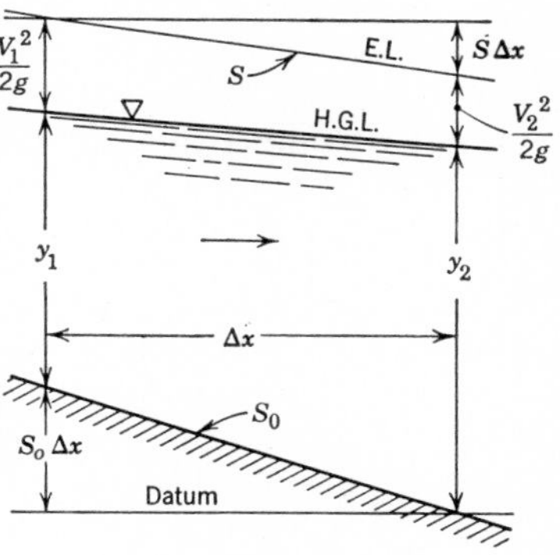

Fig. 169

The definition sketch of Fig. 169 is essentially the same as that of Fig. 167, and the analogous equation is written from the sketch; it is

$$S_o\,\Delta x + y_1 + \frac{V_1^2}{2g} = y_2 + \frac{V_2^2}{2g} + S\,\Delta x \tag{253}$$

whence, by substitution of E for $y + V^2/2g$ and solving for Δx,

$$\Delta x = \frac{E_1 - E_2}{S - S_o} \tag{254}$$

[23] V. T. Chow, "Integrating the Equation of Gradually Varied Flow," *Proc. A.S.C.E.*, Vol. 81, November 1955, and *Open-Channel Hydraulics*, McGraw-Hill Book Co., 1959.

Procedure in using this equation begins in knowing channel shape, Q, S_o, n, and one of the depths (usually at a control section); assumption of the other depth allows calculation of Δx, the horizontal distance between the two cross sections of depths y_1 and y_2—once a suitable method is devised for calculation of S. Making the assumption that the head loss between sections 1 and 2 for varied flow is equal to the head loss for a uniform flow of average velocity (V_{avg}) and hydraulic radius ($R_{h_{\text{avg}}}$), S may be calculated from the Chezy-Manning equation (228),

$$V_{\text{avg}} = (1.49/n)(R_{h_{\text{avg}}})^{\frac{2}{3}}S^{\frac{1}{2}} \tag{255}$$

in which $V_{\text{avg}} = (V_1 + V_2)/2$ and $R_{h_{\text{avg}}} = (R_{h_1} + R_{h_2})/2$.

The plausible assumption used in the calculation of S has never been precisely confirmed by experiment, but errors arising from it are small compared to those incurred in the selection of n, and over the years it has proved to be a reliable basis for design calculations. The assumption is doubtless more valid for accelerating (depth decreasing downstream) flow than for decelerating flow, but in a prismatic channel the difference between them would be very small. The disadvantage of the method is the necessity for a large number of small steps which require accurate, repetitive, and somewhat tedious numerical calculations. The modern engineer should also assimilate the more elegant integration methods of varied flow which allow rapid calculations for prismatic channels; however, for non-prismatic channels (such as natural streams) techniques akin to the step method must be used.

Illustrative Problem

A flowrate of 400 cfs occurs in a rectangular channel 20 ft wide, lined with rubble masonry and laid on a slope of 0.0001; at a point in this channel the depth is 6.00 ft. How far (upstream or downstream) from this point will the depth be 6.60 ft?

Solution. From the illustrative problem of Art. 81,

$$y_o = 8.34 \text{ ft}$$

$$y_c = \sqrt[3]{(20)^2/32.2} = 2.31 \text{ ft} \tag{239}$$

Since $y_o > 6.60 > 6.00 > y_c$, reference to Table VII indicates an M_2 surface profile, so the depth of 6.60 ft will be found upstream from that of 6.00 ft.

Set up a table for the calculations (and use equations 254 and 255).

y	A	V (V_{avg})	$V^2/2g$	E (ΔE)	R_h ($R_{h_{avg}}$)	S	$S - S_o$	Δx
6.00	120	3.335	0.173	6.173	3.75			
		(3.280)		(0.189)	(3.79)	0.000237	0.000137	1380
6.20	124	3.225	0.162	6.362	3.83			
		(3.175)		(0.190)	(3.865)	0.000216	0.000116	1640
6.40	128	3.125	0.152	6.552	3.90			
		(3.077)		(0.190)	(3.935)	0.000198	0.000098	1940
6.60	132	3.030	0.142	6.742	3.97			

The distance requested is $\Sigma(\Delta x) = 4960$ ft. The trend of the values of Δx for constant Δy shows the M_2 curve to be convex up, as expected from Table VII.

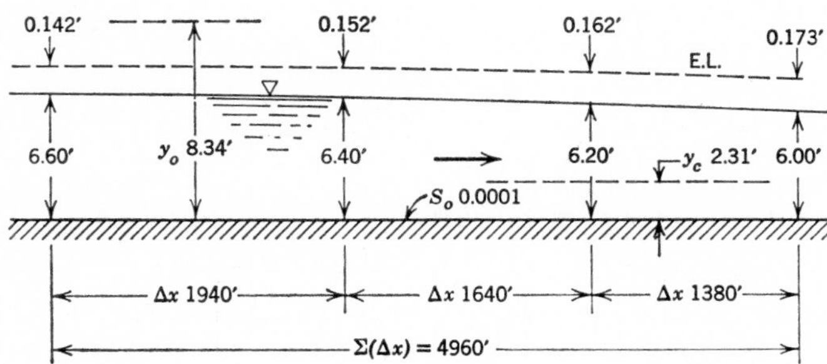

88. The Hydraulic Jump. When a change from supercritical to subcritical flow occurs in open flow a *hydraulic jump* appears, through which the depth increases abruptly in the direction of flow. In spite of the complex appearance of a hydraulic jump with its turbulence [24] and air entrainment, it may be successfully analyzed by application of the impulse-momentum principle (see Art. 44) to yield results and relationships which conform closely with experimental observations.

A hydraulic jump in an open channel of small slope is shown in Fig. 170.[25] In engineering practice the hydraulic jump frequently appears downstream from overflow structures (spillways) or underflow structures (sluice gates) where velocities are high. It may be used as an effective dissipator of kinetic energy (and thus prevent

[24] See H. Rouse, T. T. Siao, S. Nageratnam, "Turbulence Characteristics of the Hydraulic Jump," *Trans. A.S.C.E.*, Vol. 124, 1959.

[25] The jump is not so steep as shown in the figure; the length of the jump is usually (4 to 5)y_2.

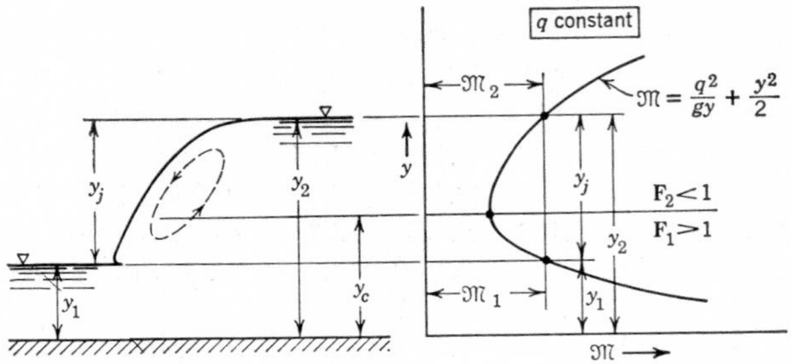

FIG. 170

scour of channel bottom) or as a mixing device in water or sewage treatment designs where chemicals are added to the flow. In design calculations the engineer is concerned mainly with prediction of existence, size, and location of the jump.

In Art. 44 the basic equation for the jump in a rectangular [26] open channel was derived; it is

$$\frac{q^2}{gy_1} + \frac{y_1^2}{2} = \frac{q^2}{gy_2} + \frac{y_2^2}{2} \tag{256}$$

the solution of which may be written

$$\frac{y_2}{y_1} = \frac{1}{2}\left[-1 + \sqrt{1 + \frac{8V_1^2}{gy_1}}\right] \tag{257}$$

in which V_1^2/gy_1 is recognized as $\mathbf{F}_1^2$; this equation shows that $y_2/y_1 > 1$ only when $\mathbf{F}_1 > 1$, thus proving the necessity of supercritical flow for jump formation. Another way of visualizing this is by defining a quantity $\mathfrak{M}$ by

$$\mathfrak{M} = \frac{q^2}{gy} + \frac{y^2}{2}$$

and plotting $\mathfrak{M}$ as a function of y (Fig. 170) for constant flowrate, whereupon the solution of the equation occurs when $\mathfrak{M}_1 = \mathfrak{M}_2$; the curve obtained features a minimum value of $\mathfrak{M}$ at the critical depth [27] and thus superficially resembles (but must not be confused with) the specific energy diagram. After construction of this curve and with one depth known, the corresponding, or *conjugate*, depth may be

[26] Here the analysis is confined to the rectangular channel for both mathematical simplicity and practical application. The same methods may be applied to channels of nonrectangular cross section; for example, see G. H. Hickox, "Graphical Solution for Hydraulic Jump," *Civil Eng.*, Vol. 4, No. 5, p. 270, 1934.

[27] This may be easily proved by setting $d\mathfrak{M}/dy$ equal to zero.

found by passing a vertical line through the point of known depth. Since a vertical line is a line of constant $\mathfrak{M}$, the intersection of this line and the other portion of the curve gives a point where $\mathfrak{M}_1$ is equal to $\mathfrak{M}_2$, and allows the conjugate depth and height of jump y_j to be taken directly from the plot. The shape of the $\mathfrak{M}$-curve also shows clearly and convincingly that hydraulic jumps can take place only across the critical depth (i.e., from supercritical to subcritical flow).

Because of eddies (rollers), air entrainment, and flow decelerations in the hydraulic jump, large head losses are to be expected; they may be calculated (as usual) from the fall of the energy line,

$$h_{L_j} = \left(y_1 + \frac{V_1^2}{2g} + z_1\right) - \left(y_2 + \frac{V_1^2}{2g} + z_2\right)$$

in which $z_1 = z_2$ if the channel bottom is horizontal. For very small jumps, the eddies and air entrainment disappear, the form of the jump changes to that of a smooth standing wave, known as an

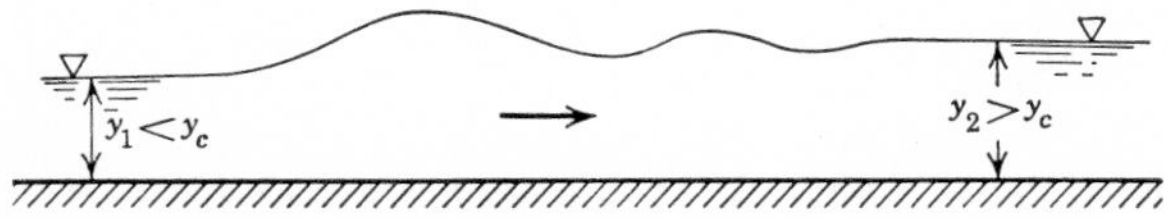

Fig. 171. Undular hydraulic jump.

undular jump (Fig. 171), with very small head loss. Laboratory tests show that undular jumps are to be expected for $\mathbf{F}_1 \gtrless 2$.

A simple problem in hydraulic jump location is shown in Fig. 172. Here a rectangular channel of steep slope ($S_o > S_c$) with uniform flow discharges into a channel of mild slope ($S_o < S_c$) of sufficient length to produce uniform flow. For the given flowrate the normal

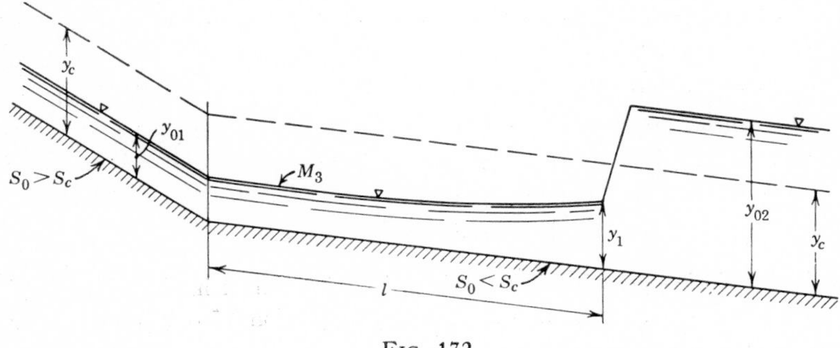

Fig. 172

depths y_{o_1} and y_{o_2} may be calculated from the Chezy-Manning equation (228), and the critical depth from equation 239.

In view of the variety of phenomena produced by combinations of uniform flow, varied flow, and hydraulic jump, the sketch of Fig. 172 must first be validated. The depth at the break in slope must be y_{o_1} since the flow in channel 1 is supercritical (with flow velocity larger than wave velocity); accordingly any influence from channel 2 cannot be transmitted upstream into channel 1. The varied flow surface profile over the reach l in channel 2 is identified as M_3 from Table VII and may be calculated (and plotted) starting with y_{o_1} and using the methods of Art. 87. The depth y_1 may be obtained from equation 257 of the hydraulic jump, with $y_2 = y_{o_2}$. The validity of using $y_2 = y_{o_2}$ may be confirmed by assuming $y_2 > y_{o_2}$ and $y_2 < y_{o_2}$; for either of these conditions no varied flow surface profile would be possible (see Table VII) between y_2 and y_{o_2}. There are two other possible flow situations in this problem: (1) $l = 0$ with the hydraulic jump at the break in slope; and (2) hydraulic jump in channel 1 (Fig. 173). The first of these will obtain when solution of the hy-

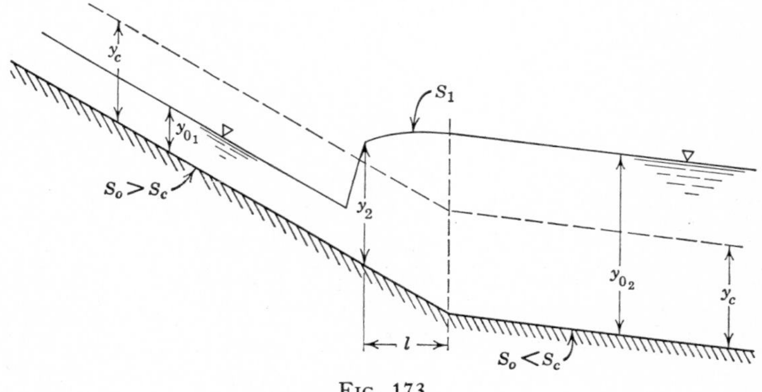

FIG. 173

draulic jump equation yields $y_{o_1} = y_1$ and $y_{o_2} = y_2$; the second will result when the hydraulic jump equation is solved (with $y_1 = y_{o_1}$) for y_2, and y_2 found to be less than y_{o_2}. The hydraulic jump will then be found to be followed by an S_1 profile from y_2 to y_{o_2} at the downstream end of channel 1.

ILLUSTRATIVE PROBLEM

A hydraulic jump occurs in a V-shaped open channel having sides sloping at 45°. Derive an equation (analogous to equation 256) relating the two depths and flowrate.

Solution.

$$F_1 - F_2 = Q\frac{\gamma}{g}(V_2 - V_1) \tag{102}$$

$$F_1 + \frac{Q\gamma V_1}{g} = F_2 + \frac{Q\gamma V_2}{g}$$

$$F = \gamma h_c A = \gamma \frac{y}{3} y^2, \quad V = \frac{Q}{y^2} \qquad (27) \quad (53)$$

Substituting these values above and canceling γ,

$$\frac{y_1^3}{3} + \frac{Q^2}{gy_1^2} = \frac{y_2^3}{3} + \frac{Q^2}{gy_2^2}$$

REFERENCES

B. A. Bakhmeteff, *Hydraulics of Open Channels*, McGraw-Hill Book Co., 1932.

S. M. Woodward and C. J. Posey, *The Hydraulics of Steady Flow in Open Channels*, John Wiley and Sons, 1941.

A. T. Ippen, "Channel Transitions and Controls," Chapter VIII, *Engineering Hydraulics* (H. Rouse, Ed.), John Wiley and Sons, 1950.

C. J. Posey, "Gradually Varied Open Channel Flow," Chapter IX, *Engineering Hydraulics* (H. Rouse, Ed.), John Wiley and Sons, 1950.

V. T. Chow, *Open-Channel Hydraulics*, McGraw-Hill Book Co., 1959.

E. A. Elevatorski, *Hydraulic Energy Dissipators*, McGraw-Hill Book Co., 1959.

PROBLEMS

739. Water flows uniformly at a depth of 4 ft in a rectangular canal 10 ft wide, laid on a slope of 1 ft per 1000 ft. What is the mean shear stress on the sides and bottom of the canal?

740. Calculate the mean shear stress over the wetted perimeter of a circular sewer 10 ft in diameter in which the depth of uniform flow is 3 ft and whose slope is 0.0001.

741. What is the mean shear stress over the wetted perimeter of a triangular flume 8 ft deep and 10 ft wide at the top, when the depth of uniform flow is 6 ft? The slope of the flume is 1 in 200, and water is flowing.

742. Calculate the Chezy coefficient which corresponds to a friction factor of 0.030.

743. Uniform flow occurs in a rectangular channel 10 ft wide at a depth of 6 ft. If the Chezy coefficient is 120 $\text{ft}^{\frac{1}{2}}$/sec, calculate friction factor, Manning n, and approximate height of the roughness projections.

744. What uniform flowrate will occur in a rectangular planed timber flume 5 ft wide and having a slope of 0.001 when the depth therein is 3 ft?

745. Calculate the uniform flowrate in an earth-lined ($n = 0.020\ \text{ft}^{\frac{1}{6}}$) trapezoidal canal having bottom width 10 ft, sides sloping 1 (vert.) on 2 (horiz.), laid on a slope of 0.0001, and having a depth of 6 ft.

746. A rough timber flume in the form of an equilateral triangle (apex down) of 4-ft sides is laid on a slope of 0.01. Calculate the uniform flowrate which occurs at a depth of 3 ft.

747. What uniform flowrate will occur in this canal cross section if it is laid on a slope of 1 in 2000 and has $n = 0.017$ ft$^{\frac{1}{6}}$?

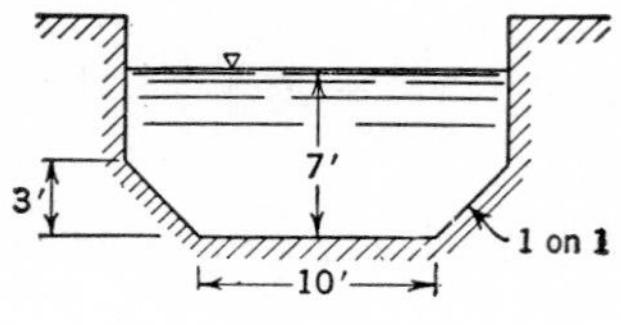

PROBLEM 747

748. A semicircular canal of 4-ft radius is laid on a slope of 0.002. If n is 0.015 ft$^{\frac{1}{6}}$, what uniform flowrate will exist when the canal is brim full?

749. A flume of planed timber has its cross section an isosceles triangle (apex down) of 8-ft base and 6-ft altitude. At what depth will 180 cfs flow uniformly in this flume if it is laid on a slope of 0.01?

750. At what depth will 150 cfs flow uniformly in a rectangular channel 12 ft wide lined with rubble masonry and laid on a slope of 1 in 4000?

751. At what depth will 400 cfs flow uniformly in an earth-lined ($n = 0.025$ ft$^{\frac{1}{6}}$) trapezoidal canal of base width 15 ft, having side slopes 1 on 3, if the canal is laid on a slope of 1 in 10,000?

752. Calculate the depth of uniform flow for a flowrate of 400 cfs in the open channel of problem 747.

753. An earth-lined ($n = 0.020$ ft$^{\frac{1}{6}}$) trapezoidal canal of base width 10 ft and side slopes 1 (vert.) on 3 (horiz.) is to carry 100 cfs uniformly at a mean velocity of 2 fps. What slope should it have?

754. What slope is necessary to carry 400 cfs uniformly at a depth of 5 ft in a rectangular channel 12 ft wide, having $n = 0.017$ ft$^{\frac{1}{6}}$?

755. A trapezoidal canal of side slopes 1 (vert.) on 2 (horiz.) and having $n =$ 0.017 ft$^{\frac{1}{6}}$ is to carry a uniform flow of 1300 cfs on a slope of 0.005 at a depth of 5 ft. What base width is required?

756. A rectangular channel 5 ft wide has uniform sand grain roughness of diameter 0.25 in. To observe the variation of n with depth, calculate values of n for depths of 0.1, 1.0, 2.0, and 3.0 ft.

757. A flowrate of 0.1 cfs of oil (per foot of width) is to flow uniformly down an inclined glass plate at a depth of 0.05 ft. Calculate the required slope. The viscosity and specific gravity of the oil are 0.009 lb-sec/sqft and 0.90, respectively.

758. Water (68°F) flows uniformly in a channel at a depth of 0.03 ft. Assuming a critical Reynolds number of 500, what is the largest slope on which laminar flow can be maintained? What mean velocity will occur on this slope?

759. What flowrate (per foot of width) may be expected for water (68°F) flowing in a wide rectangular channel at a depth of 0.02 ft if the channel slope is 0.0001? Assume laminar flow and confirm by calculating Reynolds number.

760. At what depth will a flowrate of water (68°F) of 0.003 cfs/ft occur in a wide open channel of slope 0.00015? Assume laminar flow and confirm by calculating Reynolds number.

761. Plot curves similar to those of Fig. 154 for an equilateral triangle (apex up). Find the maximum points of the curves mathematically.

762. Plot curves similar to those of Fig. 154 for a square laid with diagonal vertical. Find the maximum points of the curves mathematically

763. What uniform flowrate occurs in a 5-ft circular brick conduit laid on a slope of 0.001 when the depth of flow is 3.5 ft? What is the mean velocity of this flow?

764. Calculate the depth at which 25 cfs will flow uniformly in a smooth cement-lined circular conduit 6 ft in diameter, laid on a slope of 1 in 7000.

765. Rectangular channels of flow cross section 50 sqft have dimensions (width $\times$ depth) of (*a*) 25 ft by 2 ft, (*b*) 12.5 ft by 4 ft, (*c*) 10 ft by 5 ft, and (*d*) 5 ft by 10 ft. Calculate the hydraulic radii of these sections.

766. A channel flow cross section has an area of 200 sqft. Calculate its best dimensions if (*a*) rectangular, (*b*) trapezoidal with 1 (vert.) on 2 (horiz.) side slopes, (*c*) V-shaped.

767. Calculate the required width of a rectangular channel to carry 1600 cfs uniformly at best hydraulic conditions on a slope of 0.001 if n is 0.035 ft$^{\frac{1}{6}}$.

768. What are the best dimensions for a trapezoidal canal having side slopes 1 (vert.) on 3 (horiz.) and n of 0.020 ft$^{\frac{1}{6}}$ if it is to carry 1400 cfs uniformly on a slope of 0.009?

769. What is the minimum slope at which 200 cfs may be carried uniformly in a rectangular channel (having a value of n of 0.014 ft$^{\frac{1}{6}}$) at a mean velocity of 3 fps?

770. What is the minimum slope at which 1000 cfs may be carried uniformly at a mean velocity of 2 fps in a trapezoidal canal having n = 0.025 ft$^{\frac{1}{6}}$ and sides sloping 1 (vert.) on 4 (horiz.)?

771. Prove that the best form for a V-shaped open channel section is one of vertex angle 90°.

772. Calculate the specific energies when 225 cfs flow in a rectangular channel of 10-ft width at depths of (*a*) 1.5 ft, (*b*) 3 ft, and (*c*) 6 ft.

773. At what depths may 30 cfs flow in a rectangular channel 6 ft wide if the specific energy is 4 ft?

774. Eight hundred cubic feet per second flow in a rectangular channel of 20-ft width having n = 0.017 ft$^{\frac{1}{6}}$. Plot accurately the specific energy diagram for depths from 0 to 10 ft, using the same scales for y and E. Determine from the diagram (*a*) the critical depth, (*b*) the minimum specific energy, (*c*) the specific energy when the depth of flow is 7 ft, and (*d*) the depths when the specific energy is 8 ft. What type of flow exists when the depth is (*e*) 2 ft, (*f*) 6 ft; what are the channel slopes necessary to maintain these depths? What type of slopes are these, and (*g*) what is the critical slope, assuming the channel to be of great width?

775. Flow occurs in a rectangular channel of 20-ft width and has a specific energy of 10 ft. Plot accurately the q-curve. Determine from the curve (*a*) the critical depth, (*b*) the maximum flowrate, (*c*) the flowrate at a depth of 8 ft, and (*d*) the depths at which a flowrate of 1000 cfs may exist, and the flow condition at these depths.

776. Five hundred cubic feet per second flow in a rectangular channel 15 ft wide at depth of 4 ft. Is the flow subcritical or supercritical?

777. If 400 cfs flow in a rectangular channel 18 ft wide with a velocity of 5 fps, is the flow subcritical or supercritical?

778. If 300 cfs flow uniformly in a rectangular channel 12 ft wide having n = 0.015 ft$^{\frac{1}{6}}$ and laid on a slope of 0.005, is the flow subcritical or supercritical? What is the critical slope for this flowrate, assuming the channel to be of great width.

779. A uniform flow in a rectangular channel 12 ft wide has a specific energy of 8 ft; the slope of the channel is 0.005 and the Chezy coefficient 120 ft$^{\frac{1}{2}}$/sec. Predict all possible depths and flowrates.

780. What is the maximum flowrate which may occur in a rectangular channel 8 ft wide for a specific energy of 5 ft?

781. One thousand cubic feet per second flow uniformly in a rectangular channel, 15 ft wide (n is 0.018 ft$^{\frac{1}{6}}$, S_o is 0.002), at best hydraulic conditions. Is this flow subcritical or supercritical? What is the critical slope for this flowrate, assuming the channel to be of great width?

782. The flowrate in a rectangular channel 5 ft wide is 100 cfs. Calculate the angle of small surface waves with the wall of the channel for depths of 1 ft and 3 ft.

783. In a wide rectangular channel n is 0.017 ft$^{\frac{1}{6}}$ and does not vary over a depth range from 1 to 5 ft. To confirm the trend of Fig. 160, calculate values of the critical slope for depths of 1, 3, and 5 ft. What flowrates are associated with these depths? What is the mean value of the critical slope for this depth range? What is the maximum (percent) deviation from this mean value?

784. Calculate the specific energy when 300 cfs flow at a depth of 4 ft in a trapezoidal channel having base width 8 ft and sides sloping at 45°.

785. Calculate the specific energy when 100 cfs flow at a depth of 3 ft in a V-shaped flume, if the width at the water surface is 4 ft.

786. At what depths may 800 cfs flow in a trapezoidal channel of base width 12 ft and side slopes 1 (vert.) on 3 (horiz.) if the specific energy is 7 ft?

787. Four hundred cubic feet per second flow in a trapezoidal channel of base width 15 ft and side slopes 1 (vert.) on 3 (horiz.). Calculate the critical depth and the ratio of critical depth to minimum specific energy. If n = 0.020 ft$^{\frac{1}{6}}$, what is the critical slope?

788. Uniform flow occurs in a trapezoidal canal of base width 5 ft and side slopes 1 (vert.) on 2 (horiz.) laid on a slope of 0.002. The depth is 6 ft. Is this flow subcritical or supercritical?

789. A sewer line of elliptical cross section 10 ft high by 7 ft wide carries 200 cfs of water at a depth of 5 ft. Is this flow subcritical or supercritical?

790. What is the critical depth for a flowrate of 500 cfs in a channel having the cross section of problem 747?

791. For best hydraulic cross section in a trapezoidal channel of side slopes 2 (vert.) on 1 (horiz.), a uniform flowrate of 1000 cfs occurs at critical depth. If n is 0.017 ft$^{\frac{1}{6}}$, calculate the critical slope.

792. Derive an expression for critical depth for an open channel of V-shaped cross section with side slopes of 45°. What is the ratio between critical depth and minimum specific energy for this channel?

793. Solve the preceding problem for a channel of parabolic cross section defined by $y = x^2$.

794. Calculate more exact values of critical slopes for problems (*a*) 774, (*b*) 778, (*c*) 781 by including the effect of the sidewalls.

795. Solve problem 783 for a rectangular channel of 15-ft width.

796. What theoretical flowrate will occur over a high broad-crested weir 30 ft long when the head thereon is 2 ft?

797. The elevation of the crest of a high broad-crested weir is 100.00 ft. If the length of this weir is 12 ft and the flowrate over it 200 cfs, what is the elevation of the water surface upstream from the weir?

798. If the energy line is 3 ft above the crest of a frictionless broad-crested weir of 4-ft height, what is the water depth just upstream from the weir?

799. When the depth of water just upstream from a frictionless broad-crested weir 2 ft high is 3 ft, what flowrate per foot of crest length can be expected?

800. A dam 4 ft high and having a broad horizontal crest is built in a rectangular channel 15 ft wide. For a depth of water *on* the crest of 2.00 ft, calculate the flowrate and the depth of water just upstream from the dam.

801. The velocity in a rectangular channel of 10-ft width is to be reduced by installing a smooth broad-crested rectangular weir. Before installation the mean velocity is 5 fps and the water depth 1 ft; after installation these quantities are to be 1 fps and 5 ft, respectively. What height of weir is required?

802. Flow occurs over and under this control gate as shown. Calculate the flowrate (per foot of gate width), assuming streamlines straight and parallel at section 2.

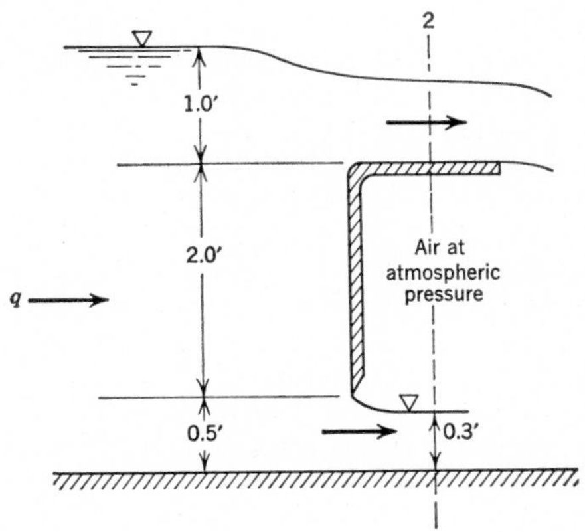

PROBLEM 802

803. An open rectangular channel 5 ft wide and laid on a mild slope ends in a free outfall. If the brink depth is measured as 0.865 ft, what flowrate exists in the channel?

804. Fifty cubic feet per second flow in a rectangular channel 8 ft wide. If this channel is laid on a mild slope and ends in a free outfall, what depth at the brink is to be expected?

805. A horizontal pipe of 2-ft diameter discharges as a free outfall. The depth of water at the brink of the outfall is 0.5 ft. Calculate the flowrate.

806. The critical depth is maintained at a point in a rectangular channel 6 ft wide by building a gentle hump 1 ft high in the bottom of the channel. When the depth over the hump is 2.20 ft, what water depths are possible just upstream from the hump?

807. If 150 cfs flow uniformly in a rectangular channel 10 ft wide, laid on a slope of 0.0004 and having $n = 0.014 \text{ ft}^{\frac{1}{6}}$, what is the minimum height of the hump that may be built across this channel to create critical depth over the hump? Sketch the energy line and water surface, showing all vertical dimensions. Neglect head losses caused by the hump.

808. Uniform flow in a rectangular channel occurs at a depth of 1.50 ft and velocity of 25 fps. When a smooth frictionless hump 2 ft high is built in the floor of the channel, what water depth can be expected on the hump? What hump height would be required to produce critical depth on the hump? What would happen if a taller hump than this were installed?

809. Solve the preceding problem for a depth of 10 ft and velocity of 3.75 fps.

810. Uniform flow at 6.00-ft depth occurs in a rectangular open channel 10 ft wide having Manning n of $0.017 \text{ ft}^{\frac{1}{6}}$ and laid on a slope of 0.0036. There is a flexible hump in the floor of the channel which can be raised or lowered. Neglecting losses caused by the hump: (*a*) How large may the hump height be made without changing the water depth just upstream from the hump? (*b*) How large should the hump height be to make the water depth 7.00 ft just upstream from the hump?

811. The water depth over a hump of 1-ft height in a rectangular open channel is 3.50 ft. Just upstream and just downstream from the hump the water depths are 5.00 ft. What depths would be expected on and just upstream from the hump when the channel downstream from the hump is removed? There will be no change of flowrate.

812. A long rectangular channel 10 ft wide carries a flowrate of 372 cfs uniformly at a depth of 2.00 ft and ends in a free outfall. When a smooth hump 1.50 ft high is installed at the very end of this channel, what depth is to be expected on the hump? How large a hump would cause the depth there to be critical?

813. A rectangular channel 12 ft wide is narrowed to a 6-ft width to cause critical flow in the contracted section. If the depth in this section is 3 ft, calculate the flowrate and the depth in the 12-ft section, neglecting head losses in the transition. Sketch energy line and water surface, showing all pertinent vertical dimensions.

814. One hundred and fifty cubic feet per second flow uniformly in a rectangular channel 10 ft wide having $n = 0.014 \text{ ft}^{\frac{1}{6}}$, and laid on a slope of 0.0004. This channel is to be narrowed to cause critical flow in the contracted section. What is the maximum width of contracted section which will accomplish this? For this width and neglecting head losses, sketch the energy line and water surface, showing vertical dimensions. If the contraction is narrowed to 4-ft width, what depths are to be expected in and just upstream from the contraction?

815. A uniform flow of 750 cfs occurs in a rectangular channel 15 ft wide at a depth of 10 ft. A hump 2 ft high is built in the bottom of the channel, and at the same point the width is reduced to 12 ft. When friction is neglected, what is the water depth over the hump?

816. A flow of 50 cfs per foot of width occurs in a wide rectangular channel at a depth of 6 ft. A bridge to be constructed across this channel requires piers spaced 20 ft on centers. Assuming that the noses of the piers are well streamlined and frictionless, how thick may they be made without causing "backwater effects" (i.e., deepening of the water) upstream from the bridge?

817. A flow of 300 cfs occurs in a long rectangular channel 10 ft wide, laid on a slope of 0.0016 and having $n = 0.018 \text{ ft}^{\frac{1}{6}}$. There is a smooth gradual constriction in the channel to 6-ft width. What depths may be expected in and just upstream from the constriction?

818. A long rectangular channel 15 ft wide has $n = 0.014 \text{ ft}^{\frac{1}{6}}$ and is laid on a slope of 0.00090. In this channel there is a smooth constriction to 13.5-ft width containing a hump 1.50 ft high. Predict the water depths in and just upstream from the constriction for a flowrate of 1000 cfs.

819. In a trapezoidal canal of 10-ft base width and sides sloping at 45°, 800 cfs flow uniformly at a depth of 10 ft. The channel is constricted by raising the sides to a vertical position. Calculate the depth of water in the constriction, neglecting local head losses. What is the minimum height of hump which may be installed in the constriction to produce critical depth there?

820. A long rectangular channel of 15-ft width, slope of 0.001, and n of $0.015 \text{ ft}^{\frac{1}{6}}$ reduces to 11-ft width as it passes through a culvert in a highway embankment. Below what flowrate can depths greater than normal depth be expected upstream from the embankment?

821. For a rectangular open channel of 8-ft width, $n = 0.015 \text{ ft}^{\frac{1}{6}}$, and slope 0.0035, the following data are given. Approximately how far apart are sections 1 and 2?

Section	Depth, ft	Velocity, fps	Hydraulic Radius, ft	Specific Energy, ft
1	3.00	15.00	1.715	6.49
2	3.20	14.06	1.775	6.26

822. The channels of Fig. 163 are of 12-ft width and have $n = 0.020 \text{ ft}^{\frac{1}{6}}$. Their slopes are 0.00500 and 0.0100 and the depth of water at the break in slope is 5.00 ft. Considering this depth to be critical depth, how far upstream and downstream from this point will depths of 5.40 and 4.60 ft be expected? Use depth intervals of 0.20 ft. If the respective slopes were slightly decreased, would the calculated distances be lengthened or shortened?

823. A uniform flow at 5-ft depth occurs in a long rectangular channel 15 ft wide having $n = 0.017 \text{ ft}^{\frac{1}{6}}$ and laid on a slope of 0.0010. A hump is built across the floor of the channel of such height that the water depth just upstream from the hump is raised to 6.00 ft. Identify the surface profile upstream from the hump and calculate the distance from the hump to where a depth of 5.60 ft is to be expected. Use depth intervals of 0.20 ft.

824. Water discharges through a sluice gate into a long rectangular open channel 8 ft wide having $n = 0.017 \text{ ft}^{\frac{1}{6}}$. The depth and velocity at the vena contracta are 2.00 ft and 25.0 fps, respectively. Identify the surface profiles if the channel slope is (*a*) favorable, (*b*) zero, (*c*) adverse. If these slopes are (*a*) 0.002, (*b*) 0.000, and (*c*) 0.002, how far downstream from the sluice will a depth of 2.20 ft be expected? Use depth intervals of 0.10 ft.

825. Eight hundred cubic feet per second flow in a rectangular channel of 20-ft width. Plot the $\mathfrak{M}$-curve of hydraulic jumps on the specific energy diagram of problem 774. From these curves determine (*a*) the depth after a hydraulic jump has taken place from a depth of 1.5 ft, (*b*) the height of this jump, (*c*) the specific energy before the jump, (*d*) the specific energy after the jump, (*e*) the loss of head in the jump, and (*f*) the total horsepower lost in the jump.

826. A hydraulic jump occurs in a rectangular open channel. The water depths before and after the jump are 2 ft and 5 ft, respectively. Calculate the critical depth.

827. For a rectangular open channel, prove that hydraulic jumps can occur only across the critical depth. Prove this for a nonrectangular channel.

828. If the maximum $\mathbf{F}_1$ for an undular hydraulic jump is $\sqrt{3}$, what is the maximum value of y_2/y_1 for such jumps? Show that the head lost in a hydraulic jump is given by $h_L/y_1 = (y_2/y_1 - 1)^3/4(y_2/y_1)$. Note that such head losses will be very small (usually negligible) for low values of y_2/y_1.

829. A supercritical flow of 100 cfs occurs at 3-ft depth in a V-shaped open channel of side slopes 45°. Calculate the depth just downstream from a hydraulic jump in this flow.

830. A hydraulic jump occurs in a V-shaped open channel with sides sloping at 45°. The depths upstream and downstream from the jump are 3 ft and 4 ft, respectively. Estimate the flowrate in the channel.

831. A hydraulic jump occurs in a horizontal storm sewer of circular cross section 4 ft in diameter. Before the jump the water depth is 2 ft and just downstream from the jump the sewer is full with a gage pressure of 1.0 psi at the top. Predict the flowrate.

832. A rectangular channel 10 ft wide carries a flowrate of 500 cfs uniformly at 2 ft depth. The channel is constricted at the end to produce a hydraulic jump in the channel. Calculate the width of constriction required for the jump to be just upstream from the constriction.

833. A very long rectangular channel of 8 ft width and n of 0.017 $\text{ft}^{\frac{1}{6}}$, laid on a slope of 0.03, carries a flowrate of 200 cfs and ends in a free outfall. Across the downstream end of the channel a smooth frictionless hump is constructed. Calculate the depths (*a*) on the hump, (*b*) just upstream from the hump, and (*c*) far up the channel, and sketch and identify the surface profiles. If a hydraulic jump occurs calculate the depths y_1 and y_2. The hump is 2.5 ft high.

834. Neglecting wall, bottom, and hump friction (but not losses in the jump), what height of hump will produce this flow picture?

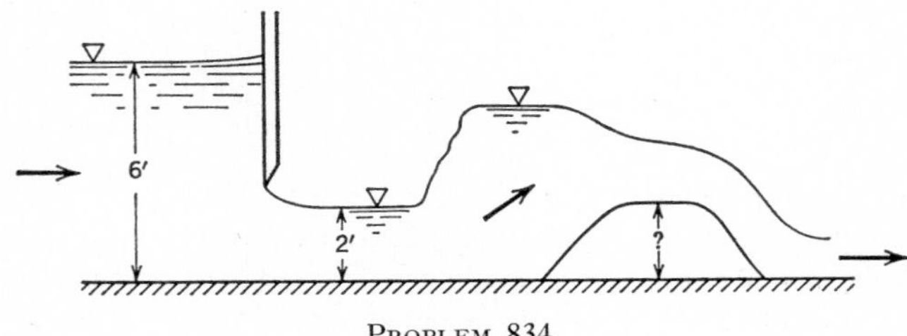

PROBLEM 834

835. The sketch shows a plan view of a frictionless Venturi flume of rectangular cross section with horizontal floor. A hydraulic jump from 1 ft to 3 ft occurs at section 3. Calculate the depth of water at sections 1, 2, and 4. Section 1 may be assumed to be infinitely wide.

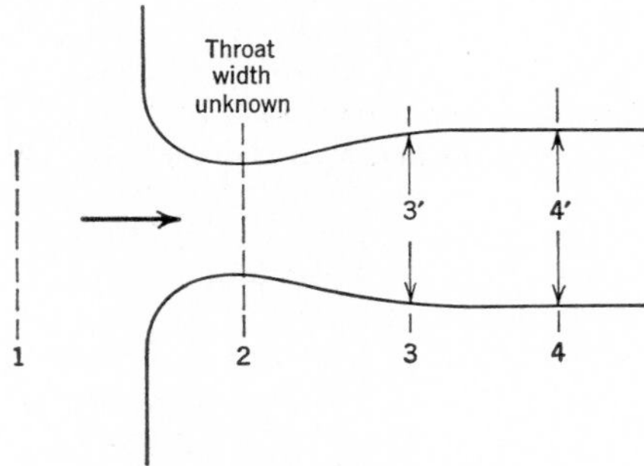

PROBLEM 835

836. The flowrate in Fig. 172 is 543 cfs. If the channels are 12 ft wide with n = 0.017 $\text{ft}^{\frac{1}{6}}$ and the downstream channel laid on a slope of 0.00228, what depth must exist in this channel for a hydraulic jump to occur to uniform flow? Calculate the length l if y_{o_1} = 2.60 ft. Use three equal depth intervals through the zone of varied flow.

837. The flowrate in Fig. 173 is 543 cfs, the channels are 12 ft wide, and n = 0.017 $\text{ft}^{\frac{1}{6}}$. The downstream channel is laid on a slope of 0.0015. If y_{o_1} = 2.60 ft, calculate l, using two equal depth intervals.

chapter 11

Fluid Measurements

In engineering and industrial practice one of the fluid mechanics problems most frequently encountered by the engineer is the measurement of many of the variables and properties discussed in the foregoing chapters. Efficient and accurate measurements are also absolutely essential for correct conclusions in the various fields of fluid mechanics research. Whether the necessity for precise measurements is economic or scientific, the engineer of today must be well equipped with a knowledge of the fundamentals and existing methods of measuring various fluid properties and phenomena. It is the purpose of this chapter to indicate only the principles and phenomena of fluid measurements; the reader will find available in the abundant engineering literature [1] the details of installation and operation of the various measuring devices. Although many of the following devices frequently appear in engineering practice as appurtenances in various designs where they are not used for measuring purposes, a study of them as measuring devices will make obvious their applications in other capacities.

89. Measurement of Fluid Properties. Of the fluid properties density, viscosity, elasticity, surface tension, and vapor pressure, the engineer is usually called upon to measure only the first two. Since measurements of elasticity, surface tension, and vapor pressure are normally made by physicists and chemists, the various experimental techniques for measuring these properties will not be reviewed here.

Density measurements of liquids may be made by the following methods, listed in approximate order of their accuracy: (1) weighing a known volume of liquid, (2) hydrostatic weighing, (3) Westphal balance, (4) hydrometer, and (5) U-tube.

[1] See references in and at end of this chapter.

To weigh accurately a known volume of liquid a device called a *pycnometer* is used. This is usually a glass vessel whose weight, volume, and variation of volume with temperature have been accurately determined. If the weight of the empty pycnometer is W_1, and the weight of the pycnometer, when containing a volume $\mathbf{V}$ of liquid at temperature t, is W_2, the specific weight of the liquid, γ_t, at this temperature may be calculated directly from

$$\gamma_t \mathbf{V} = W_2 - W_1$$

Density determination by hydrostatic weighing consists essentially in weighing a plummet of known volume (1) in air, and (2) in the liquid whose density is to be determined (Fig. 174*a*). If the weight of the plummet in air is W_a, its volume, $\mathbf{V}$, and its weight when suspended in the liquid, W_l, the equilibrium of vertical forces on the plummet gives

$$W_l + \gamma_t \mathbf{V} - W_a = 0$$

from which the specific weight, γ_t, at the temperature t, may be calculated directly.

Like the method of hydrostatic weighing, the Westphal balance (Fig. 174*b*) utilizes the buoyant force on a plummet as a measure of specific gravity. Balancing the scale beam with special riders placed at special points allows direct and precise reading of specific gravity.

Probably the most common means of obtaining liquid densities is with the hydrometer (Fig. 174*c*), whose operation is governed by the fact that a weighted tube will float with different immersions in liquids of different densities. To create a great variation of immersion for small density variation, and, thus, to give a sensitive instrument, changes in the immersion of the hydrometer occur along a slender tube, which is graduated to read the specific gravity of the liquid at the point where the liquid surface intersects the tube.

The unknown density of a liquid 1 may be obtained approximately from the known density of a liquid 2 (if the liquids are not miscible) by placing them in an open U-tube and measuring the lengths of liquid columns, l_1 and l_2 (Fig. 174*d*). From manometer principles,

$$\rho_1 l_1 = \rho_2 l_2$$

thus allowing ρ_1 to be calculated. This method is usually not precise because the various menisci prevent accurate measurement of the lengths of the liquid columns.

Viscosity measurements are made with devices known as *viscosimeters* or *viscometers*, which may be classified as *rotational*, *falling-*

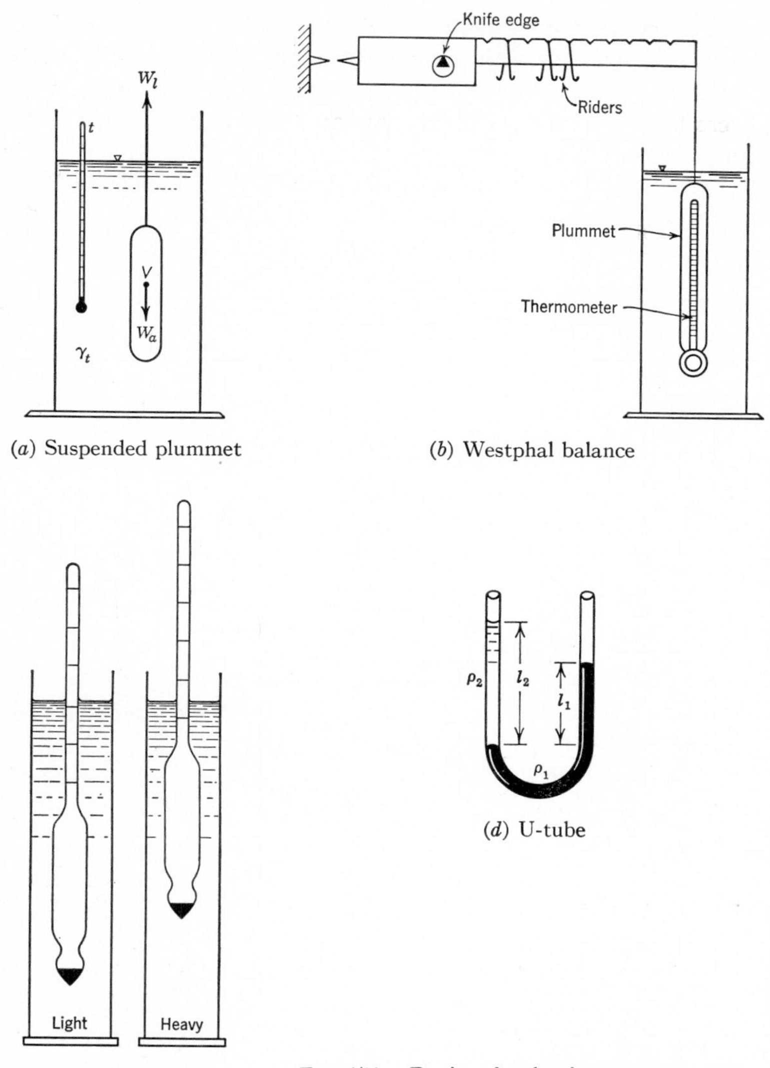

(*a*) Suspended plummet

(*b*) Westphal balance

(*c*) Hydrometers

(*d*) U-tube

FIG. 174. Devices for density measurements.

sphere, or *tube* devices according to their construction or operation. The operation of all these viscometers depends upon the existence of laminar flow (which has been seen in foregoing chapters to be dominated by viscous action) under certain controlled and reproducible conditions. In general, however, these conditions involve too many complexities to allow the constants of the viscometer to be calculated analytically, and they are, therefore, usually obtained by calibration with a liquid of known viscosity. Because of the variation of viscosity with temperature all viscometers must be immersed in constant-temperature baths and provided with thermometers for taking the temperatures at which the viscosity measurements are made.

Two instruments of the rotational type are the MacMichael and Stormer viscometers, whose essentials are shown diagrammatically in Fig. 175. Both consist of two concentric cylinders, with the

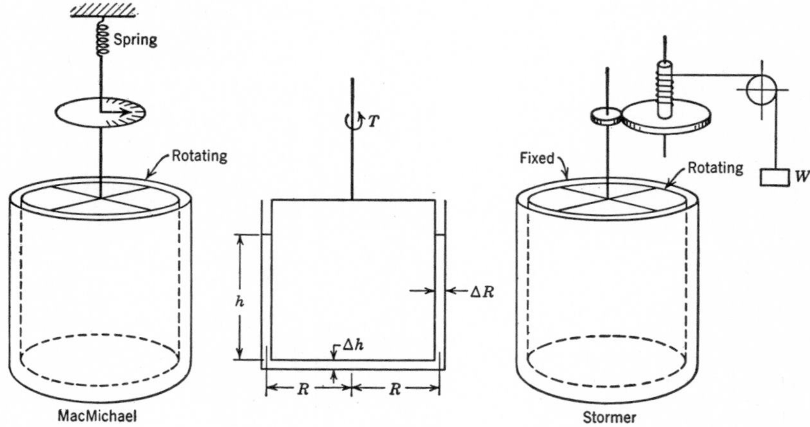

FIG. 175. Rotational viscometers (schematic).

space between containing the liquid whose viscosity is to be determined. In the MacMichael type, the outer cylinder is rotated at constant speed, and the rotational deflection of the inner cylinder (accomplished against a spring) becomes a measure of the liquid viscosity. In the Stormer instrument, the inner cylinder is rotated by a falling-weight mechanism, and the time necessary for a fixed number of revolutions becomes a measure of the liquid viscosity.

The measurement of viscosity by the above-mentioned variables may be justified by a simplified mechanical analysis, using the dimensions of Fig. 175. Assuming ΔR and Δh small, and the peripheral velocity of the moving cylinder to be V, the torque, T, may be calculated from the principles and methods of Art. 5. The result is

$$T = \frac{2\pi R^2 h \mu V}{\Delta R} + \frac{\pi R^3 \mu V}{2\,\Delta h}$$

in which the first term represents the torque due to viscous shear in the space between the cylinder walls, and the second term that between the ends of the cylinders. With R, h, ΔR, and Δh constants of the instrument, and the rotational speed (N) proportional to V, the equation may be written

$$T = K\mu N \quad \text{or} \quad \mu = T/KN$$

in which the constant K depends upon the foregoing factors. For the MacMichael instrument the torque is proportional to the torsional deflection θ ($T = K_1\theta$) with the result

$$\mu = K_1\theta/KN$$

which shows that the viscosity may be obtained from deflection and speed measurements. In the Stormer viscometer the torque is constant since it is proportional to the weight W, which is a constant of the instrument; also, the time t required for a fixed number of revolutions is inversely proportional to N ($t = K_2/N$) with the result

$$\mu = (T/KK_2)t$$

and the time required for a fixed number of revolutions produced by the same falling weight thus becomes a direct measure of viscosity.

The falling-sphere type of viscometer is shown in Fig. 176. In this type of viscometer the time t required for a small sphere to fall at constant velocity through a distance l in a liquid becomes a direct measure of the liquid's viscosity. Here again a simple analysis will confirm the foregoing statement. From Stokes' law (Art. 118), the drag D of a sphere of diameter d, moving under laminar conditions ($Vd/\nu \lesssim 0.1$) at a constant velocity V through a fluid of *infinite* extent, is given by

$$D = 3\pi\mu Vd \tag{328}$$

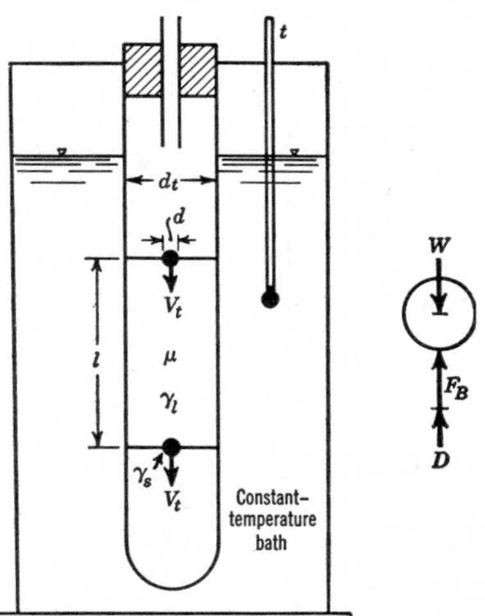

Fig. 176. Falling-sphere viscometer.

The weight of the sphere in terms of its specific weight and size is given by $W = \pi d^3\gamma_s/6$ and its buoyant force (Art. 13) by $F_B = \pi d^3\gamma_l/6$. After the sphere has acquired constant velocity, these forces are in equilibrium, giving

$$D - W + F_B = 3\pi\mu Vd - \pi d^3\gamma_s/6 + \pi d^3\gamma_l/6 = 0$$

and, solving for μ,

$$\mu = d^2(\gamma_s - \gamma_l)/18V \tag{258}$$

and thus it appears that viscosity can be obtained from simple measurements of size, density, and velocity. In practice this equation usually requires a large correction, because the extent of the fluid is far from infinite and the *wall effect* produced by the walls surrounding the fluid is surprisingly large. Wall effect is obtained experimentally, it depends on the ratio of sphere diameter to tube diameter, and it is usually expressed as a correction to the velocity by [2]

$$\frac{V}{V_t} = 1 + \left(\frac{9}{4}\frac{d}{d_t}\right) + \left(\frac{9}{4}\frac{d}{d_t}\right)^2$$

in which V_t is the observed velocity of fall in the tube. Applying this correction to equation 258 allows accurate viscosities to be computed from easily measured quantities.

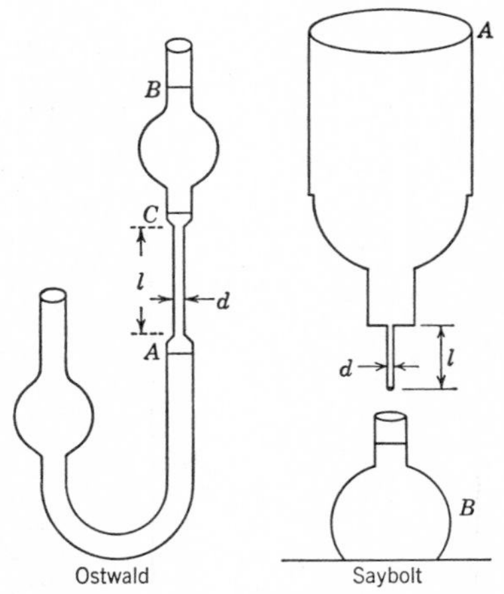

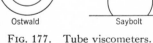

FIG. 177. Tube viscometers.

Typical tube-type viscometers are the Ostwald and Saybolt instruments of Fig. 177. Similar to the Ostwald is the Bingham type, and similar to the Saybolt are the Redwood and Engler viscometers. All these instruments involve the *unsteady laminar flow* of a fixed volume of liquid through a small tube under standard head conditions. The time for the quantity of liquid to pass through the tube becomes a measure of the kinematic viscosity of the liquid.

The Ostwald viscometer is filled to level A, and the meniscus of the liquid in the right-hand tube is drawn up to a point above B and then released. The time for the meniscus to fall from B to C becomes a measure of the kinematic viscosity. In the Saybolt viscometer the outlet is plugged, and the reservoir filled to level A; the plug is then removed, and the time required to collect a fixed quantity of liquid in the vessel B is measured. This time then becomes a direct measure of the kinematic viscosity of the liquid.

[2] J. S. McNown, H. M. Lee, M. B. McPherson, and S. M. Engez, "Influence of Boundary Proximity on the Drag of Spheres," *Proc. 7th Intern. Congr. Appl. Mech.*, 1948.

The relation between time and kinematic viscosity for the tube type of viscometer may be indicated approximately by applying the Hagen-Poiseuille law for laminar flow in a circular tube (Art. 67). The approximation involves the application of a law of steady established laminar motion to a condition of unsteady flow in a tube which may be too short for the establishment of laminar flow and, therefore, cannot be expected to give a complete or perfect relationship between efflux time and kinematic viscosity; it will serve, however, to indicate elementary principles. From equation 191,

$$Q = \pi d^4 \gamma h_L / 128 \mu l \qquad (259)$$

for steady laminar flow in a circular tube. But $Q = \boldsymbol{V}/t$, in which $\boldsymbol{V}$ is the volume of liquid collected in time t. Substituting this in equation 259 and solving for μ,

$$\mu = (\pi d^4 h_L / 128 \boldsymbol{V} l)\gamma t$$

The head loss, h_L, however, is nearly constant since it is nearly equal to the imposed head which varies between fixed limits. Since d, l, $\boldsymbol{V}$, and h_L are constants of the instrument the equation reduces to $\mu \cong K\gamma t = K\rho g t$, from which $\mu/\rho = \nu \cong Kgt$, and kinematic viscosity is seen to depend almost linearly upon measured time. A more exact (but empirical) equation relating ν and t for the Saybolt Universal viscometer is (for $t > 32$ sec)

$$\nu \text{ (sqft/sec)} = 0.000002365t - 0.001935/t$$

in which t is the time in seconds (called *Saybolt seconds*). It is to be noted that this equation approaches the linear one predicted by the approximate analysis for large values of the time, t. The familiar S.A.E. numbers used for motor oils are indices of kinematic viscosity, as shown in Table VIII.

TABLE VIII [a]

S.A.E. Viscosity No.	Saybolt Seconds at 210°F	ν (sqft/sec) at 210°F
20	45 to 58	0.000063 to 0.000104
30	58 to 70	0.000104 to 0.000137
40	70 to 85	0.000137 to 0.000178

[a] From *S.A.E. Handbook*, Soc. Auto. Engrs., 1959.

Of the tube viscometers, the Saybolt, Engler, and Redwood are built of metal to rigid specifications and hence may be used without calibration. Since the dimensions of the glass viscometers such as the Bingham and Ostwald cannot be so perfectly controlled, these instruments must be calibrated before viscosity measurements are made.

90. Measurement of Static Pressure. The accurate measurement of pressure in a fluid at rest may be accomplished with comparative ease since it depends only upon the accuracy of the gage or manometer used to record this pressure and is independent of the details of the connection between fluid and recording device. To measure the static pressure within a moving fluid with high accuracy is quite another matter, however, and depends upon painstaking attention to the details of the connection between flowing fluid and measuring device.

For perfect measurement of static pressure in a flowing fluid a device is required which fits the streamline picture and causes no flow disturbance; it should contain a small smooth hole whose axis is normal to the direction of motion at the point where the static pressure is to be measured; to this opening is connected a manometer or pressure gage. Meeting all these requirements is a virtual impossibility, but it is evident that the device must be as small as possible and constructed with great care. However, the most troublesome point in measuring the static pressure in a flowfield is the proper orientation or alignment of the device with the flow direction, which usually is not known in advance. Two basic designs (there are many adaptations) have solved this problem successfully. One of these is the thin disk of Fig. 178*a* containing separate piezometer openings in each side which lead to a differential manometer; this device is inserted in the flowfield and turned (about its stem) until the connected differential manometer reads zero, which shows the pressure

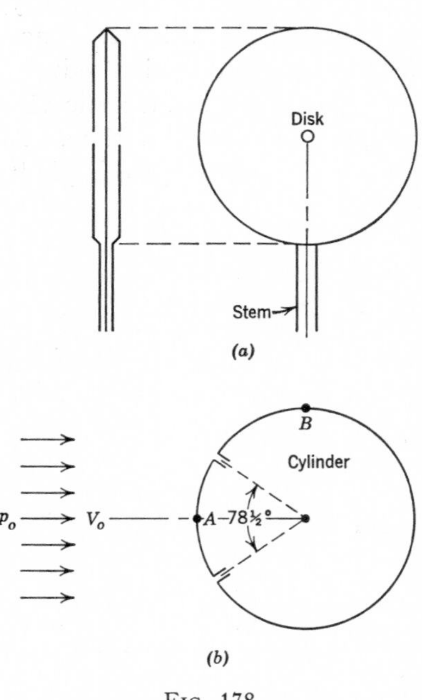

FIG. 178

on both sides of the disk to be the same and the disk thus aligned with the flow. After alignment is secured either pressure may be separately measured and taken to be the static pressure at the piezometer opening. A second device (for two-dimensional flows) is the cylinder of Fig. 178*b* containing two separate piezometer openings connected to a differential manometer; the cylinder is turned about its own axis until the manometer balances, showing the direction of flow to be along the bisector of the angle between the openings. At the stagnation point (A), $p_A > p_o$ and, at B, $p_B < p_o$ with continuous fall of pressure from p_A to p_B; accordingly, at some point between A and B on the surface of the cylinder, a local pressure equal to the static pressure will be found and may be measured on a separate manometer. Fechheimer [3] has shown that the appropriate angle between the piezometer openings should be 78.5° for incompressible flow, and Thrasher and Binder [4] have confirmed this for compressible flows of velocity up to one-fourth of the sonic (at higher velocities a larger angle is required). Once the flow direction has been found and the static pressure recorded, the cylinder may be rotated through the appropriate angle so that one of the piezometer openings is at the stagnation point, whence the stagnation pressure may be measured and the velocity V_o obtained. Thus this simple and compact device serves a three-fold purpose in the measurement of static pressure, flow direction, and velocity. A spherical counterpart of this device has been developed for use in three-dimensional flowfields.

In fields where the flow direction is known to fair accuracy, the *static tube* (Fig. 179) is the usual means of measuring static pressure. Such a tube is merely a small smooth cylinder with rounded or pointed upstream end: in the side of the cylinder are piezometer holes or a circumferential slot through which pressure is transmitted to gage or manometer. Assuming perfect alignment with the flow, the flow past the tube will be symmetrical and of mean velocity slightly larger than V_o; hence the pressure at the piezometer openings may be expected to be slightly less than p_o. This error is minimized by making the tube as small as possible and is usually negligible in engineering work. In experimental work or in the use of

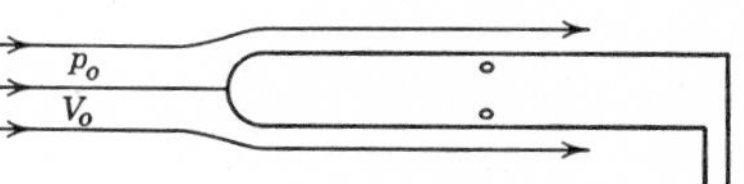

FIG. 179. Static tube.

[3] C. J. Fechheimer, "The Measurement of Static Pressure," *Trans. A.S.M.E.*, Vol. 48, p. 965, 1926.

[4] L. W. Thrasher and R. C. Binder, "Influence of Compressibility on Cylindrical Pitot-Tube Measurements," *Trans. A.S.M.E.*, Vol. 72, p. 647, 1950.

the static tube on aircraft some misalignment of the tube with flow direction is to be expected; when this occurs the pressure on one side of the tube becomes larger than p_o, on the other side less than p_o, and some flow through the tube (from opening to opening) results. With such complexities, the pressure carried to gage or manometer cannot be exactly predicted but will be close to p_o for small angles of misalignment. A static tube which is insensitive to misalignment is desired by the experimentalist since, with larger (and incurable) errors of alignment, accurate values of static pressure may still be obtained.

The static pressures in the fluid passing over a solid surface (such as a pipe wall or the surface of an object, Fig. 180) may be measured successfully by small smooth piezometer holes drilled normal to the surface; such surfaces fit the flow perfectly since (assuming no separation) they are streamlines of the flow. These piezometer openings can measure only the local static pressures at their locations on the solid surface and cannot, in general, measure the pressures at a distance from this surface since such pressures differ from those at the surface owing to flow curvatures and accelerations. Where no flow curvatures exist, as in a straight passage, a wall piezometer opening will allow pressures throughout the cross section of the passage to be predicted.

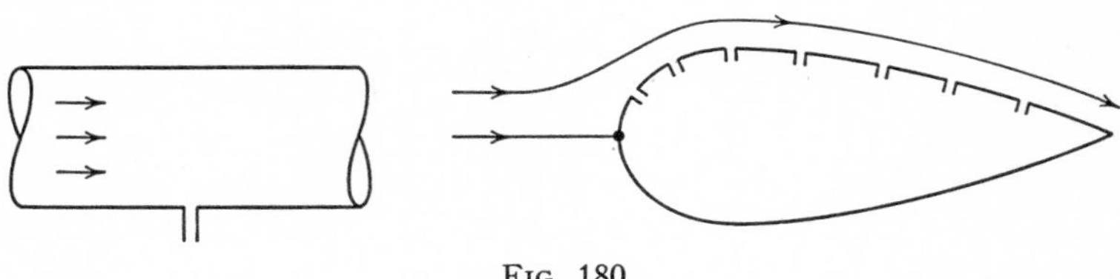

Fig. 180

Frequently in pipe-line practice a large number of piezometer openings are drilled in the pipe wall at the same cross section and led into a *piezometer ring* (Fig. 181), whence the pressure is transferred to a recording device. The pressure taken from the piezometer ring is considered more reliable than that obtained from a single piezometer opening, since probability dictates that the errors incurred by the inevitable imperfections of single openings will tend to cancel if numerous openings are used and the results averaged; the piezometer ring automatically approximates such an average.

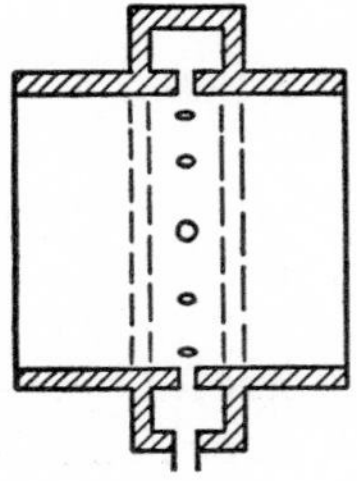

Fig. 181. Piezometer ring.

91. Measurement of Surface Elevation. The elevation of the surface of a liquid at rest may be determined by manometer, piezometer column, or pressure-gage readings (Arts. 8–10).

The same methods may be applied to flowing liquids if the foregoing precautions for static pressure measurement (Art. 90) are followed and if the piezometer method is used only where the streamlines of flow are *essentially* straight, and parallel. Correct and incorrect measurements of a liquid surface by piezometer openings are illustrated in Fig. 182.

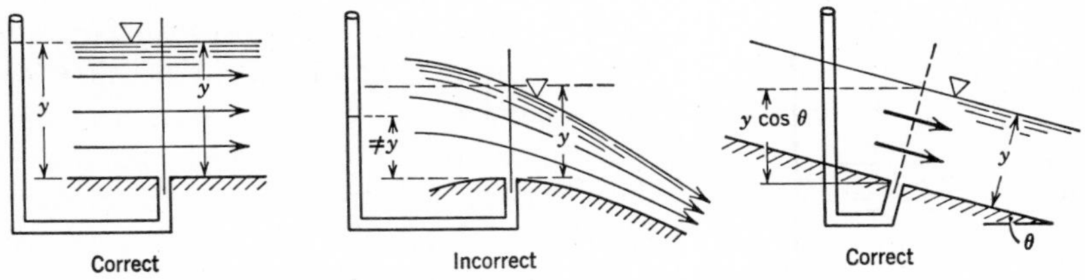

FIG. 182. Measurement of surface elevation with piezometer columns.

For direct measurements to a liquid surface the hook gage and point gage (Fig. 183) are common. The hook gage is generally used in a stilling well connected to the liquid at the point where its surface elevation is to be measured, but it may be used directly on the liquid surface if velocities are small. To set the point of the hook in the liquid surface, it is first placed below the surface and then raised until a small pimple appears on the surface. When the hook is lowered so that the pimple just disappears, its point is accurately (± 0.001 ft) at the same elevation as the liquid surface. From a graduated scale and vernier on the hook gage shaft, the surface elevation of the liquid may be read precisely.

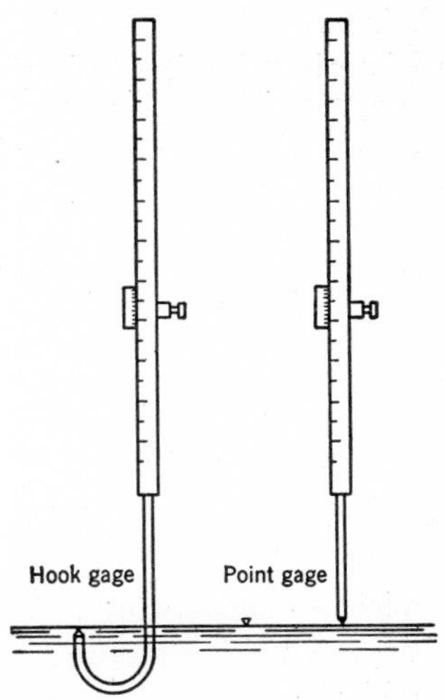

FIG. 183. Gages for measurement of surface elevation.

A point gage is suitable for swiftly flowing liquids in which the presence of a hook below the liquid surface would cause local disturbances. In measuring, the point gage is lowered until it just contacts the liquid surface (noted by slight disturbances of the surface) and then read by scale and vernier located on the gage shaft.

Floats are often used in connection with chronographic water-level recorders for measuring liquid-surface elevations. The arrangement of such floats is indicated schematically in Fig. 184. As the liquid level varies, the motion of the cable is measured on a scale or plotted automatically on a chronographic record sheet.

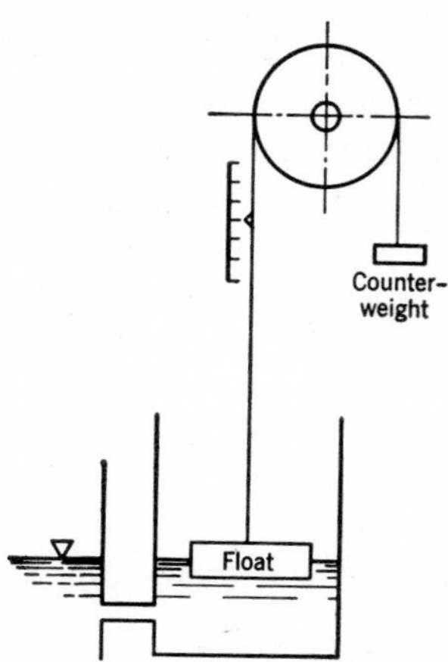

FIG. 184. Float for measurement of surface elevation (schematic).

An electrical method of liquid surface measurement has been moderately successful. It consists of using a small fixed metal plate and the liquid surface as a condenser (Fig. 185). Variation of liquid-surface elevation varies the capacitance [5] of the condenser, which may be measured electrically, and (after calibration of the device) electrical measurements become a measure of surface elevation.

Staff gages give comparatively crude but direct measurements of liquid-surface elevation. From casual observation, the reader is familiar with their use as tide gages, in the measurement of reservoir levels, and in registering the draft of ships.

92. Measurement of Stagnation Pressure. The stagnation [6] pressure (also called *total pressure*) may be measured accurately by placing in the flow a small solid object having a small piezometer hole at the stagnation point. The piezometer opening may be easily located at the stagnation point if the hole is drilled along the axis of a symmetrical object such as a cylinder, cone, or hemisphere; with the axis of the object properly aligned with the direction of flow the piezometer opening is automatically located at the stagnation point, and the pressure there may be transferred through the opening to a recording device. Theoretically, the upstream end of solid objects for this purpose may be of any shape, since the shape of the object does not affect the magnitude of the stagnation pressure; in laboratory practice, however, the upstream end is usually made convergent (conical or hemispherical) in order to fix more precisely the location of the stagnation point.

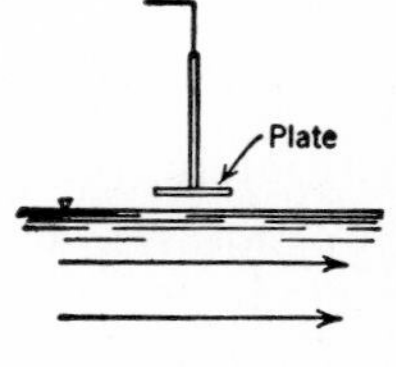

FIG. 185

[5] See R. H. Black, "Capacitance Method of Measuring Water Film Thickness," *Proc. A.S.C.E.*, No. 2524, June, 1960.

[6] See Arts. 25 and 33.

When used on airplanes or in experimental work, misalignments of tube with flow are inevitable, but reliable measurements of stagnation pressure are still required; laboratory research [7] has led to the development of a tube shielded by a special jacket which gives reliable measurements at angles of misalignment up to 45°.

The early experimental work of Henri Pitot (1732) provided the basis for the measurement of stagnation pressure by showing that a small tube with open end facing upstream (Fig. 186) provided a means for measuring velocity. He found that when such tubes

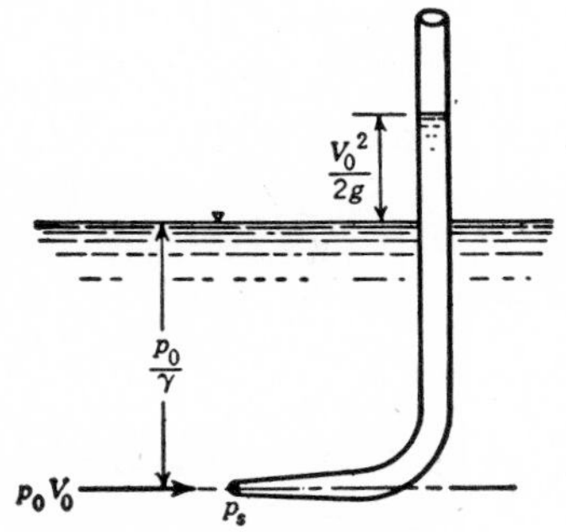

Fig. 186. Pitot tube.

(later called *pitot tubes*) were placed in an open flow where the velocity was V_o the liquid in the tube rose above the free surface a distance $V_o^2/2g$. From the sketch the expression for stagnation pressure,

$$\frac{p_s}{\gamma} = \frac{p_o}{\gamma} + \frac{V_o^2}{2g} \quad \text{or} \quad p_s = p_o + \tfrac{1}{2}\rho V_o^2 \tag{66}$$

may be written directly, thus confirming the equations of Art. 25.

In the use of pitot tubes for the measurement of velocity profiles in shear flows, experimentalists have noted [8] an error due to the asymmetry of the flow near the tip of the tube. This error is small where the velocity gradient is low but increases near solid boundaries where the velocity gradient is high. The sense of the error is to move the effective center of the tube toward the region of lower

[7] W. R. Russell, W. Gracey, W. Letko, and P. G. Fournier, "Wind Tunnel Investigation of Six Shielded Total-Pressure Tubes at High Angles of Attack," *N.A.C.A. Tech. Note* 2530, 1951.

[8] See, for example, A. D. Young and J. N. Maas, "The Behavior of a Pitot Tube in a Tranverse Pressure Gradient," *Aero. Res. Comm. R. & M.* 1770, 1936; and F. A. McMillan, "Experiments on Pitot Tubes in Shear Flow," *Aero. Res. Comm. R. & M.* 3028, 1956.

velocity gradient; this means that the velocity being measured by stagnation pressure will be found at a point farther from the wall than the axis of the pitot tube; the magnitude of this error also depends upon the detailed geometry of the tube.

MEASUREMENT OF VELOCITY

93. Pitot-Static Tube in Incompressible Flow. From the stagnation pressure equation for an incompressible fluid may be obtained

$$V_o = \sqrt{2(p_s - p_o)/\rho} = \sqrt{2g(p_s - p_o)/\gamma} \qquad (260)$$

from which it is apparent that velocities may be calculated from measurements of stagnation and static pressures. It has been shown

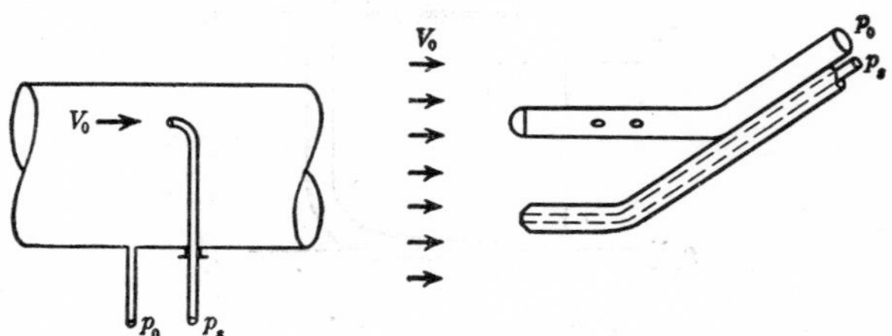

FIG. 187. Pitot-static tubes (separate).

that stagnation pressures may be measured easily and accurately by a pitot tube, and static pressures by various methods such as tubes, flat plates, and wall piezometer openings. Any combination of stagnation- and static-pressure-measuring devices is known as a *pitot-static tube.*

Pitot-static tubes may be divided into two classes: (1) those in which static and stagnation pressure connections are *separate,* and (2) those in which these are *combined.*

Separate types of pitot-static tubes are shown in Fig. 187 as used in obtaining the velocity profile in a pipe or as an air-speed indicator in aeronautical practice. Such tubes are simple in construction, but they cause inconvenience even in parallel-walled passages because of the necessity of two connections to the pipe and the difficulties of obtaining correct static pressure by a single piezometer opening. They are useless in curved flow, where the transverse pressure gradient renders $(p_s - p_o)$ uninterpretable in terms of V_o.

Modern practice favors the combined type of pitot-static tube, two types of which (for general and aircraft use) are illustrated in Fig. 188. Here the static tube jackets the stagnation pressure tube,

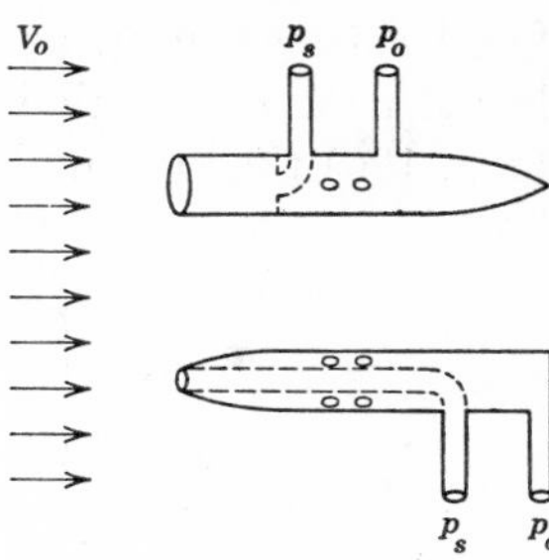

Fig. 188. Pitot-static tubes (combined).

resulting in a compact, efficient, velocity-measuring device. When connected to a differential pressure-measuring instrument, the pressure difference $(p_s - p_o)$ may be read, and it is seen from equation 260 to be a direct measure of the velocity V_o.

A static tube has been shown to record a pressure slightly less than the true static pressure, owing to the increase in velocity past the tube (Art. 90). This means that equation 260 must be modified by an experimentally determined instrument coefficient, C_I, to

$$V_o = C_I\sqrt{2(p_s - p_o')/\rho} = C_I\sqrt{2g(p_s - p_o')/\gamma} \qquad (261)$$

in which p_o' is the actual pressure measured by the static tube. Since p_o' is less than p_o, it is evident that C_I will always be less than unity. However, for most engineering problems the value of C_I may be taken as 1.00 for the conventional types of pitot-static tubes (Fig. 189), since the differences between p_o and p_o' are very small. Prandtl has designed a pitot-static tube in which the dif-

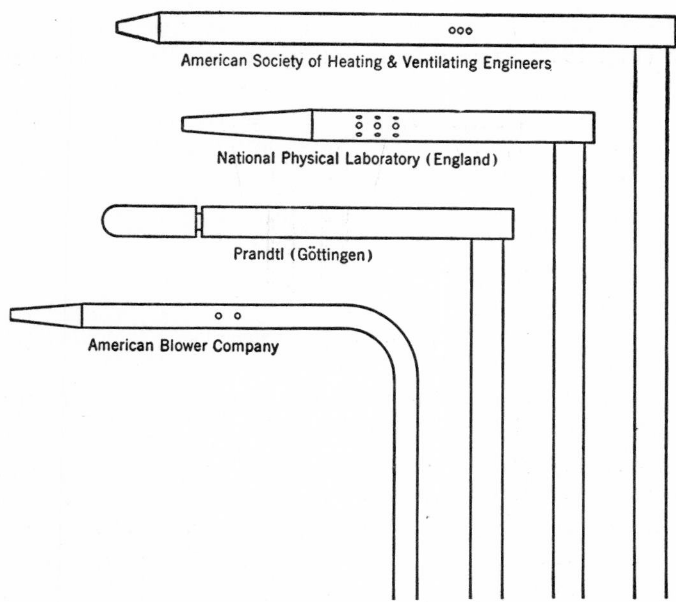

Fig. 189. Pitot-static tubes (to scale).

ference between p_o and p_o' is completely eliminated by ingenious location of the static-pressure opening. The opening is so located (Fig. 190) that the underpressure caused by the tube is exactly com-

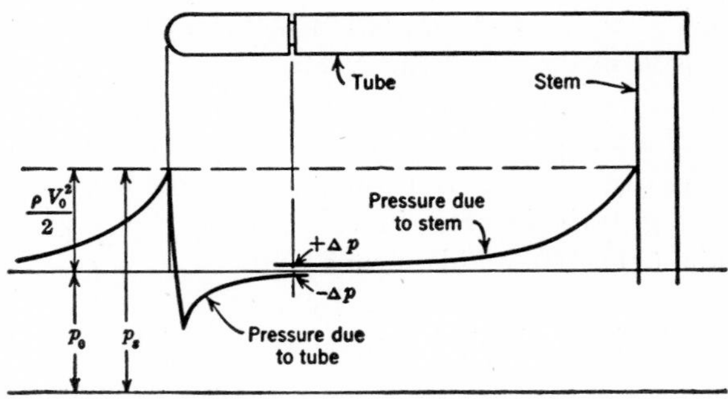

FIG. 190. Prandtl's pitot-static tube.

pensated by the overpressure due to the stagnation point on the leading edge of the stem, thus giving the true static pressure at the piezometer opening.

There are many variations of the pitot tube idea which result in devices of various shapes and coefficients. Probably the one most widely used in the United States is the Cole pitometer, which consists of two similar pitot tubes, one facing upstream, the other downstream (Fig. 191). The tube facing upstream measures the stagna-

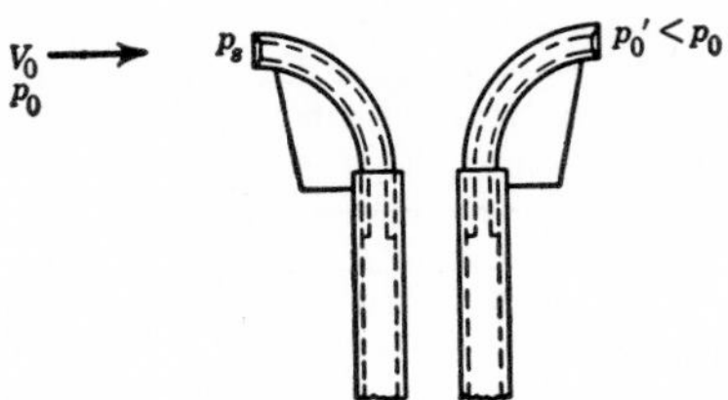

FIG. 191. Cole pitometer.

tion pressure properly, but the one facing downstream measures the pressure in the turbulent wake behind itself, which is considerably less (Art. 118) than the true static pressure, p_o. The coefficient of the pitometer is, therefore, much less than unity; experiments [9] have shown it to have values between 0.86 and 0.89. The advantages of the pitometer are its ruggedness and a compactness which allows it to be slipped easily through a cock in the wall of a pipe line.

[9] E. S. Cole, "Pitot Tube Practice," *Trans. A.S.M.E.*, Vol. 57, p. 281, 1935.

A practical aspect of velocity-measuring devices is their sensitivity to yaw or misalignment with the flow direction. Since perfect alignment is virtually impossible it is advantageous for a pitot-static tube to produce minimum error when perfect alignment does not exist. Prandtl's pitot-static tube, designed to be insensitive to small angles of yaw, gives a variation of only 1 percent in its coefficient at an angle of yaw of 19°. For the same percentage variation in coefficient the American Society of Heating and Ventilating Engineers' pitot-static tube may have an angle of yaw of 12°, and that of the National Physical Laboratory only 7°.[10]

Illustrative Problem

A pitot-static tube having a coefficient of 0.98 is placed at the center of a pipe line in which benzene is flowing. A manometer attached to the pitot-static tube contains mercury and benzene and shows a reading of 3 in. Calculate the velocity at the centerline of the pipe.

Solution. Using manometer principles,

$$\frac{p_s - p_o'}{\gamma} = \frac{3}{12} \times \frac{13.55 - 0.88}{0.88} = 3.60 \text{ ft of benzene}$$

$$v_c = 0.98\sqrt{2g(3.60)} = 14.9 \text{ fps} \tag{261}$$

94. Pitot-Static Tube in Compressible Flow. For velocity measurements in compressible flow, separate measurements of static pressure, stagnation pressure, and stagnation temperature are required. For subsonic flow ($\mathbf{M}_o < 1$) equation 83 may be used directly:

$$\frac{V_o^2}{2g} = c_p T_s \left[1 - \left(\frac{p_o}{p_s}\right)^{\frac{k-1}{k}}\right] \tag{83}$$

with p_s obtained by pitot tube, p_o by static tube, and T_s by temperature probe. The temperature probe consists of a small thermocouple surrounded by a jacket with open upstream end and small holes at the rear; a stagnation point exists on the upstream end of the probe, and a temperature close to the stagnation temperature is measured by the thermocouple.[11]

[10] Data from K. G. Merriam and E. R. Spaulding, "Comparative Tests of Pitot-Static Tubes," *N.A.C.A. Tech. Note* 546, 1935.

[11] The measured temperature is not exactly the stagnation temperature, and calibration of the instrument is required.

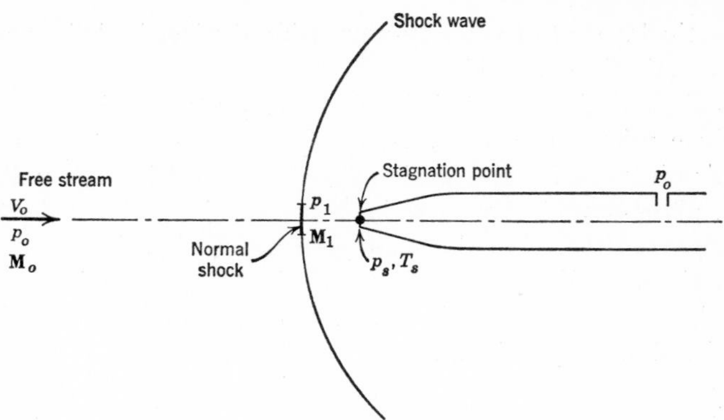

FIG. 192. Pitot tube in a supersonic flow.

For supersonic flow ($\mathbf{M}_o > 1$) a short section of normal shock wave (Art. 46) will be found upstream from the stagnation point (Fig. 192) and velocity calculations are considerably more complicated. Applying equation 115 through the shock wave,

$$\frac{p_1}{p_o} = 1 + \frac{2k}{k+1}(\mathbf{M}_o^2 - 1) \tag{262}$$

Applying equation 81 between the downstream side of the shock wave and the stagnation point,

$$\frac{p_s}{p_1} = \left(1 + \frac{k-1}{2}\mathbf{M}_1^2\right)^{\frac{k}{k-1}} \tag{263}$$

Multiplying these two equations together and substituting the relation between $\mathbf{M}_o$ and $\mathbf{M}_1$ of equation 114 yields

$$\frac{p_s}{p_o} = \frac{k+1}{2}\mathbf{M}_o^2\left[\frac{(k+1)^2\mathbf{M}_o^2}{4k\mathbf{M}_o^2 - 2k + 2}\right]^{\frac{1}{\kappa-1}} \tag{264}$$

which shows that measurements of p_s and p_o will allow prediction of the Mach number $\mathbf{M}_o$ of the undisturbed stream. Using the fact that the stagnation temperature is the same on both sides of the shock wave,

$$V_o^2/2g = c_p(T_s - T_o) \tag{265}$$

However, from $a_o^2 = kgRT_o$ and $\mathbf{M}_o = V_o/a_o$, $T_o = V_o^2/kgR\mathbf{M}_o^2$, which when substituted in the foregoing equation yields

$$\frac{V_o^2}{2g}\left(1 + \frac{2c_p}{kR\mathbf{M}_o^2}\right) = c_pT_s \tag{266}$$

Use of equations 264 and 266 allows V_o to be calculated from measurements of T_s, p_s, and p_o.

From measurements of p_s and p_o the experimentalist may easily identify subsonic and supersonic flow and thus select the proper equation for velocity calculation. Insertion of $\mathbf{M}_o = 1$ in equation 81 or 264 will yield

$$\frac{p_s}{p_o} = \left(\frac{k+1}{2}\right)^{\frac{k}{k-1}} \tag{267}$$

Thus, for $\mathbf{M}_o < 1$, $p_s/p_o < [(k+1)/2]^{k/(k-1)}$ and, for $\mathbf{M}_o > 1$, $p_s/p_o > [(k+1)/2]^{k/(k-1)}$.

It is of interest to compare equation 85 (Art. 35) with equation 267 and to note that $p_1/p_2 = p_s/p_o$ when $\mathbf{M}_2 = \mathbf{M}_o = 1$. Equation 86 was developed for an accelerated flow with increase of velocity from zero to the sonic; equation 267 characterizes decelerated motion from sonic velocity to zero. Since both flows are assumed isentropic, the equations are identical.

Illustrative Problems

A pitot-static tube in an air duct indicates a stagnation pressure of 25 in. of mercury abs; the static pressure in the duct is 18 in. of mercury abs; a temperature probe shows a stagnation temperature of 150°F. What is the local velocity just upstream from the pitot-static tube?

Solution.

$$\frac{25}{18} = 1.39 < \left(\frac{1.4+1}{2}\right)^{3.5} = 1.893 \tag{267}$$

Therefore equation 83 should be used for velocity calculation:

$$V_o^2/2g = 186.5(460+150)\left[1 - \left(\frac{18}{25}\right)^{0.286}\right] = 10220 \text{ ft of air}; \quad V_o = 810 \text{ fps} \tag{83}$$

The instruments of a high-speed airplane flying at high altitude show stagnation pressure 20 in. of mercury abs, static pressure 5 in. of mercury abs, and stagnation temperature 150°F. Calculate the speed of this airplane.

Solution.

$$\frac{20}{5} = 4 > \left(\frac{1.4+1}{2}\right)^{3.5} = 1.893 \tag{267}$$

therefore equations 264 et seq. should be used for velocity calculation:

$$4 = \frac{1.4+1}{2}\mathbf{M}_o^2\left[\frac{(2.4)^2\mathbf{M}_o^2}{5.6\mathbf{M}_o^2 - 2.8 + 2}\right]^{2.5} \tag{264}$$

Solving (by trial),

$$\mathbf{M}_o^2 = 2.71, \quad \mathbf{M}_o = 1.645$$

$$\frac{V_o^2}{2g}\left(1 + \frac{2 \times 186.5}{53.3 \times 1.4 \times 2.71}\right) = 186.5(610); \quad V_o = 1600 \text{ fps} \qquad (266)$$

95. Anemometers and Current Meters. Mechanical devices of similar characteristics are utilized in the measurement of velocity in

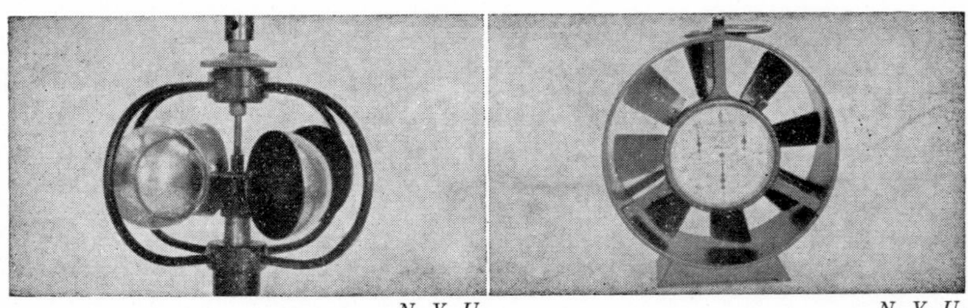

N. Y. U. *N. Y. U.*

Cup type Vane type

Anemometers

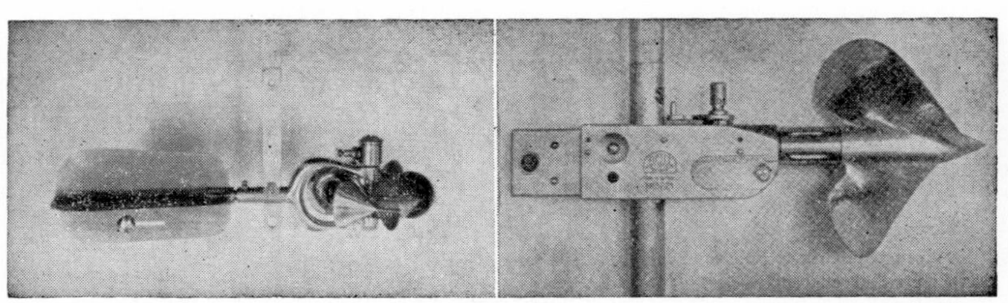

N. Y. U. *A. S. M. E.*

Cup type Vane type

Current meters

FIG. 193

air and water flow. Those for air are called *anemometers*; those for water, *current meters*. These devices consist essentially of a rotating element whose speed of rotation varies with the local velocity of flow, the relation between these variables being found by calibration. Anemometers and current meters fall into two main classes, depending upon the design of the rotating elements; these are the cup type and vane (propeller) type, as illustrated in Fig. 193. Anemometers and current meters differ slightly in shape, ruggedness, and appurtenances because of the different conditions under which they are used. The cup-type anemometer for the measurement of wind velocity is usually mounted on a rigid shaft; the vane-type anemometer is held in the hand while readings are taken. The current meters are usually suspended in a river or canal by a cable and hence must have empennages and weights to hold them in fixed positions in the flow.

Another type of anemometer which has been very successful in the field of aeronautical research is the *hot-wire anemometer*, one type of which is shown schematically in Fig. 194. The device consists of a fine platinum wire exposed to the velocity V_o which is to be measured. The fact that various velocities will have various cooling effects upon the hot wire, which will change its resistance, allows (by calibration) relating the velocity V_o to certain electrical measurements. The hot-wire anemometer of Fig. 194 is of the constant-voltage type,[12] and during its operation the drop in electrical potential across its terminals is maintained constant. Variation of velocity will change the resistance of the wire and, thus, the ammeter reading; the ammeter reading thereby becomes, after calibration, a measure of the velocity. The advantage of the hot-wire anemometer lies in the fact that it may be built in extremely small sizes and so may be employed in measuring the velocity variations in turbulent flow, velocity profiles in boundary layers, etc., where a pitot-static tube cannot be used. It must always be calibrated before use, and calibration is generally made against pitot-static tube measurements of velocity.

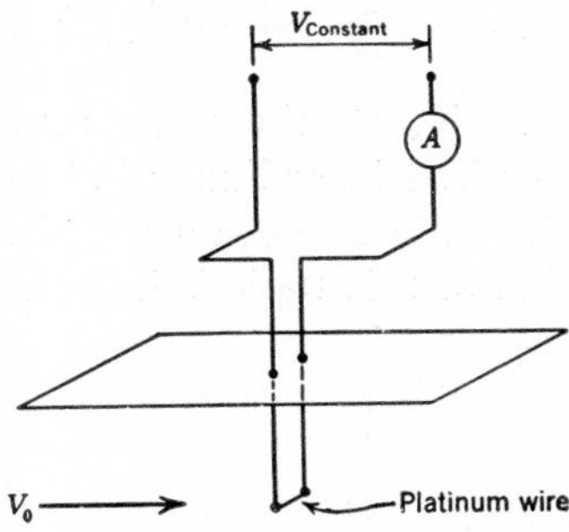

FIG. 194. Hot-wire anemometer.

Development of a hot-wire anemometer for similar measurements in liquids has proved less successful because of gas bubbles and solid particles fouling the wire and invalidating the calibration. However, considerable progress has recently been made in the development of a *hot-film anemometer*[13] which permits measurement of turbulent velocity fluctuations in liquids. Eujen[14] also reports a successful device, based on a variation of the hot-wire principle, for the measurement of small liquid velocities.

[12] Constant-current and constant-resistance types are also used.

[13] P. G. Hubbard, "Operating Manual for I.I.H.R. Hot Wire and Hot Film Anemometers," *State University of Iowa Studies in Engineering*, No. 37, 1957.

[14] See *Appl. Mech. Revs.*, p. 105, February 1951.

Kolin [15] has developed a velocity-measuring device based on electromagnetic principles which may be used successfully in liquid flows. Here the flowing liquid is utilized as a conductor which develops a voltage as it passes through a magnetic field; after calibration, measurement of this voltage allows computation of the velocity. The device may be used to obtain the mean velocity in a pipe flow or, when built in very small sizes, to obtain local velocities within a flowing liquid. Elrod and Fouse [16] discuss the application of such meters to the flow measurement of liquid metals.

MEASUREMENT OF SHEAR

Although the measurement of shear in flowing fluids is seldom required in ordinary industrial or engineering practice, it is of considerable importance to the research experimentalist. Some discussion of this topic is included here so that the reader may have some appreciation of the difficulties of measuring this simple quantity, whose importance has been repeatedly emphasized as basic to all problems involving the flow of real fluids. Also it offers some opportunity to review some devices and principles previously discussed.

96. Shear Measurements. No device has yet been invented which is capable of measuring the stress between moving layers of fluid. Shear measurements consist entirely of measurements of wall shear (τ_o) from which the shear between moving layers may be deduced from certain equations of fluid mechanics; such deductions may be of high or low accuracy, depending upon the equations used and the approximations necessary for solving them.

The wall shear for a cylindrical pipe of uniform roughness and with established flow may be obtained easily and accurately from pressure measurements through the use of equation 163, which may be expressed as

$$\tau_o = \frac{\gamma d}{4l}\left[\frac{p_1}{\gamma} + z_1 - \frac{p_2}{\gamma} - z_2\right] \tag{268}$$

in which the bracketed term is recognized as the head loss between points 1 and 2. With all details of the flow axisymmetric and no

[15] A. Kolin, "Electromagnetic Velometry, I. A Method for the Determination of Fluid Velocity Distribution in Space and Time," *Jour. Appl. Phys.*, Vol. 15, No. 2, p. 150, 1944.

[16] H. G. Elrod, Jr., and R. R. Fouse, "An Investigation of Electromagnetic Flowmeters," *Trans. A.S.M.E.*, Vol. 74, No. 4, p. 589, 1952. See also V. Cushing, "Induction Flowmeter," *Rev. Sci. Instruments,* Vol. 29, 1958, and V. P. Head, "Electromagnetic Flow Meter Primary Elements," *Trans. A.S.M.E.* (*Series D*), December 1959.

variation of wall roughness, it may be safely assumed that τ_o is the same at all points on the pipe wall and its value deducible from equation 268. The same procedure may be applied to any prismatic conduit and the same equation may be used with d replaced by $4R_h$; here, however, the flow is not axisymmetric and the shear stress must be interpreted as the *mean value.* Although on any longitudinal element of such a conduit the wall shear may be presumed constant, the equation provides no information on whether the wall shear at any point is larger or smaller than the mean value.

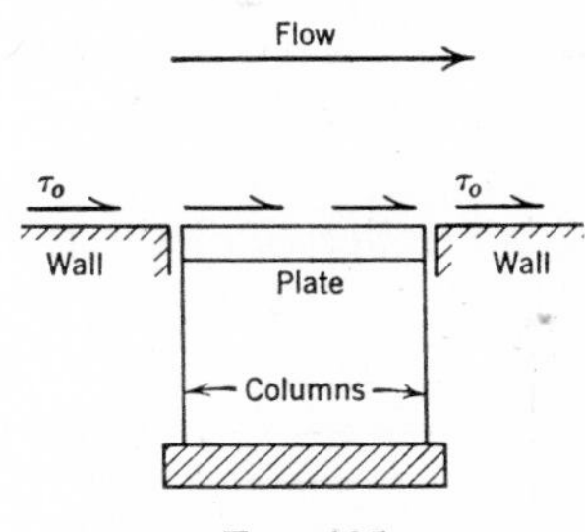

FIG. 195

Because of the foregoing limitations and for applications to problems of more complex boundary geometry, a more basic type of shear meter has been developed. This consists (Fig. 195) in replacing a small section of the wall by a movable plate mounted on elastic columns fastened to a rigid support. The columns are deflected slightly by the shearing force of fluid on plate, the small deflection measured by strain gages, and the shear stress deduced from this deflection. Although the device is basically simple, it is costly, unwieldy, and by no means easy to operate and interpret because of the relatively small shear force to be measured and the

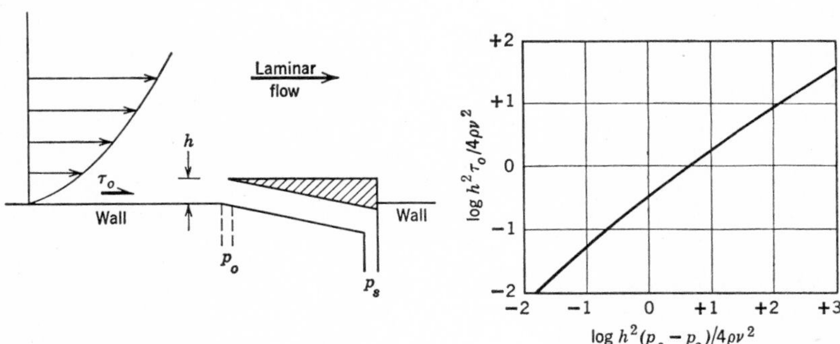

FIG. 196. Stanton tube.

relatively large extraneous forces which also contribute to the deflection of the columns.

Wall pitot tubes have been used successfully for the measurement of wall shear. Stanton [17] first used the design shown in Fig. 196,

[17] T. E. Stanton, D. Marshal, (Mrs.) C. N. Bryant, "On the Conditions at the Boundary of a Fluid in Turbulent Motion," *Proc. Roy. Soc.*, (A), Vol. 97, 1920.

the wall forming one side of the pitot tube; calibration in laminar pipe flow and in the laminar film of turbulent pipe flow showed τ_o to be a function of $(p_s - p_o)$. Its use for turbulent flows is thus restricted to measurements in the laminar film or boundary layer covering the wall, where the velocity profile is essentially the same as that in the pipe for the same fluid viscosity and wall shear. Taylor [18] presented a dimensionless calibration for the Stanton tube which may be expressed by $\tau_o h^2/4\rho\nu^2$ as a function of $(p_s - p_o)h^2/4\rho\nu^2$; this relationship is shown in Fig. 196, which allows τ_o to be predicted from measurements of $(p_s - p_o)$, ρ, ν, and h, providing that h is smaller than the thickness of the laminar region.

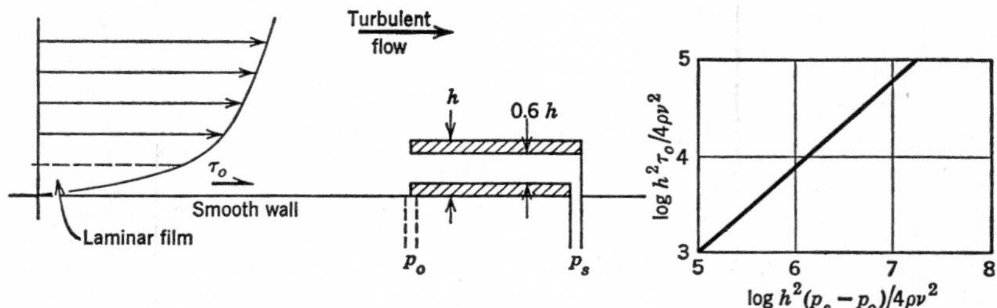

FIG. 197. Preston tube.

More recently, Preston [19] has applied the foregoing idea to turbulent flow over smooth surfaces with a tube of simpler design. The Preston tube (Fig. 197) is not submerged in the laminar film, and its performance depends upon the similarity of the velocity profiles through the buffer [20] zone between the laminar film and the turbulent region. The single calibration curve of Fig. 197 validates this similarity over the range indicated.

MEASUREMENT OF FLOWRATE

97. Total-Quantity Methods. In steady flow the flowrate (Q or G) may be obtained by measurement of the total quantity of fluid collected in a measured time. Such collections may be made by weight or volume and are the primary means of measuring fluid flow, but usually such measurements can be employed for only comparatively small flows under laboratory conditions.

[18] G. I. Taylor, "Measurements with Half-Pitot Tubes," *Proc. Roy. Soc.*, (A), Vol. 166, 1938.

[19] J. H. Preston, "The Determination of Turbulent Skin Friction by Means of Pitot Tubes," *Journ. Roy. Aero. Soc.*, Vol. 58, 1954.

[20] See Fig. 125 of Art. 69.

Measurement of flowrate by weighing consists in collecting the flowing liquid in a container placed on a scales and measuring the weight of liquid accumulated in a certain time. There are, of course, many commercial variations of this method, and many automatic devices are applied to it, but the principle remains the very simple one indicated.

Volumetric measurements of flowrate are carried out by allowing the liquid to collect in a container whose internal dimensions have been accurately determined. By noting the rise in the liquid surface in a measured time, or by noting the number of fillings of the container in a measured time, the rate of flow may be easily calculated. The accuracy of volumetric measurements is not, in general, as high as that of weighing methods because of the larger number of variables to be measured.

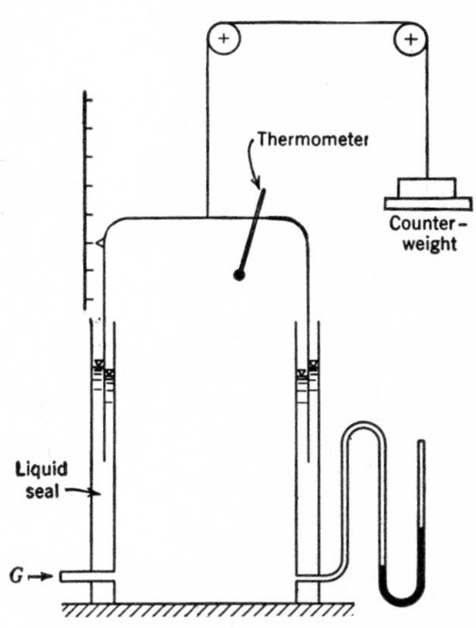

FIG. 198. Gasometer.

Gases may be measured volumetrically with a *gasometer* (Fig. 198), a device in which constant pressure and temperature are maintained while a volume of gas is collected. With pressure and temperature constant, the rate of rise of the movable top becomes a measure of the rate of flow into the gasometer, and after pressure and temperature are noted the flowrate may easily be calculated. A correction or automatic compensation must be made for the variable buoyant force, due to varying immersion, acting on the top.

98. Venturi Meters. A constriction in a streamtube has been seen (Arts. 25 and 36) to produce an accelerated flow and fall of hydraulic grade line or pressure which is directly related to flowrate, and thus is an excellent meter in which rate of flow may be calculated from pressure measurements. Such constrictions used as fluid meters are obtained by Venturi meters, nozzles, and orifices.

A Venturi meter is shown in Fig. 199. It consists of a smooth entrance cone of angle about 20°, a short cylindrical section, and a diffuser of 5° to 7° cone angle in order to minimize head loss (Art. 76). For satisfactory operation of the meter the flow should be an

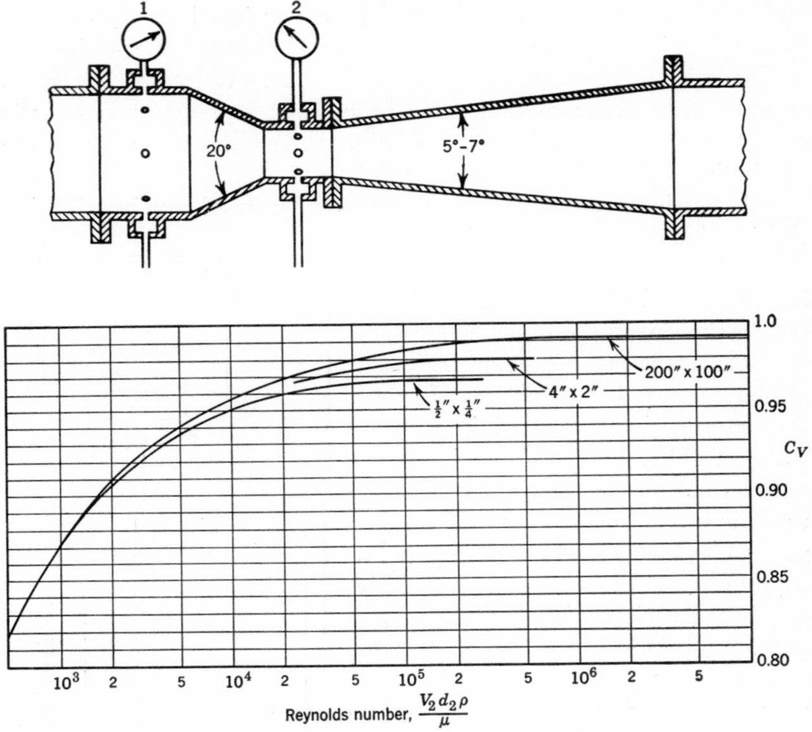

FIG. 199. Venturi meter and coefficients.

established one as it passes section 1. To ensure this the meter should be installed downstream from a section of straight and uniform pipe, free from fittings, misalignments, and other sources of large-scale turbulence, and having a length of at least 30 (preferably 50) pipe diameters. Straightening vanes may also be placed upstream from the meter for reduction of rotational motion in the flow.

The pressures at the *base* of the meter (section 1) and at the *throat* (section 2) are obtained by piezometer rings, and the fall of the hydraulic grade line between these sections is usually measured by differential manometer.[21] For the metering of gases, separate measurements of pressure and temperature are required at the base of the meter, but for liquids the differential manometer reading alone will allow computation of the flowrate.

For ideal incompressible flow the flowrate may be obtained by

[21] In industrial practice, this manometer difference is frequently carried to a mechanical device which calculates and plots a chronographic record of the flowrate.

solving simultaneous equations 53 and 64 of Art. 25 to yield

$$Q = \frac{A_2}{\sqrt{1 - (A_2/A_1)^2}} \sqrt{2g\left(\frac{p_1}{\gamma} + z_1 - \frac{p_2}{\gamma} - z_2\right)} \tag{269}$$

However, for real-fluid flow and the same $(p_1/\gamma + z_1 - p_2/\gamma - z_2)$ the flowrate will be expected to be less than that given by equation 269 because of frictional effects and consequent head loss between sections 1 and 2; in metering practice it is customary to account for this by insertion of an experimentally determined coefficient C_v in equation 269,[22] which then becomes

$$Q = \frac{C_v A_2}{\sqrt{1 - (A_2/A_1)^2}} \sqrt{2g\left(\frac{p_1}{\gamma} + z_1 - \frac{p_2}{\gamma} - z_2\right)} \tag{270}$$

Since C_v is merely a convenient means of expressing the head loss $h_{L_{12}}$, an exact relation between these variables is to be expected; this may be found by substituting $A_2 V_2$ for Q in the foregoing equation and rearranging it to the form of the conventional Bernoulli equation. The result is

$$h_{L_{1-2}} = \left[\left(\frac{1}{C_v^2} - 1\right)\left(1 - \left(\frac{A_2}{A_1}\right)^2\right)\right] \frac{V_2^2}{2g} \tag{271}$$

The bracketed quantity is the conventional minor loss coefficient, K_L, for the entrance cone of the meter and may be calculated from C_v.[23] Loss coefficients and pipe friction factors have been seen to depend on Reynolds number and to diminish with increasing Reynolds number; from the structure of the bracketed quantity it can be predicted that C_v will increase with increasing Reynolds number. This prediction is borne out by the typical experimental results of Fig. 199 [24] for Venturi meters of different size but of the same diameter ratio.[25] It should be observed in view of the principles of similitude (Chapter 8) that geometrically similar meters could be expected to give results falling on a single line on the plot, when installed in pipe lines with established flow.

[22] For compressible flow C_v is inserted in equations 88, 89, 90, and 92 in the same way.

[23] Nominal values of K_L and C_v in turbulent flow are 0.04 and 0.98, respectively. See Art. 76.

[24] Test results for Simplex Valve and Meter Co. and Builders Iron Foundry Venturi meters. *Fluid Meters: Their Theory and Application*, 4th Edition, A.S.M.E., 1937.

[25] Experiments at other diameter ratios (d_2/d_1) show a decrease of coefficient with increase of diameter ratio.

Illustrative Problem

Air flows through a 6-in. by 3-in. Venturi meter. The gage pressure is 20.0 psi and the fluid temperature 60°F at the base of the meter, and the differential manometer shows a reading of 6 in. of mercury. The barometric pressure is 14.70 psia. Calculate the flowrate.

Solution.

$$\frac{p_2}{p_1} = \frac{20 + 14.7 - 6(14.7/29.92)}{20 + 14.7} = 0.915$$

$$\gamma_1 = 34.7 \times 144/53.3 \times 520 = 0.180 \text{ lb/cuft} \qquad (3)$$

$$A_2/A_1 = 0.25, \quad A_2 = 0.0491 \text{ sqft}$$

From Appendix VIII, $Y = 0.949$ and, assuming $C_v = 0.98$,

$$G = \frac{0.949 \times 0.98 \times 0.0491 \times 0.180}{\sqrt{1 - (0.25)^2}} \sqrt{2g(6 \times 0.491 \times 144)/0.180} \qquad (92)$$

$$G = 3.31 \text{ lb/sec}$$

Checking $\mathbf{R}_2$ gives a value of 1,250,000, which (from Fig. 199) gives a better value of C_v of 0.981. The true flowrate is, therefore, $G = 3.31(0.981/0.980) = 3.31$ lb/sec.

99. Nozzles. Nozzles are used in engineering practice for the creation of jets and streams for all purposes as well as for fluid metering; when placed in or at the end of a pipe line as metering devices they are generally termed *flow nozzles*. Since a thorough study of flow nozzles will develop certain general principles which may be applied to other special problems, only the flow nozzle will be treated here.

A typical flow nozzle is illustrated in Fig. 200. Such nozzles are designed to be clamped between the flanges of a pipe, generally possess rather abrupt curvatures of the converging surfaces, terminate in short cylindrical tips, and are essentially Venturi meters with the diffuser cones omitted. Since the diffuser cone exists primarily to minimize the head losses caused by the meter, it is obvious that larger head losses will result from flow nozzles than occur in Venturi meters and that herein lies a disadvantage of the flow nozzle; this disadvantage is somewhat offset, however, by the lower initial cost of the flow nozzle.

Extensive research on flow nozzles, sponsored by the American Society of Mechanical Engineers and the International Standards Association, has resulted in the accumulation of a large amount of reliable data on nozzle installation, specifications, and experimental coefficients. Only the barest outline of these results can be presented here; the reader is referred to the original papers of these societies for more detailed information.

The A.S.M.E. flow nozzle installation of Fig. 200 is typical of those employed in American practice, section 1 being taken one pipe

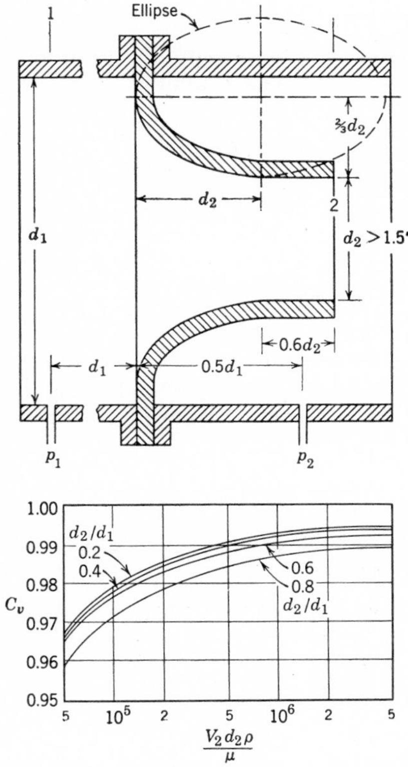

FIG. 200. A.S.M.E. flow nozzle and coefficients.[26]

diameter upstream and section 2 at the nozzle tip. It has been found that a pressure representative of that at the latter point may be obtained by a wall piezometer connection which leads, fortunately, to the simplification of the nozzle installation, since a wall piezometer is easier to construct than a direct connection to the tip of the nozzle. Pressures obtained in this manner are not, of course, the exact pres-

[26] Data from *Fluid Meters: Their Theory and Application*, 4th Edition, A.S.M.E., 1937; and H. S. Bean, S. R. Beitler, and R. E. Sprenkle, "Discharge Coefficients of Long Radius Flow Nozzles When Used with Pipe Wall Pressure Taps," *Trans. A.S.M.E.*, p. 439, 1941.

sures existing in the live stream of flowing fluid passing section 2, but the slight deviations incurred are of no consequence since they are absorbed in the experimental coefficient, C_v. The variation of C_v with area ratio and Reynolds number is typical of the geometrically similar conditions specified: the constancy of C_v at high Reynolds number and decrease of C_v with decreasing Reynolds number is observed for the flow nozzle as for the Venturi meter. Coefficients for standardized flow nozzle installations at Reynolds numbers below 50,000 are not yet available, but the reader will find [27] reliable information on special installations when confronted with a low Reynolds number metering problem.

The flow nozzle being essentially equivalent to the entrance cone of the Venturi meter, flowrates may be computed by equations 92 and 270.

Illustrative Problem

An A.S.M.E. flow nozzle of 3-in. diameter is installed in a 6-in. water line. The attached differential manometer contains mercury and water and shows a reading of 6 in. Calculate the flowrate through the nozzle and the head loss caused by its installation. The water temperature is 60°F.

Solution. Selecting tentatively $C_v = 0.99$ from Fig. 200 and calculating A_2 as 0.0491 sqft,

$$Q = \frac{0.99 \times 0.0491}{\sqrt{1 - (0.25)^2}} \sqrt{2g(\tfrac{6}{12})(13.55 - 1)} = 1.010 \text{ cfs} \qquad (270)$$

Calculating $\mathbf{R}_2$ gives 423,000, which yields a better value of $C_v = 0.988$. Using this value in place of 0.99, $Q = (0.988/0.99)1.010 = 1.008$ cfs.

Precise calculation of head loss caused by the nozzle installation is not possible or necessary, but adequate values may be obtained by computing $h_{L_{1-2}}$ from equation 271 (giving 0.11 ft) and adding it to the head loss caused by the flow deceleration downstream from the nozzle. Treating this as an abrupt enlargement, the head loss may be computed from equation 217 as $h_L = (V_2 - V_1)^2/2g = 3.7$ ft. Thus the total head loss caused by the nozzle installation is about 3.8 ft of water.

100. Orifices. Like nozzles, orifices serve many purposes in engineering practice other than the metering of fluid flow, but the study of the orifice as a metering device will allow the application of principles to other problems, some of which will be treated subsequently.

[27] See, for example, J. F. Downie Smith and S. Steel, "Rounded Approach Orifices," *Mech. Eng.*, December 1935; and B. O. Buckland, "Fluid Meter Nozzles," *Trans. A.S.M.E.*, Vol. 56, p. 827, 1934.

The orifice for use as a metering device in a pipe line consists of a concentric square-edged circular hole in a thin plate which is clamped between the flanges of the pipe (Fig. 201). The flow characteristics of the orifice differ from those of the nozzle in that the minimum section of the streamtube occurs not within the orifice but downstream from it, owing to the formation of a *vena contracta* at section 2. The cross-sectional area at the vena contracta, A_2, is charac-

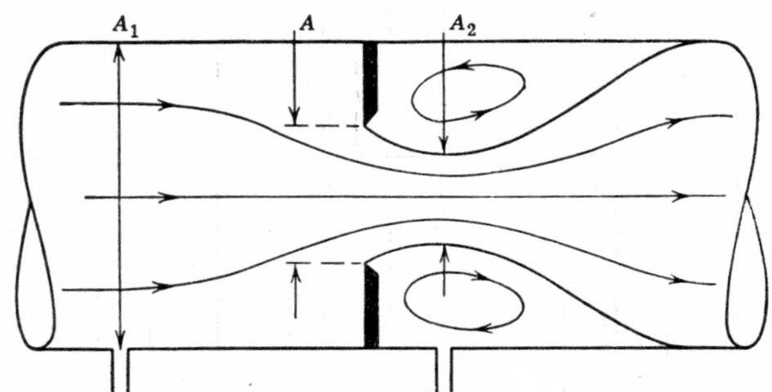

FIG. 201. Definition sketch for orifice meter.

terized by a coefficient of contraction,[28] C_c, and given by C_cA. Substituting this into equation 270,

$$Q = \frac{C_v C_c A}{\sqrt{1 - C_c^2 (A/A_1)^2}} \sqrt{2g\left(\frac{p_1}{\gamma} + z_1 - \frac{p_2}{\gamma} - z_2\right)} \qquad (272)$$

which is customarily written

$$Q = CA\sqrt{2g\left(\frac{p_1}{\gamma} + z_1 - \frac{p_2}{\gamma} - z_2\right)} \qquad (273)$$

thus defining the orifice coefficient C as

$$C = \frac{C_v C_c}{\sqrt{1 - C_c^2 (A/A_1)^2}} \qquad (274)$$

which is thus dependent not only on C_v and C_c but on the shape of the installation (defined by A/A_1) as well.

In practice it is not feasible to locate the downstream pressure connection at the vena contracta because the location of the vena contracta depends upon both Reynolds number and A/A_1. Accordingly, it is frequently located (as for the flow nozzle) at a fixed pro-

[28] The Weisbach values of Art. 76 may be used as nominal at high Reynolds numbers.

portion of the pipe diameter downstream from the orifice plate, and the other connection one pipe diameter upstream. Any coefficient, C, will thus be dependent upon, and associated with, the particular location of the pressure connections. Values of C over a wide range of Reynolds numbers may be obtained from Fig. 202; it is convenient

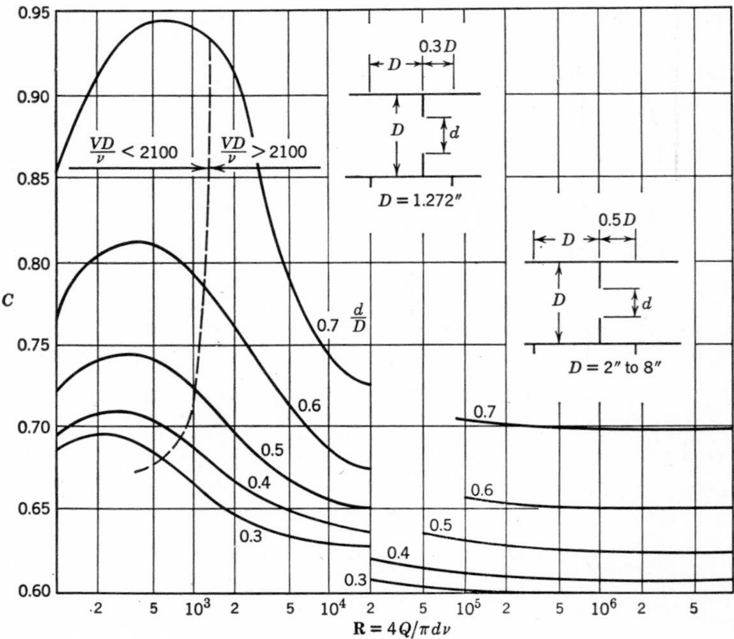

Fig. 202. Orifice meter coefficients.[29]

and standard practice to define the Reynolds number on the basis of flowrate and orifice diameter as

$$\mathbf{R} = \frac{Qd}{(\pi d^2/4)\nu} = \frac{4Q}{\pi d \nu}$$

which is used as the abscissa for the plot. The trend of the coefficient with Reynolds number is of interest when compared with that of the Venturi meter and flow nozzle. The constancy of C at high Reynolds number is again noted, reflecting the substantial constancy of both C_v and C_c in this range. At lower Reynolds numbers an increase of

[29] Data from G. L. Tuve and R. E. Sprenkle, "Orifice Discharge Coefficients for Viscous Liquids," *Instruments*, Vol. 6, p. 201, 1933; Vol. 8, pp. 202, 225, 232, 1935; and *Fluid Meters: Their Theory and Application*, 4th Edition, A.S.M.E., 1937.

C is noted, in spite of the expectation of a decrease of C_v in this region; evidently, increased viscous action not only lowers C_v but also raises C_c (by increasing the size of the vena contracta), and the latter effect is predominant. In the range of very low Reynolds numbers the effect of viscous action on the vena contracta remains at a maximum (with C_c around 1), and the decrease of C with further decrease of Reynolds number reflects the steady decrease of C_v produced by viscous resistance.

In metering the flow of compressible fluids at high pressure ratios, equation 88 may be used but it strictly applies only when the downstream pressure connection is at the vena contracta. Values of Y from Appendix VIII may generally be used only as a first approximation; accurate values of Y for various locations of the pressure taps will be found in the A.S.M.E. Fluid Meters report cited in footnote 24.

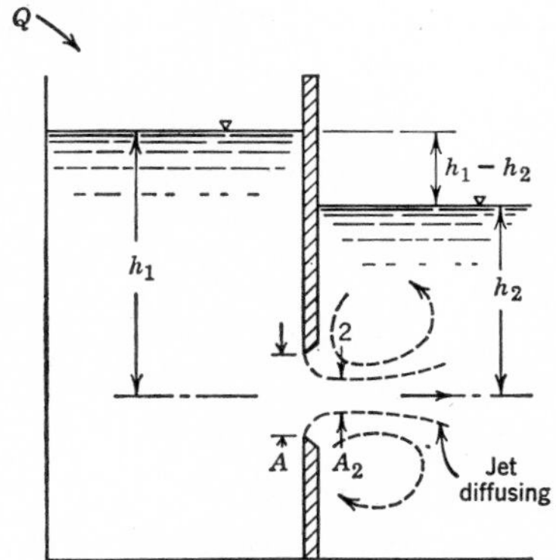

FIG. 203. Submerged orifice.

An extension of the pipe-line orifice of Fig. 201 is the *submerged orifice* of Fig. 203 featured by orifice discharge from one large reservoir into another. Here with A/A_1 virtually zero, C (of equation 274) becomes C_vC_c. Assuming a perfect fluid and applying the Bernoulli equation between the upstream reservoir and section 2,

$$h_1 = h_2 + V_2^2/2g \quad \text{or} \quad V = \sqrt{2g(h_1 - h_2)}$$

providing that the pressure distribution may be considered hydrostatic [30] in the downstream reservoir. For the real fluid, frictional effects will prevent the attainment of this velocity and C_v is intro-

[30] This is a valid assumption if h_2 is large compared to orifice size.

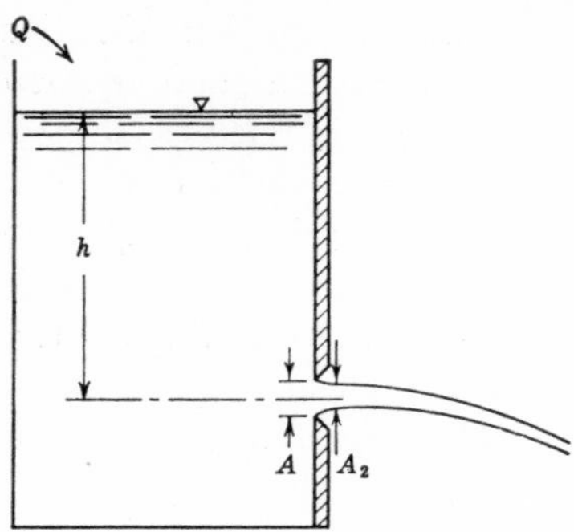

FIG. 204. Orifice discharging freely.

duced as before, so that

$$V_2 = C_v\sqrt{2g(h_1 - h_2)}$$

The flowrate may be calculated from A_2V_2, in which A_2 is replaced by C_cA:

$$Q = A_2V_2 = C_cC_vA\sqrt{2g(h_1 - h_2)} = CA\sqrt{2g(h_1 - h_2)} \qquad (275)$$

in which C_vC_c is defined as the coefficient (of discharge) of the orifice. When the orifice discharges freely into the atmosphere (Fig. 204), h_2 becomes zero and the equation reduces to

$$Q = C_cC_vA\sqrt{2gh} = CA\sqrt{2gh}$$

The dependence of the various orifice coefficients upon shape of orifice is illustrated by Fig. 205. The coefficients given are nominal

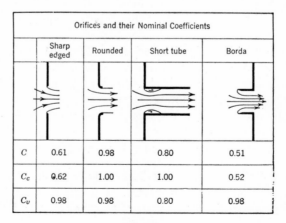

Orifices and their Nominal Coefficients				
	Sharp edged	Rounded	Short tube	Borda
C	0.61	0.98	0.80	0.51
C_c	0.62	1.00	1.00	0.52
C_v	0.98	0.98	0.80	0.98

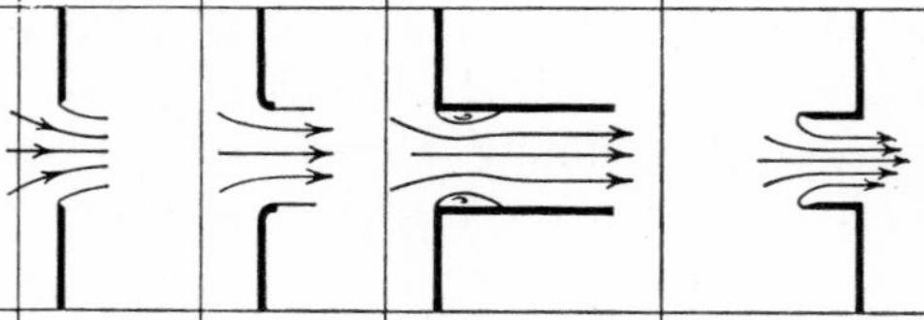

FIG. 205

values for large orifices ($d > 1$ in.) operating under comparatively large heads of water ($h > 4$ ft). Above these limits of head and size, various experiments have shown that the coefficients are practically constant. Coefficients foı sharp-edged orifices over a wide range of Reynolds numbers are given in Fig. 206, which shows the same trend of values (for the same reasons) as that of Fig. 202. The plot of Fig. 206, although convenient and applicable to the flow of all

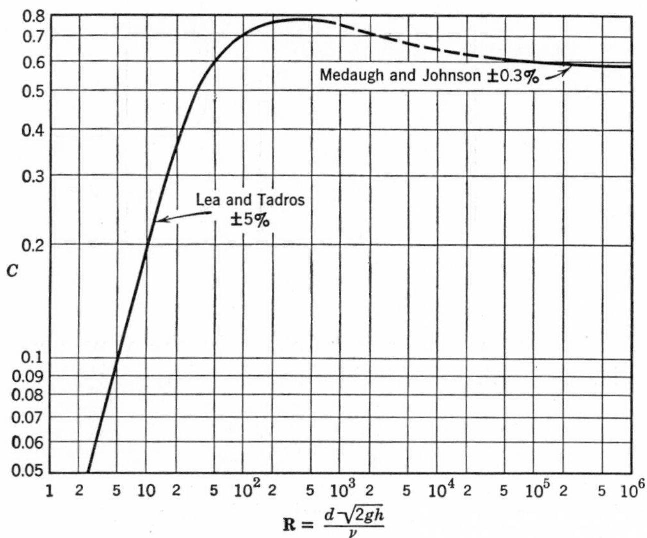

FIG. 206. Coefficients for sharp-edged orifices under static head[31] ($h/d > 5$).

fluids, has a certain limitation in orifice size caused by the action of surface tension. Surface-tension effects (although impossible to predict except in idealized situations) will increase with decreasing orifice size; the plotted values are valid only where such effects are negligible and, thus, cannot be applied to very small orifices.

The head lost between the reservoir and section 2 in an orifice operating under static head may be calculated from the coefficient of velocity and flowrate by equation 271. Since $A_2/A_1 \cong 0$ this equation reduces to

$$h_L = \left(\frac{1}{C_v^2} - 1\right)\frac{V_2^2}{2g}$$

[31] Data from F. C. Lea, *Hydraulics*, 6th Edition, p. 87, Edward Arnold and Co., 1938; and F. W. Medaugh and G. D. Johnson, *Civil Eng.*, Vol. 10, No. 7, p. 422, July, 1940.

A special problem of orifice flow is that of the two-dimensional sluice gate of Fig. 207 in which jet contraction occurs only on the

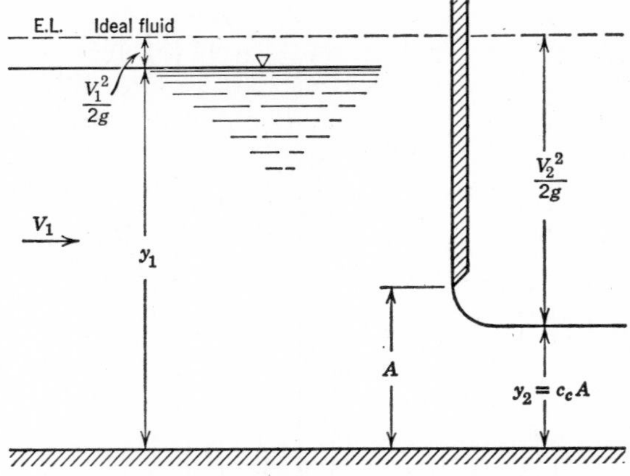

FIG. 207. Sluice gate.

top of the jet and pressure distribution in the vena contracta is hydrostatic. Assuming an ideal fluid,

$$y_1 + \frac{V_1^2}{2g} = y_2 + \frac{V_2^2}{2g}$$

and substituting $V_2 y_2 / y_1$ for V_1, and solving for V_2,

$$V_2 = \frac{1}{\sqrt{1 - (y_2/y_1)^2}} \sqrt{2g(y_1 - y_2)}$$

The actual velocity (allowing for head loss) is obtained by multiplying the above by C_v, and the flowrate by multiplying the result by $C_c A$. The flowrate through the sluice is, therefore,

$$q = \frac{C_v C_c A}{\sqrt{1 - (y_2/y_1)^2}} \sqrt{2g(y_1 - y_2)}$$

from which it is noted that the effective head on the sluice is $(y_1 - y_2)$, that the equation is analogous to equation 272, and that it reduces to equation 275 as the depth y_1 becomes large compared to y_2.

101. Elbow Meters. The orifice, nozzle, and Venturi meter as applied in the measurement of pipe-line flow have been seen to be fundamentally methods of producing a regular and reproducible fall of the hydraulic grade line which is related to flowrate. Another meter of this type is the elbow meter of Fig. 208 which utilizes the fall of hydraulic grade line between the outside and inside of a regular pipe bend (see Fig. 53 of Art. 28). Analytical solutions of such problems are not feasible, and such devices are calibrated by determining experimentally the relation between fall of hydraulic grade line and flowrate. Lansford [32] has done this for a variety of

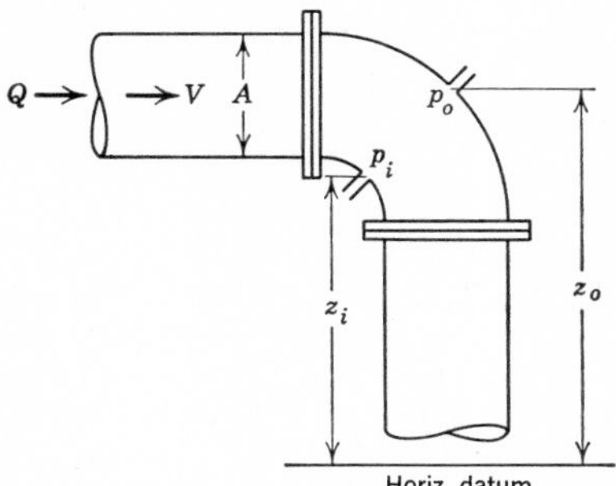

FIG. 208. Elbow meter.

90° flanged elbows, and it allows their use as accurate and economical flow meters; for a basic equation he proposes

$$\left(\frac{p}{\gamma} + z\right)_o - \left(\frac{p}{\gamma} + z\right)_i = C_k \frac{V^2}{2g}$$

with coefficient C_k ranging between 1.3 and 3.2, the magnitudes depending upon the size and shape of the elbow. This equation may be solved for V and multiplied by A to obtain the flowrate, Q. If the coefficient of the meter is then defined as $\sqrt{1/C_k}$ the resulting equation has the same form as that for nozzles and orifices,

$$Q = CA\sqrt{2g\left(\frac{p_o}{\gamma} + z_o - \frac{p_i}{\gamma} - z_i\right)}$$

in which C will have values between 0.56 and 0.88.

102. Pitot Tube Methods. The flowrate in pipe lines may be measured by means of pitot-static tube or pitometer. These devices have been seen (Art. 93) to be velocity-measuring instruments which may be employed in pipe lines to establish the distribution of velocity; their use in measuring flowrate is essentially an integration of the product of velocity and the area through which it occurs.

One method of obtaining flowrate from velocity measurements is to divide the pipe cross section into a number of *equal annular*

[32] W. M. Lansford, "The Use of an Elbow in a Pipe Line for Determining the Flow in the Pipe," *Eng. Exp. Sta. Univ. Ill., Bull.* 289, 1936.

areas (B, C, D, Fig. 209) and to measure the mean velocities through these areas. The point of mean velocity is assumed to be at the midpoint of the annular area, i.e., at points where circles divide these areas in half. This is really assuming the velocity to vary linearly over the areas considered, which (in turbulent flow) is obviously more true near the center of the pipe than near the walls; this assumption does not cause serious errors, however, if a large number of annular areas are taken. In general, the velocity profile is not symmetrical about the centerline of the pipe, and the mean

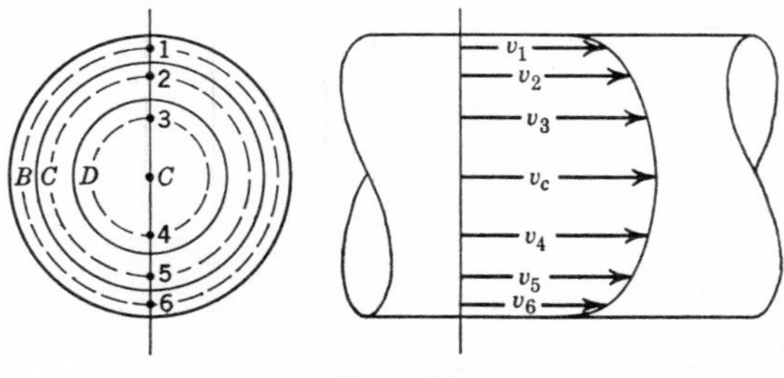

FIG. 209

velocity through an annular area is taken to be the numerical average of the two velocities measured in this area. Thus (Fig. 209)

$$Q_B = \left(\frac{V_1 + V_6}{2}\right)\frac{A}{3}, \quad Q_C = \left(\frac{V_2 + V_5}{2}\right)\frac{A}{3}, \quad Q_D = \left(\frac{V_3 + V_4}{2}\right)\frac{A}{3}$$

but the flowrate, Q, in the line is given by

$$Q = Q_B + Q_C + Q_D$$

or, by substitution of the above values,

$$Q = A\left(\frac{V_1 + V_2 + V_3 + V_4 + V_5 + V_6}{6}\right)$$

Thus the mean velocity in the pipe line is a simple numerical average of the velocities existing at certain predetermined points on the diameter of the pipe.

Another method of obtaining flowrate from velocity distribution is by graphical integration. For a cylindrical pipe there are two convenient ways of accomplishing this, and both begin with the basic equation

$$Q = \int^A v\, dA \tag{56}$$

For a cylindrical passage, $dA = 2\pi r\,dr = \pi\,d(r^2)$, so the expression for Q may be written

$$Q = \pi \int_0^{R^2} v\,d(r^2) \quad \text{or} \quad 2\pi \int_0^R (vr)\,dr$$

in which R is the radius of the pipe. These integrals may be evaluated by plotting v against r^2 or vr against r and measuring the areas under the resulting curves, the usual forms of which are shown in Fig. 210.

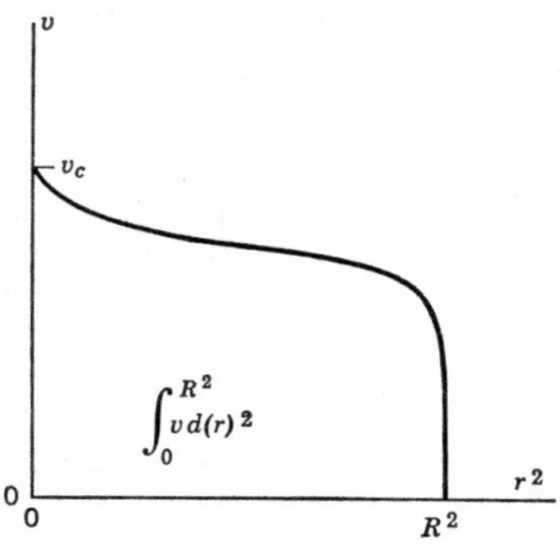

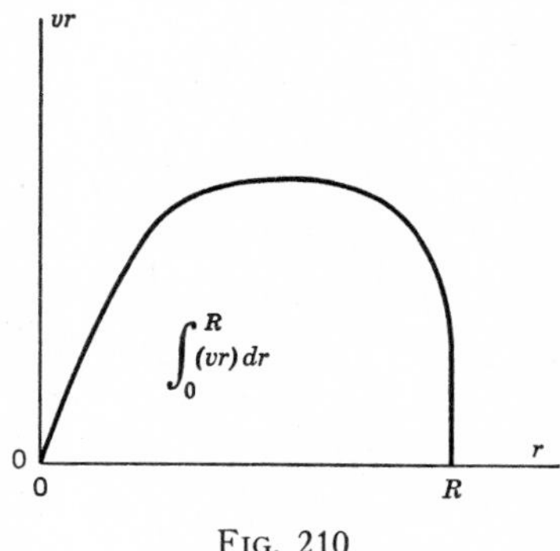

FIG. 210

103. Dilution Methods.[33] Dilution methods for measuring flowrate consist essentially in introducing (at a steady rate) a concentrated foreign substance to the flow, measuring the concentration of the substance after thorough mixing has taken place, and calculating from the dilution of the substance the flowrate which has brought about this dilution.

A concentrated salt solution has been used in Europe in applying this method to the calculation of flow in small mountain streams, and in this country to the calculation of the flow through the turbines of hydroelectric power plants. If the rate of flow of salt solution into the unknown flow is Q_S, and the concentration of salt in this solution is C_1 lb/cuft, the number of pounds per second of salt added to the unknown flow is given by $Q_S C_1$. If the concentration of salt in the unknown flow after mixing has occurred is C_2, the number of pounds of salt flowing in the stream per second is also given by $(Q + Q_S)C_2$. Therefore

$$Q_S C_1 = (Q + Q_S) C_2$$

Since Q_S is so small compared to Q, it is usually neglected on the right-hand side of the equation and the flowrate computed from

[33] See W. A. Cawley and J. W. Woods, "An Improved Dilution Method for Flow Measurement," *Trans. A.S.C.E.*, Vol. 123, 1958.

$Q = Q_S(C_1/C_2)$. Hence by control and measurement of Q_S and with C_1 and C_2 easily obtainable by titration methods, the unknown flowrate Q may be found, providing that complete mixing has been attained.

104. Salt-Velocity Method. An ingenious method of flow measurement, which has met with success in the measurement of flowrate in large pipe lines, is the salt-velocity method developed by Allen and Taylor.[34] In this method a quantity of concentrated salt solution is introduced suddenly to the flow, and the mean velocity is obtained by measuring the velocity of the salt solution as it moves with the flow.

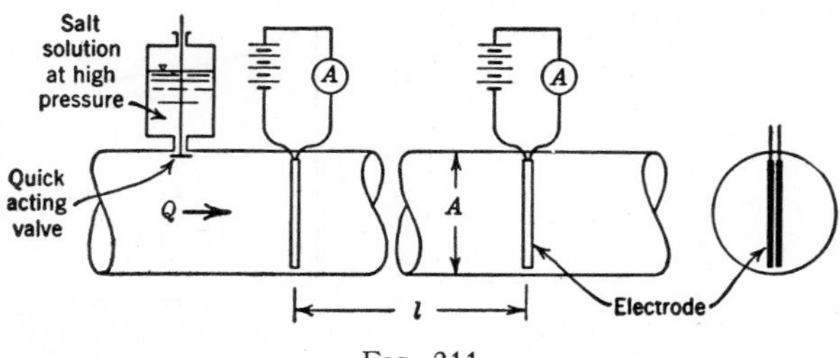

FIG. 211

The mechanism for the method (illustrated schematically in Fig. 211) is essentially a device for introducing the salt solution and two similar electrodes and circuits. The passage of the salt solution between the plates of an electrode is recorded by a momentary increase in the ammeter reading due to the greater conductivity of the salt solution. By noting the time, t, between the deflections of the two ammeter needles and knowing the distance l between electrodes, the mean velocity in the pipe may be calculated from $V = l/t$ and the flowrate Q from $Q = AV$.

The testing techniques involved in the salt-velocity method are somewhat more complex than the above statement of principles implies, owing primarily to the use of a chronographic device to record automatically (1) the variation of electrical current with the passage of the salt solution, and (2) the time of passage of the solution between electrodes.

105. Weirs. For measuring large and small open flows in field or laboratory, the weir finds wide application. A weir may be defined in a general way as "any regular obstruction *over* which flow occurs." Thus, for example, the overflow section (spillway) of a

[34] C. M. Allen and E. A. Taylor, "The Salt Velocity Method of Water Measurement," *Trans. A.S.M.E.*, Vol. 45, p. 285, 1923.

dam is a special type of weir and may be utilized for flow measurement. However, weirs for measuring purposes are usually of more simple and reproducible form, consisting of smooth, vertical, flat plates with upper edges sharpened. Such weirs, called *sharp-crested weirs*, appear in a variety of forms, the most popular of which is the rectangular weir; this type has a straight, horizontal crest and extends over the full width of the channel in which it is placed. The flow picture produced by such a weir is essentially two-dimensional and for this reason it will be used as a basis for the following discussion.

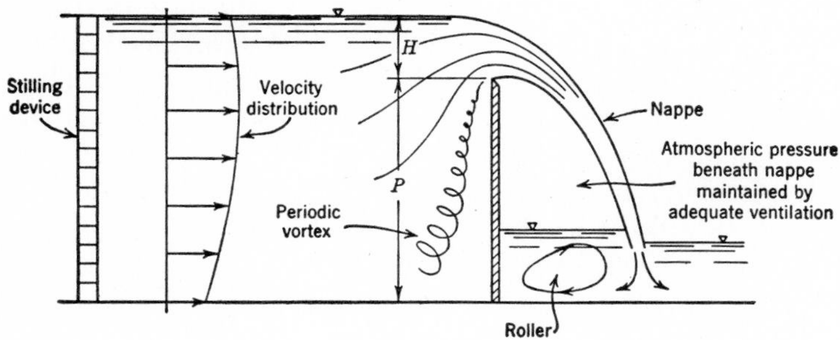

FIG. 212. Weir flow (actual).

The flow of liquid over a sharp-crested weir is at best an exceedingly complex problem and impossible of rigorous analytical solution.[35] An appreciation for the complexities, however, is necessary to an understanding of experimental results and of the deficiencies of simplified weir formulas. These complexities may be discovered by considering the flow over the sharp-crested weir of Fig. 212. Although it is obvious at once that the head, H, on the weir is the primary factor causing the flow (Q) to occur, no simple relationship between these two variables can be derived, for two fundamental reasons: (1) the shape of the flow picture, and (2) the effect of turbulence and frictional processes cannot be calculated. The most important factors which affect the shape of the flow picture are the head on the weir, H, the weir height, P, and the extent of ventilation beneath the nappe. Although the effect of these factors may be found experimentally, there is no simple method of predicting the flow picture from the values of H, P, and pressure beneath the nappe. The effects of turbulence and friction cannot be predicted, nor can they be isolated for experimental measurement. It may be

[35] However, the two-dimensional problem for the ideal fluid may be solved by the trial-and-error methods of Arts. 28 and 114.

noted, however, that frictional resistance at the sidewalls will affect the flowrate to an increasing extent as the channel becomes narrower and the weir length, b (normal to the paper), smaller. Fluid turbulence and frictional processes at the sides and bottom of the approach channel also contribute to the velocity distribution in an unpredictable way. The effects of velocity distribution on weir flow have been shown by Schoder and Turner [36] to be appreciable, and an effort should be made in all weir installations to provide a good length of approach channel, with stilling devices such as racks and screens for the even distribution of turbulence and the prevention of abnormal velocity distributions. Another influence of frictional processes is the creation of a periodic helical secondary flow in the corners just upstream from the weir plate, resulting in a vortex (Fig. 212), which influences the flow in an unpredictable manner. The free liquid surfaces of weir flow also bring surface-tension effects into the problem, and these forces, although small, affect the flow picture appreciably at low heads and small flows.

In the light of the foregoing complexities, the derivation of any simple weir formula obviously requires a drastic (and somewhat artificial) simplification of the problem which can only lead to an approximate result; however, by such methods the *form* of the relationship between flowrate and head can be found and an experimental coefficient defined. To derive a simple weir equation, let it be assumed that (1) velocity distribution upstream from the weir is uniform, (2) all fluid particles move horizontally as they pass the weir crest, (3) the pressure in the nappe is zero, and (4) the influence of viscosity, turbulence, secondary flows, and surface tension may be neglected. These assumptions produce the flow picture of Fig. 213. Taking section 1 in the approach channel well upstream from the weir and section 2 slightly downstream from the weir crest, Bernoulli's equation may be applied to a typical streamline to find the velocity v_2. Using streamline AB as typical and taking the weir crest as datum,

$$H + \frac{V_1^2}{2g} = (H - h) + \frac{v_2^2}{2g}$$

whence

$$v_2 = \sqrt{2g(h + V_1^2/2g)}$$

which shows v_2 to be dependent upon h. Taking dq to be the two-dimensional flowrate through a strip of height dh, $dq = v_2\, dh$ allow-

[36] E. W. Schoder and K. B. Turner, "Precise Weir Measurements," *Trans. A.S.C.E.*, Vol. 93, p. 999, 1929.

ing integration to obtain the flowrate, q,

$$q = \int_0^H v_2\, dh = \sqrt{2g} \int_0^H \left(h + \frac{V_1^2}{2g}\right) dh$$

The result is

$$q = \tfrac{2}{3}\sqrt{2g}\left[\left(H + \frac{V_1^2}{2g}\right)^{\frac{3}{2}} - \left(\frac{V_1^2}{2g}\right)^{\frac{3}{2}}\right]$$

Because of the cumbersome form of this equation[37] and because in many weir problems $P >> H$ and V_1 is small, $V_1^2/2g$ is customarily

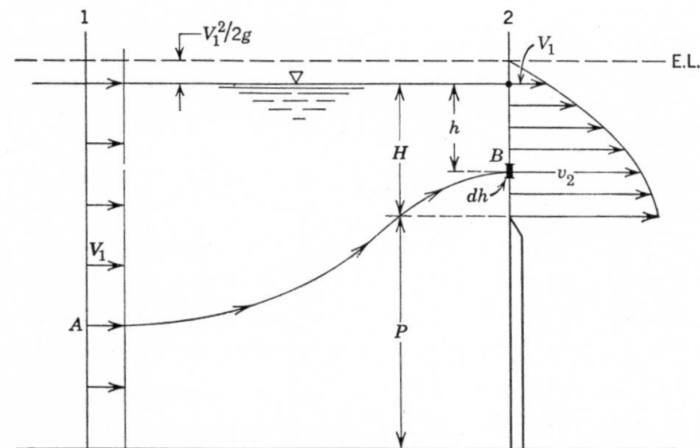

FIG. 213. Weir flow (simplified).

neglected and the equation further simplified to

$$q = \tfrac{2}{3}\sqrt{2g}\, H^{\frac{3}{2}} \tag{276}$$

which is the basic equation for rectangular weirs. Into this equation must be inserted an experimentally determined coefficient, C_w, which includes the effects of the many phenomena disregarded in the foregoing development and simplifications. For real weir flow the relation between flowrate and head then becomes

$$q = C_w \tfrac{2}{3}\sqrt{2g}\, H^{\frac{3}{2}} \tag{277}$$

showing that (to the extent C_w is constant), for rectangular weirs, $q \propto H^{\frac{3}{2}}$. The coefficient C_w is essentially a factor which transforms the simplified weir flow of Fig. 213 into the real weir flow of Fig. 212 and its magnitude is thus fixed by the most important difference between these flows—the shape of the flowfield. Thus the weir

[37] Which (knowing P and H) requires trial-and-error solution for q.

coefficient is primarily a coefficient of contraction which expresses the extent of contraction of the true nappe below that assumed in the simplified analysis. Since the size of the weir coefficient depends primarily on the shape of the flowfield, the effect of other fluid properties and phenomena may usually be discovered by examining their influence upon this shape.

Although a dimensional analysis of the weir problem must necessarily be incomplete because of the impossibility of including all the pertinent factors, it will provide a rational basis for an understanding of some of the factors affecting weir coefficients. The expressible independent variables entering the two-dimensional weir problem are q, H, P, μ, σ, ρ, and g. Application of dimensional analysis to these variables will show the existence of a relationship between four dimensionless numbers,[38]

$$q/\sqrt{g}\,H^{\frac{3}{2}} = F(\mathbf{R}, \mathbf{W}, P/H)$$

in which the Froude number on the left-hand side is a direct measure of C; dimensional analysis thus shows the dependence of the weir coefficient on **R**, **W**, and P/H. Of these numbers P/H has been found to be the most important in determining the magnitude of C_w—as would be expected since this ratio, more than any of the others, has the greatest influence on the shape of the flowfield. The effect of **W** is negligible except at low heads where surface-tension effects may be large; the effect of **R** is small (except at low heads) since water is usually involved in weir flows, **R** is high, and viscous action small.

The experimental work of Rehbock [39] led to an empirical formula for the coefficient of well-ventilated, sharp-crested rectangular weirs for water measurement, which has an accuracy of better than 1 percent if care is taken with the details of installation. Rehbock's formula for C_w is

$$C_w = 0.605 + 0.08\frac{H}{P} + \frac{1}{305H} \tag{278}$$

which shows clearly the strong influence of P/H, except where H is small and $1/305H$ of some significance. In this term H may be shown [40] (for a liquid of constant physical properties) to be a direct

[38] See Appendix IX.

[39] Th. Rehbock, "Wassermessung mit scharfkantigen Uberfallwehren," *Zeitschrift des V.d.I.*, Vol. 73, No. 24, June 15, 1929. Also see C. E. Kindsvater and R. W. Carter, "Discharge Characteristics of Rectangular Thin Plate Weirs," *Trans. A.S.C.E.*, Vol. 124, 1959.

[40] See footnote 42.

measure of **R** and **W**; this implies (as expected) that the influences of viscosity and surface tension on weir flow are strong only when H is small.

Weirs are reliable measuring devices only at heads above the range of strong action by viscosity and surface tension; in this range the formula shows that the coefficient can be expected to increase with increasing head and decreasing weir height. Although it is desirable to calibrate a new weir installation *in place*, this is frequently not possible and a formula must be selected for the weir

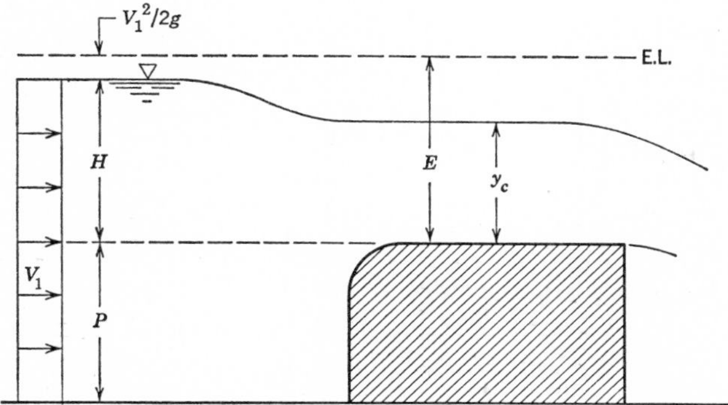

FIG. 214. Broad-crested weir.

coefficient. The Rehbock formula can be expected to give good results if such important details as adequate ventilation, stilling devices, crest sharpness, and smoothness of upstream face are not overlooked.

Broad-crested weirs (Fig. 214) have been shown (Art. 86) to be critical-depth meters; here, for ideal flow,

$$q = \sqrt{g(2E/3)^3} = (2/3)^{\frac{3}{2}}\sqrt{g}\,E^{\frac{3}{2}} \tag{279}$$

The weir coefficient for the ideal broad-crested weir may be calculated by equating equations 277 and 279 to yield

$$C_w = \frac{1}{\sqrt{3}}\left(\frac{E}{H}\right)^{\frac{3}{2}}$$

For a very high weir $P/H \to \infty$, $E \to H$, $E/H \to 1$, and $C_w \to 1/\sqrt{3} = 0.577$ as shown in Art. 86; for a lower weir $P/H < \infty$, $E > H$, $E/H > 1$, and $C_w > 0.577$. Hence the weir coefficient increases with decreasing P/H and thus exhibits the same trend as the Rehbock formula. Experimental measurements also substantiate this variation of C_w with P/H, but the values of C_w obtained from experiment are a few percent lower than those of ideal flow because of head loss accompanied by a falling energy line.

For small flowrates, *notch weirs* are widely used as measuring devices; of these the most popular is the triangular weir or V-notch (Fig. 215). A simplified analysis similar to that used on the rectangular weir (but neglecting velocity of approach) yields (after inserting the experimental coefficient) the fundamental formula

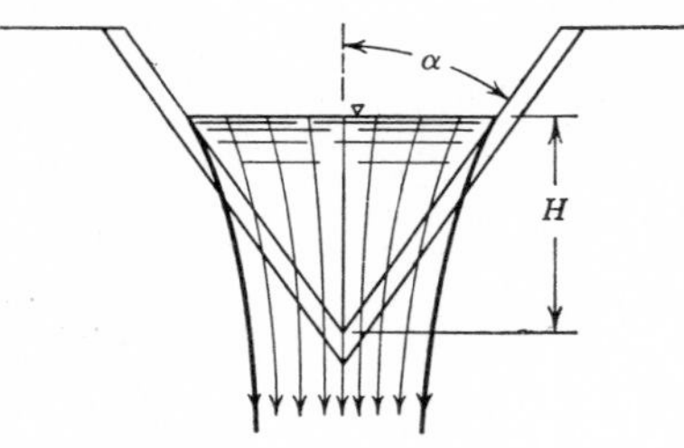

FIG. 215. Triangular weir.

$$Q = C_w \tfrac{8}{15} \tan \alpha \sqrt{2g}\, H^{\frac{5}{2}} \qquad (280)$$

Triangular weirs of 90° notch angle (2α) have coefficients (for water) near 0.59, but these are affected by viscosity, surface tension, and weir plate roughness, increases of any one of these tending to increase the coefficient. A comprehensive study of triangular weir flow has been made by Lenz,[41] who used many liquids in order to discover the effects of viscosity and surface tension on weir coefficients, thus extending the utility of the triangular weir as a reliable measuring device. For notch angles of 90° Lenz proposed [42]

$$C_w = 0.56 + \frac{0.70}{\mathbf{R}^{0.165}\mathbf{W}^{0.170}}$$

as applying to all liquids providing that the falling sheet of liquid does not cling to the weir plate and that $H > 0.2$ ft, $\mathbf{R} > 300$, and $\mathbf{W} > 300$. The work of Lenz has not only broadened the field of application of the weir as a measuring device but has also documented the increase of coefficient with decreasing **R** and **W**, a characteristic of the coefficients for all sharp-crested weirs.

The crest of a *spillway structure* is shown in Fig. 216. Major considerations in the design of such a spillway are structural stability against hydrostatic pressure and other loads, and prevention of separation (Art. 59) and reduced pressures on the surface of the structure. The rectangular weir equation may be applied to the spillway, the coefficient C_w ranging from 0.60 to 0.75. The relatively high value of C_w may be explained by a comparison of a sharp-crested

[41] A. T. Lenz, "Viscosity and Surface-Tension Effects on V-Notch Weir Coefficients," *Trans. A.S.C.E.*, Vol. 108, 1943.

[42] $\mathbf{R} = H\sqrt{gH}/\nu$, $\mathbf{W} = \rho g H^2/\sigma$

weir (Fig. 212) and a spillway crest designed exactly to fit the curvature of the lower side of the nappe of this weir for a certain design head, H_D. In spite of the greater friction of the spillway, with a fixed reservoir surface the flow over the two structures will be

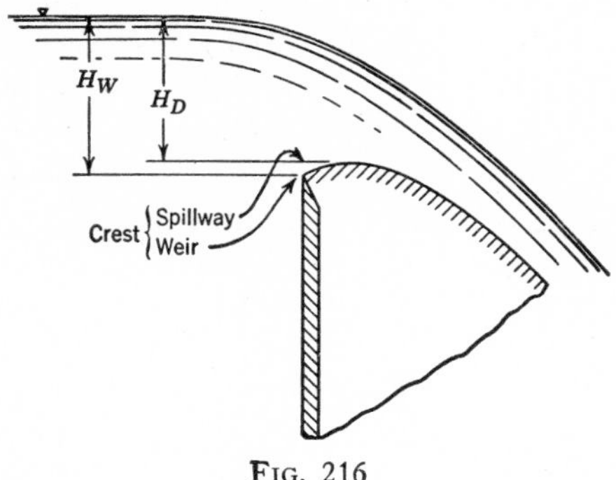

FIG. 216

approximately the same, but the heads for each structure will be measured from their respective crests and will, therefore, be quite different, the head on the weir being greater than the head on the spillway. Since for (about) the same flowrate the smaller head will be associated with the larger coefficient, the spillway coefficient is seen to be larger than that of the sharp-crested weir.

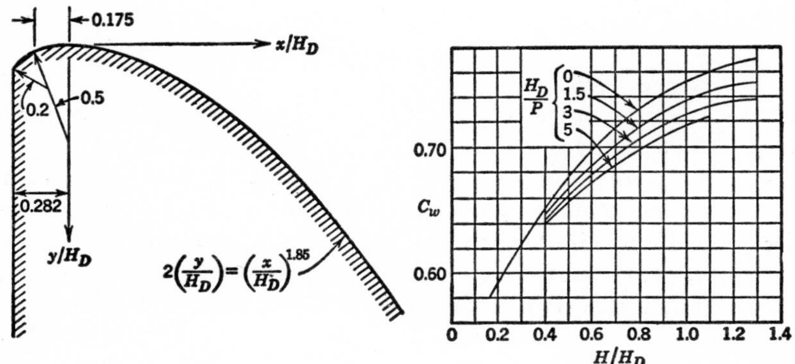

FIG. 217. Spillway profiles and coefficients.[43]

A spillway profile which will fit the flow of a sharp-crested weir and thus prevent harmful discontinuities of pressure is shown with its coefficients in Fig. 217. The results are both useful and instruc-

[43] F. R. Brown, "Hydraulic Models as an Aid to the Development of Design Criteria," *Waterways Expt. Sta., Corps of Engrs., Bull.* 37, Vicksburg, Miss., June, 1951.

tive, since in defining the coefficients the head is taken as the vertical distance between spillway crest and energy line upstream from the structure and thus contains the velocity head of the approaching flow. After the design head and height of structure have been determined, the profile of the structure may be laid out and discharge coefficients accurately predicted for heads between 40 percent and 130 percent of the design head. For any weir height, P, a steady increase of C_w with H is noted as for the sharp-crested weir (equation 278). For the sharp-crested weir this trend was due to change in the overall shape of the flow picture; for the spillway crest it is

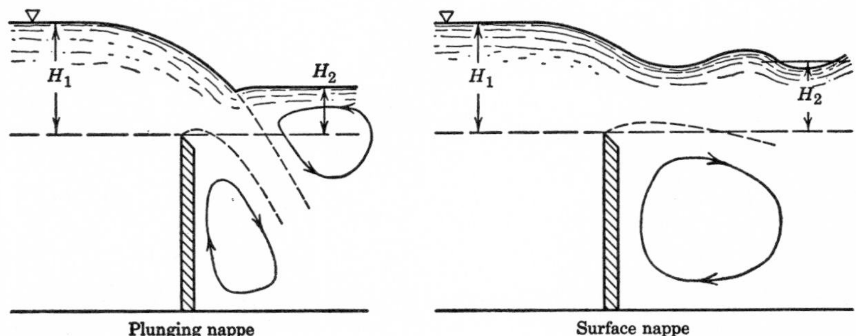

FIG. 218. Submerged weirs.

due mostly to the steady decline in pressure over the crest with increasing head, which increases the effective head and is reflected in an increase of coefficient. This trend is beneficial up to the point where cavitation or separation occurs, inasmuch as an increase in C_w may be interpreted as an increase in the efficiency of the spillway.

The effects of submergence (Fig. 218) on weir flow are of some theoretical and practical interest. Sharp-crested weirs cannot be considered to be precise measuring devices when operating submerged, but the effect of submergence on broad-crested weirs is surprisingly small, making this type of weir a reliable measuring device even for high submergence. This reliability is due primarily to the straighter streamlines and essentially hydrostatic pressure distribution on the crest of the broad-crested weir which change little as the downstream water level rises above the crest of the weir. For the sharp-crested weir discharging freely the pressure distribution through the nappe is featured by zero pressure on top and bottom, and thus is far from hydrostatic; raising the downstream water level above the crest of the weir will drastically change this pressure distribution and immediately affect the whole flow picture. In all submerged weir problems the two flow situations (Fig. 218) of *plunging nappe* at low submergence and *surface nappe* at higher submergence are observed. Approximate results[44] of investigations on submerged rectangular weirs may be seen on the sketch of Fig. 219 in which submergence ratio H_2/H_1 is plotted against the ratio of measured flowrate (Q_S) to that which would have existed with free flow (Q_F) for a head H_1. Engineers require accurate information on weir submergence when flowrate measurements are needed in times of flood, in spillway

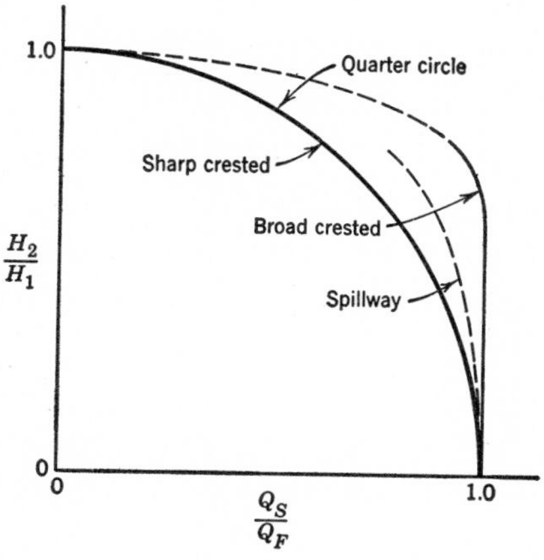

FIG. 219. Approximate effects of submergence.

design, and in situations where adequate vertical drop is unavailable for use of a (more precise) freely flowing weir.

106. Current-Meter Measurements. The construction of a weir for measuring the flowrate in large canals, streams, or rivers is impractical for many obvious reasons, but existing spillways whose coefficients are known may frequently serve as measuring devices. The standard method of river flow measurement is to measure the velocity by means of a current meter (Art. 95) and integrate the results in a manner similar to that used for pitot tube measurements (Art. 102).

Fundamental to the use of a current meter is some knowledge of the general properties of velocity distribution in open flow. As in pipes, the velocities are reduced at the banks and bed of the channel,

[44] For accurate and detailed information see J. G. Woodburn, "Tests of Broad Crested Weirs," *Trans. A.S.C.E.*, Vol. 96, 1932; J. K. Vennard and R. F. Weston, "Submergence Effect on Sharp-Crested Weirs," *Eng. News-Record*, June 3, 1943; J. R. Villemonte, "Submerged Weir Discharge Studies," *Eng. News-Record*, December 25, 1947.

but it must be realized that in open flow the roughnesses and turbulences are of such great and irregular magnitudes that the velocity distribution problem cannot be placed on the precise basis which it enjoys in pipe flow. However, from long experience and thousands of measurements, the United States Geological Survey has established certain average characteristics of velocity distribution in streams and rivers which serve as a basis for current-meter measurements. These characteristics of velocity distribution in a vertical are shown in Fig. 220 and may be amplified by the following state-

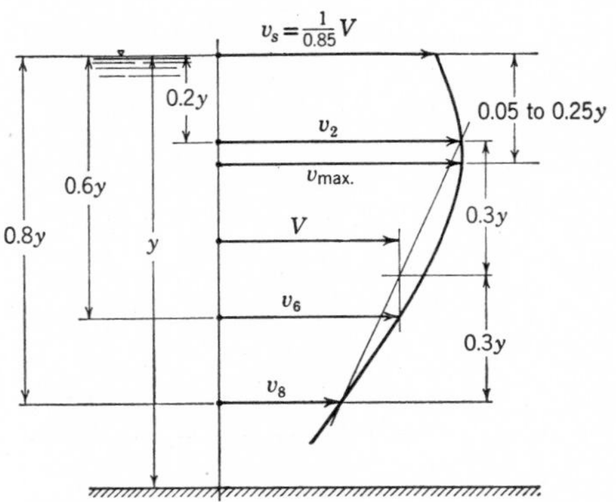

FIG. 220. Standard velocity distribution in a vertical in open flow.

ments: (1) the curve may be assumed parabolic;[45] (2) the location of the maximum velocity is from $0.05y$ to $0.25y$ below the water surface; (3) the mean velocity occurs at approximately $0.6y$ below the water surface; (4) the mean velocity is approximately 85 percent of the surface velocity; (5) a more accurate and reliable method of obtaining the mean velocity is to take a numerical average of the velocities at $0.2y$ and $0.8y$ below the water surface. These average values will, obviously, not apply perfectly to a particular stream or river, but numerous measurements with the current meter will tend toward accurate results since deviations from the average values will tend to compensate, thus giving a greater accuracy than can be obtained from individual measurements.

Current-meter measurements for calculation of flowrate may be taken in the following manner. A reach of river having a fairly regular cross section is selected. This cross section is measured accurately by soundings. It is then divided into vertical strips

[45] This is merely a convenient approximation; it does not imply laminar flow.

(Fig. 221), the current meter is suspended, and velocities are measured at the two-tenths and eight-tenths points in each vertical (1, 2, 3, etc., Fig. 221). From these measurements the mean velocities (V_1, V_2, V_3, etc.) in each vertical may be calculated. The mean velocity through each vertical strip is taken as the average of the mean velocities in the two verticals which bound the strip, and thus the rates of flow (Q_{12}, Q_{23}, etc.) through the strips may be calculated from

$$Q_{12} = b_{12}\left(\frac{y_1 + y_2}{2}\right)\left(\frac{V_1 + V_2}{2}\right)$$

$$Q_{23} = b_{23}\left(\frac{y_2 + y_3}{2}\right)\left(\frac{V_2 + V_3}{2}\right)$$

and the flowrate in the stream may be calculated by totaling the flowrates through the various strips.

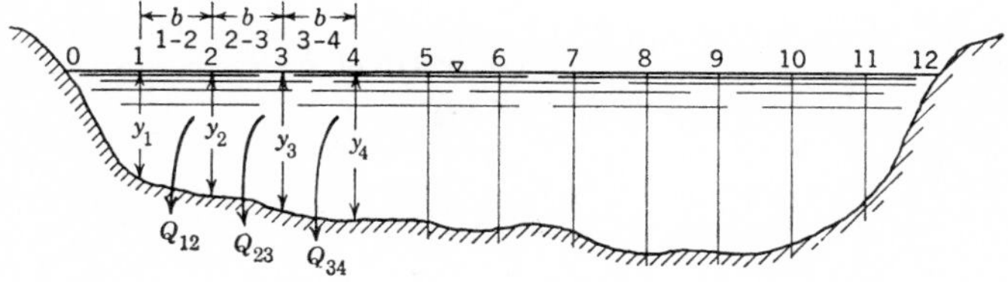

FIG. 221. Division of river cross section for current-meter measurements.

107. Float Measurements. The velocities of surface floats may sometimes be used under satisfactory flow conditions to obtain rough measurements of river flow, but floats of this type are subject to the vagaries of winds and local surface currents which may drive them far off their courses. The reach of river selected for float measurements should be straight and uniform and should have a minimum of surface disturbances; measurements should be taken on a windless day. The time that it takes the floats to travel a certain distance may be measured easily, and from this the surface velocities may be computed and the average velocities approximated by using the relationships of Fig. 220. It should not be inferred, however, that even under ideal conditions the accuracy of float measurements is high. This is due to the above-mentioned general difficulties and to the fact that the ratio of mean velocity to surface velocity, although having an average value of 0.85, may be as low as 0.80 or as high as 0.95 and quite unpredictable for a given reach of river.

REFERENCES

H. Diederichs and W. C. Andrae, *Experimental Mechanical Engineering*, Vol. I, John Wiley and Sons, 1930. (Out of print.)

K. H. Beij, "Aircraft Speed Instruments," *N.A.C.A. Rept.* 420, 1932.

"Density Determinations," Part 16 of *A.S.M.E. Power Test Codes on Instruments and Apparatus*, A.S.M.E., 1931.

"Determination of the Viscosity of Liquids," Part 17 of *A.S.M.E. Power Test Codes on Instruments and Apparatus*, A.S.M.E., 1931.

"Head Measuring Apparatus," Part 4 of *A.S.M.E. Power Test Codes on Instruments and Apparatus*, A.S.M.E., 1933.

"Pressure Measurement," Part 2 of *A.S.M.E. Power Test Codes on Instruments and Apparatus*, A.S.M.E., 1938–1942.

"Standards for Discharge Measurement with Standardized Nozzles and Orifices," *N.A.C.A. Tech. Mem.* 952, 1940.

D. M. Corbett and others, "Stream Gaging Procedure," *U.S. Geological Survey Water Supply Paper* 888, 1943.

H. Addison, *Hydraulic Measurements*, 2nd Edition, John Wiley and Sons, 1946. (Out of print.)

H. Schaefer, "Machmeters for High-Speed Flight Research," *Jour. Aero. Sci.*, Vol. 15, No. 6, p. 351, 1948.

E. Ower, *The Measurement of Air Flow*, Chapman and Hall, 1949.

J. W. Howe, "Flow Measurement," Chapter III of *Engineering Hydraulics* (H. Rouse, Ed.), John Wiley and Sons, 1950.

R. F. Stearns, R. R. Johnson, R. M. Johnson, and C. A. Larson, *Flow Measurement with Orifice Meters*, D. Van Nostrand, 1951.

F. W. Fleming and R. C. Binder, "Study of Linear Resistance Flow Meters," *Trans. A.S.M.E.*, July 1951.

H. W. Stoll, "The Pitot-Venturi Flow Element," *Trans. A.S.M.E.*, October 1951.

A. L. Jorissen and H. T. Newton, "Discharge Measurement by Means of Cylindrical Nozzles," *Trans. A.S.M.E.*, July 1952.

W. M. Schulze, G. C. Ashby, Jr., and J. R. Erwin, "Several Combination Probes for Surveying Static and Total Pressure and Flow Direction," *N.A.C.A. Tech. Note* 2830, December 1952.

Water Measurement Manual, U.S. Bureau of Reclamation, 1953.

L. K. Spink, *Principles and Practice of Flow Meter Engineering*, Foxboro Co., 1958.

W. F. Coxon, *Flow Measurement and Control*, The Macmillan Co., 1959.

"Flow Measurement by Means of Standardized Nozzles and Orifice Plates," Chapter 4 of Part 5 of *A.S.M.E. Power Test Codes on Instruments and Apparatus*, A.S.M.E., 1959.

Fluid Meters—Their Theory and Application, 5th Ed., A.S.M.E., 1959.

PROBLEMS

838. A pycnometer weighs 0.2207 lb when empty and 0.9256 lb when filled with liquid. If its volume is 0.007639 cuft, calculate the specific gravity of the liquid.

839. A plummet weighs 0.900 lb in air and 0.667 lb in a liquid. If the volume of liquid displaced by the plummet is 0.00456 cuft, what is the specific gravity of the liquid?

840. A cylindrical plummet weighing 0.10 lb, of 1-in. diameter, and having a specific gravity of 7.70, is suspended in a liquid from the end of a balance arm, 6 in. from the knife edge. The arm is balanced by a weight of 0.12 lb, 4 in. from the knife edge. What is the specific gravity of the liquid? Neglect the weight of the balance arm.

841. A crude hydrometer consists of a cylinder of $\frac{1}{2}$-in. diameter and 2-in. length, surmounted by a cylindrical tube of $\frac{1}{8}$-in. diameter and 8 in. long. Lead shot in the cylinder brings the hydrometer's total weight to 0.30 oz. What range of specific gravities can be measured by this hydrometer?

842. To what depth will the bottom of the hydrometer of the preceding problem sink in a liquid of specific gravity 1.10?

843. A hydrometer weighing 0.25 lb and having a volume of 0.0035 cuft is placed in a liquid of specific gravity 1.60. What percent of the volume remains above the liquid surface?

844. Mercury is placed in an open U-tube and liquid is poured into one of the legs. A liquid column 10 in. high balances a mercury column 1.5 in. high. What is the specific gravity of the liquid?

845. Water is placed in an open U-tube, and oil (s.g. < 1) is poured into each leg. The water column is 6 in. high; one oil column is 3 in. and the other 10 in. What is the specific gravity of the oil?

846. If the torque required to rotate the inner cylinder of problem 56 at a constant speed of 4 rpm is 2 ft-lb, calculate the approximate viscosity of the oil.

847. A Stormer-type viscometer consists of two cylinders, one of 3.0-in. outside diameter, the other of 3.1-in. inside diameter; both are 10 in. high. A 1-lb weight falls 5 ft in 10 sec, its supporting wire unwinding from a spool of 2-in. diameter on the main shaft of the viscometer. If the space between the cylinders is filled with oil to a depth of 8 in. and the space between the ends of the cylinders is 0.05 in., calculate the viscosity of the oil.

848. A steel sphere (s.g. = 7.8) of 0.5-in. diameter falls at a constant velocity of 0.2 fps through an oil (s.g. = 0.90). Calculate the viscosity of the oil, assuming that the fall occurs in (*a*) a large tank, (*b*) through a tube of 3-in. diameter.

849. What constant speed will be attained by a lead (s.g. = 11.4) sphere of 0.5-in. diameter falling freely through an oil of kinematic viscosity 0.12 sqft/sec and s.g. 0.95, if the fall occurs (*a*) in a large tank, (*b*) in a tube of 2-in. diameter.

850. Assuming a critical Reynolds number of 0.1, calculate the approximate diameter of the largest air bubble which will obey Stokes' law while rising through a large tank of oil of viscosity 0.004 lb-sec/sqft and s.g. 0.90.

851. Glass spheres of 0.1-in. diameter fall at constant velocities of 0.1 and 0.05 fps through two different oils (of the same specific gravity) in very large tanks. If the viscosity of the first oil is 0.002 lb-sec/sqft, what is the viscosity of the second?

852. A Saybolt Universal viscometer has tube diameter and length of 0.0693 in. and 0.482 in., respectively. The internal diameter of the cylindrical reservoir is 1.17 in., and the height from the tube outlet to rim of reservoir is 4.92 in. Assuming that the loss of head may be taken as the average of the total heads on the tube outlet at the beginning and end of the run, derive an approximate relationship between ν (sqft/sec) and t (Saybolt seconds), and compare with the exact equation relating these quantities. Volume collected is 60 cc.

853. If in 150 sec a Saybolt viscometer discharges a standard volume of one oil, how long will it take to discharge the same volume of a second oil which is 10 percent denser and has a 10 percent smaller absolute viscosity than the first?

854. This viscometer is to be used for liquids of kinematic viscosity between 0.00002 and 0.0005 sqft/sec. Calculate the minimum tube length which will hold all Reynolds numbers below 1500. Consider tube friction only.

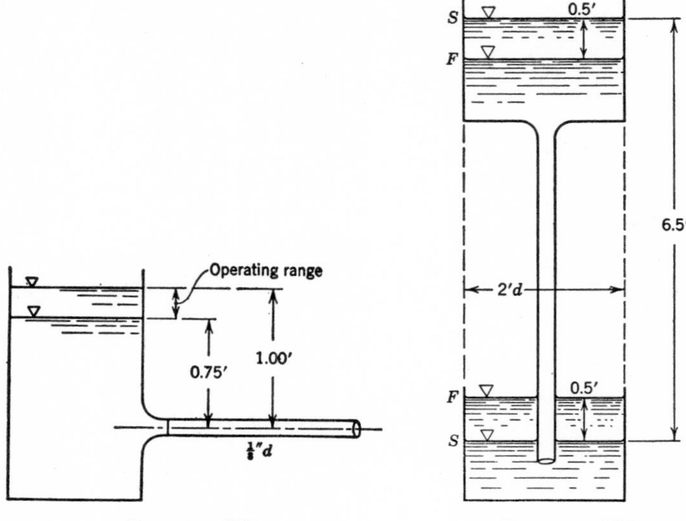

PROBLEM 854 PROBLEM 855

855. In this viscosity test the time from start (S) to finish (F) is 200 sec. The oil has s.g. 0.87, the pipe is 7 ft long and of 0.5-in. diameter, K_L for the entrance is 1.2, and for the exit 1.0. Calculate the approximate viscosity of the oil.

856. The disk of Fig. 178*a* and a pitot tube are placed in an air stream aligned properly with the flow, and connected to a U-tube containing water. If the difference of water elevation in the legs of the manometer is 4 in., calculate the air velocity, assuming a specific weight of 0.0763 lb/cuft.

857. A pitot-static tube installed in an air duct shows a differential manometer reading of 2 in. of water. If the pressure and temperature in the duct are 15.0 psia and 80°F, respectively, what velocity is indicated? Neglect compressibility effects.

858. A differential manometer containing water is attached to the pitot-static tube on an airplane flying close to the ground through the ICAO Standard Atmosphere (Appendix IV) at 150 mph. What manometer reading can be expected? What velocity would be indicated if this manometer reading occurred at altitude 20,000 ft?

859. A pitot-static tube is placed at the center of a 6-in. smooth pipe in which there is an established flow of carbon tetrachloride at 68°F. The attached differential manometer containing mercury and carbon tetrachloride shows a difference of 3 in. What flowrate exists in the line?

860. In a laminar flow in a 12-in. pipe two pitot tubes are installed, one on the centerline, the other 3 in. from the centerline. If the specific weight of the liquid flowing is 50 lb/cuft and the flowrate is 10 cfs, calculate the reading of a differential manometer connected to the two tubes, if there is mercury in the bottom of the manometer.

861. The friction factor of this pipe is 0.025. Calculate the manometer reading when the mean velocity is 10 fps.

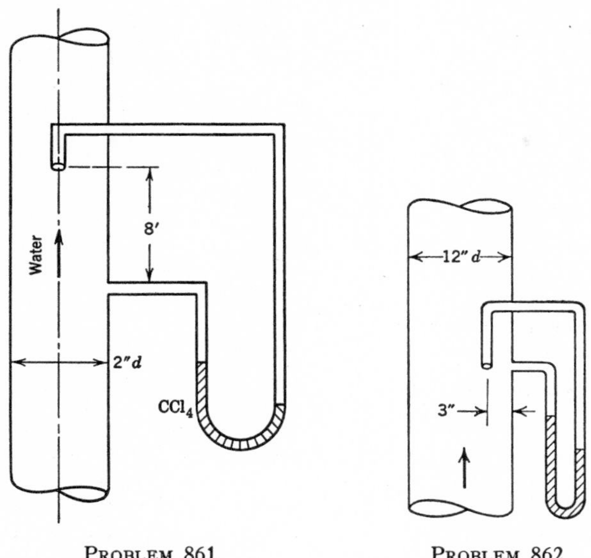

PROBLEM 861 PROBLEM 862

862. If the mean velocity is 8.02 fps and the fluid flowing has specific gravity and kinematic viscosity of 0.80 and 0.00401 sqft/sec, respectively, calculate the manometer reading.

863. A liquid flows from left to right in this 6-in. diameter clean cast iron pipe at $\mathbf{R} > 10^6$. Calculate the head lost per 100 ft of pipe.

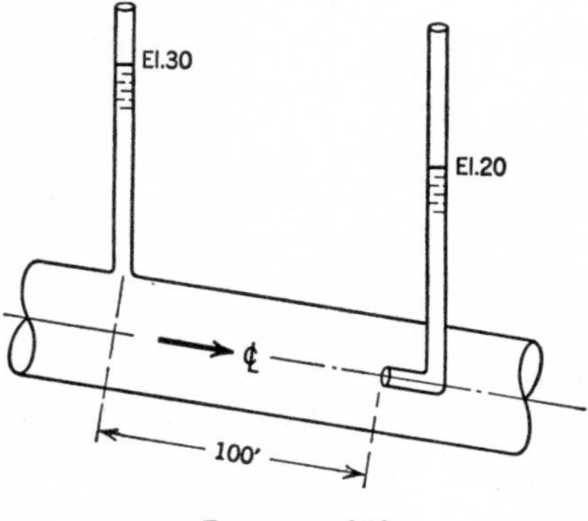

PROBLEM 863

864. A pitometer ($C_I = 0.87$) is placed at a point in a water line. If the attached differential manometer containing mercury and water shows a reading of 5 in., what is the velocity at the point?

865. The flowrate is 10 cfs of water, the pitometer coefficient 0.87, and the friction factor of the pipe 0.020. Calculate the manometer readings.

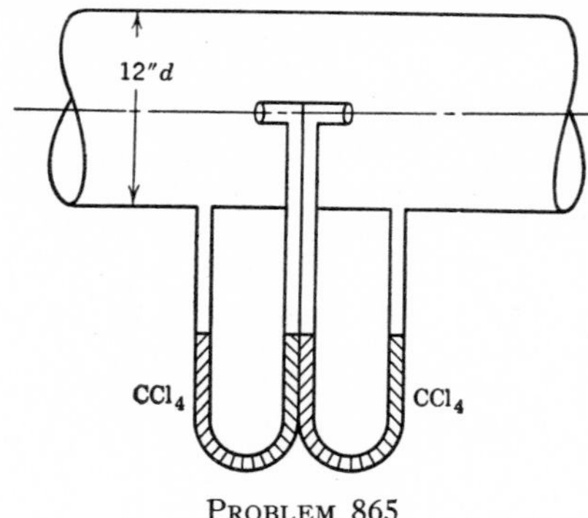

PROBLEM 865

866. Calculate the velocity at a point in an air duct if the stagnation temperature is 150°F and the separate stagnation and static pressure manometers show readings of 8 in. and 15 in. of mercury vacuum, respectively. The barometric pressure is 28 in. of mercury.

867. Carbon dioxide flows in a pipe. At a point in the flow the stagnation temperature is 100°F, and the static pressure 14.0 psia. The differential manometer attached to the pitot-static tube shows a reading of 30 in. of mercury. Calculate the velocity at this point.

868. A pitot-static tube and temperature probe are installed in a duct where nitrogen is flowing. The stagnation pressure and temperature are 15 in. of mercury gage and 200°F, respectively. The static pressure is 3 in. of mercury vacuum. If the barometric pressure is 30 in. of mercury, what velocity is indicated?

869. Calculate the pressure and temperature on the nose of a projectile moving near the ground at 2000 fps through the ICAO Standard Atmosphere. (Appendix IV.)

870. A Stanton tube of height h of 0.015 in. is installed in the wall of a 12-in. smooth pipe in which a fluid is flowing with a mean velocity of 10 fps. The fluid has viscosity and density of 0.002 lb-sec/sqft and 2 slugs/cuft, respectively. What pressure difference is to be expected? If this pressure difference is interpreted (as for a pitot-static tube) as $\rho v^2/2$, how far from the wall will this velocity be located?

871. A laminar flow occurs in a 12-in. smooth pipe at a Reynolds number of 1000. In the pipe wall a Stanton tube having h = 0.04 in. is installed. Calculate the ratio between the expected pressure difference and the wall shear. If this pressure difference is presumed to equal $\rho v^2/2$, how far from the wall will this velocity be located?

872. A Preston tube of 0.5-in. outside diameter is attached to the hull of a ship to measure the local shear. When the ship moves through fresh water (68°F) the pressure difference is found to be 75 psf. Calculate the local shear.

873. The gasometer of Fig. 198 is used to measure the flow of hydrogen at 65°F. The internal diameter of the cover is 5 ft, and it rises 1.87 ft in 34.6 sec. The manometer reads 16 in. of water, and the barometric pressure is 14.66 psia. Calculate the weight flowrate. If the counterweight weighs 600 lb, what is the weight of the cover?

874. A 12-in. by 6-in. Venturi meter is installed in a horizontal water line. The pressure gages read 30 and 20 psi. Calculate the flowrate if the water temperature is 68°F. Calculate the head lost between the base and throat of the meter. Calculate the total head lost by the meter if the diffuser tube has cone angle 7°. Calculate the flowrate if the pipe is vertical and the throat of the meter 2 ft below the base.

875. Linseed oil flows through a horizontal 6-in. by 3-in. Venturi meter. What is the difference in pressure head between the base and throat of the meter when 120 gpm flow at (*a*) 80°F, (*b*) 120°F? What is the head loss for each?

876. The maximum flowrate in a 9-in. water line is expected to be 5 cfs. To the Venturi meter is attached a mercury-under-water manometer 36-in. long. Calculate the minimum throat diameter which should be specified.

877. If the head lost between base and throat of a 3-in. by 6-in. Venturi meter is neglected, what coefficient of velocity would be required to allow for the change in velocity distribution between these points if α_1 and α_2 are 1.06 and 1.00, respectively (see Fig. 96)?

878. A pitot tube is installed at the center of the base of a 4-in. by 2-in. Venturi meter through which 1.00 cfs of water is flowing. The pitot tube is connected to one side of a differential manometer (containing mercury and water); the other side of this manometer is connected to the throat piezometer ring. Calculate the manometer reading if V/v_c at sections 1 and 2 are 0.82 and 1.00, respectively.

879. Air (γ = 0.08 lb/cuft) flows through a 4-in. by 2-in. frictionless Venturi meter. The pressures in the 4-in. and 2-in. sections are 0.25 psi and 0.300 in. of mercury vacuum, respectively. Calculate the flowrate, neglecting compressibility of the air.

880. Carbon dioxide flows through a 6-in. by 3-in. Venturi meter. Gages at the base and throat read 20 psi and 14 psi and temperature in the fluid at the base of the meter is 80°F. Calculate the weight flowrate, assuming standard barometer and $C_v = 0.99$.

881. An A.S.M.E. flow nozzle of 3-in. diameter is installed in a 6-in. water (68°F) line. The attached manometer contains mercury and water and registers a difference of 15 in. Calculate the flowrate through the nozzle. Calculate the head lost by the nozzle installation.

882. A 3-in. nozzle is installed at the end of a 6-in. air duct in which the specific weight is 0.0763 lb/cuft. A differential manometer connected to a piezometer opening 6 in. upstream from the base of the nozzle and to a pitot tube in the jet shows a reading of 0.25 in. of water. Calculate the flowrate, assuming uniform velocity distribution in the jet and C_v of 0.97. Assume the air incompressible.

883. If air flows through the pipe and nozzle of problem 881 and if open mercury manometers at points 1 and 2 show positive gage pressures of 30 in. and 20 in., and if the temperature of the air at point 1 is 60°F, calculate the weight flowrate, assuming standard barometric pressure.

884. A 1-in. nozzle has C_v of 0.98 and is attached to a 3-in. hose. What flowrate (water) will occur through the nozzle when the pressure in the hose is 60 psi? What is the velocity of the jet at the nozzle tip? How much head is lost through the nozzle? To what maximum height will this jet rise (neglect air friction)?

885. A sharp-edged orifice with conventional pressure connections is to be installed in a 12-in. water line. For a flowrate of 10 cfs the maximum allowable head loss is 25 ft. What is the smallest orifice that may be used? Since calculations are approximate, assume $Cv = 1$.

886. A 4-in. orifice at the end of a 6-in. line discharges 5.30 cfs of water. A pressure gage upstream from the orifice reads 58.0 psi and a gage connected to a pitot tube in the vena contracta reads 60.0 psi. Calculate C_c and C_v for this orifice assuming $\alpha_2 = 1.00$.

887. A 6-in. flow nozzle is installed in a 12-in. water line. An orifice of what diameter will produce the same head loss as the nozzle? Assume C_v the same for nozzle and orifice.

888. If the coefficient of an orifice of 3-in. diameter installed in a 6-in. line is approximately 0.65 for the conventional piezometer connections of Fig. 201, what approximate coefficient can be expected if the downstream connection is made at a point where the expanding jet has a 4-in. diameter?

889. Predict the location of the water surface in the middle piezometer tube relative to one of the other water surfaces, C_v is 0.97.

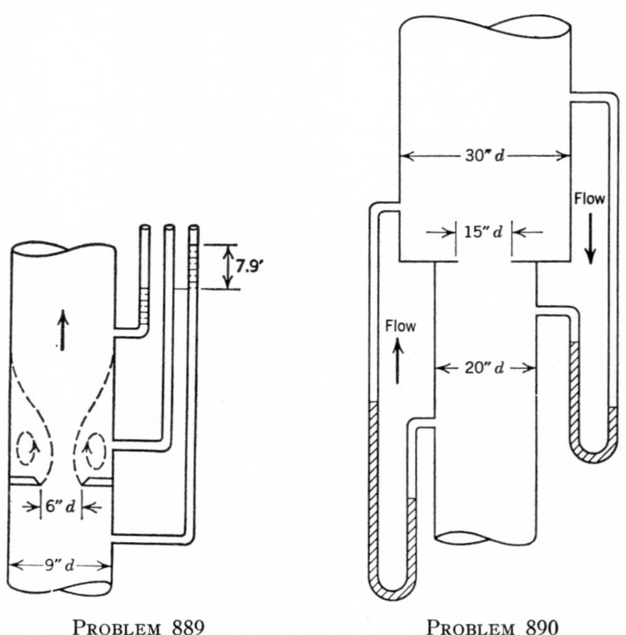

PROBLEM 889 PROBLEM 890

890. Find the ratio of the manometer readings for upward and downward flow of the same flowrate. The manometer liquids are the same and each downstream pressure connection is opposite the vena contracta.

891. A 4-in. sharp-edged orifice at the end of a 6-in. water line has C_v of 0.97. Calculate the flowrate when the pressure in the line is 40 psi.

892. A conical nozzle of 2-in. tip diameter and having C_c of 0.85 and C_v of 0.97 is attached to the end of a 4-in. water line. A manometer (containing carbon tetrachloride and water) is connected to a pitot tube in the vena contracta and to a piezometer ring at the base of the nozzle. Taking $\alpha_2 = 1.0$, calculate the flowrate and the pressure at the base of the nozzle. The manometer reads 24 in.

893. A 2-in. conical nozzle having C_v of 0.98 and C_c of 0.80 is attached to a 4-in. pipe line and delivers water to an impulse turbine. The pipe line is 1000 ft long and leaves a reservoir of surface elevation 450 at elevation 420. The nozzle is at elevation 25. Assuming a square-edged pipe entrance and a friction factor of 0.02, calculate (*a*) the flowrate through the pipe and nozzle, (*b*) the horsepower of the nozzle stream, (*c*) the horsepower lost in line and nozzle.

894. This *inlet orifice* is used to meter the flow of air into the pipe. Assuming the downstream pressure connection to be in the plane of the vena contracta, predict the flowrate. Consider the air incompressible.

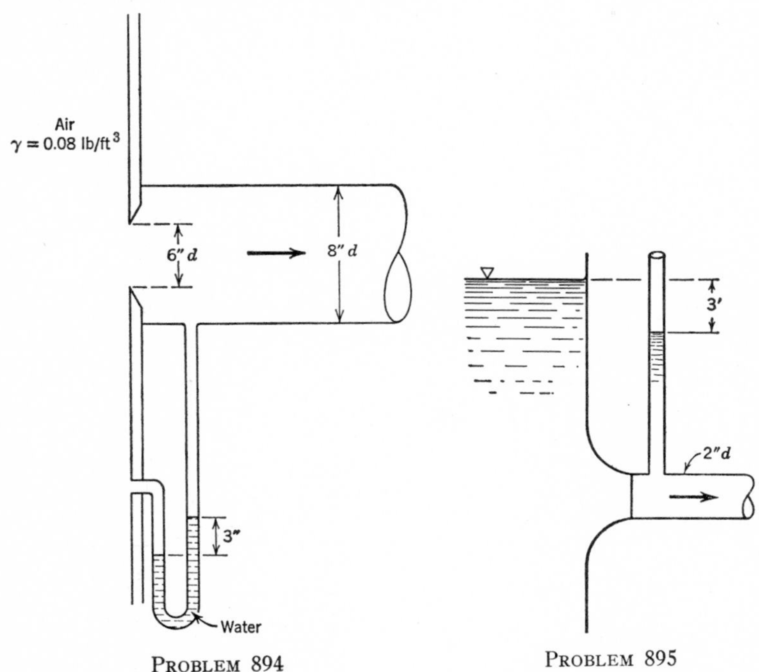

PROBLEM 894 PROBLEM 895

895. Calculate the flowrate if C_v for this entrance nozzle is 0.96.

896. Water (68°F) discharges into the atmosphere from a 1.5-in. sharp-edged orifice under a 5-ft head. Calculate the flowrate. Repeat the calculation for linseed oil at 68°F.

897. Water flows from one tank to an adjacent one through a 3-in. sharp-edged orifice. The head of water on one side of the orifice is 6 ft and on the other 2 ft. Taking C_c as 0.62 and C_v as 0.95, calculate the flowrate.

898. A jet discharges vertically upward from a 2-in. sharp-edged orifice located in a horizontal plane. If the head (on the vena contracta) is 20 ft and the jet rises to a height of 19 ft above the vena contracta, what is the flowrate if the diameter of the vena contracta is 1.6 in.? Air friction on the jet may be neglected.

899. A conventional sharp-edged orifice of 2-in. diameter discharges into the atmosphere from a large tank. At a point in the jet the height of the energy line is measured by pitot tube and found to be 0.30 ft below the free surface level in the tank. Calculate the flowrate and the head on the orifice. Air friction on the jet may be neglected.

900. A 3-in. sharp-edged orifice discharges vertically upward. At a point 10 ft above the vena contracta, the diameter of the jet is 3 in. Under what head is the orifice discharging?

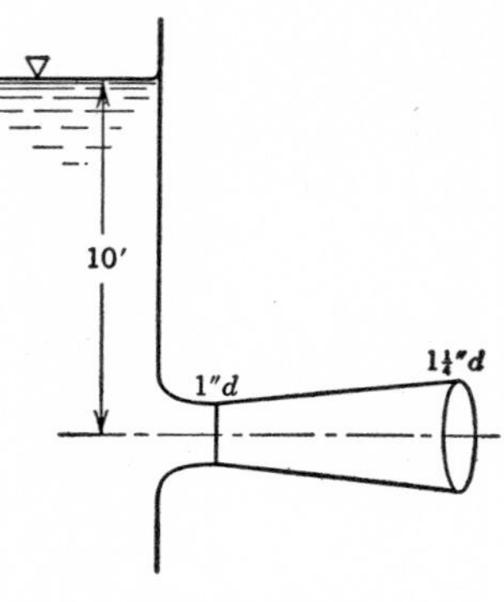

PROBLEM 901

901. The flowrate is 0.19 cfs, and the head lost in the diffuser is 0.48 ft. Predict the flowrate when the diffuser is removed.

902. Under a 4.42-ft head, 0.056 cfs of water discharges from a 1-in. sharp-edged orifice in a vertical plane; 3.30 ft outward horizontally from the vena contracta the jet has dropped 0.65 ft below the centerline of the orifice. Calculate C, C_v, and C_c.

903. A 2-in. sharp-edged orifice discharges with a 20-ft head on its vena contracta. To what height will the jet rise (above the vena contracta) if the jet discharges (*a*) vertically upward, (*b*) upward at an angle of 45°? Neglect air friction on the jet.

904. Water discharges through a 1-in. diameter sharp-edged orifice under a 3-ft head. At what head will the same flowrate occur through a horizontal pipe of 1-in. diameter and 12 in. long (friction factor 0.020) having a square-edged entrance?

905. A short tube of 1-in. diameter and 1.5-in. length may flow full or not full at its exit. Calculate the approximate ratio between the respective flowrates for the same head.

906. Predict the discharge coefficients of the standard short tube and re-entrant short tube from the loss coefficients of Fig. 134. Assume that the tubes are 4 diameters in length with friction factors of 0.020.

907. Predict the coefficient of velocity for this short tube with restricted entrance if the friction factor for the tube is 0.020.

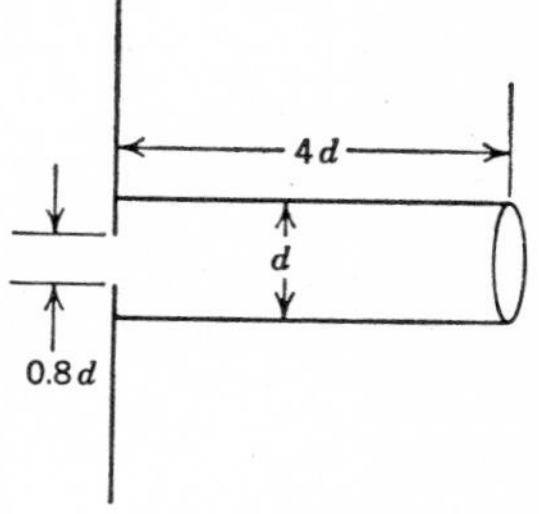

PROBLEM 907

908. A sluice gate 4 ft wide is open 3 ft and discharges onto a horizontal surface. If the coefficient of contraction is 0.80 and the coefficient of velocity 0.90, calculate the flowrate if the upstream water surface is 4 ft above the top of the gate opening.

909. This sluice gate extends the full width of a rectangular channel 5 ft wide. Assuming C_v 0.96 and C_c 0.75, estimate the flowrate, neglecting the dynamics of the roller.

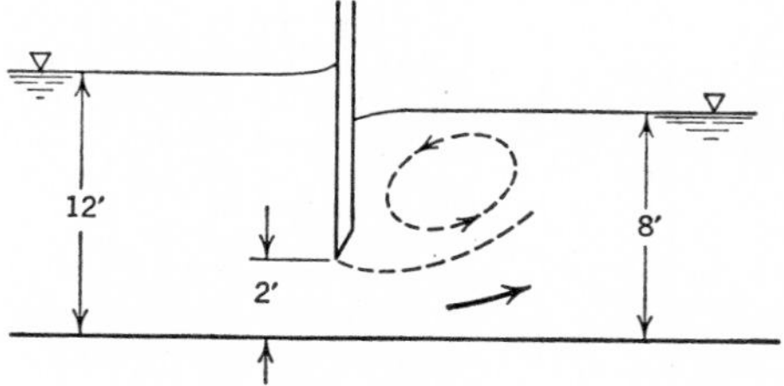

PROBLEM 909

910. When this sluice gate is open 2 ft its vena contracta is in the plane of the brink of the outfall. If its coefficient of contraction is 0.75, channel and gate widths 8 ft, and the flowrate 175 cfs, what is its coefficient of velocity?

911. An elbow meter of 4-in. diameter has a coefficient of 0.815. What flowrate of water occurs through this meter when the attached manometer (containing mercury and water) shows a difference of 10 in.?

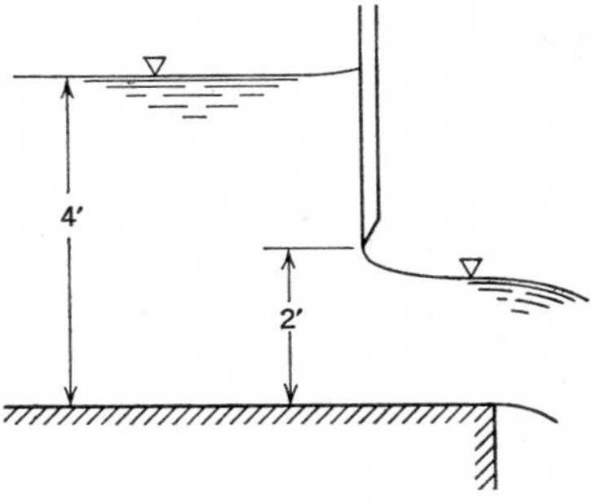

PROBLEM 910

912. A pitometer ($C_I = 0.87$) is installed in a 6-in. water line. A manometer containing carbon tetrachloride and water, connected to the pitometer, shows the following readings when the tip of the instrument is placed at the points specified for the numerical average method for calculating flowrate.

Pitometer location	1	2	3	C	4	5	6
Manometer reading, in.	1.20	2.04	2.83	3.03	2.89	2.10	1.26

Calculate the flowrate, V/v_c, and the distance from centerline to station 2.

913. A pitot-static tube is placed at various points along a diameter of a 20-in. pipe in which water is flowing. The pressure difference is measured on a manometer containing mercury and water. If the following manometer readings are taken, calculate the flowrate and V/v_c by the graphical method.

Distance from pitot-static tube location to pipe centerline, in.	0	2	4	6	8	9	9.5
Manometer reading, in.	6	5.95	5.64	5.07	4.12	3.25	2.45

914. The flow in a brook is measured by the salt-dilution method; 0.20 gpm of salt solution having a concentration of 20 lb of salt per gallon is introduced and mixes with the flow. A sample extracted below the mixing point shows a concentration of 0.00008 lb/gal. Calculate the flowrate in the brook.

915. The salt-velocity method is to be used in a 24-in. pipe line, and electrodes are installed 100 ft apart. The time between deflection of the ammeter needles is 12.0 sec. Calculate the flowrate in the line.

916. The head on a sharp-crested rectangular weir 4 ft long and 3 ft high is 4 in. Calculate flowrate and velocity of approach. Repeat the calculation for a weir of 1-ft height.

917. A certain flowrate passes over a sharp-crested rectangular weir 2 ft high under a head of 1 ft. Calculate the head on a similar weir 1 ft high for the same flowrate.

918. What depth of water must exist behind a rectangular sharp-crested weir 5 ft long and 4 ft high, when a flow of 10 cfs passes over it? What is the velocity of approach?

919. A rectangular channel 18 ft wide carries a flowrate of 50 cfs. A rectangular sharp-crested weir is to be installed near the end of the channel to create a depth of 3 ft upstream from the weir. Calculate the necessary weir height.

920. Across one end of a rectangular tank 3 ft wide is a sharp-crested weir 4 ft high. In the bottom of the tank is a sharp-edged orifice of 3-in. diameter. If 2.0 cfs flow into the tank, what depth of water will be attained?

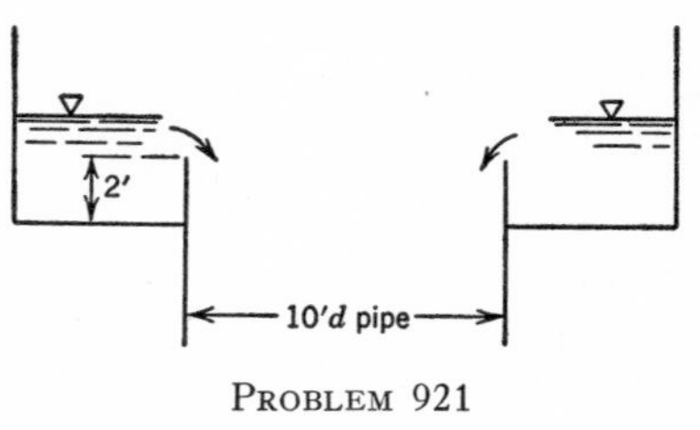

PROBLEM 921

921. Treating the upper edge of the pipe as a sharp weir crest, estimate the flowrate when the water depth in the basin is 2.5 ft.

922. Derive the theoretical flow equation for the triangular weir.

923. Calculate the flowrate of water (68°F) over a smooth sharp-crested triangular weir of 90° notch angle when operating under a head of 6 in. Repeat the calculation for linseed oil at the same temperature.

924. A triangular weir of 90° notch angle is to be used for measuring water flowrates up to 1.5 cfs. What is the minimum depth of notch which will pass this flowrate?

925. A 90° triangular weir discharges water at a head of 0.5 ft into a tank with a 2.5-in. sharp-edged orifice in the bottom. Predict the depth of water in the tank.

926. Calculate the approximate flowrate to be expected through this sharp-edged opening. Assume a weir coefficient of 0.62.

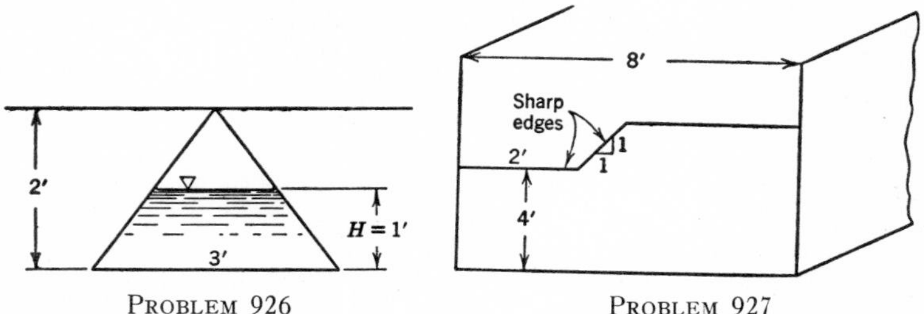

PROBLEM 926 PROBLEM 927

927. The depth of water behind the weir plate is 4.8 ft. Predict the flowrate over the weir.

928. Estimate the flowrate over this sharp-crested weir, assuming a coefficient of 0.62.

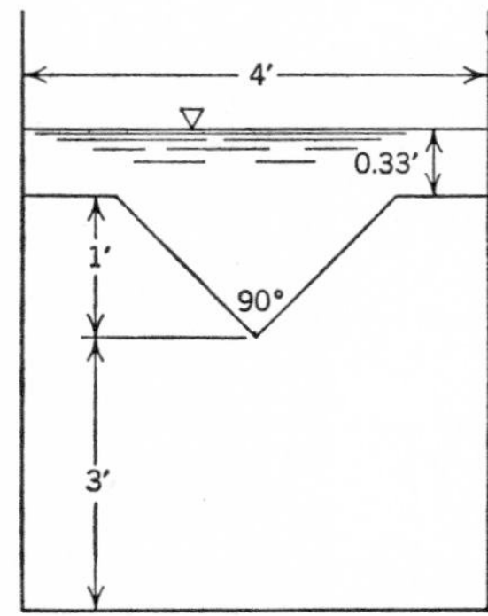

PROBLEM 928

929. A rectangular channel 20 ft wide carries 100 cfs at a depth of 3 ft. What height of broad-crested rectangular weir must be installed to double the depth? Assume a weir coefficient of 0.56.

930. A broad-crested weir 3 ft high has a flat crest and a coefficient of 0.55. If this weir is 20 ft long and the head on it is 1.5 ft, what flowrate will occur over it? What maximum flowrate could be expected if the flow were frictionless?

931. A frictionless broad-crested weir 4 ft high is built across a channel 8 ft wide. If the energy line is 3 ft above the weir crest, calculate the head, flowrate, and weir coefficient.

932. In order to justify the form of the relation between C and P/H of the Rehbock formula, calculate C for frictionless broad-crested weirs having values of P/H of 1, 5, and 9. Compare these with values obtained from the Rehbock formula, disregarding $1/305H$.

933. What flowrate will occur over a spillway of 500-ft length when the head thereon is 4 ft, if the coefficient (referenced to the static head) of the spillway is 0.72?

934. A spillway 1000 ft long is found by model experiments to have a coefficient (referenced to the static head) of 0.68. It has a crest elevation of 100.00. When a flood flow of 50,000 cfs passes over the spillway, what is the elevation of the water surface in the reservoir just upstream from the spillway?

935. A spillway structure 20 ft high is designed by the methods of Fig. 217 for a head of 10 ft. Calculate the coefficients and flowrates for heads of 4, 10, and 12 ft. Calculate the corresponding coefficient for the design head when this is referenced to water surface rather than energy line.

936. Upstream and downstream from a sharp-crested rectangular weir 3 ft high the water depths are 4 ft and 3.5 ft, respectively. Calculate the approximate flowrate.

937. The drop in the water surface in passing a submerged sharp-crested rectangular weir 4 ft high is 0.25 ft. Calculate the approximate flowrate if the depth upstream from the weir is 5 ft.

938. The following data are collected in a current-meter measurement at the river cross section of Fig. 221, which is 60 ft wide at the water surface. Assume $V = 2.22 \times$ (rps), and calculate the flowrate in the river.

Station	0	1	2	3	4	5	6	7	8	9	10	11	12
Depth, ft	0.0	3.0	3.2	3.5	3.6	3.7	3.9	4.0	4.4	4.4	4.2	3.5	0.0
						Rpm of rotating element							
$0.2y$	...	40.0	53.5	58.6	63.0	66.7	61.5	56.3	54.0	52.6	50.0	45.0	...
$0.8y$	...	30.7	42.8	50.0	54.2	58.8	53.3	49.4	46.5	43.2	40.1	32.5	...

chapter 12
Elementary Hydrodynamics

Although the word *hydrodynamics* means (literally) *water-motion* to the layman, the early scientists and mathematicians used it in a more definite way to define the study of the flowfields of any ideal fluid by mathematical methods—and thus gave it the specific meaning used here. In this sense the flow of an ideal fluid about an airfoil would be considered a problem in hydrodynamics, whereas the established flow of a real fluid in a pipe would not. Although modern hydrodynamics now includes problems with viscous action, none of these can be included here and the word is used in the classical sense described above.

The aim of this chapter is to provide the beginner with a modest introduction to classical hydrodynamics so that he may see its possibilities and limitations, and thus widen his horizons. Modern hydrodynamics is such a vast and complex field that only a small portion of it may be presented here. For a mathematically rigorous approach to the subject, considerable facility with partial differential equations is needed; since it cannot be assumed that the beginner has acquired such experience and space does not permit its development here, a compromise has been effected in the writing of Appendix V, which summarizes the basic mathematical operations necessary to the chapter. To confine the treatment within realistic limits it is restricted to the incompressible ideal fluid, steady flow, and two-dimensional flowfields mostly in the horizontal plane.

The methods of higher mathematics are being used increasingly in the solution of engineering problems. For steady one-dimensional flow such methods are not needed, but in flowfield problems they become necessary and often provide a clear, elegant, and efficient mode of attack. Because of simplifying assumptions, however, these methods usually do not yield the final solutions to engineering problems; instead they provide a theoretical result from which the final solution may, with good judgment, be deduced. Such methods thus lead to insight and understanding rather than complete engineering answers. In hydrodynamics an ideal fluid is assumed as in Chapters 4 and 5, accompanied by the same limitations discussed there. Since then the reader has learned that real fluids flow as ideal ones if outside the zone of viscous influence, or may be considered ideal as an

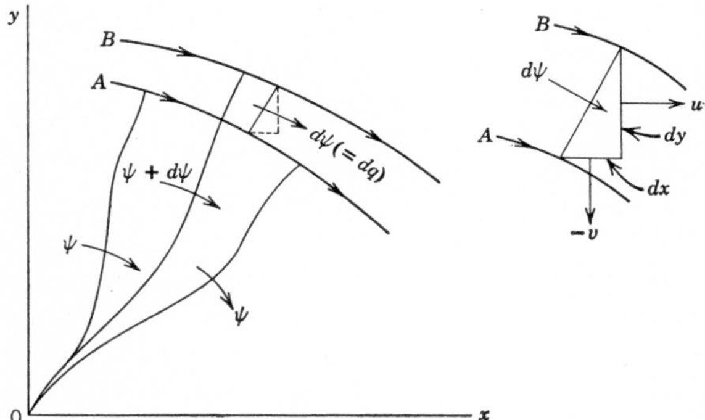

FIG. 222. Definition of the streamfunction.

approximation when viscous action is small. The flowfields about a submerged object or in a short smooth nozzle or pipe entrance are examples of this. Other problems in which viscous action predominates or triggers the separation of fluid from boundary obviously cannot be treated with the ideal fluid; however, hydrodynamical methods can be employed in the study of flowfields in which separation points are not determined by viscous action. For example, the prediction of the complete flowfield about an elliptical object in a free stream would be impossible with an ideal fluid because of the separation caused by viscous action in the boundary layer, but for a disk normal to the flow, where the separation at the edges is independent of viscous action, mathematical methods and the ideal fluid may be effectively used.

108. The Streamfunction. Definition of the streamfunction, a concept based on the continuity principle and the properties of the streamline, provides a mathematical means of plotting and interpreting flowfields. Consider the streamline A of Fig. 222. By its definition no flow crosses it and thus the flowrate ψ across all lines

OA [1] is the same. Accordingly, ψ is a constant of the streamline and, if ψ can be found as a function of x and y, the streamline may be plotted. Similarly, the flowrate between O and a closely adjacent streamline (B) will be $\psi + d\psi$, and the flowrate between the streamlines (i.e., in the streamtube) will be $d\psi$. As the flowrates into and out of the elemental triangle are equal from continuity considerations,

$$d\psi = -v\,dx + u\,dy$$

However, if $\psi = F(x, y)$, mathematics shows (see Appendix V) that

$$d\psi = \frac{\partial\psi}{\partial x}\,dx + \frac{\partial\psi}{\partial y}\,dy \tag{281}$$

and by comparison of these equations [2]

$$u = \frac{\partial\psi}{\partial y} \quad \text{and} \quad v = -\frac{\partial\psi}{\partial x} \tag{282}$$

Thus, if ψ is a known function of x and y, the components of velocity at any point may be obtained by taking the appropriate partial derivative of ψ. Conversely, if u and v are known as functions of x and y, ψ may be obtained by integration of equation 281, yielding

$$\psi = \int\left(\frac{\partial\psi}{\partial x}\right)dx + \int\left(\frac{\partial\psi}{\partial y}\right)dy + C$$

The equation of continuity,

$$\frac{\partial u}{\partial x} + \frac{\partial v}{\partial y} = 0 \tag{58}$$

may be easily (and usefully) expressed in terms of ψ by substituting the relations of equations 282; this leads to the identity

$$\frac{\partial}{\partial x}\left(\frac{\partial\psi}{\partial y}\right) = \frac{\partial}{\partial y}\left(\frac{\partial\psi}{\partial x}\right) \quad \text{or} \quad \frac{\partial^2\psi}{\partial x\,\partial y} = \frac{\partial^2\psi}{\partial y\,\partial x}$$

which expresses the fact that if $\psi = F(x, y)$ the derivatives taken in either order yield the same result.

[1] Point O need not be the origin of coordinates.

[2] Analogous relations for polar coordinates are

$$v_r = \frac{\partial\psi}{r\,\partial\theta} \quad \text{and} \quad v_t = -\frac{\partial\psi}{\partial r}$$

in which v_r is positive radially outward from the origin; v_t and θ are positive counterclockwise.

The equation for vorticity,

$$\xi = \frac{\partial v}{\partial x} - \frac{\partial u}{\partial y} \tag{62}$$

may also be expressed in terms of ψ by similar substitutions:

$$\xi = -\frac{\partial^2 \psi}{\partial x^2} - \frac{\partial^2 \psi}{\partial y^2}$$

However, for irrotational flows, $\xi = 0$, and the classic Laplace equation,

$$\frac{\partial^2 \psi}{\partial x^2} + \frac{\partial^2 \psi}{\partial y^2} = 0$$

results. This means that the streamfunctions of all irrotational flows must satisfy the Laplace equation and that such flows may be identified in this manner; conversely, flows which do not satisfy the Laplace equation in ψ are rotational ones. Since both rotational and irrotational flowfields are physically possible, the satisfaction of the Laplace equation is no criterion of the physical existence of a flowfield.

Illustrative Problem

A flowfield is described by the equation $\psi = y - x^2$. Sketch the streamlines $\psi = 0$, $\psi = 1$, and $\psi = 2$. Derive an expression for the velocity V at any point in the flowfield. Calculate the vorticity of this flow.

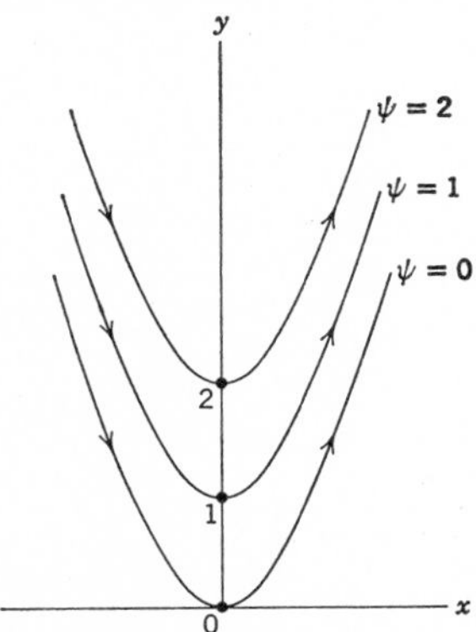

Solution. From the equation for ψ, the flowfield is a family of parabolas symmetrical about the y-axis with the streamline $\psi = 0$ passing through the origin of coordinates.

$$u = \frac{\partial}{\partial y}(y - x^2) = 1 - 0 = 1 \text{ fps} \tag{282}$$

$$v = -\frac{\partial}{\partial x}(y - x^2) = 0 + 2x = (2x) \text{ fps} \tag{282}$$

which allows the directional arrows to be placed on the streamlines as shown. The velocity V may be calculated from

$$V = \sqrt{u^2 + v^2} = \sqrt{1 + 4x^2}$$

and the vorticity from

$$\xi = \frac{\partial}{\partial x}(2x) - \frac{\partial}{\partial y}(1) = 2 \text{ sec}^{-1} \text{ (counterclockwise)} \tag{62}$$

Since $\xi \neq 0$, this flowfield is seen to be a rotational one.

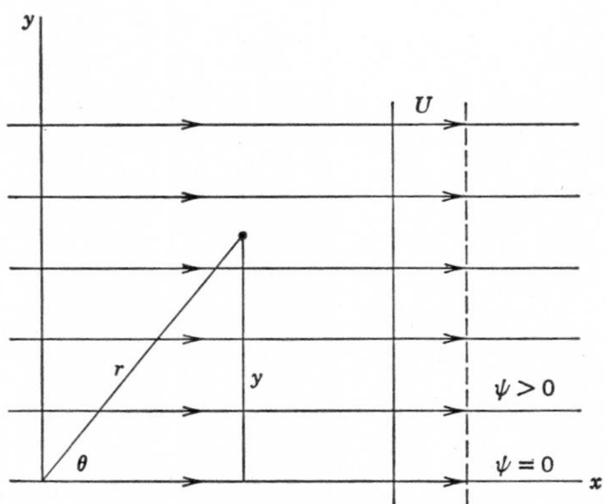

FIG. 223. Horizontal rectilinear flow.

109. Basic Flowfields. (*a*) *Rectilinear Flow.* In dealing with the flow about solid objects immersed in a stream, the approaching flow is frequently of practically infinite extent and possesses straight and parallel streamlines, and uniform velocity distribution. Such a horizontal flow of velocity U is shown in Fig. 223. Clearly, $u = U$ and $v = 0$ for this flowfield; its streamfunction may be obtained from

$$\psi = \int U\, dy + \int (0)\, dx = Uy + C$$

If the streamline coincident with the x-axis is designated by $\psi = 0$, the constant vanishes and the streamfunction of the flowfield becomes

$$\psi = Uy \tag{283}$$

or, expressed in polar coordinates (where $y = r \sin \theta$),

$$\psi = Ur \sin \theta \tag{284}$$

(*b*) *Source and Sink.* Imagine now a symmetrical flowfield consisting of radial streamlines directed outward from a common point (Fig. 224). Continuity considerations show that the velocities will diminish as the streamlines spread, and symmetry will require that all velocities be the same at the same radial distance from the origin. Across all circles of radius r will pass the same flowrate q, and thus the velocity at any point in the flowfield may be determined from $v_r = q/2\pi r$ and $v_t = 0$. Here a singularity occurs at the center of

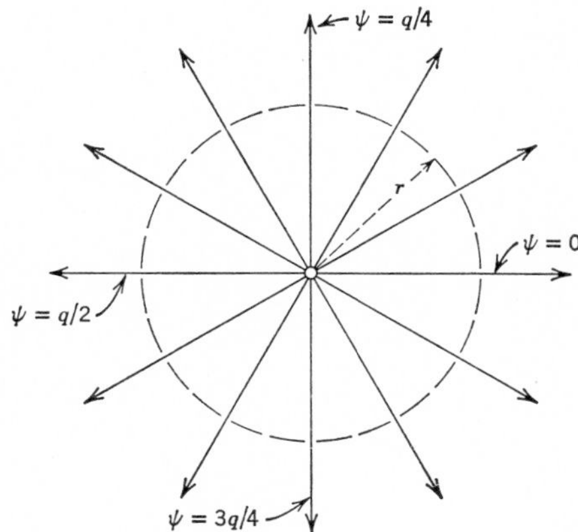

FIG. 224. Source flow.

the source, since, as $r \rightarrow 0$, $v_r \rightarrow \infty$; this singularity is of no importance in practical problems which are always concerned with the flowfield outside of the center of the source.

The streamfunction for a source flow can be easily found from

$$\psi = \int \left(\frac{q}{2\pi r}\right) r\, d\theta + \int (0)\, dr = \frac{q\theta}{2\pi} + C$$

and if the streamline $\psi = 0$ is chosen along the radial line $\theta = 0$ the constant vanishes and

$$\psi = \frac{q\theta}{2\pi} \tag{285}$$

The exact opposite of the *outward* flow from a source is the *inward* flow to a *sink*, whose streamfunction may be written (without separate proof) as

$$\psi = -\frac{q\theta}{2\pi} \tag{286}$$

(*c*) *Free Vortex.* Another useful basic flow is the free vortex described by concentric circular streamlines (Fig. 225) and velocity distribution such that the flowfield is irrotational; in this flowfield the radial component of velocity will be everywhere zero. To find the velocity distribution and streamfunction for this flowfield, select a small convenient differential element and calculate the circulation $d\Gamma$ around it, the circulation to be zero if the flow is to be irrotational. Proceed around the element in counterclockwise direction

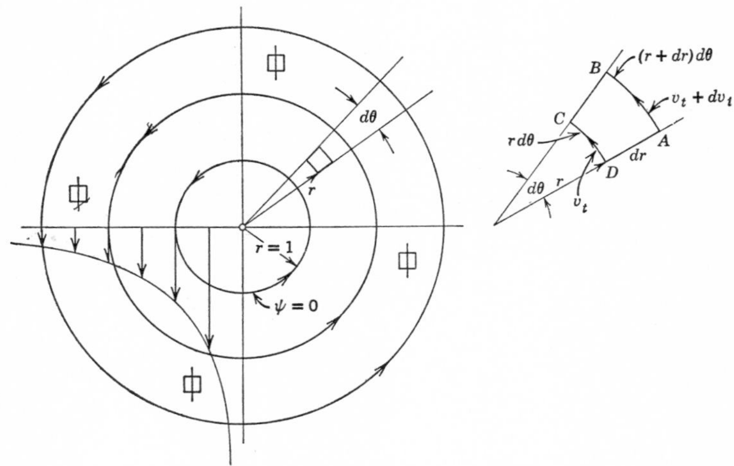

FIG. 225. Free (irrotational) vortex.

from A to B, B to C, C to D, and D to A, writing the components of $d\Gamma$ in that order:

$$d\Gamma = 0 = (v_t + dv_t)(r + dr)\,d\theta + 0 - v_t r\,d\theta + 0$$

whence (canceling $d\theta$ and neglecting $dv_t\,dr$)

$$v_t\,dr + r\,dv_t = d(rv_t) = 0$$

Thus the velocity distribution is characterized by $v_t r$ = constant, showing that with $r \to 0$, $v_t \to \infty$, a singularity is to be expected at the origin.

The circulation Γ along any closed curve coincident with any streamline may be calculated as

$$\Gamma = (2\pi r)v_t$$

in which $v_t r$ = constant. Thus the circulation Γ is a constant of the vortex, being the same along all streamlines. However, if the circulation is finite along a streamline at an infinitesimal distance from the origin, the vorticity within the infinitesimal area enclosed by this streamline cannot be zero. Thus the free vortex is a flowfield which is everywhere irrotational except at the singularity in the core of the vortex. In practical problems, however, this paradox causes no difficulties since such problems always involve the irrotational flowfield outside the vortex core.

The irrotationality of a free vortex may be easily visualized by imagining such a vortex with a free surface on which small particles of solid material are floating. As these particles move with the fluid they will be found not to rotate about their own axes, a directional line on a particle remaining parallel to itself during the motion (see Fig. 225). This may be convincingly demonstrated in the laboratory with the "drain hole vortex" (Art. 28) which closely approximates the free vortex in spite of viscous action and radial component of velocity.

The streamfunction for the vortex may be found from

$$v_t = \frac{\Gamma}{2\pi r} = -\frac{\partial \psi}{\partial r} \quad \text{and} \quad v_r = 0$$

and

$$\psi = \int \left(-\frac{\Gamma}{2\pi r}\right) dr + \int (0) r\, d\theta + C$$

which gives $\psi = -(\Gamma/2\pi) \ln r + C$ and, with ψ taken as 0 for the streamline passing through $r = 1$, the constant of integration vanishes and

$$\psi = -\frac{\Gamma}{2\pi} \ln r \tag{287}$$

in which Γ is known as the *vortex strength.* From the sense of the foregoing derivation it can be seen without proof that the streamfunction of a *clockwise* vortex of strength Γ will be given by

$$\psi = +\frac{\Gamma}{2\pi} \ln r \tag{288}$$

(*d*) *Forced Vortex.* The so-called forced vortex is defined by the velocity distribution $v_t = \omega r$ and is thus another name for the problem (of rotation of a body of fluid about an axis) discussed in Art. 14. Although the forced vortex has fewer applications in engineering than the free vortex, it is of considerable interest to study it here for further understanding of vorticity.

The streamfunction for this clockwise vortex (Fig. 226) may be found from $v_t = -\omega r$, $v_r = 0$, and

$$\psi = \int (\omega r)\, dr + \int (0) r\, d\theta + C$$

which yields $\psi = \omega r^2/2 + C$; if ψ is then taken to be zero where $r = 0$, the constant disappears and $\psi = \omega r^2/2$. This could also have been obtained directly from the original definition of the stream-

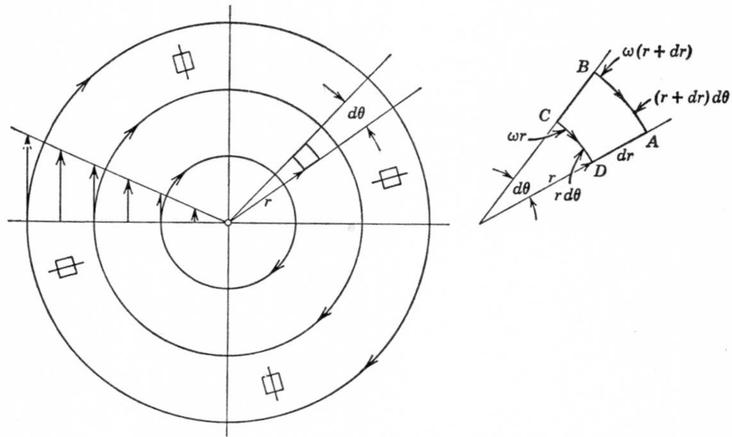

FIG. 226. Forced (rotational) vortex.

function (Fig. 222) as a flowrate between origin and (circular) streamline by writing

$$\psi = q = (\text{mean velocity})(\text{area}) = \left(\frac{\omega r}{2}\right) r = \frac{\omega r^2}{2} \tag{289}$$

Consider now the differential circulation around any differential element of fluid in the forced vortex. Starting from A and proceeding in a counterclockwise direction,

$$d\Gamma = -\omega(r + dr)(r + dr)\, d\theta + 0 + \omega r(r\, d\theta) + 0 = -2\omega r\, dr\, d\theta$$

However, the vorticity ξ is defined by $d\Gamma/dA$, in which $dA = r\, d\theta\, dr$, and thus $\xi = -2\omega$, which shows the vorticity to be directly related to the angular velocity of the fluid mass. From this it may be concluded that fluid elements in the forced vortex rotate about their own axes, whereas in the free (irrotational) vortex they do not. This may be generalized to conclude that vorticity is a measure of the rotational aspects of the fluid particles as they move through the flowfield.

110. Combining Flows by Superposition. By combination of rectilinear, vortex, source, and sink flows, flowfields of considerable engineering importance may be developed. Analytical and graphical procedures will now be developed for accomplishing this (see Fig. 227). At a point P in a flowfield ψ_A there will be some velocity V_A tangent to the streamline passing through this point; *at the same point* in a

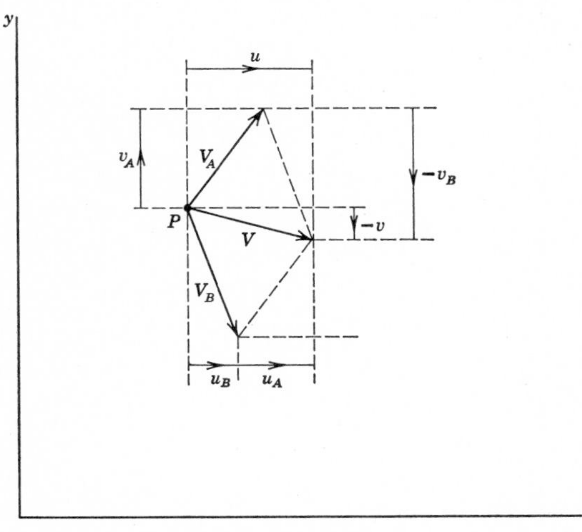

FIG. 227

flowfield ψ_B there will be some velocity V_B. When these flows are superposed the resulting velocity V at the point must be the vector sum (resultant) of the two velocities V_A and V_B. At other points the same vector addition of velocities occurs and a third flowfield ψ results. Although the principle of superposing flows is seen to be a basically simple process, its application on a point-by-point basis over a flowfield would be excessively tedious and time-consuming. A simpler procedure may be deduced from Fig. 227, where it is easily seen that

$$u = u_A + u_B$$

$$-v = -v_A - v_B$$

and, using the relations of equations 282,

$$\frac{\partial \psi}{\partial y} = \frac{\partial \psi_A}{\partial y} + \frac{\partial \psi_B}{\partial y} = \frac{\partial}{\partial y}(\psi_A + \psi_B)$$

$$\frac{\partial \psi}{\partial x} = \frac{\partial \psi_A}{\partial x} + \frac{\partial \psi_B}{\partial x} = \frac{\partial}{\partial x}(\psi_A + \psi_B)$$

it is observed that

$$\psi = \psi_A + \psi_B$$

From this it may be concluded that to obtain a flowfield by superposition of other flowfields merely involves algebraic addition of the streamfunctions of the latter; the sum of these streamfunctions will be the streamfunction of the resultant (combined) flowfield.

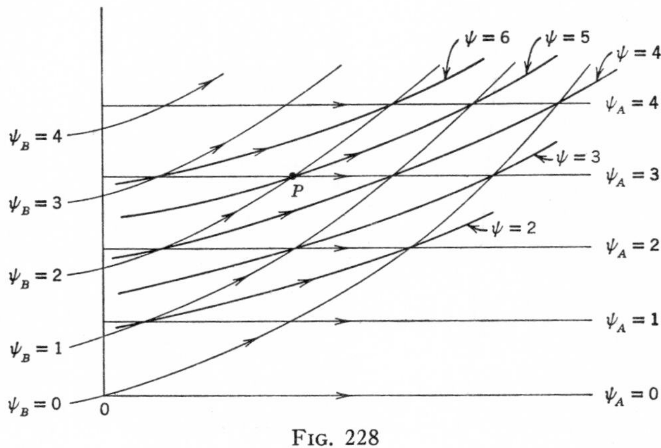

FIG. 228

The graphical approach to such problems is to return to the definition of the streamfunction and consider the flowrates between streamline and origin for both the separate and combined flowfields (Fig. 228). Across any line drawn between O and P will pass a flowrate of 3 cfs/ft in flowfield A, and across the same line in flowfield B will pass a flowrate of 2 cfs/ft; therefore, for the combination of flowfields, the flowrate will be 5 cfs/ft across any line OP. Thus for the combined flowfield the streamline passing through point P has $\psi = 5$; by similar reasoning this streamline must also pass through all other points at which $\psi_A + \psi_B = 5$. Because there are many such points at appropriate streamline intersections, this and other streamlines may be sketched to complete the combined flowfield. After ψ is obtained, the velocities at any point in the flowfield may be obtained from $\partial\psi/\partial x$ and $\partial\psi/\partial y$ and, from these, pressures may be found by application of the Bernoulli equation. When the streamfunctions of a flowfield are unknown (as a function of x and y) the graphical technique is the only one which may be used to combine these flowfields; however, with ψ a known function of x and y, either the analytical or graphical approach may be used in combining the flowfields.

ILLUSTRATIVE PROBLEM

Sketch the streamline $\psi = 2$ for the combination of the flowfields $\psi_A = y - x^2$ and $\psi_B = x$, and derive an expression for the velocity V anywhere on this streamline.

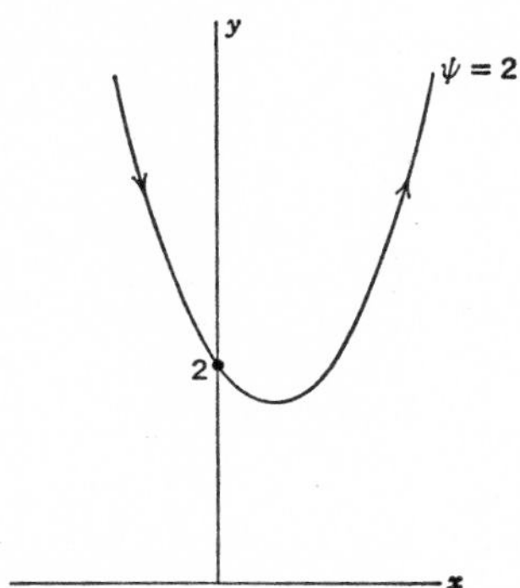

Solution. Refer to Art. 108 for a description of flowfield A. Flowfield B features straight vertical streamlines with $u_B = 0$ and $v_B = -(\partial/\partial x)x = -1$ fps. The superposition of these two flows will yield the streamfunction $\psi = y - x^2 + x$ and the equation of the required streamline will be $y - x^2 + x = 2$, allowing the streamline to be plotted or sketched. The velocity components and velocity V (anywhere in the combined flowfield) are

$$u = \frac{\partial \psi}{\partial y} = \frac{\partial}{\partial y}(y - x^2 + x) = 1 - 0 + 0 = 1 \text{ fps} \tag{282}$$

$$v = -\frac{\partial \psi}{\partial x} = -\frac{\partial}{\partial x}(y - x^2 + x) = 0 + 2x - 1 = (2x - 1) \text{ fps} \tag{282}$$

and

$$V^2 = (1)^2 + (2x - 1)^2$$

By inserting values of x for points on the streamline, the respective velocities V may be determined at these points.

111. Some Useful Combined Flowfields. (*a*) *Source in a Rectilinear Flow.* Superposition of a source and rectilinear flow will yield, if the source is at the origin of coordinates, the streamfunction

$$\psi = Ur \sin \theta + \frac{q\theta}{2\pi} \tag{290}$$

from which the streamlines may be plotted, using polar coordinates. This will yield the flowfield of Fig. 229, which is the streamline picture about the nose of a solid object (of special form) in a free stream. Although the reader may easily confirm this flowfield in a mathematical plotting exercise, certain physical features are worthy of discussion and necessary to a full understanding of the problem.

Near the source at O the velocities are very high, but they vary inversely with the radial distance from the source. Accordingly, the effect of the source will be nil at large (theoretically infinite) dis-

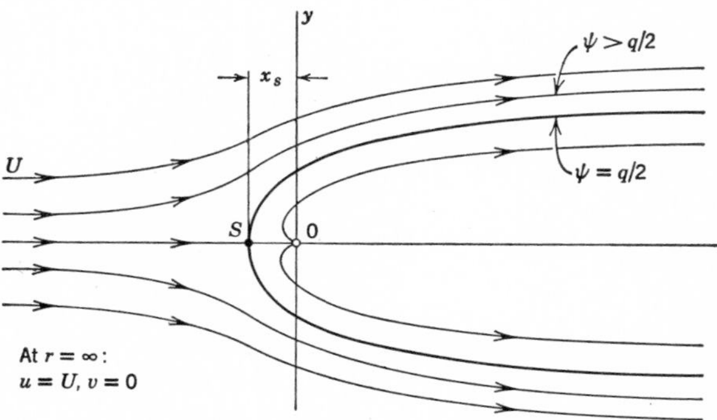

FIG. 229. Source in horizontal rectilinear flow.

tances from O and there the free stream velocity will be found to be unchanged by the presence of the source. However, at some point (on the x-axis) to the left of O, the velocity U of the free stream must exactly equal and cancel the source velocity, yielding a net velocity of zero for the combined flowfield. This condition defines the stagnation point s whose distance x_s from the origin may be computed from $U = q/2\pi x_s$, showing x_s to be dependent upon source strength q and free stream velocity U. Fluid particles issuing from the source and moving to s cannot proceed further to the left, so they must move up or down from s and be carried to the right along a streamline which separates the source flow from that of the free stream. This streamline may be considered to be the contour of a solid body around which the original free stream flow is forced to pass. The equation of the body contour may be completely determined once the value of ψ for its (coincident) streamline is found. Using the coordinates of the stagnation point, which is a known point on the body contour,

$$\psi = U\left(\frac{q}{2\pi U}\right)\sin\pi + \frac{q\pi}{2\pi} = \frac{q}{2}$$

and the body contour is thus described by the equation

$$Ur \sin \theta + \frac{q\theta}{2\pi} = \frac{q}{2}$$

Other principal dimensions of the body contour exist at $\theta = \pi/2$ and $\theta = 0$. For $\theta = \pi/2$, $r = y = q/4U$; for $\theta = 0$, r is infinite but $r \sin \theta = y = q/2U$, showing that the body contour approaches this line as an asymptote.

Anywhere in the flowfield the velocities v_r and v_t may be obtained from

$$v_r = \frac{\partial \psi}{r\,\partial \theta} = \frac{1}{r}\frac{\partial}{\partial \theta}\left(Ur \sin \theta + \frac{q\theta}{2\pi}\right) = U \cos \theta + \frac{q}{2\pi r}$$

$$v_t = -\frac{\partial \psi}{\partial r} = -\frac{\partial}{\partial r}\left(Ur \sin \theta + \frac{q\theta}{2\pi}\right) = -U \sin \theta$$

and the resultant velocity V from $V^2 = v_t^2 + v_r^2$.

The body contour formed by a source in a rectilinear flow is unique in that it has a head (or nose) but no tail; for this reason it is known classically as a *half body*. Half bodies may be used effectively in studying the pressure and velocity distributions near the upstream end of symmetrical objects such as airplane stabilizers, struts, and bridge piers, or torpedo and aircraft fuselages for the axisymmetric three-dimensional case. A generalization of this scheme is the use of many sources of different strengths arrayed along the x-axis to produce a boundary streamline coincident with a prescribed body contour; by this method, the velocity and pressure field about the upstream end of an object may be accurately predicted without costly experimentation.

(*b*) *Source and Sink of Equal Strength.* In the flowfield produced by a source and sink of equal strength the total flowrate passes from one to the other and is thus featured by a family of streamlines originating at the source and ending in the sink. A source and sink on the x-axis are shown in Fig. 230. The streamfunction of the flowfield will be

$$\psi = \frac{q\theta_1}{2\pi} - \frac{q\theta_2}{2\pi} = \frac{q}{2\pi}(\theta_1 - \theta_2) \tag{291}$$

Consider any point P in the flowfield; from geometrical considerations it will be noted that the angle between the radii r_1 and r_2 is given by $\alpha = \theta_2 - \theta_1$. Thus the equation of any streamline is

$$\psi = -\frac{q}{2\pi}\alpha \tag{292}$$

and, for ψ constant along any streamline, α will be constant, showing (again from geometry) that all streamlines of the flowfield are circles which pass through source and sink. The similarity of this flowfield to (1) the flux lines between the poles of a magnet and (2)

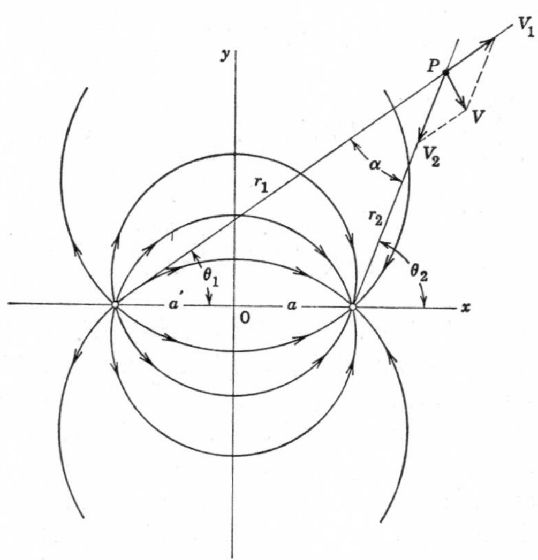

FIG. 230. Source and sink of equal strength.

the flow of electrical current through a homogeneous conductor between points of high and low potential will be observed.

In Cartesian coordinates the streamfunction for this flowfield may be written by replacing the θ's with the appropriate trigonometric function. This yields

$$\psi = \frac{q}{2\pi}\left(\arctan\frac{y}{x+a} - \arctan\frac{y}{x-a}\right) \tag{293}$$

No simple expressions for u and v may be derived from this, but the velocity at any point P will be the vector sum of those (V_1 and V_2) caused separately by the source and sink; these are easily calculated from $q/2\pi r_1$ and $q/2\pi r_2$.

In view of the simple streamline configuration, it is hardly necessary to resort to the graphical technique for this combined flowfield; nevertheless, it proves instructive to consider this (see Fig. 231).

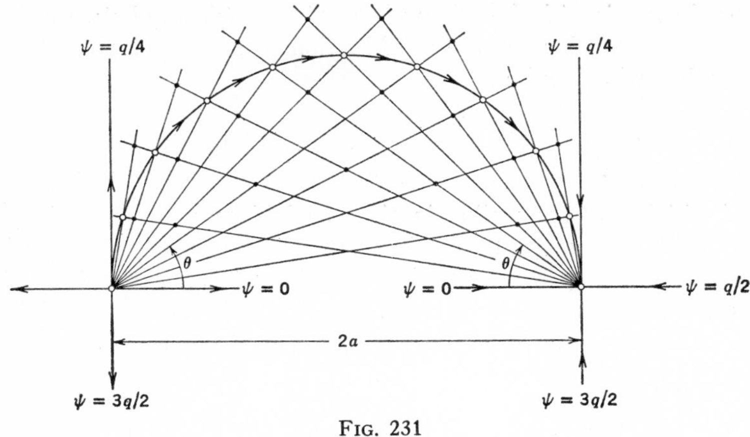

FIG. 231

For the graphical addition of these flows the sink is treated as a negative source and the angle θ for the sink is therefore measured in a clockwise direction. The right angle between the axes may then be divided into a large number of small equal angles, thus defining specific streamlines emanating from the source or entering the sink, each streamline carrying a specific value of ψ. Through streamline intersections which have the same total ψ a streamline may be drawn; the streamline $\psi = q/4$ is shown on Fig. 231.

(*c*) *Source and Sink of Equal Strength in a Rectilinear Flow.* To close the downstream end of the half body cited in section (*a*) of this article, a source and sink of equal strength may be used, resulting in the classic *Rankine oval* of Fig. 232. Let the source and sink

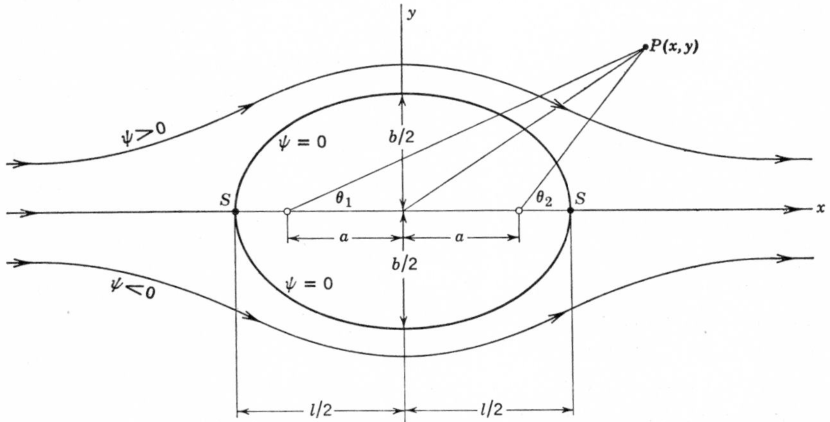

FIG. 232. Source and sink (of equal strength) in horizontal rectilinear flow.

be on the x-axis with the origin of coordinates midway between them. The streamfunction for the combination of source, sink, and rectilinear flow will then be

$$\psi = Uy + \frac{q}{2\pi}\left(\arctan\frac{y}{x+a} - \arctan\frac{y}{x-a}\right) \tag{294}$$

from which any streamline in the flowfield may be plotted. The body contour is of particular interest, and its principal dimensions ($l/2$ and $b/2$) may be obtained as in section (a). Stagnation points (s) are to be expected to the left of the source and to the right of the sink. Here the resultant velocity from source, sink, and rectilinear flow will be zero, so

$$0 = -\frac{q}{2\pi(l/2 - a)} + \frac{q}{2\pi(l/2 + a)} + U$$

whence

$$l/2 = a\sqrt{1 + q/a\pi U}$$

The stagnation points will be at $y = 0$ and $x \neq a$, so ψ (from equation 294) for the body contour streamline will be zero, from which the body contour may be plotted. The other principal dimension $b/2$ may be obtained by substituting $x = 0$, $y = b/2$, and $\psi = 0$ into equation 294, which will reduce to

$$0 = U\frac{b}{2} + \frac{q}{\pi}\arctan\frac{b}{2a} - \frac{q}{2}$$

from which $b/2$ may be obtained by trial.

The flowfields about certain oval bodies may thus be predicted and the pressure distributions on the body contour deduced from them.[3] To the engineer the results will be somewhat unrealistic downstream from the midsection of the oval because of separation, wake formation, and lack of stagnation point in real-fluid flow. However, on the upstream end of the object, where boundary layers are thin and flow accelerating, the theoretical predictions will be found to agree very closely with experimental measurements.

A generalization of this basic scheme (but with considerably greater mathematical difficulties) is the distribution of sources and sinks of *zero net strength* along the x-axis. In combination with a rectilinear flow a closed body contour is to be expected, but the

[3] The top half of such a flow picture is a mathematical representation of the flow of the wind over an oval building (such as a dirigible hangar or quonset hut); the wind pressure on such structures may be studied in this manner.

form of the body may be made by this technique into a more useful streamlined one than that of the Rankine oval.

(*d*) *The Doublet*. The flowfield for a *doublet* is produced by reducing the distance a of Fig. 230 to zero, i.e., superposing the source and sink of equal strength. However, if this operation were carried out without some restriction the result would be no flowfield at all, since source and sink would exactly cancel each other! To obtain a useful flowfield a certain strategem must be employed to prevent this cancellation. Consider now (on Fig. 233) the same arrangement

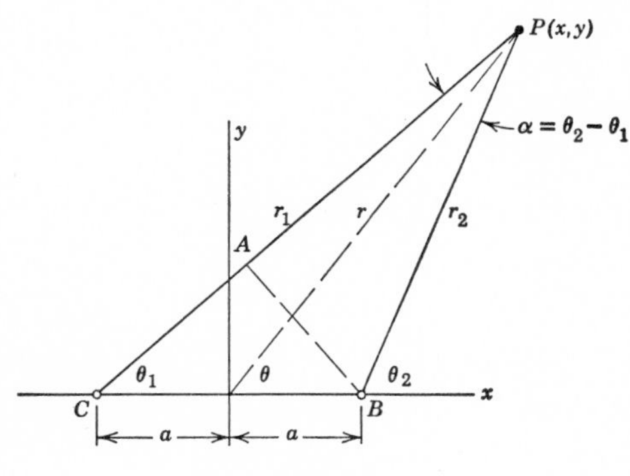

FIG. 233

of source, sink, and point P as that of Fig. 230. The common side AB of the two triangles ABP and ABC is seen to be

$$AB = r_2 \sin \alpha = 2a \sin \theta_1$$

Now as $a \to 0$: $\alpha \to 0$, $\sin \alpha \to \alpha$, $r_2 \to r_1 \to r$, and $\theta_1 \to \theta_2 \to \theta$. The foregoing equation thus reduces to (for $a \to 0$)

$$r\alpha = 2a \sin \theta$$

Substituting this in equation 292, the streamfunction for the doublet is obtained as

$$\psi = -\frac{q}{2\pi}\left(\frac{2a \sin \theta}{r}\right)$$

This form of the equation is quite useless since, with $a = 0$, $\psi = 0$. Suppose, however, that as a approaches zero, q is increased in such a manner that $2qa$ remains a constant, m, known as the *strength* of the doublet. Then the streamfunction of the doublet becomes

$$\psi = -\frac{m \sin \theta}{2\pi r} \tag{295}$$

For any streamline, ψ is a constant and $r = C \sin \theta$, showing the streamlines to be circles tangent to the x-axis at the origin. The flowfield produced by a doublet is therefore that of Fig. 234.

(*e*) *Doublet in Rectilinear Flow.* When the doublet is combined with the rectilinear flow, a limiting case of the Rankine oval results.

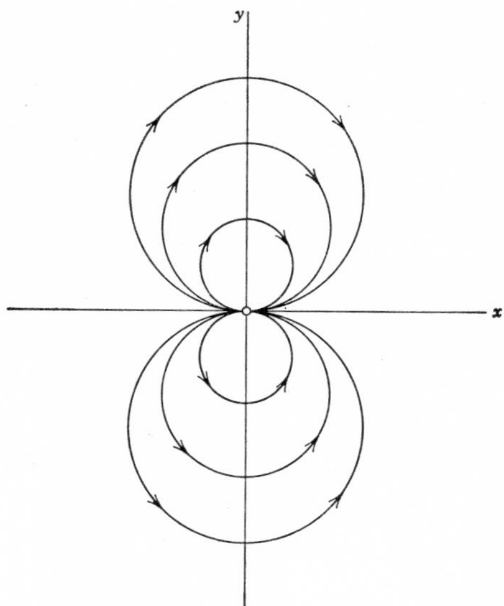

FIG. 234. Doublet.

From experience with the latter a closed body contour will be expected and from the streamfunction

$$\psi = Ur \sin \theta - \frac{m \sin \theta}{2\pi r} \tag{296}$$

the form of the body may be deduced once the value of ψ for the body streamline is determined. From experience with the Rankine oval, the value of ψ for the body contour is zero and, with $\psi = 0$ in equation 296, the dimension R of the body may be obtained as $R = \sqrt{m/2\pi U}$. This shows R to be a constant and the body contour a circle (Fig. 235).

The streamfunction for this flowfield may now be expressed in terms of R instead of the doublet strength by substitution of $m = 2\pi UR^2$ into equation 296. This gives

$$\psi = U\left(r - \frac{R^2}{r}\right) \sin \theta \tag{297}$$

and simple expressions for radial and tangential velocity components anywhere in the flowfield are obtained; these are

$$v_r = \frac{\partial \psi}{r\, \partial \theta} = U\left(1 - \frac{R^2}{r^2}\right)\cos\theta \tag{298}$$

$$v_t = -\frac{\partial \psi}{\partial r} = -U\left(1 + \frac{R^2}{r^2}\right)\sin\theta \tag{299}$$

Since the radial velocity component on the body contour (where $r = R$) is zero, $V = v_t = -2U\sin\theta$, from which the pressure varia-

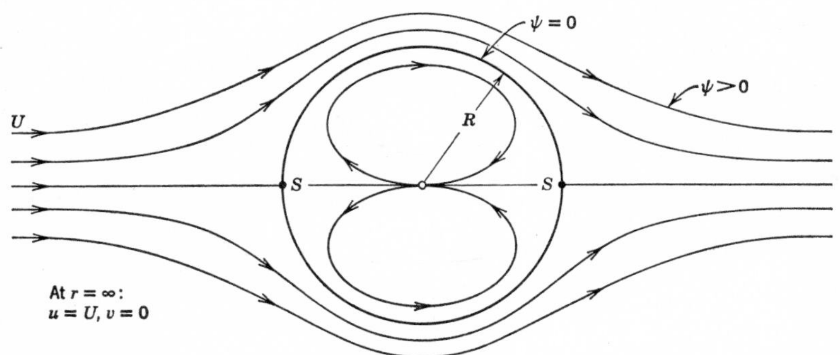

FIG. 235. Doublet in horizontal rectilinear flow.

tion over the body may be deduced. Of particular interest are the facts that (1) the velocity on the surface of the body at its midsection is exactly twice[4] that of the free stream and (2) on the body surface at 30° from the x-axis,[4] the velocity and the pressure are exactly equal to those in the undisturbed free stream.

(f) Doublet in Rectilinear Flow with Circulation. Another useful flowfield may be constructed by superposing[5] free vortex, doublet, and rectilinear flowfields. The streamfunction is (for the clockwise vortex)

$$\psi = U\left(r - \frac{R^2}{r}\right)\sin\theta + \frac{\Gamma}{2\pi}\ln r \tag{300}$$

The use of the first term, representing the doublet in the rectilinear flow, implies that the body contour, a circle of radius R, is unchanged by the superposition of the vortex flow; this is easily justified by

[4] For the corresponding three-dimensional flowfield about a sphere these values are $3U/2$ and 42°, respectively.

[5] With vortex and doublet both at the origin of coordinates.

recalling that the velocities induced by the vortex are wholly tangential, therefore possess no component normal to the body, and thus cannot change its form. When the flowfield is plotted it will be found (for small Γ) to be as shown in Fig. 236 with the stagna-

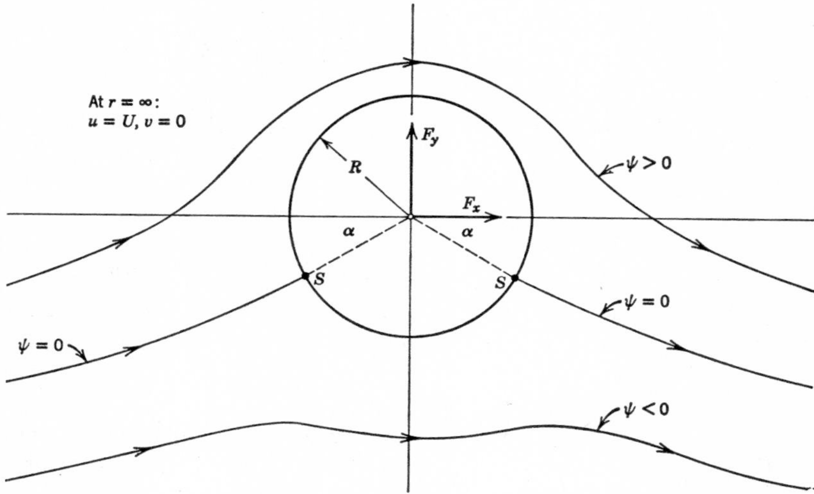

FIG. 236. Doublet and free vortex in horizontal rectilinear flow.

tion points at angle α below the horizontal diameter of the circle. The velocity components anywhere in the flowfield are

$$v_r = \frac{\partial \psi}{r\,\partial \theta} = U\left(1 - \frac{R^2}{r^2}\right)\cos\theta$$

$$v_t = -\frac{\partial \psi}{\partial r} = -U\left(1 + \frac{R^2}{r^2}\right)\sin\theta - \frac{\Gamma}{2\pi r}$$

On the body contour, $r = R$, $v_r = 0$, and $v_t = -2U\sin\theta - \Gamma/2\pi R$, from which the stagnation points may be located. At the stagnation points, $v_t = 0$ and $\theta = -\alpha$; therefore $\sin\alpha = \Gamma/4\pi UR$. For the largest value of α, which is seen to be $\pi/2$, the two stagnation points will merge. Here $\sin\alpha = 1$, $\Gamma/4\pi UR = 1$ and the flowfield will be that of Fig. 237*a*. For $\Gamma/4\pi UR > 1$, the stagnation point will be found at some point below the circle and on its vertical axis as in Fig. 237*b*.

For the flowfield of Fig. 236 the forces F_x and F_y exerted by fluid on the circle may be computed by integrating the appropriate differential force components over the perimeter of the circle. Apply-

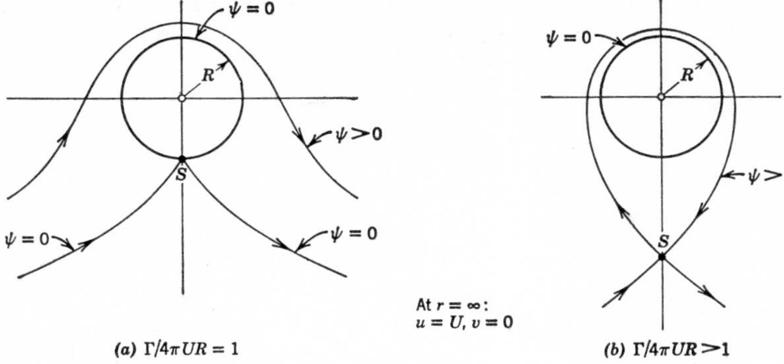

FIG. 237

ing the Bernoulli equation between a point in the free stream (where velocity is U and pressure p_o) and another point on the surface of the body where pressure is p and velocity is $(-2U \sin \theta - \Gamma/2\pi R)$,

$$p = p_o + \tfrac{1}{2}\rho U^2 - \tfrac{1}{2}\rho(-2U \sin \theta - \Gamma/2\pi R)^2 \qquad (301)$$

The pressure p (see Fig. 238) will be normal to the perimeter of the circle, will act on a differential area $R\, d\theta$ at the point (R, θ), and will

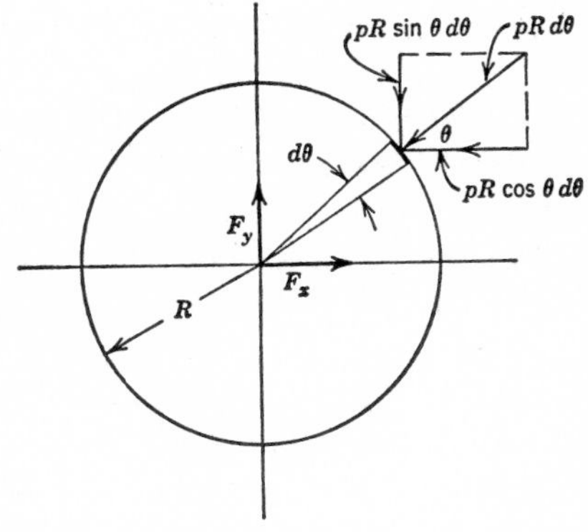

FIG. 238

produce a differential force $pR\, d\theta$ with horizontal and vertical components of $pR \cos \theta\, d\theta$ and $pR \sin \theta\, d\theta$, respectively. Evidently, then,

$$-F_x = \int_0^{2\pi} pR \cos \theta\, d\theta$$

$$-F_y = \int_0^{2\pi} pR \sin \theta\, d\theta$$

The integrations may be carried out after inserting the expression for p given by equation 301. The results are

$$F_x = 0 \quad \text{and} \quad F_y = \rho U \Gamma \tag{302}$$

The first of these might have been anticipated from symmetry of the flowfield, and it shows that the drag force exerted by an ideal fluid on the body contour is zero. This may be generalized to apply to bodies of any shape and is known as the *d'Alembert paradox*.[6] The expression for F_y is of more far-reaching consequence in modern fluid mechanics as it shows that a circulation Γ and a velocity U are both necessary to the existence of a transverse (*lift*) force. The

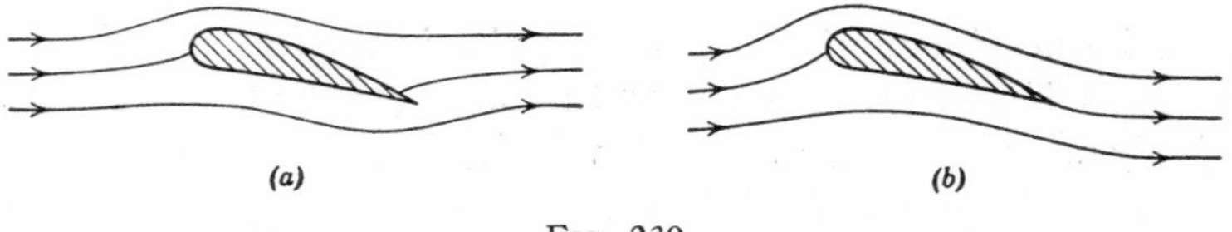

FIG. 239

foregoing derivation, when generalized to apply to a body contour of any shape, is called the *Kutta-Joukowsky theorem* after those who first developed it. It serves to explain certain familiar phenomena in which bodies spinning in real fluids create their own circulation and when exposed to a rectilinear flow are acted upon by a transverse force; some examples are the forces on the rotating cylinders of a "rotorship" and the transverse force which causes a pitched baseball to curve. However, its most important engineering application is in the theory of the lift force exerted on airfoils, hydrofoils, and the blades of turbines, pumps, propellers, windmills, etc. The theory does not, of course, explain the origin of circulation about a nonspinning object but does demonstrate that a circulation is required for the generation of a transverse force.

By the use of the mathematics of the complex variable, *conformal transformations* of these flows may be constructed. Using the Joukowsky transformation, Fig. 235 may be distorted into an airfoil or hydrofoil in a flowfield as shown in Fig. 239*a*, which yields no transverse force and (the unrealistic) flow around the trailing edge of the foil. However, the transformation of Fig. 236 yields the more realistic flow picture of Fig. 239*b*; here, with circulation, and a flow

[6] In d'Alembert's time the discrepancy between theory and observation was considered to be a contradiction; today it is completely explained by the action of viscosity.

tangent to the surfaces at the trailing edge, a lift is produced and experimental results are closely approximated.

(*g*) *Vortex in Rectilinear Flow.* The streamfunction of a clockwise vortex of strength Γ in a rectilinear flow of velocity U is

$$\psi = Ur \sin \theta + \frac{\Gamma}{2\pi} \ln r \tag{303}$$

from which

$$v_r = \frac{\partial \psi}{r \, \partial \theta} = U \cos \theta$$

$$v_t = -\frac{\partial \psi}{\partial r} = -U \sin \theta - \frac{\Gamma}{2\pi r}$$

The resulting flowfield will appear as in Fig. 240, resembling that of Fig. 237*b*. The closed streamline may be considered to be a body

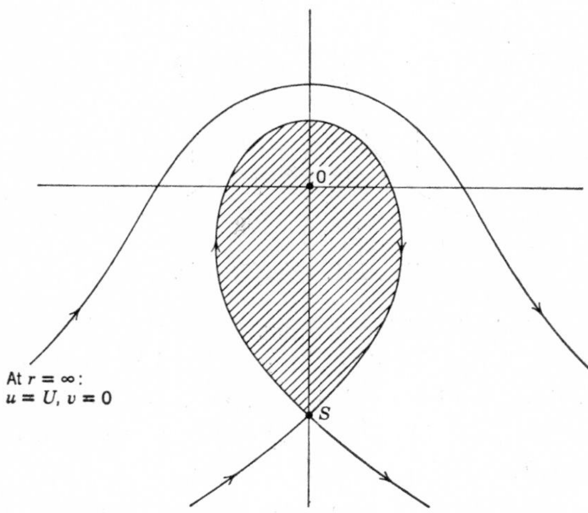

FIG. 240. Free vortex in horizontal rectilinear flow.

contour on which an upward force (by the generalized Kutta-Joukowsky theorem) $\rho U\Gamma$ may be expected. Such a flowfield, therefore, becomes a mathematical representation of the large flowfield about a small wing or blade element. When used in company with another vortex (of the same sense) the flowfield may be made to represent that about a biplane; an infinite number of vortices in line and uniformly spaced will yield the flowfield near an infinite series of blades such as encountered in propeller or turbine theory.

(*h*) *Vortex Pair.* Two vortices of equal strength but opposite sense in close proximity are known as a *vortex pair*. If these vortices are on the x-axis (Fig. 241) the streamfunction of the combined flow will be

$$\psi = \frac{\Gamma}{2\pi}\ln r_1 - \frac{\Gamma}{2\pi}\ln r_2 = \frac{\Gamma}{2\pi}\ln\frac{r_1}{r_2} \tag{304}$$

and, when plotted, a symmetrical flowfield will result as shown. Since the velocities induced by the vortices in the zone between $-a$ and $+a$ are larger than elsewhere in the flowfield, a region of lower pressure is to be expected there and the tendency noted for such vortices to move toward each other.

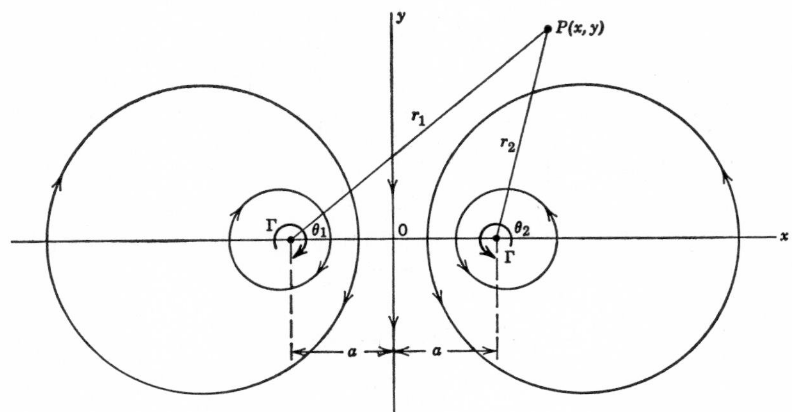

FIG. 241. Vortex pair.

(*i*) *Vortex Pair in Rectilinear Flow.* When combined with a rectilinear flow the vortex pair will form a closed body contour (Fig. 242) analogous to the Rankine oval of Fig. 232. If the vortex pair is on the x-axis and the rectilinear flow upward and parallel to the y-axis, the streamfunction of the flowfield will be

$$\psi = \frac{\Gamma}{2\pi}\ln\frac{r_1}{r_2} - Vx \tag{305}$$

From symmetry of the flow, two stagnation points will be expected on the y-axis and may be located by setting the resultant velocity equal to zero:

$$2(\Gamma/2\pi r_s)\cos\alpha - V = 0$$

and with $r_s^2 = y_s^2 + a^2$ and $\cos\alpha = a/r_s$,

$$y_s^2 = \Gamma a/\pi V - a^2$$

from which y_s may be obtained.[7] The equation for the body contour

[7] For the existence of stagnation points on the y-axis, $\Gamma a/\pi V$ must be larger than a^2.

may be obtained by noting that, at the stagnation point, $r_1 = r_2$ and $x = 0$ and therefore $\psi = 0$. The equation of the body contour (see Fig. 242) is then (from equation 305)

$$x = \frac{\Gamma}{4\pi V} \ln \frac{(x + a)^2 + y^2}{(x - a)^2 + y^2}$$

from which it may be plotted; the half-width $b/2$ may be obtained by solving (by trial) for x with $y = 0$.

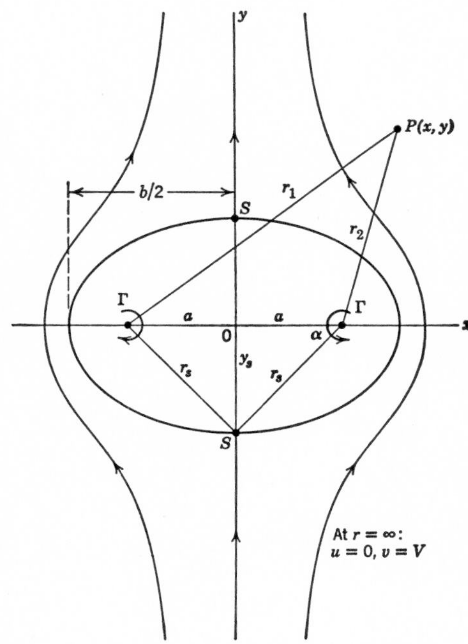

FIG. 242. Vortex pair in vertical rectilinear flow.

(*j*) *Source and Vortex.* Superposition of a source and vortex produces a flowfield of some utility in the analysis of flow through fluid machinery. The streamfunction of this flow will be (for a clockwise vortex)

$$\psi = \frac{q\theta}{2\pi} + \frac{\Gamma}{2\pi} \ln r \tag{306}$$

The equation of the streamline, $\psi = 0$, will be

$$r = e^{-q\theta/\Gamma}$$

showing the flowfield to be a family of logarithmic spirals (Fig. 243). The center of the source-vortex combination contains the double singularity of infinite tangential and radial velocity components; this point is to be avoided in practical calculations. However, the flow between circles 1 and 2 is of considerable engineering interest. Through the zone between circles 1 and 2 the velocity diminishes

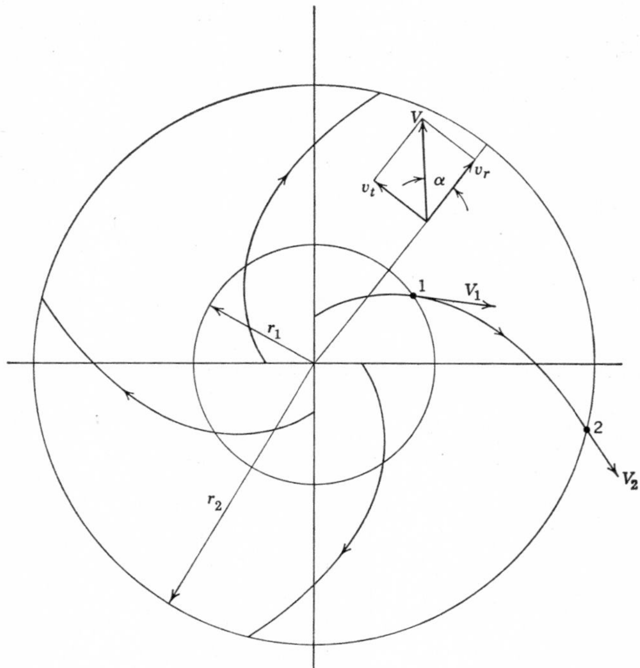

FIG. 243. Source and vortex.

from V_1 to V_2, showing the zone to be a *diffuser* through which a rise of pressure is to be expected. The velocity components anywhere in the flowfield are

$$v_r = \frac{\partial \psi}{r\, \partial \theta} = \frac{q}{2\pi r} \quad \text{and} \quad v_t = -\frac{\partial \psi}{\partial r} = -\frac{\Gamma}{2\pi r}$$

With Γ constant,

$$(v_t r)_1 = (v_t r)_2$$

which means that the moments of momentum (see Art. 51) at sections 1 and 2 are the same; since no torque is exerted on the fluid between sections 1 and 2, this is to be expected. Also it may be noted (from the velocity triangle) that tan α is constant throughout the flowfield, confirming a well-known property of the logarithmic spiral. Frequently in engineering this flow is encountered in the annular passage (called *vortex chamber or vaneless diffuser*) just downstream from the rotating impeller of a centrifugal pump or supercharger. Similarly a sink-vortex combination is a mathematical representation of the flowfield (if two-dimensional) between the guide vanes and runner of a hydraulic turbine. (See Fig. 85.)

112. The Velocity Potential. Suppose now that another function $\phi(x, y)$ be defined in such a way that for Cartesian coordinates

$$u = -\frac{\partial \phi}{\partial x} \quad \text{and} \quad v = -\frac{\partial \phi}{\partial y} \tag{307}$$

or, in polar coordinates,

$$v_r = -\frac{\partial \phi}{\partial r} \quad \text{and} \quad v_t = -\frac{\partial \phi}{r\,\partial \theta} \tag{308}$$

The function ϕ is known as the *velocity potential*, and it has some significant properties which may now be examined in general terms.

The continuity equation

$$\frac{\partial u}{\partial x} + \frac{\partial v}{\partial y} = 0 \tag{58}$$

may be written in terms of ϕ by substitution of the above definitions, to yield the Laplacian differential equation,

$$\frac{\partial^2 \phi}{\partial x^2} + \frac{\partial^2 \phi}{\partial y^2} = 0 \tag{309}$$

Thus all practical flows (which must conform to the continuity principle) must satisfy the Laplacian equation in terms of ϕ.

Similarly the equation for vorticity,

$$\xi = \frac{\partial v}{\partial x} - \frac{\partial u}{\partial y} \tag{62}$$

may be put in terms of ϕ to give

$$\xi = \frac{\partial}{\partial x}\left(-\frac{\partial \phi}{\partial y}\right) - \frac{\partial}{\partial y}\left(-\frac{\partial \phi}{\partial x}\right) = -\frac{\partial^2 \phi}{\partial x\,\partial y} + \frac{\partial^2 \phi}{\partial y\,\partial x}$$

from which a valuable conclusion may be drawn: Since $\partial^2\phi/\partial x\,\partial y = \partial^2\phi/\partial y\,\partial x$, *the vorticity must be zero for the existence of a velocity potential.* From this it may be deduced that only irrotational ($\xi = 0$) flowfields can be characterized by a velocity potential ϕ; for this reason *irrotational* flows are also known as *potential* flows.

Illustrative Problem

Calculate the velocity potential ϕ and sketch the equipotential lines of the flowfield produced by a doublet and rectilinear flow.

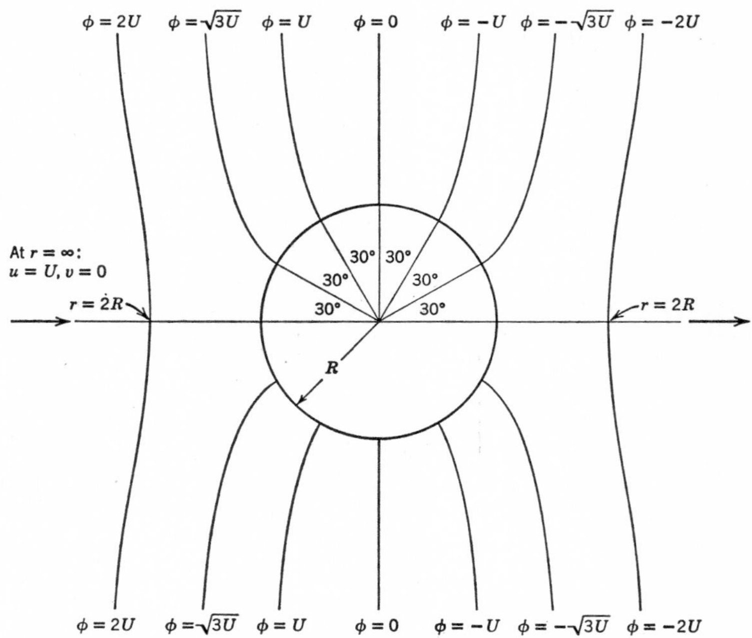

Solution. From section (*e*) of Art. 111,

$$v_r = U\left(1 - \frac{R^2}{r^2}\right)\cos\theta, \qquad v_t = -U\left(1 + \frac{R^2}{r^2}\right)\sin\theta \tag{298 and 299}$$

but

$$v_r = -\frac{\partial\phi}{\partial r} \quad \text{and} \quad v_t = -\frac{\partial\phi}{r\,\partial\theta} \tag{308}$$

and

$$\phi = \int\left(\frac{\partial\phi}{\partial r}\right)dr + \int\left(\frac{\partial\phi}{r\,\partial\theta}\right)r\,d\theta + C$$

Therefore

$$\phi = \int -U\left(1 - \frac{R^2}{r^2}\right)\cos\theta\,dr + \int U\sin\theta\left(1 + \frac{R^2}{r^2}\right)r\,d\theta + C$$

Expanding and rearranging,

$$\phi = -U\left[\int(\cos\theta)\,dr - \int(r\sin\theta)\,d\theta\right] + UR^2\left[\int\left(\frac{\cos\theta}{r^2}\right)dr + \int\left(\frac{\sin\theta}{r}\right)d\theta\right] + C$$

The bracketed quantities may be shown to be (respectively) equivalent to the integrals indicated in

$$\phi = -U\int d(r\cos\theta) + UR^2\int d\left(\frac{-\cos\theta}{r}\right) + C$$

and thus

$$\phi = -U\left(r + \frac{R^2}{r}\right)\cos\theta + C$$

Assuming for convenience that the constant of integration is zero, it is seen that $\phi = 0$ when $\theta = \pi/2$ or $3\pi/2$; thus the line $\phi = 0$ coincides with y-axis. For $\theta \to \pi$ and $r \to \infty$, $\theta \to +\infty$, and, for $\theta \to 0$ and $r \to \infty$, $\phi \to -\infty$. For $\theta = \pi/3$ and $r = R$, $\phi = -U$, and when plotted or sketched the equipotential lines must appear as shown. It should be noted that these lines appear to be perpendicular (orthogonal) to the streamlines of Fig. 235.

113. Relation between Streamfunction and Velocity Potential. A geometric relationship between streamlines and equipotential lines may be derived from the foregoing equations and restatement of certain mathematical definitions; the latter are (with definitions of u and v inserted)

$$d\psi = \frac{\partial\psi}{\partial x}dx + \frac{\partial\psi}{\partial y}dy = -v\,dx + u\,dy$$

$$d\phi = \frac{\partial\phi}{\partial x}dx + \frac{\partial\phi}{\partial y}dy = -u\,dx - v\,dy$$

However, along any streamline ψ is constant and $d\psi = 0$, so $dy/dx = v/u$; also along any equipotential line ϕ is constant and $d\phi = 0$, so $dy/dx = -u/v$. The geometric significance of this is seen in Fig. 244 (and has been suggested in the preceding illustrative problem): *The equipotential lines are normal to the streamlines.* Thus the streamlines and equipotential lines (for an irrotational flow) [8] form a net, called a *flownet*, of mutually perpendicular families of lines, a fact of great significance for the study of flowfields where formal mathematical expressions of ϕ and ψ are unobtainable. Another

[8] There can be no equipotential lines for rotational flows; see Art. 112.

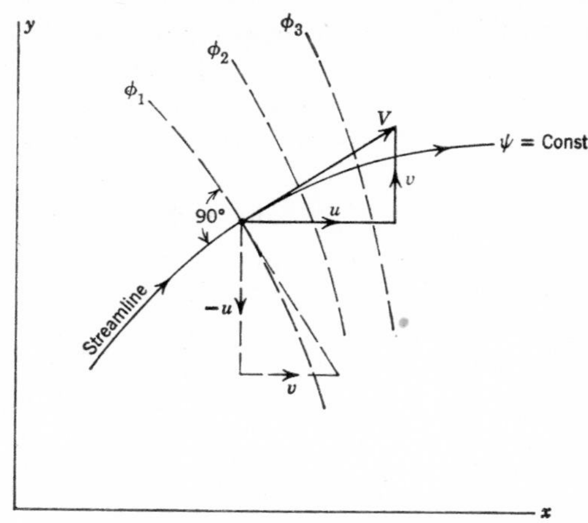

FIG. 244

feature of the velocity potential which may be deduced from Fig. 244 and equations 307 is the decline of the value of ϕ *along the direction of the flow*, i.e., $\phi_3 < \phi_2 < \phi_1$.

114. The Flownet.[9] The fact that streamlines and equipotential lines are orthogonal offers a graphical technique for the solution of two-dimensional irrotational flow problems. In Fig. 245 it is seen that, with $\Delta s = \Delta n$ and both of these dimensions approaching zero,

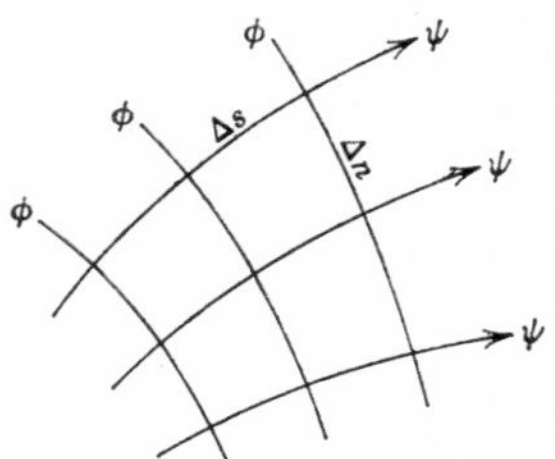

FIG. 245. Flownet.

the figures approach tiny squares; if streamlines and equipotential lines can be constructed to satisfy this requirement, the only possible flowfield may be found, the velocities determined by the streamline spacing, and the pressures from the Bernoulli equation. Consider-

[9] Another trial-and-error technique for plotting the streamlines of a flowfield is the *relaxation method* developed by R. V. Southwell. See J. S. McNown, E.-Y. Hsu, and C.-S. Yih, "Application of the Relaxation Technique in Fluid Mechanics," *Trans. A.S.C.E.*, Vol. 120, 1955. Space does not permit the presentation of this method here.

able patience is needed to apply the technique because of trial and adjustment of streamline and equipotential line positions, but results are invariably obtained more quickly and less expensively than from an experimental program. The accuracy of such results depends upon the fineness of the flownet, and it is obvious at once that greater accuracy will require more time and patience since there are many more lines to adjust; however, the engineer will usually have no trouble reaching a suitable compromise between available time and desired accuracy of results.

Illustrative Problems

Determine the distribution of pressure over the cylindrical surface BC of the two-dimensional nozzle for a flowrate q.

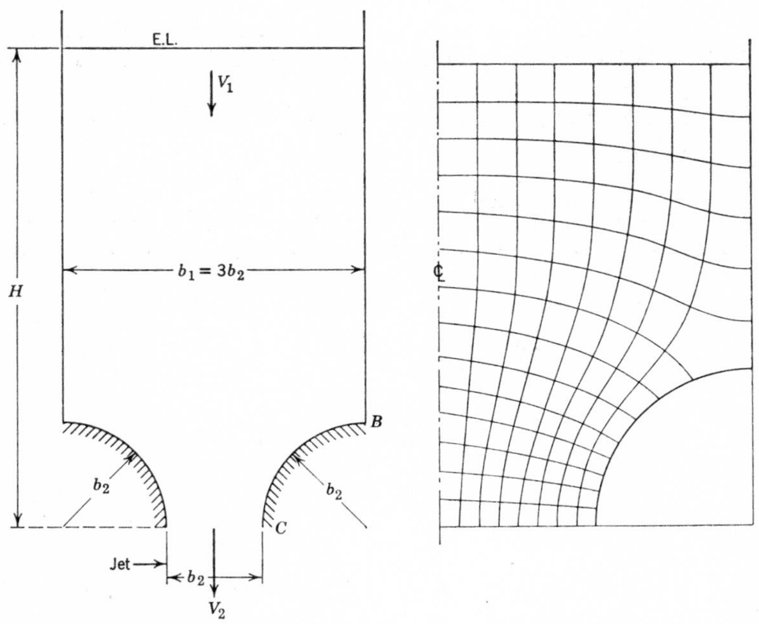

Solution. Recognizing B as a stagnation point, it is evident at once that the pressure along BC will fall from stagnation pressure at B to zero at C. Calculating velocities, $V_2 = q/b_2$, $b_1 = 3b_2$, $V_1 = q/3b_2$. Distance between nozzle exit and energy line, H, is $V_2^2/2g$. With position of energy line known, any point (where pressure is p and velocity V) in the flowfield is a distance $(p/\gamma + V^2/2g)$ vertically from the energy line. Thus, the vertical position of all points on streamline BC being known, the pressures at these points may be obtained once the velocities are determined; this will require a flownet analysis as follows. The velocity distribution at the nozzle exit and well upstream from the nozzle can be expected to be uniform from one-dimensional flow considerations; thus this becomes a relatively easy problem with simple boundary conditions upstream and downstream from the nozzle. First draw the central streamline, noting that it is a line of symmetry and that only one side of the flowfield need be investigated. Divide the distance between centerline and wall (at nozzle exit and in passage) into the same number of streamtubes of equal width. Evidently these streamtubes are straight and parallel upstream from the nozzle and at the nozzle exit, and the flowrate in each will be $q/2n$ if n is the number of streamtubes between centerline and boundary walls. Next sketch in curved streamlines connecting the straight ones and then draw equipotential lines normal to the streamlines to produce a network of squares; here it should be noted that all equipotential lines must be normal to the centerline and the solid boundary of passage or nozzle; this is a good guide to the construction of these lines. After the streamlines and equipotential lines have been readjusted to conform to the squares of the flownet, the width of the streamtube adjacent to BC may be scaled, the velocities obtained from the continuity principle, and the pressures calculated as indicated above.

Predict the flowrate q from the large reservoir over this outflow structure for a given head H.

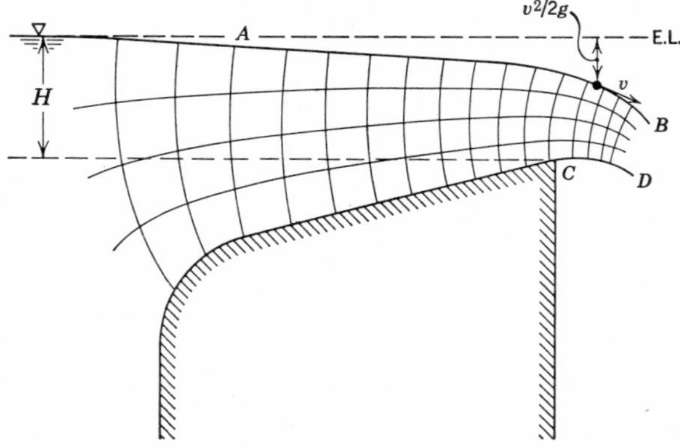

Solution. First sketch a plausible flow picture about as shown. From the position of free streamlines AB and CD the velocity at any point on them may be determined by the vertical distance from the energy line; this distance is $v^2/2g$. At a section downstream from BD the sheet of liquid will be freely falling and a velocity distribution and total flowrate may be estimated. At this section the streamlines may be spaced in such a manner that the flowrate is the same in all the streamtubes and the streamlines may be sketched back to the resevoir. The flownet is then drawn by trial and error to satisfy the condition of orthogonality of streamlines and equipotential lines. Once the flownet is established the velocities obtained from streamline spacing must conform to those obtained from the energy line. If they do not, a new flow picture must be drawn and the process repeated until this conformance is obtained; when it is, the total flowrate q will be the sum of the flowrates in the separate streamtubes.

The flownet technique may also be applied to the flow of real fluids under some restrictions which (although severe) are frequently encountered in engineering practice. Consider the one-dimensional flow of an incompressible real fluid in a streamtube. The Bernoulli equation written in differential form is

$$d\left(\frac{p}{\gamma} + \frac{V^2}{2g} + z\right) = -dh_L$$

Suppose now that V is small (so that $d(V^2/2g)$ may be neglected) and the head loss dh_L given by

$$dh_L = \frac{1}{K} V\, dl \tag{310}$$

in which dl is the differential length along the streamtube and K is a constant. The Bernoulli equation above then reduces to

$$d\left(\frac{p}{\gamma} + z\right) = -\frac{1}{K} V\, dl \quad \text{or} \quad V = -\frac{d}{dl} K\left(\frac{p}{\gamma} + z\right)$$

and, if this may be extended [10] to the two-dimensional case,

$$u = -\frac{\partial}{\partial x} K\left(\frac{p}{\gamma} + z\right) = -\frac{\partial \phi}{\partial x}$$

$$v = -\frac{\partial}{\partial y} K\left(\frac{p}{\gamma} + z\right) = -\frac{\partial \phi}{\partial y}$$

and $K(p/\gamma + z)$ is seen to be the velocity potential of such a flow-field.

The conditions of the foregoing hypothetical problem are almost exactly satisfied when a fluid flows in laminar condition through a homogeneous porous medium; even though the flow in such problems is rotational, the flownet technique may be applied to obtain useful engineering information for the "seepage flow" of water through or under structures, to wells and under-drains, or for the flow of petroleum through the porous materials of subsurface "reservoirs"; K in such problems is known as the coefficient of permeability and has the dimensions (from equation 310) of velocity.

[10] Validation of this extension will be found in any treatise on flow through porous media.

Illustrative Problems

Fine-grained uniform sand is packed in this vertical permeameter tube. Calculate the coefficient of permeability, K, if the flowrate is 0.00010 cfs.

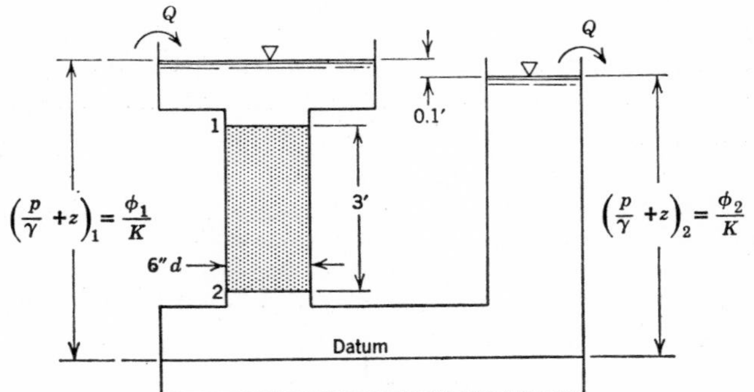

Solution. Neglecting the velocities in the reservoirs, $(p/\gamma + z)$ at 1 and 2 are as shown for the top and bottom of the sand column, respectively. Therefore

$$\frac{\Delta\phi}{\Delta l} = \frac{K \times 0.1}{3} = V = \frac{0.00010}{(\pi/4)(\frac{6}{12})^2}; \quad K = 0.00153 \text{ fps}$$

However, the foregoing calculation is valid only if the flow is laminar. This will be the case only if the Reynolds number Vd/ν is less than 1 (approx.). With d the effective diameter of the sand grain and ν the kinematic viscosity of the liquid, it may be inferred that a valid K depends on both of these quantities.

Show how to calculate the seepage flowrate from reservoir to reservoir beneath the sheet piling.

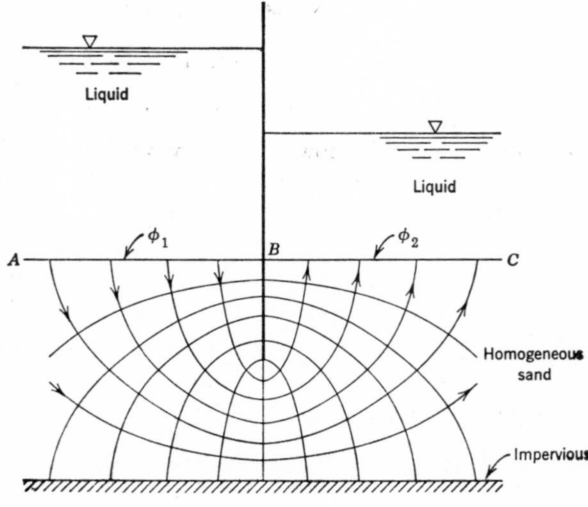

Solution. Line AB is recognized as an equipotential line having $\phi_1 = K(p/\gamma + z)_1$; likewise for line BC, $\phi_2 = K(p/\gamma + z)_2$. Sketch streamlines and equipotential lines about as shown; note that the former must all be normal to the top of the sand bed and the latter normal to the sheet piling and impervious layer. Refine these into a net of squares as accurately as possible. Assign values of ϕ to all the equipotential lines so that $\Delta\phi$ is the same between any consecutive pair of lines. Divide $(\phi_1 - \phi_2)$ by the number of spaces between the equipotential lines to obtain $\Delta\phi$, and assign values to the equipotential lines $\phi_1 - \Delta\phi$, $\phi_1 - 2\Delta\phi$, $\phi_1 - 3\Delta\phi$, etc., in order downstream. The flowrate q through any square (of side Δl) may be computed from $\Delta q = V\,\Delta l = -(\Delta\phi/\Delta l)\,\Delta l = -\Delta\phi$, which establishes the flowrate through the streamtube containing that square. Use the same procedure for squares in the other streamtubes and obtain the total flowrate from $q = \Sigma(\Delta q)$. When the number of streamlines does not equal the number of equipotential lines the flownet will contain small rectangles instead of squares. Through any rectangle of length Δl (along the flow) and Δn (across the flow), the flowrate Δq is $\Delta q = V\,\Delta n = -(\Delta\phi/\Delta s)\,\Delta n$; then $q = \Sigma(\Delta q)$ as before.

REFERENCES

H. Lamb, *Hydrodynamics*, 6th Ed., Cambridge University Press, 1932, and Dover Publications, 1945.

L. Prandtl and O. G. Tietjens, *Fundamentals of Hydro- and Aeromechanics*, McGraw-Hill Book Co., 1934.

W. F. Durand and M. M. Munk, *Aerodynamic Theory*, Vol. I (W. F. Durand, Ed.), Julius Springer, Berlin, 1935.

H. Glauert, *The Elements of Aerofoil and Airscrew Theory*, Cambridge University Press, 1937.

S. L. Green, *Hydro- and Aero-Dynamics*, Sir Isaac Pitman and Sons, 1937.

V. L. Streeter, *Fluid Dynamics*, McGraw-Hill Book Co., 1948.

G. Birkhoff, *Hydrodynamics*, Princeton University Press, 1950.

G. Temple, *An Introduction to Fluid Dynamics*, Oxford University Press, 1958.

D. H. Wilson, *Hydrodynamics*, Edward Arnold, Ltd., 1959.

D. E. Rutherford, *Fluid Dynamics*, Interscience Publishers, Inc., 1959.

H. R. Vallentine, *Applied Hydrodynamics*, Butterworths Scientific Publications, 1959.

L. M. Milne-Thomson, *Theoretical Hydrodynamics*, 4th Ed., The Macmillan Co., 1960.

PROBLEMS

939. Determine the streamfunctions for the flowfields of problem 237 and plot the streamline $\psi = 2$.

940. Determine the streamfunction for a parabolic velocity profile (as in laminar flow) between parallel plates separated by a distance b. Take the origin of coordinates midway between the plates.

941. A flowfield is characterized by the streamfunction $\psi = 3x^2y - y^3$. Is this flow irrotational? If not, calculate the vorticity. Show that the magnitude of velocity at any point in the flowfield depends only upon its distance from the origin of coordinates. Plot the streamline $\psi = 2$.

942. Plot the body contour formed by a source (at the origin) of 40π cfs/ft in a uniform horizontal stream (from left to right) of velocity 10 fps. Calculate the velocities at the body contour for values of θ of 180°, 150°, 120°, 90°, 60°. Calculate the velocities at radial distances of 2, 3, and 4 ft for $\theta = 180°$. Calculate the pressures at the above eight points if water is flowing and the pressure in the undisturbed free stream is 0 psi. At what point on the body contour will the velocity be a maximum?

943. The nose of a two-dimensional strut having a 6-in. thickness is made in the shape of a half body. If the strut is designed to operate at a speed of 30 fps, find the source strength required to simulate the nose of the strut, the distance between source and stagnation point, and the pressure on the body at a point 90° from the stagnation point.

944. A bluff overlooking a plain approximates a section of a half body as shown. For use as a guide in soaring, derive an equation for lines of constant vertical component of velocity for a wind velocity of 30 mph.

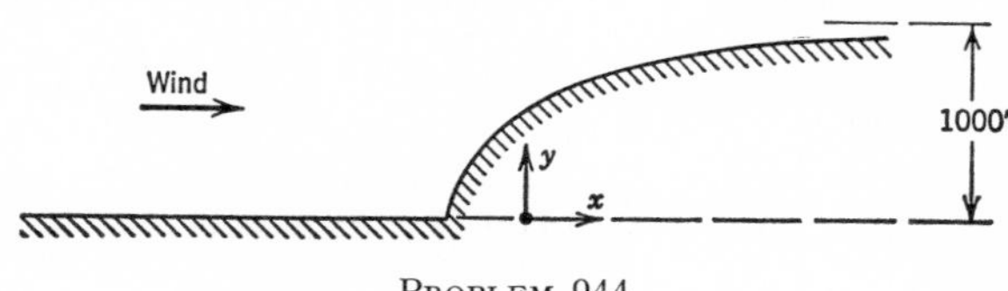

PROBLEM 944

945. Devise a graphical scheme for plotting the contour of a half body.

946. A sink of strength 100 sqft/sec located at the origin of coordinates is combined with a horizontal rectilinear flow (from left to right) having a velocity of 10 fps. Determine the streamfunction of the combined flowfield. Locate any stagnation points. Derive the equation for the line of separation between the flow entering the sink and that which passes it by.

947. Two sources of equal strength are located on the x-axis at a distance a from the origin of coordinates. Determine the streamfunction of the flowfield and locate any stagnation points. If this flowfield is combined with one of streamfunction Uy, sketch the flowfield and locate the stagnation points.

948. Two sources each of strength 100 sqft/sec are located on the y-axis one foot above the origin, the other one foot below. Their flowfields are combined with one of streamfunction $10y$. Determine the streamfunction of the combined flowfield. Sketch the resulting body contour. Determine the location of any stagnation points. Find the point where the body contour intersects the y-axis and calculate the velocity there. Sketch this flowfield when the uniform velocity is 30 fps.

949. A source and sink each of strength 60 sqft/sec are located on the x-axis (see Fig. 230) with $a = 3$ ft. Calculate the velocity at the origin of coordinates. Calculate the value of ψ for the streamline passing through the point (0, 4) and calculate the velocity there.

950. If the sink of the preceding problem is replaced by a source, calculate the streamfunction for the flowfield and the other quantities requested.

951. A Rankine oval 8 ft long and 4 ft wide is in a uniform rectilinear flow of velocity 10 fps. Calculate the velocity along the body contour at the midsection.

952. A source and a sink of equal strengths 6.28 sqft/sec are placed in a uniform stream moving vertically upward. The source is at $(-1, 0)$ and the sink at $(1, 0)$. Derive the equation for the streamfunction and sketch the flowfield. Locate the stagnation points if any. Calculate the pressure at the origin relative to that in the undisturbed uniform stream, which has a velocity of 1 fps.

953. What strength of doublet will be needed to produce a velocity of 10 fps at the point (0, 5)? What is the value of ψ for the streamline passing through this point?

954. Calculate the velocities and pressures on the surface of a cylinder of 4-ft diameter in a horizontal uniform flow of velocity 10 fps at values of θ of 165°, 140°, 115°, and 90°. Also calculate the velocities (at the foregoing angles) along the streamline $\psi = 5$. Assume water flowing and pressure zero in the free stream.

955. Estimate the resultant dynamic force on the upstream half of a cylindrical bridge pier 5 ft in diameter when the water approaching it is 10 ft deep and moving at a mean velocity of 8 fps.

956. For a doublet in a rectilinear flow there will be a line in the flowfield at all points of which the velocity and pressure are the same as those of the undisturbed free stream. Show that the equation of this line is $(R/r)^2 = 2 \cos 2\theta$.

957. Calculate the velocities and pressures at 90° and 270° on the surface of the cylinder of problem 954 after a clockwise vortex of strength 150 sqft/sec has been imposed on the flowfield at the center of the cylinder. Also calculate these velocities and pressures for a vortex strength of 300 sqft/sec.

958. A rotorship has two cylindrical rotors, each of 30 ft height and 10 ft diameter. The ship moves due north at 18 mph when a 30-mph east wind is blowing. If the circulation induced by rotating each cylinder is 3000 sqft/sec, calculate the propulsive force on the ship. Assume that specific weight of the air is 0.0765 lb/cuft.

959. A uniform horizontal flow (from left to right) of velocity 10 fps is combined with a clockwise vortex whose center is at the origin of coordinates. The stagnation point is located at (0, −5). Plot the streamline passing through the stagnation point in the region between $x = -5$ and $x = 5$. What is the velocity in this flowfield at (0, 5)?

960. Clockwise vortices of equal strength are located on the y-axis at points $(0, a)$ and $(0, -a)$. Derive the streamfunction of the resulting flowfield and sketch the streamlines. Calculate the velocity at the origin of coordinates.

961. Solve the preceding problem if the upper vortex is clockwise and the lower one counterclockwise.

962. Sketch the flowfield suggested by Fig. 242 when: (*a*) $\Gamma a/\pi V = a^2$ and (*b*) $\Gamma a/\pi V < a^2$. Locate the stagnation points for (*a*) and for $\Gamma a/\pi V = a^2/4$.

963. The two-dimensional vaneless diffuser of a centrifugal pump has entrance and exit diameters of 2 ft and 3 ft, respectively. Water enters at a velocity of 40 fps and at an angle of 60° with a radial line. Calculate the streamfunction for this flowfield and the expected pressure rise through the diffuser.

964. The exit circle for the guide vanes of a hydraulic turbine (see Fig. 85) is of 14-ft diameter. The entrance to the runner blades is of 10-ft diameter. If water leaves the guide vanes at a velocity of 80 fps and at an angle of 75° with a radial line, calculate the velocity at the runner entrance, the streamfunction for the flowfield, and the pressure drop between guide vanes and runner.

965. Determine (if possible) the velocity potentials of the flowfields of problem 237.

966. Determine (if possible) the velocity potentials of the four basic flowfields of Art. 109.

967. Determine the velocity potentials of the flowfields of: (*a*) Fig. 229, (*b*) Fig. 234, (*c*) Fig. 240, and (*d*) Fig. 243.

968. For a frictionless flow of water through this U-bend find the pressures at points A and B by the flownet method. The two-dimensional flowrate is 40 cfs/ft, the bend is in a horizontal plane, and the pressure in the straight passage is 10 psi. Draw the bend to a scale of 1 in. = 1 ft and divide the flow into 10 streamtubes.

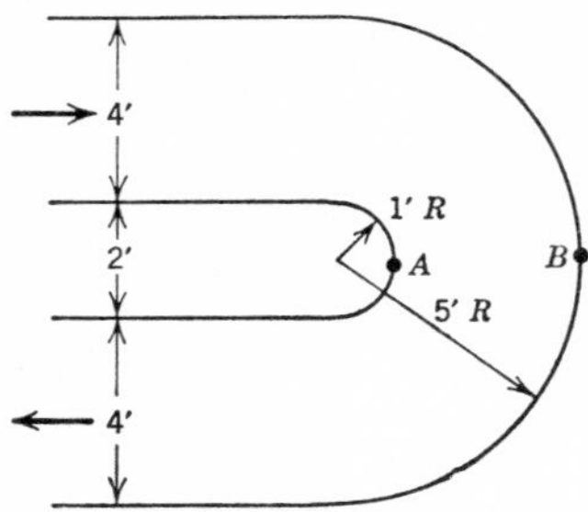

PROBLEM 968

969. Predict the seepage flowrate under this dam if the coefficient of permeability of the pervious layer is 50 ft/day. Use a scale of 1 in. = 4 ft and divide the flow into 10 streamtubes.

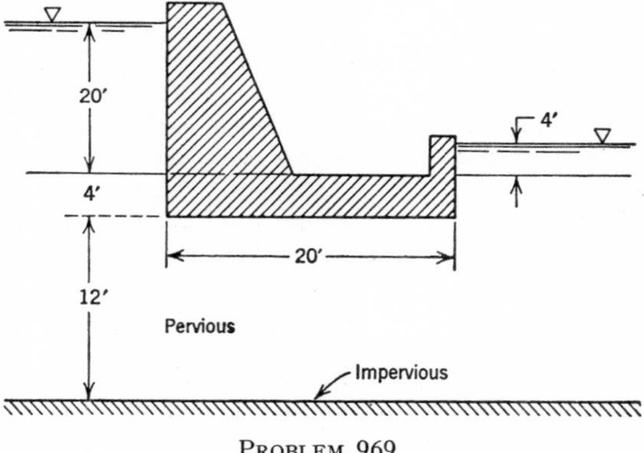

PROBLEM 969

970. Predict the seepage flowrate through this dike on an impervious foundation if the dike material is homogeneous and has a coefficient of permeability of 40 ft/day. Draw the dike to a scale of 1 in. = 1 ft and use 10 streamtubes.

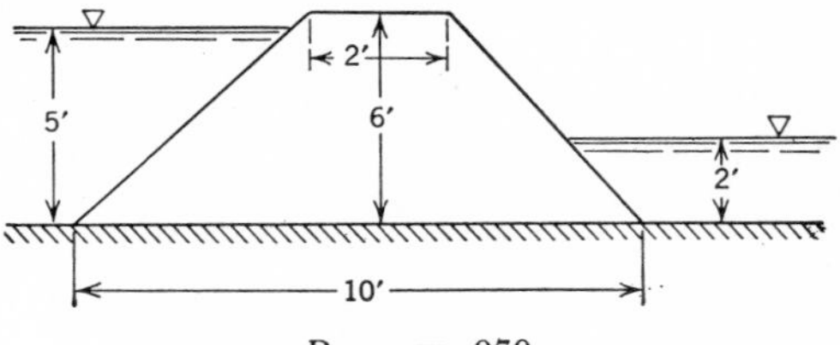

PROBLEM 970

chapter 13

Fluid Flow about Immersed Objects

Problems involving the forces exerted on a solid body when fluid flows by it no longer belong exclusively to the naval architect and aeronautical engineer. In the design of buildings, bridges, automobile bodies, and trains, a knowledge of fluid resistance or drag has become of increasing importance; in the design of propellers, turbines, and centrifugal pumps, the principles of lift are being applied with increasing success. As these principles find wide application, it becomes necessary for all engineers to become familiar with the fundamental mechanics of the fluid motion. It is the purpose of this chapter to outline the fundamental and elementary aspects of this subject for both incompressible and compressible flow.

115. Fundamentals and Definitions. In general, when flow occurs about an object which is either unsymmetrical or whose axis is not aligned with the flow, the flowfield will be unsymmetrical, the local velocities and pressures on either side of the object will be different, and a force normal to the oncoming flow will be exerted. Accompanying this, the action of frictional stress in the boundary layers over the surface of the object will produce a force along the direction of the oncoming flow. These forces are known (from their aeronautical backgrounds) as *lift* and *drag*, respectively. The classic and most useful example is the airfoil (or hydrofoil) of Fig. 246, on which the lift L, drag D, resultant force F, angle of attack α, and chord, c, are shown; the length of the foil perpendicular to the plane of the paper is termed the *span*. The force F, which is seen to be the resultant of L and D, is also the resultant of all forces of pressure and friction exerted by fluid on foil. However, the contribution of the frictional stresses to the lift may be neglected since such stresses are small compared to the pressure and act in a direction roughly normal

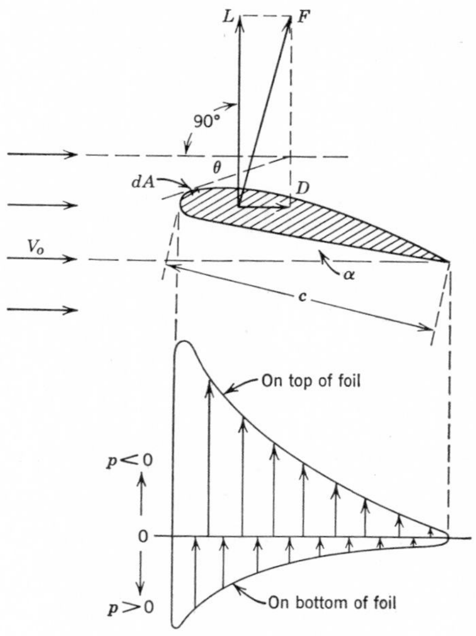

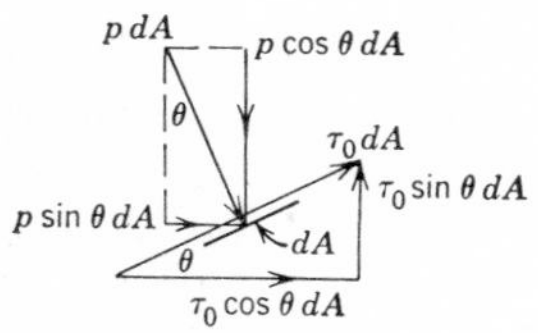

FIG. 246

to L. This is an important simplification of the problem in that it allows L to be considered to be the result of pressure variation alone and thus permits the use of the ideal fluid and the methods of the preceding chapter for analytical predictions of lift; this is accomplished by using the method of Art. 12, plotting the pressure distribution over top and bottom of the foil, and planimetering the resulting area as shown on Fig. 246.

Prediction of the drag force on immersed objects is much more difficult than that of lift, since usually no simplifications are possible and both pressure and frictional forces must be considered. However, both lift and drag may be predicted from experimental measurements on small models in wind tunnels, water tunnels, or towing basins.

The foregoing preliminary discussion becomes more meaningful through formal application to a surface element of an immersed body followed by appropriate integration. Consider the element dA of Fig. 246 on which a pressure p and frictional stress τ_o act. The differential drag and lift on this element are seen to be

$$dD = p\,dA \sin\theta + \tau_o\,dA \cos\theta$$

$$dL = p\,dA \cos\theta - \tau_o\,dA \sin\theta$$

which may be integrated to yield

$$D = \int^s p\,dA \sin\theta + \int^s \tau_o\,dA \cos\theta$$

$$L = \int^s p\,dA \cos\theta - \int^s \tau_o\,dA \sin\theta$$

in which $\int^s$ designates *integral over the surface of the object.* The second integral of the lift expression usually being negligible, the lift is expressed by

$$L = \int^s p\,dA \cos\theta$$

which is a formal mathematical expression for the graphical integration of Fig. 246. The integrals in the drag equation are equally important; the first one is called the *pressure drag* D_p, and the second the *frictional drag* D_f. The former will depend upon the form of the object and flow separation, the latter upon the extent and character of the boundary layer. Although the prediction of separation points is generally a very complex problem dependent upon both body form and boundary layer properties, the breakdown of total drag into pressure drag and frictional drag proves of great value in studying these separately in further detail. Frictional drag may be isolated by considering the flow past a thin flat plate parallel to the oncoming flow (Fig. 247*a*); here $\sin\theta = 0$, $\cos\theta = 1$, $D_p = 0$, and $D = D_f = \int^s \tau_o\,dA$. Pressure drag may be isolated by studying the flow about a flat plate normal to the oncoming flow (Fig. 247*b*); here $\cos\theta = 0$, $\sin\theta = 1$, $D_f = 0$, and $D = D_p = \int^s p\,dA$. Although these prob-

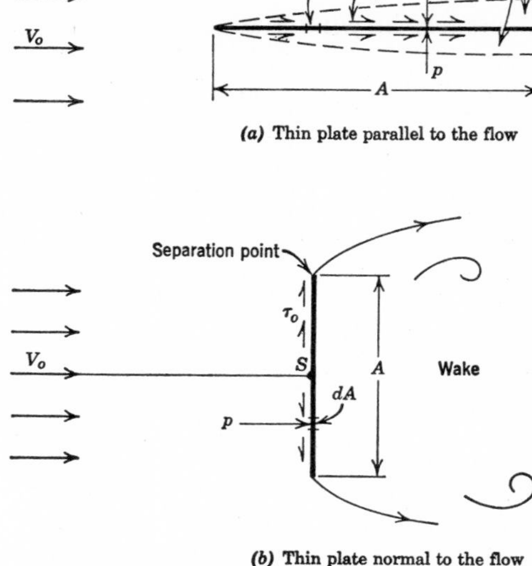

(a) Thin plate parallel to the flow

(b) Thin plate normal to the flow

FIG. 247

lems (and others) will be examined more intensively in Arts. 117 and 118, a descriptive preview will prove useful in setting up certain guideposts.

Consider the circular disk, sphere, and streamlined body of Fig. 248, all of which have the same cross-sectional area, and are immersed

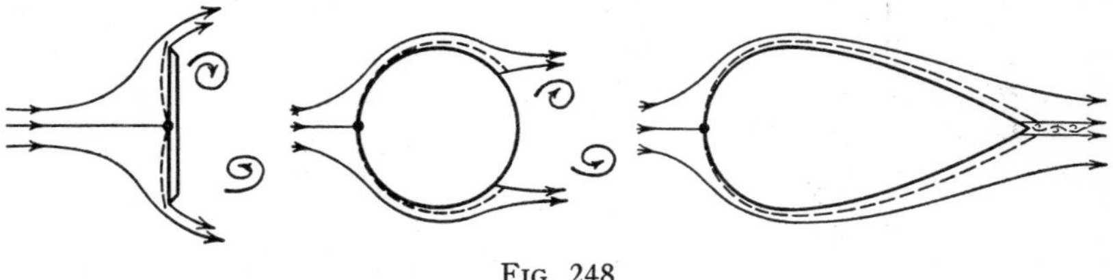

FIG. 248

in the same turbulent flow; all these flows will feature a stagnation point on the upstream side of the object and a maximum local pressure there.

For the disk, separation will be expected at the edges with high local velocity and low pressure. This reduced pressure, being adjacent to the wake, will be transmitted into it, causing the downstream side of the disk to be exposed to a mean pressure considerably below that on the upstream side. The result is a large drag force caused wholly by pressure since none of the shear forces on the disk has components in the original direction of motion.

For the sphere, the wake is smaller than that of the disk and (from the streamline picture) will contain a somewhat higher pressure, leading to the expectation that the drag of the sphere is considerably smaller than that of the disk.[1] For the sphere, the frictional drag is not zero since all shear stresses acting on the sphere will have components parallel to the oncoming flow. These shear stresses are extremely difficult to calculate, but they are small and result in a frictional drag which is negligible compared to the pressure drag for spheres and other objects of similar (blunt) form.

For the well-streamlined body, the wake may be extremely small, being only the width of the boundary layer at the tail of the object. The pressure in such a wake is comparatively large, since the gentle contour of the body allows deceleration of the flow and consequent regain of pressure without incurring separation. Thus the pressure drag of such objects is a very small fraction of that of the disk. However, the frictional drag of streamlined bodies is considerably larger than that of the sphere, since streamlining has brought more surface area in contact with the flow. For well-streamlined objects, frictional drag is usually larger than pressure drag but both are so small that their total is only about one-fortieth that of the disk.

The foregoing examples illustrate the fact that the viscosity property of a fluid is the root of the drag problem. Viscosity has been seen to cause drag either by frictional effects on the surface of an object or through pressure drag by causing separation and the creation of a low pressure wake behind the object. By streamlining an object, the size of its wake is decreased and a reduction in pressure drag is accomplished, but in general an increase in frictional drag is incurred.

For an ideal fluid in which there is no viscosity and thus no cause for frictional effects or wake formation regardless of the shape of the object about which flow is occurring, it is evident that the drag of the object is zero. Two centuries ago, d'Alembert's observation that all objects in an ideal fluid exhibit no drag was a fundamental and disturbing paradox; today this fact is a logical consequence of the fundamental reasoning presented above.

116. Dimensional Analysis of Drag and Lift. The general aspects of drag and lift forces on completely[2] immersed bodies may be

[1] Experiments show the total drag of the sphere to be about one-third that of the disk.

[2] The drag of a surface vessel, a partially submerged body, has been presented in Arts. 61 and 62.

examined advantageously by dimensional analysis before further consideration of the physical details of the problem.

The smooth object of Fig. 249 having area[3] A moves through a

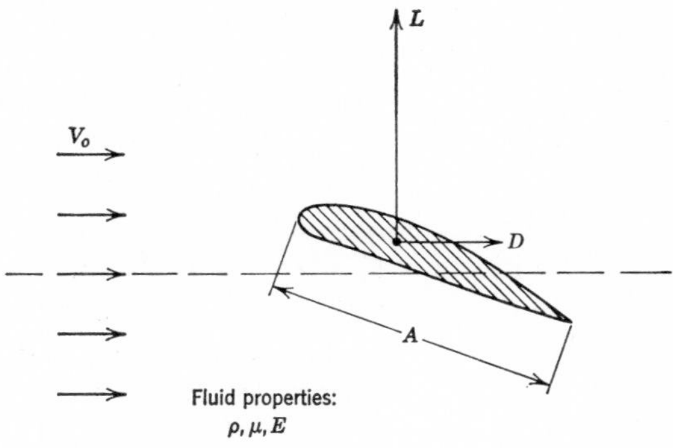

FIG. 249

fluid of density ρ, viscosity μ, and modulus of elasticity E, with a velocity V_o. If the drag force exerted on the body is D,

$$D = f_1(A, \rho, \mu, V_o, E)$$

and similarly

$$L = f_2(A, \rho, \mu, V_o, E)$$

The methods of dimensional analysis allow rearrangement[4] of these very general statements to

$$D = f_3(\mathbf{R}, \mathbf{M})A\rho V_o^2/2 \quad \text{and} \quad L = f_4(\mathbf{R}, \mathbf{M})A\rho V_o^2/2$$

If drag and lift coefficients, C_D and C_L, respectively, are defined by

$$C_D = 2D/A\rho V_o^2 \quad \text{and} \quad C_L = 2L/A\rho V_o^2 \tag{311}$$

it follows that

$$C_D = f_3(\mathbf{R}, \mathbf{M}) \quad \text{and} \quad C_L = f_4(\mathbf{R}, \mathbf{M})$$

These equations indicate: (1) that bodies of the same shape and having the same alignment with the flow (i.e., models of each other) possess the same drag and lift coefficients if their Reynolds numbers and Mach numbers are the same, or (2) that the drag and lift coefficients of bodies of given shape and alignment may be expected to depend upon their Reynolds and Mach numbers only. Thus dimensional analysis has, as in previous problems (ship resistance and pipe friction), opened the way to a comprehensive treatment of the

[3] Any convenient significant area may be selected.

[4] See Appendix IX for details.

resistance of immersed bodies by indicating the dimensionless combinations of variables upon which the drag coefficient depends. An example of this is given in Fig. 250, where experimentally determined drag coefficients for spheres are plotted as contours on a Reynolds number-Mach number graph.

From the foregoing dimensional analysis alone no conclusion can be reached on the quantitative effects of **R** and **M** on the drag and

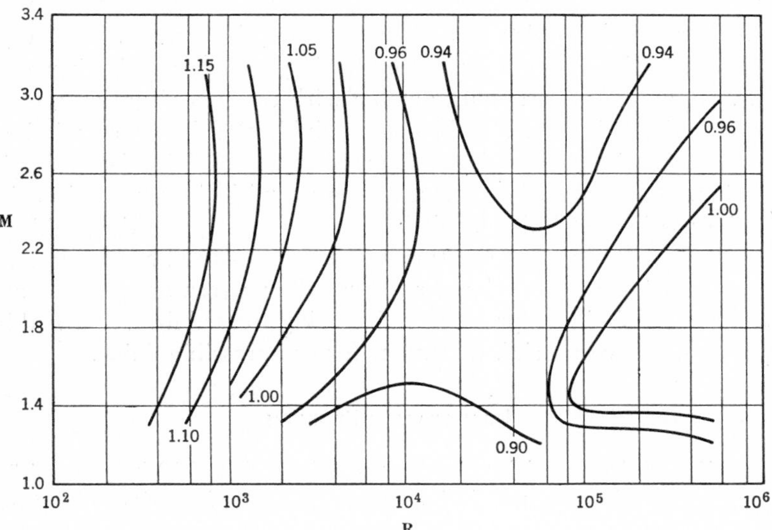

FIG. 250. Drag coefficients for spheres.[5]

lift coefficients, but theory and experiment both show that **R** is predominant when the fluid may be considered incompressible and **M** predominant [6] when compressibility effects must be considered. Usually this means that over a wide range of **R**, where **M** is small and velocities subsonic, the fluid may be considered incompressible and C_D and C_L functions of **R** only. On the other hand, when **M** approaches or exceeds unity and velocities approach or exceed that of sound, C_D and C_L are functions of **M** only, whatever the magnitude of **R**. Although such a division of flow problems is useful and convenient, it is somewhat arbitrary as there is no definite point at which the effects of compressibility begin, such effects being present at all velocities; therefore it is to be expected that there will be exceptions to this convenient division of flow problems and situations en-

[5] A. May, "Supersonic Drag of Spheres at Low Reynolds Numbers in Free Flight," *Jl. Appl. Physics*, Vol. 28, 1957, pp. 910–912.

[6] This tendency may be seen on Fig. 250.

countered where the effects of **R** and **M** are of the same order and in which neither may be ignored. Such problems are highly complex and far beyond the scope of an elementary text, but the beginner should be aware of their existence and not expect them to be classified by usual and convenient methods. Reynolds and Mach numbers have been shown (Art. 61) to be, respectively, ratios of inertia-to-viscous and inertia-to-elastic forces; from this it may be concluded directly that the flow phenomena, when governed by Reynolds number, will result from viscous action (boundary layers, etc.) but, when governed by Mach number, from elastic phenomena (shock waves, etc.). Examples of problems dealing with the first situation are the settling of solid particles through a fluid, the forces of wind on structures, and most aerodynamic problems in the field of commercial aviation. Examples of problems concerned with the second situation are the motions of bullets, missiles, rockets, propeller tips, military aircraft, and elements of gas turbines and high-speed compressors.

The remainder of the chapter follows the pattern above, drag and lift on objects in incompressible fluids being treated first, followed by a parallel, but very brief, treatment for compressible fluids.

Illustrative Problem

The lift and drag coefficients of a rectangular airfoil of 50-ft span and 7-ft chord are 0.6 and 0.05, respectively, when at an angle of attack of 7°. Calculate the horsepower required to drive this airfoil (in horizontal flight) at 150 mph through still, standard air at altitude 10,000 ft. What lift force is obtained when this power is expended? Also calculate the Reynolds and Mach numbers.

Solution. From Appendix IV,

$$\rho = 0.001756 \text{ slug/cuft}$$

Converting, 150 mph yields 220 fps.

$$D = 0.05 \times (50 \times 7) \times 0.001756(220)^2/2 = 745 \text{ lb} \qquad (311)$$

From mechanics,

$$\text{Horsepower} = 745 \times 220/550 = 298 \text{ hp}$$

$$L = 0.6 \times (50 \times 7) \times 0.001756(220)^2/2 = 8930 \text{ lb} \qquad (311)$$

From Appendix IV,

$$\mu = 3.534 \times 10^{-7} \text{ lb-sec/sqft}$$

$$\mathbf{R} = 220 \times 7 \times 0.001756/3.534 \times 10^{-7} = 7{,}650{,}000 \qquad (175)$$

$$a = \sqrt{1.4 \times 32.2 \times 53.3(23.4 + 459.6)} = 1077 \text{ fps} \qquad (11)$$

$$\mathbf{M} = 220/1077 = 0.204$$

DRAG AND LIFT—INCOMPRESSIBLE FLOW

117. Frictional Drag—Boundary Layer. As real fluid flows past a smooth immersed object the effects of viscosity produce velocity profiles featured by no velocity at the solid surface, high shear and velocity gradient, and the velocity of ideal flow at a relatively small distance from the surface. The zone in which the velocity profile is governed by frictional action is known as the *boundary layer;* if this layer is thin and there is no flow separation, the methods of Chapter 12 may be used to determine the flowfield about the object. With

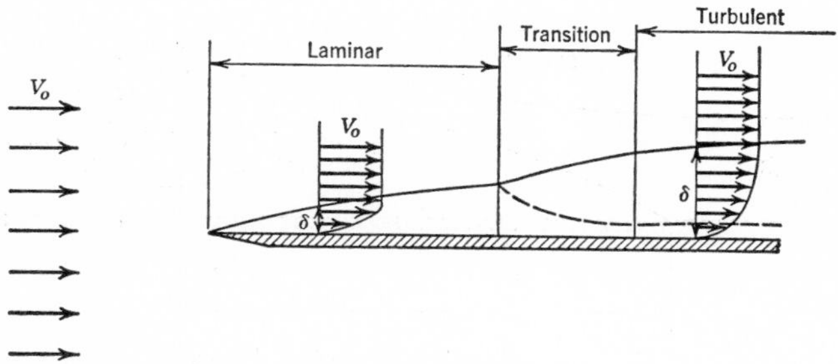

FIG. 251. Boundary layers on a flat plate.

frictional effects confined to the boundary layer, the flow outside of the boundary layer may be considered ideal (irrotational) and thus may be treated by the methods of mathematical hydrodynamics.

Boundary layer phenomena may be most easily visualized on a smooth flat plate parallel to the oncoming flow (Fig. 251). Assuming the leading edge of the plate to be smooth and even, a *laminar boundary layer* is to be expected adjacent to the upstream portions of the plate since there the boundary layer is thin, viscous action intense, and turbulence impossible.[7] The laminar nature of the boundary layer may be characterized in a Reynolds number, $\mathbf{R}_\delta = V_o\delta/\nu$, which experiments have shown to be less than a critical value of about 3900. The boundary layer must start from no thickness at the leading edge of the plate where viscous action begins and steadily increase in thickness as increased viscous action extends into the flow and slows down more and more fluid; in terms of Reynolds number, R_δ will vary over the length of the laminar boundary layer from zero at the leading edge of the plate to the critical value of about

[7] The same argument has been used in pipe flow to justify the laminar character of the film or sublayer over a smooth surface. See Art. 54.

3900 at the downstream end [8] of the layer. At this point instabilities in the boundary layer produce breakdown of the laminar structure of the flow and cause turbulence to begin. After the onset of turbulence the (transition) boundary layer thickens rapidly, developing into a turbulent boundary layer having many of the characteristics of turbulent pipe flow; as in pipe flow, a thin laminar film, or *sublayer*, will exist between the solid surface and the turbulence of the boundary layer.

An approximate analysis of the laminar boundary layer may be carried out by assuming the velocity profile in the layer to be parabolic [9] and applying the impulse-momentum principle (Fig. 252).

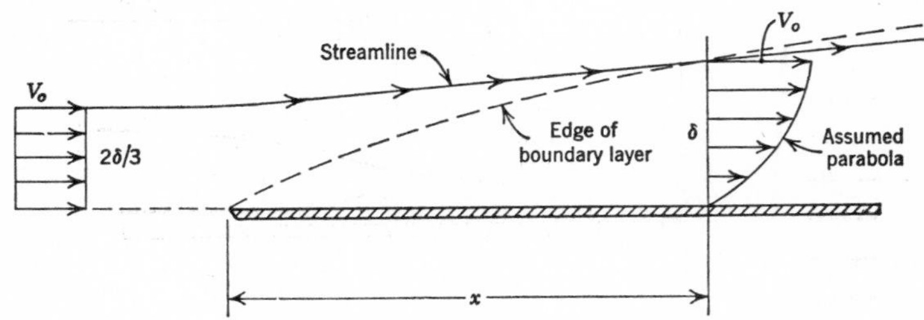

FIG. 252. Laminar boundary layer.

Taking a control volume bounded by plate and streamline (since outside of this zone there is no change of momentum) and with no pressure gradient along the flow, it is apparent that the drag force D_x exerted by one side of the plate (of length x) on the fluid is the only force producing change of momentum between sections 1 and 2. Applying equation 102,

$$-D_x = \beta_2 q \rho V_2 - \beta_1 q \rho V_1 \tag{312}$$

[8] The critical value of 3900 is a nominal one suitable for use in rough calculations and textbook problems. For excessive turbulence in the oncoming flow or rough leading edge of plate it will be considerably less than 3900; indeed, if these upstream conditions are very poor it may be essentially zero, implying a turbulent boundary layer beginning at the leading edge of the plate. The critical value of R_δ is also very sensitive to pressure gradient; a favorable gradient (see Fig. 100) stabilizes the laminar boundary layer and produces a critical value of R_δ greater than 3900. Conversely, an unfavorable pressure gradient (as in a diffuser) causes early breakdown of the laminar flow and a critical value of considerably less than 3900.

[9] The actual velocity profile is nearly parabolic but not precisely so, and this fact will necessitate subsequent adjustments in the derived equations. Comprehensive analytical solution of this boundary layer problem was accomplished by H. Blasius (*Zeit. Math. Physik*, Vol. 56, p. 1, 1908), and his predicted velocity distributions have been confirmed experimentally many times since 1924. The advanced mathematical methods used by Blasius preclude development of his exact solution in an elementary text.

in which [10] (from the parabolic velocity profile) $V_2 = 2V_o/3$, $q = 2V_o\delta/3$, and $\beta_2 = \frac{6}{5}$. From the uniform velocity distribution at section 1, $V_1 = V_o$ and $\beta_1 = 1$. Substituting these values in equation 312 yields

$$D_x \cong 2\rho V_o^2\delta/15 \tag{313}$$ [11]

However, this equation is not directly useful for drag computations since δ is usually not known. The next objective is to obtain the relationship of δ to x so that the form of the laminar boundary layer may be predicted and δ calculated from x. This may be done by considering the differential drag force dD_x acting on a differential length dx of the plate; clearly this differential force may also be expressed as $\tau_o\,dx$, and therefore

$$\tau_o = dD_x/dx$$

Since the flow is laminar, $\tau_o = \mu(dv/dy)_o$ in which $(dv/dy)_o$ may be shown (from the parabolic velocity profile) to be $2V_o/\delta$. Substituting the value of τ_o and performing the indicated differentiation gives

$$\tau_o \cong \frac{2\mu V_o}{\delta} \cong \tfrac{2}{15}\rho V_o^2 \frac{d\delta}{dx} \tag{314}$$ [11]

The variables of this equation may be separated and the equation integrated:

$$\int_0^\delta \delta\, d\delta \cong \frac{15\mu}{\rho V_o}\int_0^x dx \cong \frac{\delta^2}{2} \cong \frac{15\mu x}{\rho V_o} \tag{315}$$ [11]

giving the sought relationship between δ and x, which is conveniently expressed in terms of another Reynolds number, $\mathbf{R}_x = V_o x/\nu$.

$$\frac{\delta}{x} \cong \sqrt{\frac{30}{V_o x/\nu}} \cong \sqrt{\frac{30}{\mathbf{R}_x}} \tag{316}$$ [11]

When this is substituted into equation 313, the drag force is obtained as a function of plate length:

$$D_x \cong \sqrt{32/15\mathbf{R}_x}\, x\rho V_o^2/2 \tag{317}$$ [11]

[10] See illustrative problem, Art. 55.

[11] For agreement with the exact results of Blasius' analysis and also with experimental values, the following changes should be made in the coefficients of the equations: 2/15 to 0.1278 in equation 313, 2 to 1.723 and 2/15 to 0.1278 in equation 314, 2 × 15 to 27 in equation 315, 30 to 27 in equations 316 and 319, $\sqrt{32/15}$ to 1.328 in equations 317 and 318, $\sqrt{8/15}$ to 0.664 in equations 320 and 321.

This equation is easily compared with the standard drag equation to obtain an expression for the drag coefficient,[12] C_f; this is

$$C_f \cong \sqrt{32/15\mathbf{R}_x} \qquad (318)^{11}$$

which shows the drag coefficient to be a function of Reynolds number (for incompressible flow) as predicted by the dimensional analysis of Art. 116.

Other useful relationships may be derived from the foregoing equations. The relation between $\mathbf{R}_x$ and $\mathbf{R}_\delta$ may be obtained by mere rearrangement of equation 316; the result is

$$\mathbf{R}_x \cong \mathbf{R}_\delta^2/30 \qquad (319)^{11}$$

When the critical value of 3900 is substituted for $\mathbf{R}_\delta$, $\mathbf{R}_x \cong 500{,}000$, showing that, in terms of plate length x, a laminar boundary layer is not to be expected beyond this critical value of $\mathbf{R}_x$.

Information on shear stress and its variation along the plate may be obtained from equation 314, with $d\delta/dx$ obtained from equation 316. The result is

$$\tau_o \cong \sqrt{8/15\mathbf{R}_x}\, \rho V_o^2/2 \qquad (320)^{11}$$

This equation shows τ_o to vary *inversely* with $\sqrt{x}$; this means that the frictional stress on the upstream portions of the plate will be larger than that on the portions downstream. It is also customary[13] to express shearing stress in terms of a *local friction coefficient* c_f as $\tau_o = c_f \rho V_o^2/2$, whereupon, by comparison with equation 320,

$$c_f \cong \sqrt{8/15\mathbf{R}_x} \qquad (321)^{11}$$

Further insight into boundary layer properties may be obtained by comparing (see Fig. 253) the assumed distributions of velocity and shear with those derived by the exact analysis of Blasius (for which the velocity profile has been accurately confirmed by experiment). Experience with pipe flow has shown (Art. 67) that a parabolic velocity profile in laminar flow is accompanied by a linear shear stress distribution. However, the exact analysis shows that the velocity profile is not precisely parabolic nor the stress profile precisely linear, the most important property of both of these profiles being their asymptotic character at the outer edge of the boundary layer; this means that boundary layer thickness δ must be arbitrarily

[12] The subscript f is used here to emphasize the wholly frictional nature of the coefficient.

[13] As in pipe flow, see equation 183.

defined; in engineering work it is customary to define δ to the point at which the velocity is 99 percent of the free stream velocity V_o. Another basic feature of boundary layer flows is the deflection of the free stream away from the plate or body, which is caused by the

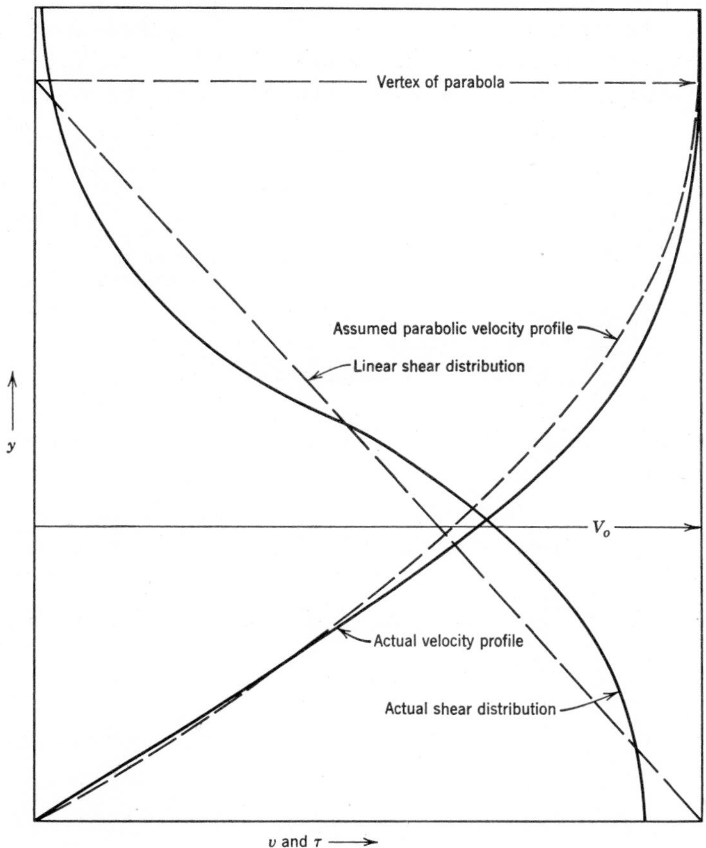

FIG. 253. Shear and velocity profiles in a laminar boundary layer.

presence of the boundary layer; from Fig. 252 the amount of the deflection is seen to be $\delta/3$ for the (assumed) parabolic velocity profile and is also very close to this value for the exact analysis. This quantity is known as the *displacement thickness* of the boundary layer, and it becomes important in precise calculations since it in effect augments the thickness of plate or body and thus alters the pressure distributions on them; in the case of the flat plate, for example, the deflection of the streamlines causes a slight increase of free stream velocity along the plate accompanied by a slight drop of pressure, and a small favorable pressure gradient. Even though these features have been ignored in Blasius' exact analysis of the laminar boundary layer, there is no appreciable effect upon the final results, all of which have been confirmed by experiment.

An approximate analysis of the turbulent boundary layer (Fig. 254) may be made assuming the seventh-root velocity profile and the accompanying shear stress expression developed by Blasius for

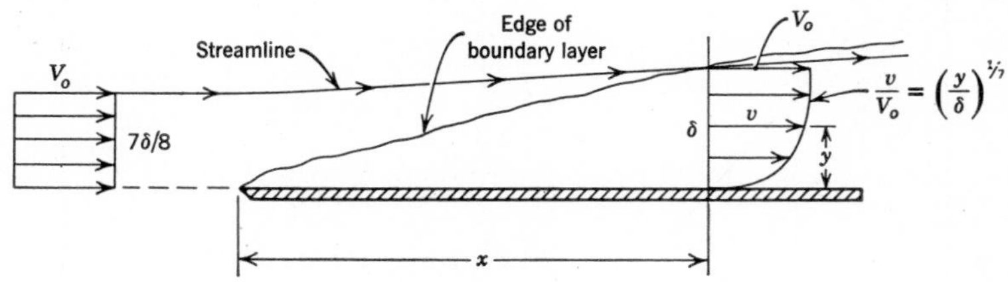

FIG. 254. Turbulent boundary layer.

established turbulent pipe flow. Here the flowrate whose momentum is changed in the distance x of Fig. 254 is

$$q = \int_0^\delta v\,dy = \frac{V_o}{\delta^{\frac{1}{7}}}\int_0^\delta y^{\frac{1}{7}}\,dy = \frac{7V_o\delta}{8}$$

Applying the impulse-momentum principle, as for the laminar boundary layer,

$$-D_x = \beta_2 q\rho V_2 - \beta_1 q\rho V_1 \tag{322}$$

in which $V_2 = 7V_o/8$, $V_1 = V_o$, $\beta_1 = 1$ and β_2 may be computed from equation 153 as follows:

$$\beta_2 = \frac{1}{\delta V_2^2}\int_0^\delta v^2\,dy$$

Substituting $7V_o/8$ for V_2 and $V_o(y/\delta)^{\frac{1}{7}}$ for v and integrating yields $\beta_2 = 64/63$. Substituting these values into equation 322 gives

$$D_x \cong \tfrac{7}{72}\rho V_o^2\delta \tag{323}$$

As for the laminar case, $\tau_o = dD_x/dx$, but here equation 207 (Art. 72) for τ_o may be used. Thus

$$\tau_o = \frac{dD_x}{dx} \cong \tfrac{7}{72}\rho V_o^2\frac{d\delta}{dx} \cong 0.0464\left(\frac{\nu}{V_o\delta}\right)^{\frac{1}{4}}\frac{\rho V_o^2}{2}$$

Separating variables and integrating,

$$\int_0^\delta \delta^{\frac{1}{4}}\, d\delta \cong 0.2385 \left(\frac{\nu}{V_o}\right)^{\frac{1}{4}} \int_0^x dx$$

Performing the integration and substituting $\mathbf{R}_x$ for $V_o x/\nu$ gives

$$\frac{\delta}{x} \cong \left(\frac{0.0079}{\mathbf{R}_x}\right)^{\frac{1}{5}} \cong \frac{0.38}{\mathbf{R}_x^{\ 0.2}} \tag{324}$$

allowing the approximate shape and size of the turbulent boundary layer to be predicted. Substitution of δ from equation 324 into equation 323 will yield expressions for D_x and C_f. These are

$$D_x \cong \tfrac{7}{72}\rho V_o^{\ 2}(0.38x)/\mathbf{R}_x^{\ 0.2} \cong \frac{0.074}{\mathbf{R}_x^{\ 0.2}}\, x\rho V_o^{\ 2}/2$$

whence, by comparison with equation 311,

$$C_f \cong 0.074/\mathbf{R}_x^{\ 0.2} \tag{325}$$

and again the drag coefficient is seen to be a function of Reynolds number only. The seventh-root law in boundary layers, as in pipe flow, can be expected to be adequate only over a limited range of Reynolds numbers which must be determined by experiment; comparison of these on Fig. 255 shows this range of $\mathbf{R}_x$ to be from 10^5 to about 10^8 and implies that computations for boundary layer thickness may be made with equation 324 in this range.

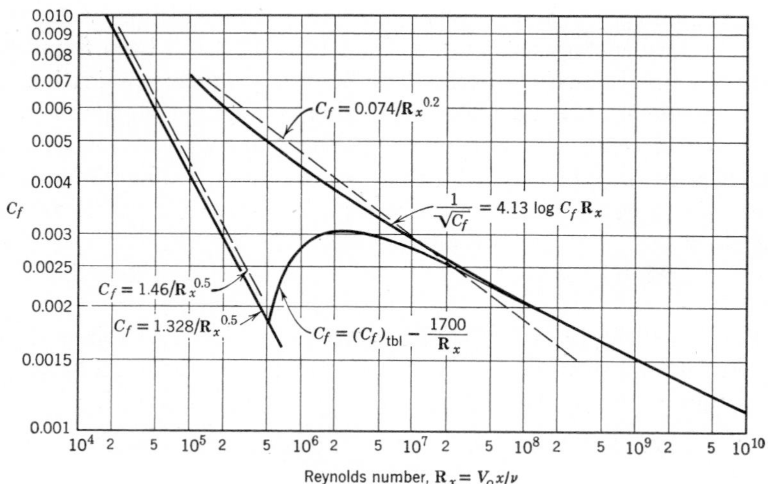

FIG. 255. Drag coefficients for smooth, flat plates.

A more refined analysis of the turbulent boundary layer has been made by von Kármán,[14] who used the logarithmic law of velocity distribution which leads, with some adjustment of coefficients, to

$$\frac{1}{\sqrt{C_f}} = 1.70 + 4.15 \log C_f \mathbf{R}_x$$

However, Schoenherr[15] found that further adjustment of the constants gave an equation which more accurately represented the results of a wide range of experiments; his equation, now used as a standard in American practice, is

$$\frac{1}{\sqrt{C_f}} = 4.13 \log C_f \mathbf{R}_x \qquad (326)$$

A plot of C_f against $\mathbf{R}_x$ for smooth flat plates (Fig. 255) with laminar and turbulent boundary layers bears a striking resemblance to the Stanton diagram for smooth circular pipes (Fig. 119). However, the critical Reynolds number is not so well defined as its counterpart in pipe flow because of flow conditions which are not so well controlled. With increased initial turbulence in the approaching flow, earlier breakdown of the laminar boundary layer occurs, thus reducing the critical Reynolds number; roughening the leading edge of the plate has also been found to decrease the critical Reynolds number by decreasing the flow stability and causing earlier breakdown of the laminar layer. On the other hand, a decrease in pressure along the flow (*favorable pressure gradient*) produced by curvature of the boundary surface or at a pipe inlet (Fig. 100) has been found to delay the breakdown of the laminar boundary layer; application of this principle has produced the so-called *laminar flow wing* for aircraft which exhibits a drag well below that of ordinary wings. The sizable reduction in drag obtained by maintenance of a laminar boundary layer is obvious at once from Fig. 255; in aerodynamic practice, suction slots and porous materials, along with smooth leading edges, and appropriately shaped wing profiles are used to accomplish this reduction.

When smooth plates feature a laminar boundary layer followed by a turbulent one (Fig. 251), experimental drag coefficients fall between the laminar and turbulent lines of Fig. 255, leaving the former

[14] Th. von Kármán, "Turbulence and Skin Friction," *Jour. Aero. Sci.*, Vol. 1, p. 1, 1934.

[15] K. E. Schoenherr, "Resistance of Flat Plates Moving through a Fluid," *Trans. Soc. Nav. Arch. and Marine Engrs.*, Vol. 40, p. 279, 1932.

abruptly and approaching the latter asymptotically. Prandtl expressed this fact mathematically by

$$C_f = (C_f)_{\text{tbl}} - 1700/\mathbf{R}_x \tag{327}$$

in which $(C_f)_{\text{tbl}}$ would be obtained from the Schoenherr equation 326; equation 327 is also plotted on Fig. 255, giving a critical Reynolds number of 537,000. This value, however, should be taken only as a typical or nominal one; the range of critical Reynolds numbers determined experimentally without a favorable pressure gradient range from 100,000 to 600,000. For favorable pressure gradients they may be considerably larger than 600,000.

Illustrative Problem

A ship model 5 ft long with draft of 0.5 ft is towed at a velocity of 1 fps in a basin containing water at 60°F. Assuming that one side of the immersed portion of the hull may be approximated by a smooth flat plate (5 ft x 0.5 ft), estimate the frictional drag of the hull and the thickness of the boundary layer at the stern of the model if the boundary layer is (*a*) laminar, (*b*) turbulent. If the measured total drag of the model is 0.04 lb, estimate the total drag of the prototype if the model scale is 1:64.

Solution. Obtaining ν from Appendix II,

$$\mathbf{R}_x = 1 \times 5/0.00001217 = 410{,}000 \tag{175}$$

From Fig. 255, (*a*) $C_f = 0.0020$; (*b*) $C_f = 0.0052$. For the laminar boundary layer,

$$D_x \text{ (2 sides of plate)} = 2 \times 0.002 \times (5 \times 0.5)1.938 \times (1)^2/2 = 0.0097 \text{ lb} \tag{311}$$

Similarly, for the turbulent boundary layer,

$$D_x \text{ (2 sides)} = 0.025 \text{ lb}$$

For the thickness of the laminar boundary layer,

$$\delta = 5 \times 5.20/\sqrt{410{,}000} = 0.040 \text{ ft} \tag{316}$$

For the turbulent boundary layer,

$$\delta = 5 \times 0.38/(410{,}000)^{0.2} = 0.145 \text{ ft} \tag{324}$$

Restudying Art. 61, the corresponding speed of the prototype is obtained from the Froude law of similitude:

$$(1)^2/5 \times 32.2 = V_o^2/(5 \times 64)32.2; \quad V_o = 8 \text{ fps} \tag{171}$$

The Reynolds number for the prototype is therefore

$$\mathbf{R} = 8 \times 5 \times 64/0.00001217 = 210{,}000{,}000 \qquad (175)$$

giving C_f (from Fig. 255) 0.00185 and allowing the frictional drag of the prototype to be estimated:

$$D_x \text{ (2 sides)} = 2 \times 0.00185(5 \times 0.5 \times \overline{64}^2)1.938 \times (8)^2/2 = 2350 \text{ lb} \qquad (311)$$

With the boundary layer turbulent for the prototype, the turbulent boundary layer assumption for the model must be used to estimate the wave drag, D_w; this is the difference between total and frictional drags:

$$D_w = 0.040 - 0.025 = 0.015 \text{ lb}$$

Since the wave drag is modeled by the Froude law and the Froude numbers are the same in model and prototype, D_w will vary with $A\rho V^2$; hence D_w for the prototype may be calculated from the proportion

$$\frac{D_w}{0.015} = \frac{(\overline{64}^2 \times 0.5 \times 5)8^2}{(0.5 \times 5)1^2}; \quad D_w = 3930 \text{ lb}$$

Finally, the total drag of the prototype (at a speed of 8 fps) is estimated to be

$$D = 3930 + 2350 = 6280 \text{ lb}$$

118. Profile Drag. Although the total drag force on any immersed object is always the sum of frictional and pressure drag, it will be seen later that this breakdown of the drag is inconvenient for objects (such as airfoils) on which a transverse (lift) force is exerted. Here the total drag is considered to be the sum of (1) that which would be developed if the airfoil had no ends (i.e., two-dimensional flow), and (2) that produced by any end effects. Since the former depends only upon the shape (profile) and orientation of the airfoil, it is called *profile drag*, whereas the latter, which depends upon the airfoil plan form and is *induced* by the lift force, is termed *induced drag*. Evidently, for objects which exhibit no lift, the induced drag will be zero and the profile drag equal to the total drag.

Pressure drag has been shown (Art. 115) to be that part of total drag resulting from pressure variation over the surface of an object and to be dependent upon wake formation downstream from the object. In general, when wakes are large, pressure drag is large and, when wake width is reduced by streamlining, pressure drag is reduced also; pressure drag is thus critically dependent upon the existence and position of flow separation which in turn depends upon shape of object and structure of the boundary layer. Consider first the case of a well-streamlined object (Fig. 256); here there is no separation, and the flowfield for ideal- or real-fluid flow is essentially the same except for boundary layer growth and lack of trailing edge stagnation point in the latter. In this case a good approximation to the pressure distribution on the object may be made by neglecting the boundary layer thickness and applying the methods of mathematical hydrodynamics (Chapter 12); such computations may be refined (with considerable difficulty) through altering the shape of the object by the displacement thickness of the boundary layer. These methods cannot, of course, yield the pressure at the trailing edge of the body, which for ideal flow is the stagnation pressure but for real flow is considerably less than this. However, the pressure may be estimated (by extrapolating the pressure distribution over the body) with sufficient accuracy that the pressure variation over the body may be appropriately integrated to allow a reasonable estimate of the pressure drag. More accurate values of pressure drag may be obtained from experimentally determined pressure distributions in the same manner.

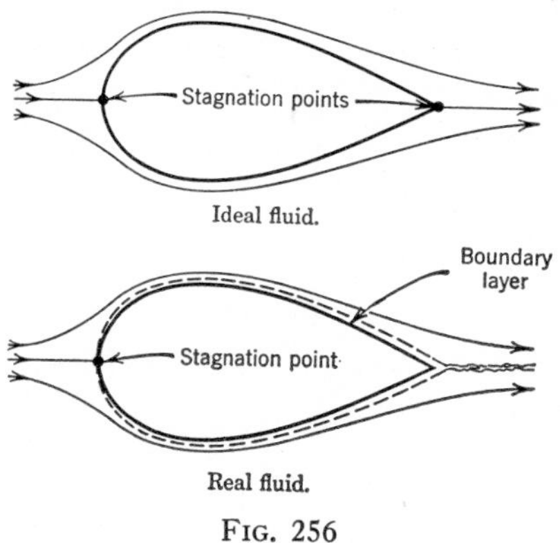

FIG. 256

For a blunt object (Fig. 257) there is a drastic difference between the ideal and real flowfields caused by flow separation and wake formation in the latter. Here the critical feature either for analytical calculations or for understanding of experimental results is the position of the separation points. Lack of resistance and energy dissipation in the ideal flow will allow fluid particles adjacent to the object to move between stagnation points, accelerating over the upstream end of the object in a favorable pressure gradient and decelerating over the downstream end through an unfavorable pressure gradient. For the real fluid, boundary layer growth will begin at the stagnation point and energy will be dissipated in overcoming resistance caused by shear stresses in the boundary layer. The momentum of fluid particles in the boundary layer will thus be considerably less than

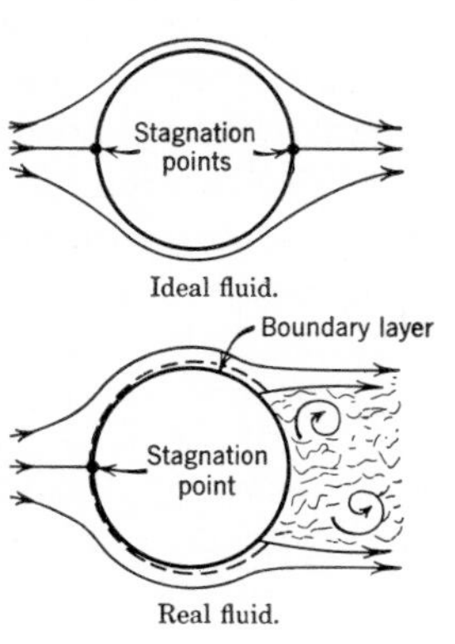

FIG. 257

those at corresponding positions in the ideal flowfield; the momentum of such particles will be further reduced by the unfavorable pressure gradient until at some point they will come to rest, accumulate, and be given a rotary motion by the surrounding flow; separation of the live flow from the object then results as the eddy increases in size. This description of the separation process applies at the inception of flow and at the beginning of wake formation. Once separation has occurred, a new flowfield is established and there is no reason to expect reattachment [16] of the live stream to the object.

With separation dependent upon boundary layer growth, it may be expected that the laminar or turbulent character of the boundary layer will be of critical importance in determining the position of separation. A simple comparison of laminar and turbulent layers of the same ρ, V_o, and δ will show their momentum fluxes ($\beta q \rho V$) to be $8\rho V_o^2\delta/15$ and $7\rho V_o^2\delta/9$, respectively—the momentum flux of the turbulent layer being nearly 50 percent greater [17] than that of the laminar one. Thus the turbulent boundary layer may be considered the "stronger" of the two and better able to survive unfavorable pressure gradient; accordingly, it may be expected that the separation point for a turbulent boundary layer will be found farther downstream than that for a laminar boundary layer.

Quantitative aspects of profile drag may be obtained from a study of the experimentally determined drag coefficients of various objects. First, consider the basic sphere, which has been exhaustively studied. At low Reynolds number (Fig. 258*a*) the flow will close behind the sphere and no wake will form; under these conditions profile drag is composed almost entirely of frictional drag. Stokes [18] has shown analytically that, in laminar flow at very low Reynolds numbers, where inertia forces may be neglected and those of viscosity alone considered, the drag of a sphere of diameter d, moving at a velocity V_o through a fluid of viscosity μ, is given by

$$D = 3\pi\mu V_o d \tag{328}$$

and this equation has been confirmed by many experiments. The drag coefficient, C_D, for the sphere under these conditions may be found by equating the preceding expression to equation 311:

$$C_D A \rho V_o^2/2 = 3\pi\mu V_o d$$

Taking A to be the area of the projection of the object on a plane,

[16] However, this may occur under special conditions for well-streamlined objects.

[17] A better comparison of laminar and turbulent boundary layers on the same object shows an even larger difference.

[18] G. G. Stokes, *Mathematical and Physical Papers*, Vol. III, p. 55, Cambridge University Press, 1901.

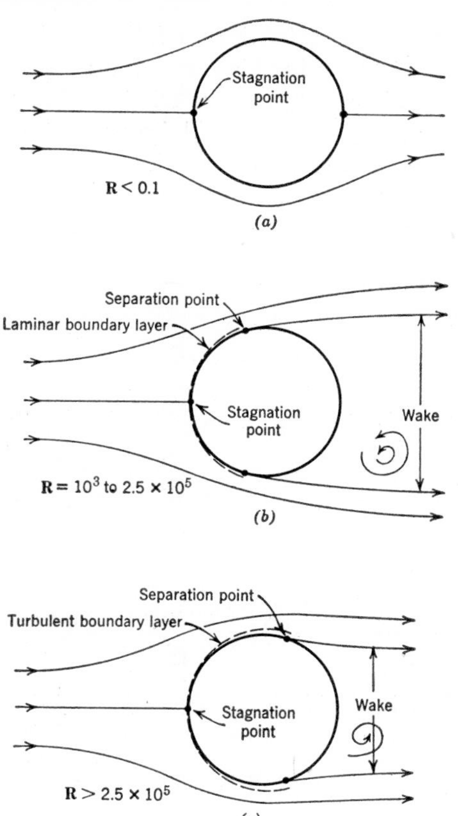

FIG. 258. Flow about a sphere at various Reynolds numbers.

normal to the direction of V_o, $A = \pi d^2/4$, and substituting this above gives

$$C_D = \frac{24\mu}{V_o d\rho} = \frac{24}{\mathbf{R}}$$

Thus the drag coefficients of spheres at low velocities are dependent only on the Reynolds number—another confirmation of the results of the dimensional analysis of Art. 116.

As the Reynolds number increases, the drag coefficients of spheres continue to depend only upon the size of this number, and a plot of experimental results over a large range of Reynolds numbers for spheres of many sizes, tested in many fluids, gives the single curve of Fig. 259.

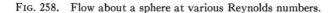

Up to a Reynolds number of 0.1, the Stokes equation applies accurately and the drag coefficient results from frictional effects. As the Reynolds number is increased to about 10, separation and weak eddies begin to form, enlarging into a fully developed wake near a Reynolds number of 1000; in this range the drag coefficient results from a combination of pressure and frictional drag, the frictional drag becoming about 5 percent of total drag as a Reynolds number of 1000 is reached (see Fig. 258). Above this figure the effects of friction become even smaller and the drag problem becomes primarily one of pressure drag.

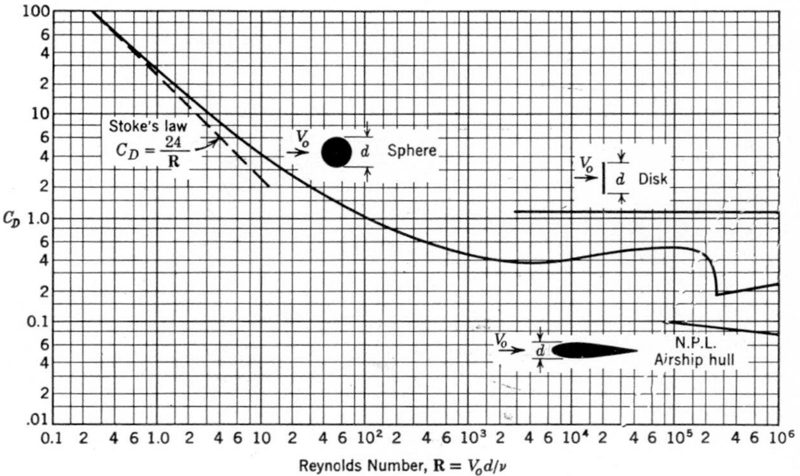

FIG. 259. Drag coefficients for sphere, disk, and streamlined body.[19]

The drag coefficient of the sphere ranges (approximately) between 0.4 and 0.5 from $\mathbf{R} \cong 1000$ to $\mathbf{R} \cong 250{,}000$, at which point it suddenly drops more than 50 percent and then increases gradually with further increase in the Reynolds number (see Figs. 258 and 259). In this range of Reynolds numbers, experiments have shown the separation point to be upstream from the midsection of the sphere, resulting in a relatively wide turbulent wake; the boundary layer on the surface of the sphere from stagnation point to separation point has been found to be laminar up to $\mathbf{R} \cong 250{,}000$. With further increase in $\mathbf{R}$ the length of the laminar boundary layer decreases, the boundary layer flow past the separation point becomes turbulent, and the separation point moves to a position downstream from the center of the sphere (Fig. 258*c*), causing a decrease in the width of the wake and consequent decrease in the drag coefficient.

[19] Data from L. Prandtl, *Ergebnisse der aerodynamischen Versuchsanstalt zu Göttingen*, Vol. II, p. 29, R. Oldenbourg, 1923; and G. J. Higgins, "Tests of the N. P. L. Airship Models in the Variable Density Wind Tunnel," *N.A.C.A. Tech. Note* 264, 1927.

The change from laminar to turbulent boundary layer on a flat plate has been seen (Art. 117) to occur at a critical Reynolds number dependent upon the turbulence of the approaching flow. It also occurs with a sphere, and with increased turbulence in the approaching flow the sudden drop in the drag coefficient curve occurs at lower Reynolds number. Thus a sphere may be used as a relative measure of turbulence by noting the Reynolds number at which a drag coefficient of 0.30 (see Fig. 259) is obtained. Before the development of the hot-wire anemometer (Art. 95), this method was used to compare the turbulence characteristics of different wind tunnels.

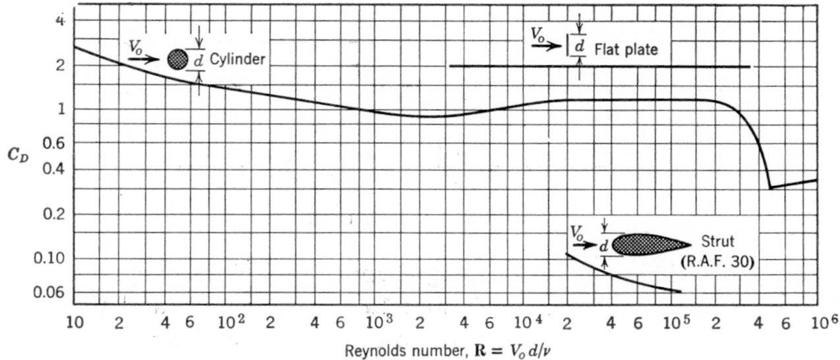

Fig. 260. Drag coefficients for circular cylinders, flat plates, and streamlined struts of infinite length.[20]

The drag coefficient of a thin circular disk placed normal to the flow shows practically no variation with the Reynolds number, since the separation point is fixed at the edge of the disk and cannot shift from this point, regardless of the condition of the boundary layer. Thus the width of the wake remains essentially constant, as does the drag coefficient. This idea may be usefully generalized and applied to all brusque or very rough objects in a fluid flow; experiment indicates that such objects have drag coefficients which are essentially constant in the range of high Reynolds numbers.[21]

The drag coefficients of circular cylinders placed normal to the flow show characteristics similar to those of spheres. The coefficients shown in Fig. 260 are for infinitely long cylinders. The drag coef-

[20] Data from L. Prandtl, *Ergebnisse der aerodynamischen Versuchsanstalt zu Göttingen*, Vol. II, p. 24, R. Oldenbourg, 1923; and B. A. Bakhmeteff, *Mechanics of Fluids*, Part II, p. 44, Columbia University Press, 1933.

[21] Compare this with the relation of the friction factor f, and Reynolds number, $\mathbf{R}$, for rough pipes, Figs. 119 and 121, and also the fact that the minor loss coefficients of pipe flow show little variation with the Reynolds number.

ficients of streamlined struts[22] and flat plates of infinite length are also shown for comparison. The total drags of the flat plate and cylinder contain negligible frictional drag at ordinary velocities, whereas the streamlined strut, because of its small turbulent wake, possesses little pressure drag. The curves are typical of those resulting from tests of brusque and streamlined objects.

Long blunt objects such as cylinders, when placed crosswise to a fluid flow, sometimes exhibit the property of shedding large eddies regularly and alternately from opposite sides (Fig. 261). Because of von Kármán's studies of the stability of these regular vortex patterns they are generally known as *Kármán vortex trails.* Experiments have

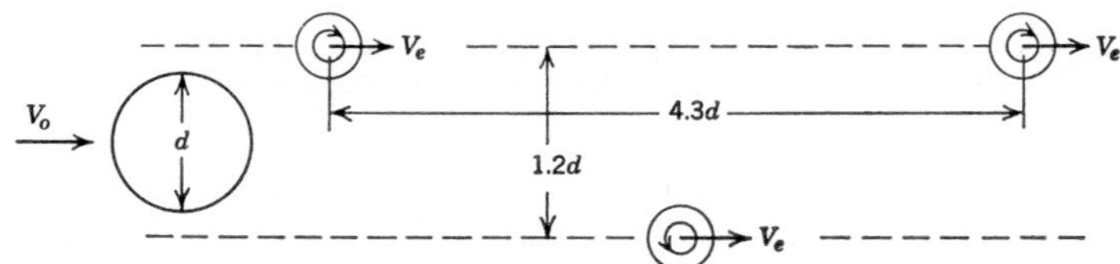

FIG. 261. Vortex trail.

shown that eddies will be shed regularly from a circular cylinder for Reynolds numbers between 10^3 and 10^5 (see Fig. 259) and that the frequency of the phenomenon is about $0.2V_o/d$ in this range. The regular periodic nature of the eddy formation produces transverse forces on the cylinder which are also periodic, and thus tend to produce transverse oscillations. These considerations are vital to the design of elastic structures such as tall chimneys and suspension bridges which are exposed to the wind; they become critical if the natural frequency of vibration is close to the frequency of the eddy formation.

ILLUSTRATIVE PROBLEM

Calculate the ratio between the drag forces on the same sphere in the same fluid stream at Reynolds numbers 400,000 and 200,000.

Solution. From Fig. 259 the drag coefficients at these Reynolds numbers are 0.20 and 0.43, respectively. With sphere size and fluid the same, Reynolds number will vary directly with velocity, and drag with $C_D\mathbf{R}^2$. Writing this as a ratio,

$$\frac{D_4}{D_2} = \frac{0.2(4)^2}{0.43(2)^2} = 1.86$$

Frequently for small changes of Reynolds number such calculations are made neglecting the change of drag coefficient. Had this been done here, the error would have been 115 percent.

[22] The area to be used in the drag equation is the projection of the body on a plane normal to the direction of flow.

119. Lift and Circulation. The Kutta-Joukowsky theorem (Art. 111) has demonstrated that circulation about an object is one of the requirements for the existence of a lift force on the object. Although it is not difficult to imagine a rotating body in a viscous fluid in-

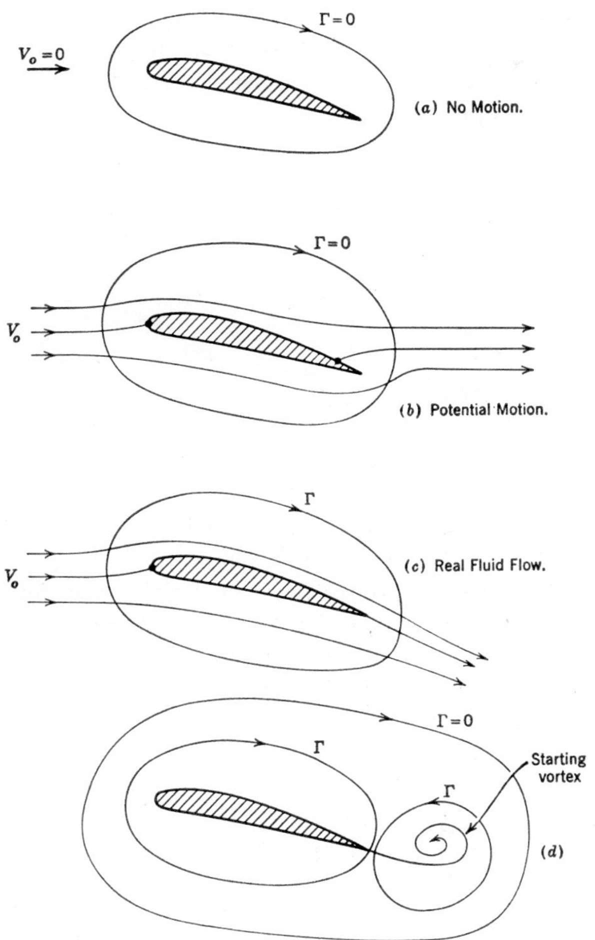

FIG. 262. Development of circulation about an airfoil.

ducing its own circulation, to explain the origin of circulation about an airfoil, or an element of a propeller or turbine blade, requires knowledge of other principles.

Consider the flow conditions about a typical airfoil as it starts to move. Before motion begins, the circulation about the foil is obviously zero (Fig. 262*a*). As motion starts, the circulation about the airfoil tends to remain zero, and the potential flow of Fig. 262*b* tends to be set up, but such a flow, which includes a stagnation point near the rear of the airfoil and flow around its sharp trailing edge, cannot be maintained in a real fluid because of separation. This momentary potential flow gives way immediately to the flow of Fig 262*c*, and in the process a circulation Γ develops about the airfoil, and a vortex, the *starting vortex* (Fig. 262*d*), is shed from the airfoil. During the creation of this vortex, however, the circulation around a closed curve, including and at some distance from the airfoil, is not

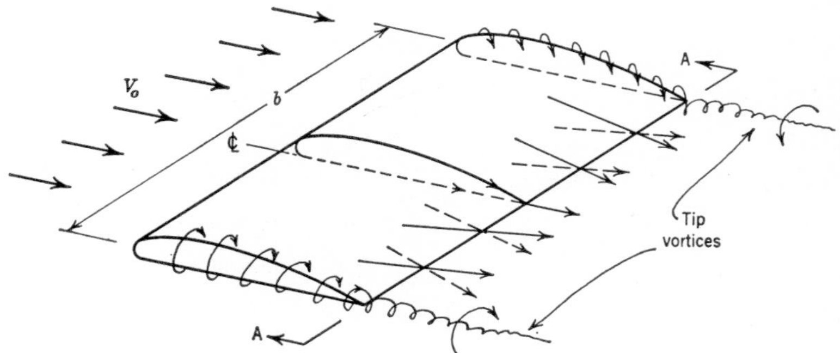

FIG. 263. Airfoil of finite length.

changed and must still be zero; thus, from the properties of circulation, the circulation about, or the strength of, the starting vortex must be equal and opposite to that about the airfoil; the existence of circulation about an airfoil is thus dependent upon the creation of a starting vortex.

120. Airfoils of Finite Length. When fluid flows about airfoils of finite length, flow phenomena result which affect both lift and drag of the airfoil; these phenomena may be understood by further investigation and application of the foregoing circulation theory of lift.

Since pressure on the bottom of an airfoil is greater than that on the top, flow will escape from below the airfoil at the ends and flow toward the top, thus distorting the general flow about the airfoil, causing fluid to move inward over the top of the airfoil and outward over the bottom (Fig. 263). As the fluid merges at the trailing edge of the airfoil, a surface of discontinuity is set up, and flows above and below this surface are, respectively, inward and outward as shown. The tendency for vortices to form from these velocity components is apparent, and, in fact, this surface of discontinuity is a sheet of vortices. However, such a vortex sheet is unstable, and the rotary motions contained therein combine to form two large vortices trailing from the tips of the airfoil (Fig. 263); these are called tip vortices and are often visible when an airfoil passes through dust-laden air or as vapor trails produced by condensation of atmospheric moisture.

Since the pressure difference between top and bottom of an airfoil must reduce to zero at the tips, it is evident that the lift per unit length of span varies over the span (Fig. 264), being maximum at the center and reducing to zero at the tips. The total lift of the airfoil is, of course, the total resulting from this lift diagram. Since lift per unit length of span varies directly with circulation ($L = \Gamma\rho V_o$),

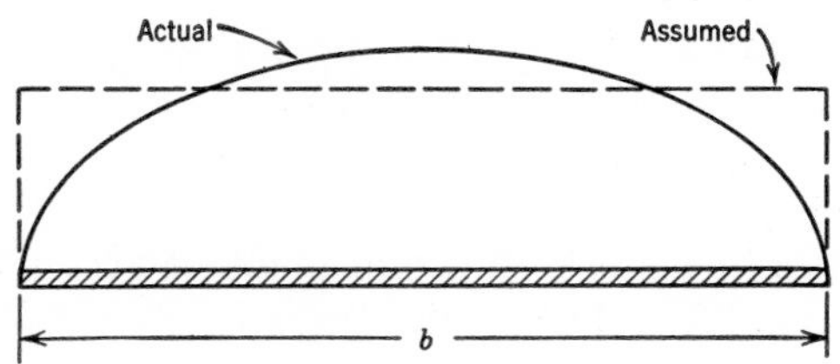

FIG. 264. Distribution of lift and circulation over an airfoil of finite length.

a diagram showing distribution of circulation over the span has the same shape as a diagram of lift distribution. The variation of lift and circulation over the span of an airfoil cannot, of course, be disregarded in a rigorous treatment of the subject, but such treatment leads to mathematical and physical complexities which are beyond the scope of this volume. A simple physical picture may be obtained, however, from the following analysis in which lift and circulation will be assumed to be distributed uniformly over the span (Fig. 264).

One of the properties of vortices is that their axes can end only at solid boundaries. Since there is no solid boundary at the end of the airfoil, the circulation Γ cannot stop here, but must continue to exist about the axes of the tip vortices (Fig. 265). The axes of the tip vortices extend rearward to the axis of the starting vortex; thus, according to the theory, the axis of the vortex having circulation Γ does not end, but is a closed curve composed of the axes of the airfoil, tip vortices, and starting vortex. In the real fluid the circulation persists only about the airfoil and portions of the tip vortices close to the airfoil; the starting vortex and remainder of the tip vortices are quickly extinguished by viscous action.

The circulations about the tip vortices induce a downward motion

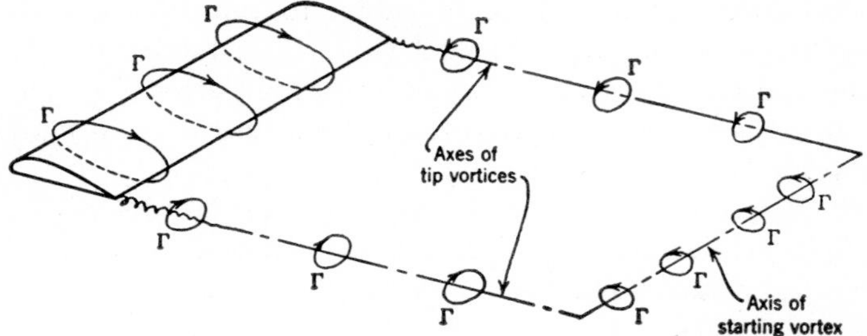

FIG. 265. Circulation about an airfoil of finite length.

in the fluid passing over an airfoil of finite length and in so doing affect both lift and drag *by changing the effective angle of attack.* The strength of this induced motion will, obviously, depend upon the proximity of the tip vortices and, thus, upon the span of the airfoil which may be expressed in terms of the *aspect ratio* b^2/A.

An airfoil of finite span is shown at angle of attack α in the horizontal flow of Fig. 266. The vertical velocity induced near the wing

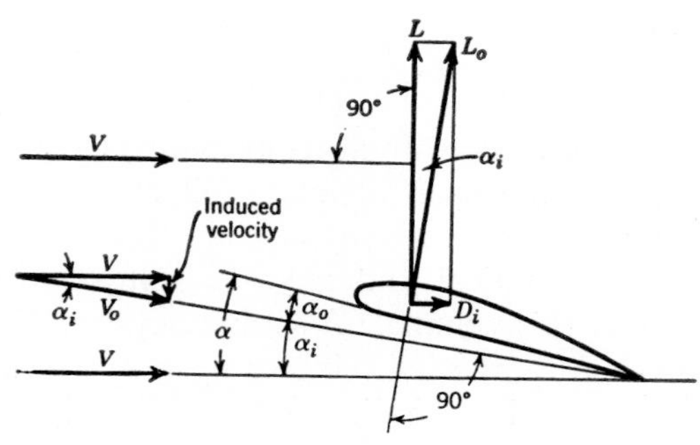

FIG. 266

by the tip vortices decreases the angle of attack by a small angle α_i, making the effective angle of attack $(\alpha - \alpha_i)$. This effective angle of attack is that for no induced velocity, or, in other words, it is the angle of attack which would be obtained if the foil had infinite span and aspect ratio. Calling this angle of attack α_o,

$$\alpha_o = \alpha - \alpha_i \tag{329}$$

Now, treating the airfoil as one of infinite span at an angle of attack α_o, the lift L_o exerted on such an airfoil is by definition normal to the direction of flow in which it is placed; therefore L_o is normal to the effective velocity V_o and at an angle α_i with the vertical. The lift,

L, on the airfoil of finite span is normal to the approaching horizontal velocity V and is the vertical component of L_o. But L_o also has a component in the direction of the original velocity V, which is a drag force, D_i, called the *induced drag* because its existence depends upon the downward velocity induced by the tip vortices. Thus an additional drag force, D_i, must be added to profile drag in computing the total drag of a body of finite length about which a circulation exists. Calling the profile drag D_o, since it is the drag of an airfoil of infinite span (which has no end effects and, therefore, no induced drag), the total drag, D, of an airfoil of finite length is given by

$$D = D_o + D_i$$

which, by dividing by $A\rho V^2/2$, may be expressed in terms of dimensionless drag coefficients as

$$C_D = C_{D_o} + C_{D_i} \tag{330}$$

From the foregoing statements and Fig. 266, it is evident that induced drag, D_i, is related to lift L, angle α, and aspect ratio b^2/A; the equations relating these variables are of great practical importance. Since α_i is small,

$$L = L_o, \quad V = V_o, \quad D_i = L\alpha_i$$

If the distribution of lift over a wing of finite span is taken to be a semiellipse [23] (see Fig. 264), it may be shown that

$$\alpha_i = \frac{C_L}{\pi b^2/A} \tag{331}$$

with drag and lift proportional to their respective coefficients, $D_i = L\alpha_i$ may be written $\alpha_i = C_{D_i}/C_L$, and by substitution of this into equation 331 there results

$$C_{D_i} = \frac{{C_L}^2}{\pi b^2/A} \tag{332}$$

which relates lift and induced drag through their dimensionless coefficients and shows that induced drag is inversely proportional to aspect ratio, becoming zero at infinite aspect ratio (infinite span) and increasing as aspect ratio and span decrease—thus offering mathematical proof of the foregoing statements on the effect of span, aspect ratio, and proximity of tip vortices on induced downward velocity and induced drag.

[23] An assumption which gives minimum induced drag and conforms well (but not perfectly) with fact.

The use of the derived expressions for α_i and C_{D_i} in equations 331 and 332 allows airfoil data obtained at one aspect ratio to be converted into corresponding conditions at infinite aspect ratio, and these data, in turn, to be reconverted to airfoils of any aspect ratio; thus extensive testing of airfoils of the same profile at various aspect ratios becomes unnecessary.

Illustrative Problem

A rectangular airfoil of 6-ft chord and 36-ft span (aspect ratio 6) has a drag coefficient of 0.0543 and lift coefficient of 0.960 at an angle of attack of 7.2°. What are the corresponding lift and drag coefficients and angle of attack for a wing of the same profile and aspect ratio 8?

Solution. For aspect ratio 8:

$$C_L = 0.960 \text{ (negligible change of lift coefficient)}$$

For aspect ratio 6:

$$C_{D_i} = (0.960)^2/6\pi = 0.0489 \qquad (332)$$

For aspect ratio ∞:

$$C_{D_o} = 0.0543 - 0.0489 = 0.0054 \qquad (330)$$

For aspect ratio 8:

$$C_D = 0.0054 + (0.960)^2/8\pi = 0.0421 \qquad (330)$$

For aspect ratio 6:

$$\alpha_i = 0.960/6\pi = 0.0509 \text{ radian} = 2.9° \qquad (331)$$

For aspect ratio ∞:

$$\alpha_o = 7.2 - 2.9 = 4.3° \qquad (329)$$

For aspect ratio 8:

$$\alpha = 4.3 + (0.960/8\pi)(360/2\pi) = 6.5° \qquad (329 \ \& \ 331)$$

121. Lift and Drag Diagrams. The relation between lift and induced drag coefficients suggests plotting lift coefficient against drag coefficient and gives the so-called *polar diagram* of Fig. 267, which is used extensively in airplane design. On this diagram equation 332 appears as a parabola passing through the origin and symmetrical about the C_D-axis, the position of the parabola depending on the aspect ratio. Since the two curves are for airfoils of the same aspect ratio, the horizontal distance between them is the profile drag coefficient C_D. However, the diagram shows much more than this. The important ratio of lift to drag is the slope of a straight line drawn between origin and the point for which this ratio is to be found; the maximum value of this ratio is the slope of a straight line tangent to the curve and passing through the origin; on the diagram are also easily seen the points of zero lift and minimum drag, and the point of maximum lift or *stall*,[24] which determines stalling angle above which lift no longer continues to increase with angle of attack; the end

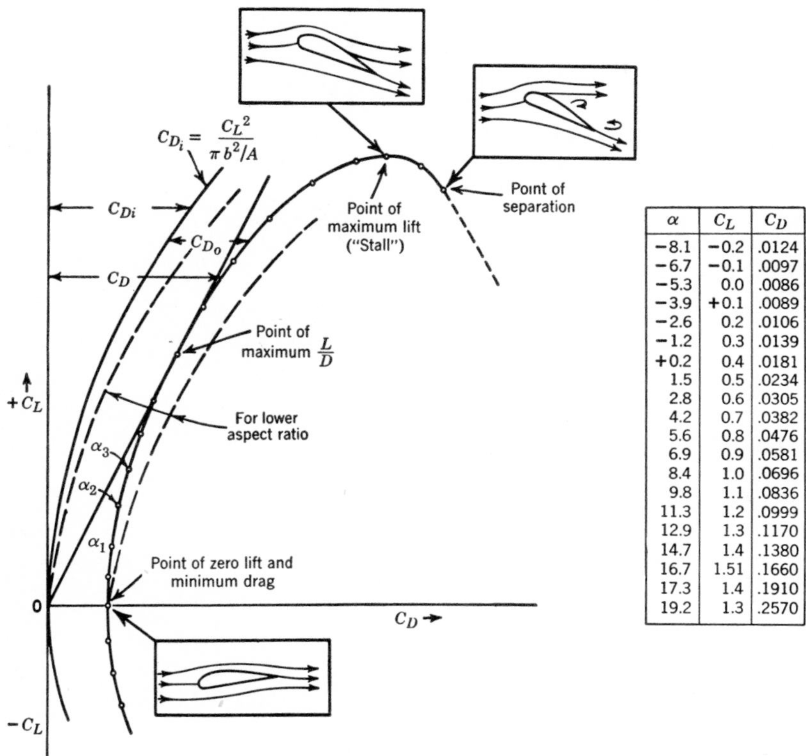

α	C_L	C_D
−8.1	−0.2	.0124
−6.7	−0.1	.0097
−5.3	0.0	.0086
−3.9	+0.1	.0089
−2.6	0.2	.0106
−1.2	0.3	.0139
+0.2	0.4	.0181
1.5	0.5	.0234
2.8	0.6	.0305
4.2	0.7	.0382
5.6	0.8	.0476
6.9	0.9	.0581
8.4	1.0	.0696
9.8	1.1	.0836
11.3	1.2	.0999
12.9	1.3	.1170
14.7	1.4	.1380
16.7	1.51	.1660
17.3	1.4	.1910
19.2	1.3	.2570

FIG. 267. Polar diagram for a typical airfoil and numerical data for the Clark-Y airfoil.[25]

of the upper solid portion of the curve is the point at which the flow separates completely from the upper side of the wing, forming a wake which increases the profile drag (and, therefore, the drag coefficient) and is accompanied by a large drop in lift and lift coefficient because

[24] The mechanism of stall has recently come under intensive study. See H. W. Emmons, R. E. Kronauer, and J. A. Rockett, "A Survey of Stall Propagation—Experiment and Theory," and S. J. Kline, "On the Nature of Stall." Both papers in *Trans. A.S.M.E.* (*Series D*), Vol. 81, 1959.

[25] A 48-ft x 8-ft rectangular airfoil tested at **R** $\sim$ 6,000,000. Data from A. Silverstein, "Scale Effect on Clark-Y Airfoil Characteristics from N.A.C.A. Full-Scale Wind-Tunnel Tests," *N.A.C.A. Rept.* 502, 1934.

of increased pressure on the upper side of the wing. Curves for other aspect ratios may be obtained [26] by the methods of the preceding illustrative problem and will be found to follow the trend indicated. Worthy of note are the equal horizontal distances C_{D_o}, between corresponding curves and the decrease of L/D ratio with decreasing aspect ratio.

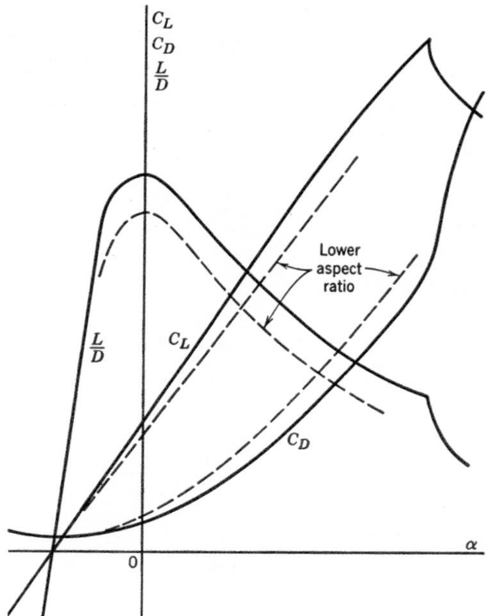

FIG. 268. Airfoil characteristics.

Another method of presenting airfoil data is to plot C_L, C_D, and L/D against angle of attack. Such a plot for the Clark-Y data of Fig. 267 is shown on Fig. 268. Of significance are the slope of the straight portion of the C_L curve, the location of the point of maximum L/D, the overall shape of the C_D curve, and the change in the position of the curves with change of aspect ratio.

The data of Fig. 267 were obtained at Reynolds numbers around 6,000,000, and from many foregoing statements it should be expected that the data will change with changing Reynolds number. The

[26] Because of deviations from the assumed semielliptical lift distribution which led to equations 331 and 332, the latter must be corrected by experimental coefficients which depend upon the plan form of the wing. The size of the corrections increases with the aspect ratio, but the order of these is 5 percent on C_{D_i} and 15 percent on α_i for rectangular wings.

following trends, which are confirmed by experiment, are of some interest in the light of foregoing principles. With increasing Reynolds number the drag coefficient at low angles of attack decreases; here the drag coefficient contains predominantly frictional effects and its variation with Reynolds number is similar to that of the flat plate (Fig. 255). With increased turbulence, due either to increased initial turbulence or increased Reynolds number, the maximum lift coefficient usually increases; in other words, higher angles of attack can be attained without causing separation. Here the momentum of the turbulent boundary layer delays separation, allowing high-velocity flow to cling to the upper side of the airfoil, causing lower pressures and greater lift.

DRAG AND LIFT—COMPRESSIBLE FLOW

The problems of flow of compressible fluids about solid objects cover a vast field and are highly complex in their physical and mathematical aspects. Analytical solutions are available for only a small number of the less difficult problems, but steady progress is being made by theoretical and experimental research. In view of the complexity and advanced nature of such problems, only the most rudimentary ideas and results can be presented in the following elementary treatment.

122. The Mach Wave. Consider the motion of a tiny source of disturbance such as a needle point or razor edge moving through a fluid (Fig. 269). If the fluid were truly incompressible its modulus of elasticity would be infinite, the velocity of propagation (equation 9, 10, or 11) of the disturbance through the fluid would also be infinite,

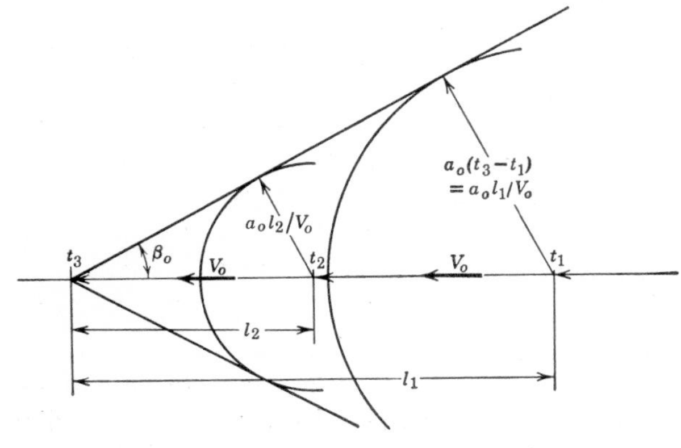

FIG. 269

and the Mach number always zero. The effects of the disturbance would be felt instantaneously through the flowfield, and the fluid in front of the object would "know" of the presence of the object and adjust itself accordingly as the object approached. This situation is closely approximated in real fluid when the flow velocities are small compared with the sonic velocity (which is the velocity of propagation of the disturbance) and produces the conventional flowfields with which the reader is now familiar.

For the opposite case in which the velocity of motion of the disturbances is appreciably greater than the sonic velocity, a very different situation develops in which fluid upstream from the object is "unaware" of its existence; as the object encounters this fluid, the fluid must suddenly change its direction, producing the sharp discontinuities known as *shock waves* (Art. 47). To investigate this, assume the point of Fig. 269 to move at a (supersonic) velocity V_o, and let it occupy positions 1, 2, 3, at times t_1, t_2, and t_3. At time t_1 the disturbance sends out an elastic wave from point 1 with a celerity a_o, and after the time $(t_3 - t_1)$ has elapsed the distance covered by the wave is $a_o(t_3 - t_1)$. In this elapsed time, however, the source of disturbance has moved to a point 3, and $l_1 = V_o(t_3 - t_1)$; eliminating $(t_3 - t_1)$, the distance covered by the wave is $a_o l_1/V_o$. Similarly the wave which started at point 2 has (when the disturbance reaches point 3) covered a distance of $a_o l_2/V_o$, and many other waves have done likewise from numerous intermediate positions. The result is a wave front (or *oblique shock wave*) which represents the line of advance of the elastic waves; the fluid to the left of this is at rest and "unaware" of the presence of the disturbance. This wave front, known as a *Mach wave*, will be conical (three-dimensional) if produced by the needle point, and wedge-shaped (two-dimensional) if produced by the razor edge. The *Mach angle*, β_o, may be seen from the figure to be defined by $\sin \beta_o = a_o/V_o$. This is useful in estimating from photographs of such wave fronts the velocity, V_o, by measuring β_o on the photograph and computing a_o from the fluid properties.

The same wave picture is, of course, produced when the fluid moves with velocity V_o past a source of disturbance at rest. On such a picture the streamlines may also be conveniently shown, and for a vanishingly small disturbance these will be straight parallel lines. For a disturbance of finite magnitude (Fig. 270) the velocity of propagation will be greater than a_o and the angle of the wave, therefore, greater than the Mach angle. Through a finite wave the streamlines will be deflected away from the object, velocities will diminish, and pressures increase (see Art. 47), these changes being comparable in sense (not in magnitude)[27] to those occurring through the normal shock wave (Art. 46). All these changes are very rapid as the shock wave is exceedingly thin; for most practical purposes it may be considered an abrupt discontinuity.

123. Phenomena and Definitions. To visualize the flow phenomena in compressible flow about an object, consider the chain of events which occurs as the free stream velocity is increased from low subsonic to high subsonic to supersonic. At low subsonic speeds the

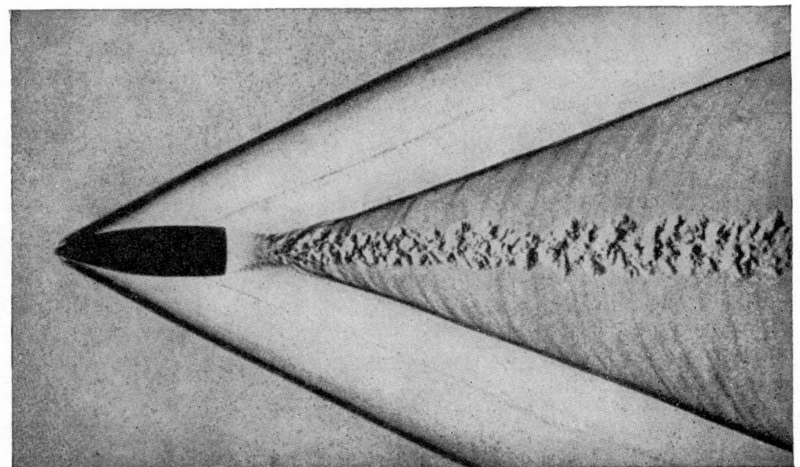

C. Cranz

FIG. 270. Small-bore bullet in flight.[28]

fluid may be considered incompressible, resulting in a conventional streamline picture featuring a stagnation point, s, on the nose of the object and a point, m, of maximum velocity and minimum pressure on the upper side (Fig. 271*a*). Increasing the velocity of the free stream will, of course, raise the stagnation pressure, increase the maximum velocity, and lower the minimum pressure, even if compressibility of the fluid is neglected; inclusion of compressibility exaggerates these changes, however, and they continue until the velocity past m becomes equal to the *local sonic velocity;*[29] this means

[27] The changes are largest in the case of the normal shock wave through which the velocity decreases from supersonic to subsonic; the velocities upstream and downstream from the oblique shock wave are both supersonic except in the case of relatively large β, where the oblique shock approaches the normal shock.

[28] From C. Cranz, *Lehrbuch der Ballistik*, Vol. I, B. G. Teubner, Leipzig, 1917.

[29] It is very important to note that the pressure, density, and temperature at m are considerably less than those properties in the free stream and that the local sonic velocity (equation 11) is thus smaller than free stream sonic velocity.

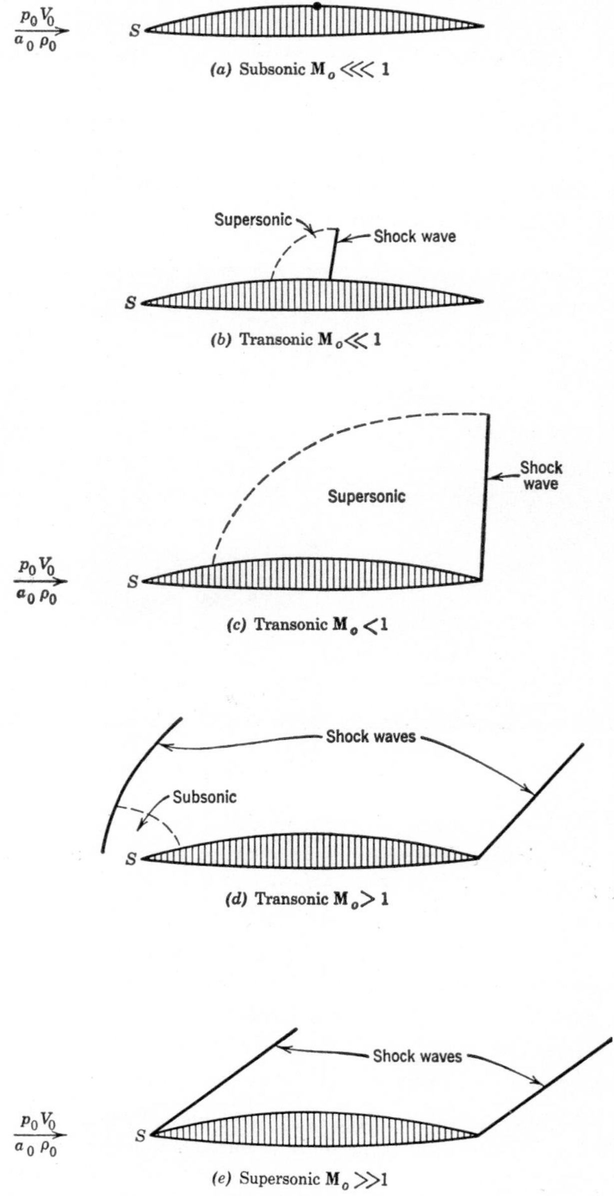

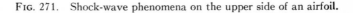

FIG. 271. Shock-wave phenomena on the upper side of an airfoil.

that the *local Mach number* at point m is now unity, although the free stream Mach number is considerably less than this. In other words, the body is moving at subsonic speed, but sonic phenomena are beginning to appear at a point on its surface. This situation marks the end of the *subsonic* flow region (where all velocities in the flow-field are subsonic) and the beginning of the *transonic* regime (in which some velocities are subsonic and others supersonic); the transonic regime ends and the supersonic begins when all velocities in the field are supersonic.

A simplified [30] picture of a transonic flow is shown in Fig. 271*b*. Of particular interest is the large region of supersonic flow and the presence of the shock wave through which the velocity decreases from supersonic to subsonic. Through the shock wave the sudden jump in pressure produces an adverse pressure gradient in the boundary layer, which promotes separation and the usual effects on lift and drag. As the free stream velocity is further increased (but V_o maintained less than a_o) the shock wave is forced rearward (Fig. 271*c*) and the major portion of the flow immediately above the surface is supersonic although the free stream Mach number $\mathbf{M}_o$ is still less than 1.

Increasing the free stream velocity V_o to slightly more than a_o will produce the wave arrangement of Fig. 271*d*, featuring a new shock wave upstream from the body and the flowfield completely supersonic except in a small region [31] between this new shock wave and the nose of the object. The supersonic regime (Fig. 271*e*) will exist at higher free stream Mach numbers when the subsonic zone becomes of negligible size or vanishes entirely, as it will for a sharp-nosed object, to which the upstream shock wave will attach itself.

124. Drag. In compressible fluid motion, drag results from energy dissipated in shock waves as well as from the skin friction and separation effects discussed in Arts. 117 and 118. Skin friction drag may be computed approximately by the methods of Art. 117 up to free stream Mach numbers around 2, but boundary layer thicknesses are much greater in compressible fluid motion and stability is greatly

[30] The complex problem of shock wave-boundary layer interaction is omitted from these simplified sketches. In the boundary layer the velocities near the surface are subsonic while the flow outside the boundary layer and upstream from the shock wave is supersonic; the subsonic region allows transmission upstream of the adverse pressure gradient produced by the shock and thus spreads out the region of shock phenomena at the boundary surface.

[31] Evidently the extent of this region depends upon the nose geometry of the body since considerations of Art. 122 have shown it to be nonexistent if the nose is sharply edged or pointed.

affected by the transfer of heat between solid surface and boundary layer. At higher Mach numbers, frictional heating in the boundary layer and adjacent surface becomes a serious problem. The approach to compressible fluid motion which assumes Mach number effects predominant and Reynolds number effects negligible implies that frictional effects and boundary layers are of little importance compared to shock phenomena. This should be recognized as an adequate working assumption but not an absolute fact, and many exceptions to this convenient rule are to be expected. For example, if methods are found for minimizing shock wave effects and preventing separation, a very large portion of total drag force will be composed of frictional drag, and Reynolds number effects will then by no means be negligible.

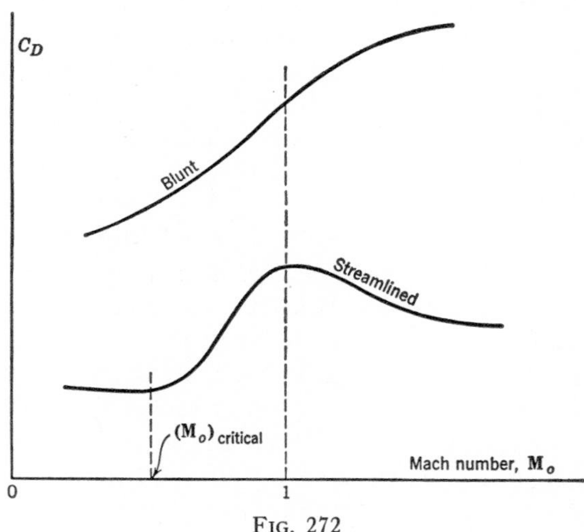

FIG. 272

The variation of drag coefficients with Mach number (Fig. 272) is central to drag problems in compressible flows and may be examined fruitfully by considering the extreme cases of a streamlined airfoil or body of revolution and a blunt body such as a flat-nosed projectile. For the latter, separation of fluid from body is fixed in position by the geometry of the body, skin friction is small, and a steady increase in drag coefficient with Mach number results from the predominant compressibility effects on or near the flat nose of the projectile. Streamlining the tail of such an object will increase rather than reduce the drag coefficient, but pointing the nose, and thus reducing the frontal area near the stagnation point, will materially reduce the drag coefficient (see Fig. 273).

For a streamlined object the variation of drag coefficient is more interesting because of the changing position of the separation point which accompanies the formation of shock wave phenomena.

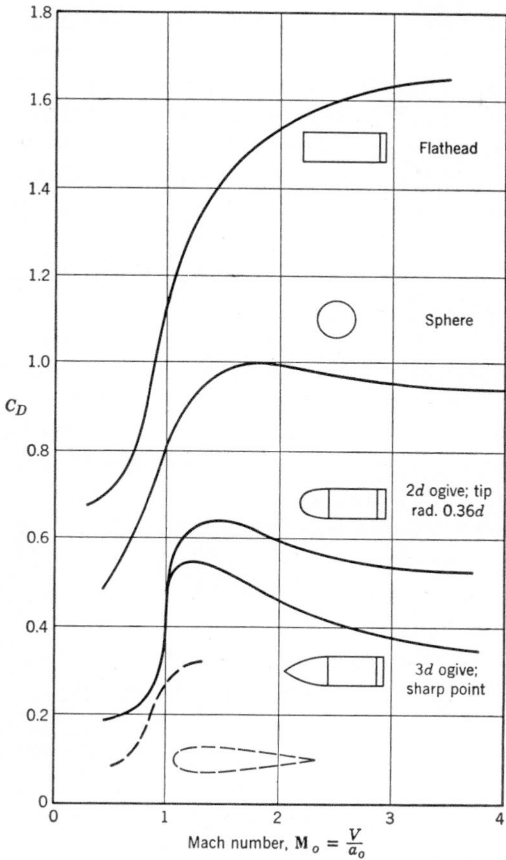

FIG. 273. Drag coefficients for bodies of revolution.[32]

Through the subsonic range (Fig. 271a) the slight decrease in drag coefficient with Reynolds number will appear on the plot (Fig. 272) of drag coefficient against Mach number since both of these numbers are directly proportional to V_o. As the transonic range is entered (Fig. 271b and accompanying discussion), shock phenomena and

[32] Data from: F. R. W. Hunt, *The Mechanical Properties of Fluids*, p. 341, Blackie and Son, 1925; and A. C. Charters and R. N. Thomas, "The Aerodynamic Performance of Small Spheres from Subsonic to High Supersonic Velocities," *Jour. Aero. Sci.*, Vol. 12, No. 4, October, 1945.

separation appear in the vicinity of point m and drag coefficient begins to increase[33] rapidly. Toward the middle of the transonic range (Fig. 271c) the separation point has moved well to the rear, reducing the size of the wake, but shock phenomena have intensified and drag coefficient continues to rise but more slowly, eventually reaching a maximum value as these opposite effects cancel each other. In the supersonic range (Fig. 271e), the drag coefficient depends primarily on the energy dissipation through the inclined shock waves and decreases steadily with further increase in Mach number.

Of great practical interest is the conversion of airfoil data obtained for incompressible flow to flow at higher velocities where the effects of compressibility are significant. The appearance of shock phenomena prevents such conversion above the critical Mach number, but below this there is a considerable range of engineering interest, covering free stream velocities between 20 and (roughly) 80 percent of the velocity of sound. The higher figure is the critical Mach number and has been found to depend on the thickness of the airfoil, being larger for thinner airfoils. The method of conversion depends only upon the Mach number and is known as the *Prandtl-Glauert rule;* the critical Mach number (the limit of application of the rule) may be established fairly reliably by application of elementary principles of compressible flow.

Pressure distribution data obtained for an airfoil at a certain angle of attack in incompressible flow are usually presented in dimensionless fashion as a plot of *pressure coefficient*, C_p, against the chord of the airfoil. The pressure coefficient is defined by

$$C_p = \frac{p - p_o}{\rho_o V_o^2/2} \tag{333}$$

in which p_o, V_o, and ρ_o are at free stream conditions and p is the pressure at any point on the airfoil. A typical plot of pressure coefficient is shown on Fig. 274, on which it is customary to plot C_p positive downward. The Prandtl-Glauert rule states[34] that such data may be corrected for use at higher Mach numbers by simply dividing C_p by $\sqrt{1 - \mathbf{M}_o^2}$, providing the critical Mach number for the wing is not exceeded. Space does not permit proof of the Prandtl-Glauert rule here, but its limit of applicability, the critical Mach number, may be easily found. Referring to the discussion of Fig. 271a, ignoring the boundary layer, and using equation 79 for an ideal

[33] Calculations show the major part of this increase to be due more to the separation than to the energy dissipated in the shock waves.

[34] A more refined (and more complicated) rule is that of von Kármán and Tsien.

compressible fluid between the free stream and point m, which is the point of minimum pressure, p_m, and minimum pressure coefficient, C_{p_m},

$$\frac{V_m^2 - V_o^2}{2g} = \frac{p_o}{\gamma_o}\frac{k}{k-1}\left[1 - \left(\frac{p_m}{p_o}\right)^{\frac{k-1}{k}}\right] \tag{79}$$

Shock phenomena begin to appear at m when V_m attains the *local sonic velocity* $\sqrt{kp_m/\rho_m}$. Substituting this for V_m, and a_o for $\sqrt{kp_o/\rho_o}$, this equation may be reduced to

$$\frac{p_m}{\rho_m}\frac{\rho_o}{p_o} - \mathbf{M}_o^2 = \frac{2}{k-1}\left[1 - \left(\frac{p_m}{p_o}\right)^{\frac{k-1}{k}}\right]$$

Substituting $(p_m/p_o)^{1/k}$ for ρ_o/ρ_m, and solving for p_m/p_o yields

$$\frac{p_m}{p_o} = \left[\frac{2 + (k-1)\mathbf{M}_o^2}{k+1}\right]^{\frac{k}{k-1}} \tag{334}$$

By dividing numerator and denominator of the right-hand side of equation 333 by p_o and k, C_{p_m} becomes

$$C_{p_m} = \frac{p_m - p_o}{\rho_o V_o^2/2} = \frac{2(p_m/p_o - 1)}{k\rho_o V_o^2/p_o k} = \frac{2(p_m/p_o - 1)}{k\mathbf{M}_o^2} \tag{335}$$

into which p_m/p_o from equation 334 may be inserted to yield the limiting C_p below which values obtained by the Prandtl-Glauert rule are invalid. The solution for C_{p_m} and the resulting (critical) $\mathbf{M}_o$ may be done by trial or by plotting as indicated on Fig. 275.

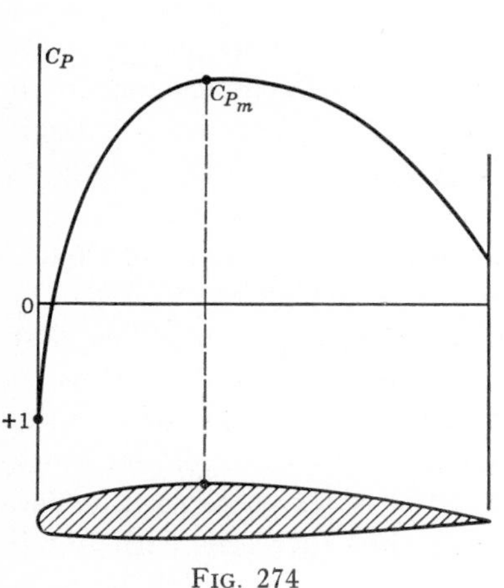

Fig. 274

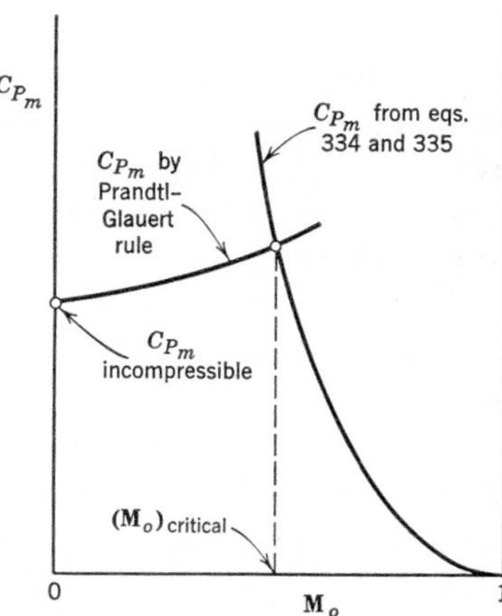

Fig. 275

125. Lift. Because of the steady decrease in pressure over the major portion of the upper side of an airfoil with increase of Mach number in the subsonic range (Fig. 271*a*), the lift coefficient at a given angle of attack will increase through this range. The Prandtl-Glauert rule shows that this may be computed approximately by dividing the lift coefficient of incompressible flow by $\sqrt{1 - \mathbf{M}_o^2}$. The result may be seen on Fig. 276 and applies only up to the point of stall.

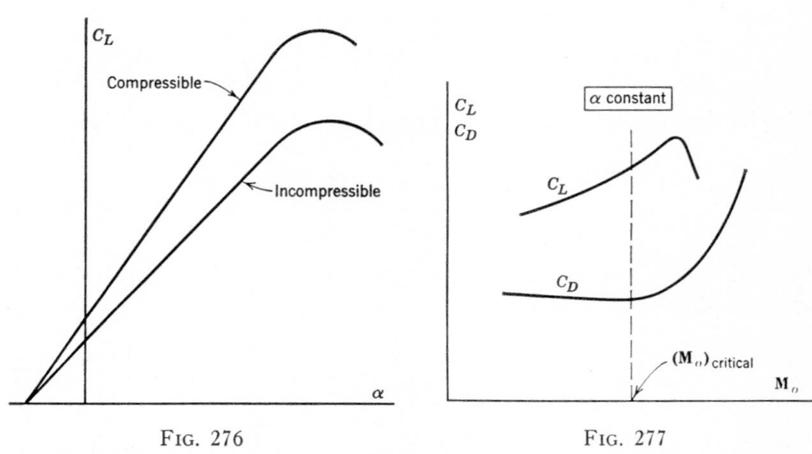

FIG. 276 FIG. 277

With increase of Mach number into the transonic range the lift coefficient continues to increase (for the same angle of attack) until the shock phenomena produce separation (Fig. 271*b*). The presence of the shock wave (through which there is a sudden rise in pressure) and the resulting separation produce a region of increased pressure on the upper side of the airfoil which will cause the lift coefficient to drop sharply. This so-called *shock-stall* usually occurs at Mach numbers slightly above the critical (Fig. 277), but reliable methods for predicting it are as yet unavailable.

The variation of the lift coefficient with Mach number through the transonic and into the supersonic range is pictured in Fig. 278 and related to the simplified flow pictures of Fig. 271. The first drop in lift coefficient brought about by the shock-stall is arrested by the formation of a shock wave and increase of pressure on the lower side of the airfoil and by the change in position of the shock wave on the upper side, which reduces the region of separation and high pressure there. The lift coefficient then increases with Mach number to another maximum until intense shock wave phenomena become predominant, after which the trend is steadily downward with increasing Mach number.

Although analytical methods are unavailable for treating problems in the transonic range, simple expressions for C_L have been worked out for the subsonic and supersonic flows about thin symmetrical airfoils of infinite span and small angle of attack, which have been

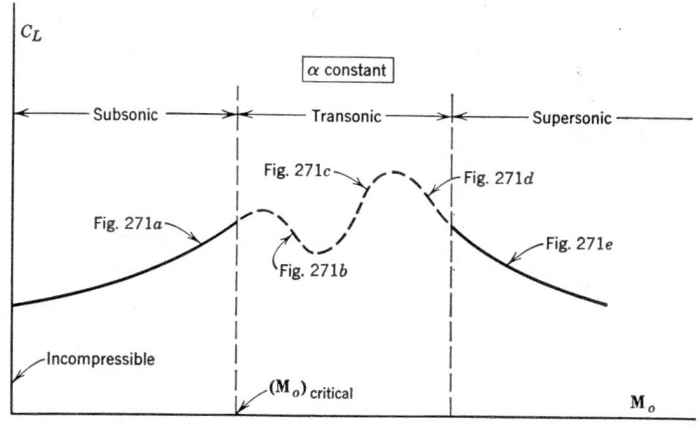

FIG. 278

confirmed by experiment and may be used by engineers as rough guides. These expressions are

$$C_L = \frac{2\pi\alpha}{\sqrt{1 - \mathbf{M}_o^2}} \quad \text{and} \quad C_L = \frac{4\alpha}{\sqrt{\mathbf{M}_o^2 - 1}}$$

for the subsonic and supersonic cases, respectively. In spite of the tremendous advances in theory, however, reliable results for airfoils of finite length and arbitrary profile cannot yet be predicted, and experimental data are considered the only safe basis for design.

REFERENCES

C. V. Drysdale (Ed.), *The Mechanical Properties of Fluids*, Blackie and Son, 1925.

H. Glauert, *Aerofoil and Airscrew Theory*, Cambridge University Press, 1932.

W. F. Durand (Ed.), *Aerodynamic Theory*, Julius Springer, 1934, six volumes.

L. Prandtl and O. G. Tietjens, *Fundamentals of Hydro- and Aeromechanics*, *Applied Hydro- and Aeromechanics*, McGraw-Hill Book Co., 1934.

S. Goldstein (Ed.), *Modern Developments in Fluid Dynamics*, Oxford University Press, 1938, two volumes.

R. von Mises, *Theory of Flight*, McGraw-Hill Book Co., 1945.

P. E. Hemke, *Elementary Applied Aerodynamics*, Prentice-Hall, 1946.

H. L. Dryden, "Recent Advances in the Mechanics of Boundary Layer Flow," *Advances in Applied Mechanics*, Vol. I, Academic Press, 1948.
A. Ferri, *Elements of Aerodynamics of Supersonic Flows*, The Macmillan Co., 1949.
O. G. Sutton, *The Science of Flight*, Penguin Books, 1949.
L. Prandtl, *Essentials of Fluid Dynamics*, Hafner Publishing Co., 1952.
A. H. Shapiro, *The Dynamics and Thermodynamics of Compressible Fluid Flow*, Vol. II, Ronald Press Co., 1953.
L. Howarth (Ed.), *Modern Developments in Fluid Dynamics (High Speed Flow)*, Oxford University Press, 1953, two volumes.
Th. von Kármán, *Aerodynamics*, Cornell University Press, 1954.
H. Schlichting, *Boundary Layer Theory*, McGraw-Hill Book Co., 1955.
Symposium on Boundary Layer Effects in Aerodynamics, National Physical Laboratory (England), 1955.
E. Carafoli, *High-Speed Aerodynamics*, Pergamon Press, 1956.
F. H. Clauser, "The Turbulent Boundary Layer," *Advances in Applied Mechanics*, Vol. IV, Academic Press, 1956.
H. W. Liepmann and A. Roshko, *Elements of Gasdynamics*, John Wiley and Sons, 1957.
S. F. Hoerner, *Fluid Dynamic Drag*, published by the author, 1958.
Th. von Kármán, "Some Significant Developments in Aerodynamics since 1946," *Journ. Aero. Sci.*, March 1959.
W. D. Hayes and R. F. Probstein, *Hypersonic Flow Theory*, Academic Press, 1959.
A. M. Kuethe and J. D. Schetzer, *Foundations of Aerodynamics*, 2nd ed., John Wiley and Sons, 1959.

PROBLEMS

971. A rectangular airfoil of 40-ft span and 6-ft chord has lift and drag coefficients of 0.5 and 0.04, respectively, at an angle of attack of 6°. Calculate the drag and horsepower necessary to drive this airfoil at 50, 100, and 150 mph horizontally through still air (40°F and 13.5 psia). What lift forces are obtained at these speeds?

972. A rectangular airfoil of 30-ft span and 6-ft chord moves horizontally at a certain angle of attack through still air at 150 mph. Calculate the lift and drag, and the horsepower necessary to drive the airfoil at this speed through air of (*a*) 14.7 psia and 59°F, and (*b*) 11.5 psia and 0°F. $C_D = 0.035$, $C_L = 0.46$. Calculate the speed and power required for condition (*b*) to obtain the lift of condition (*a*).

973. The drag coefficient of a circular disk when placed normal to the flow is 1.12. Calculate the force and horsepower necessary to drive a 12-in. disk at 30 mph through (*a*) standard air at sea level (Appendix IV), and (*b*) water.

974. The drag coefficient of an airship is 0.04 when the area used in the drag formula is the two-thirds power of the volume. Calculate the drag of an airship of this shape having a volume of 500,000 cuft when moving at 60 mph through still standard air at sea level (Appendix IV).

975. A wing model of 5-in. chord and 2.5-ft span is tested at a certain angle of attack in a wind tunnel at 60 mph using air at 14.5 psia and 70°F. The lift and drag are found to be 6.0 lb and 0.4 lb, respectively. Calculate the lift and drag coefficient for the model at this angle of attack.

976. A wing of 500 sqft plan form area is to produce a lift of 10,000 lb in level flight through standard air at sea level (Appendix IV) at 250 mph by expending 600 hp. What C_L and C_D are required?

977. A steel sphere of 0.25-in. diameter is fired at a velocity of 2000 fps at an altitude of 30,000 ft in the ICAO Standard Atmosphere (Appendix IV). Calculate the drag force on this sphere.

978. A smooth plate 10 ft long and 3 ft wide moves through still sea level air (Appendix IV) at 15 fps. Assuming the boundary layer to be wholly laminar, calculate (*a*) the thickness of the layer at 2, 4, 6, 8, and 10 ft from the leading edge of the plate; (*b*) the shear stress, τ_o, at those points; and (*c*) the total drag force on one side of the plate. (*d*) Calculate the thickness at the above points if the layer is turbulent. (*e*) Calculate the total drag for the turbulent boundary layer. (*f*) What percentage saving in drag is effected by a laminar boundary layer?

979. A smooth flat plate 8 ft long and 2 ft wide is placed in an air stream (14.7 psia and 59°F) of velocity 30 fps. Calculate the total drag force on this plate (2 sides) if the boundary layer at the trailing edge is (*a*) laminar, (*b*) transition, and (*c*) turbulent.

980. A flat-bottomed scow having a 150-ft by 20-ft bottom is towed through still water (60°F) at 10 mph. What is the frictional drag force exerted by the water on the bottom of the scow? How long could the laminar portion of the boundary layer be, using a critical Reynolds number of 537,000? What is the thickness of the laminar layer at its downstream end? What is the approximate thickness of the boundary layer at the rear end of the bottom of the scow?

981. A streamlined train 400 ft long is to travel at 90 mph. Treating the sides and top of the train as a smooth flat plate 30 ft wide, calculate the total drag on these surfaces when the train moves through still air at sea level (Appendix IV). Calculate the possible length of the laminar boundary layer and the thickness of this layer at its downstream end. What is the thickness of the boundary layer at the rear end of the train? What horsepower must be expended to overcome this resistance?

982. Estimate the frictional drag encountered by the hull of a submarine when traveling deeply submerged at a speed of 20 mph. The length of the hull is 200 ft and its surface area 18,000 sqft. Assume water density 1.99 slugs/cuft and kinematic viscosity 0.000013 sqft/sec.

983. The two rectangular smooth flat plates are to have the same drag in the same fluid stream. Calculate the required x. If the two plates are combined into the T-shape indicated, what ratio exists between the drag of the combination and that of either one? Assume laminar boundary layers in all calculations.

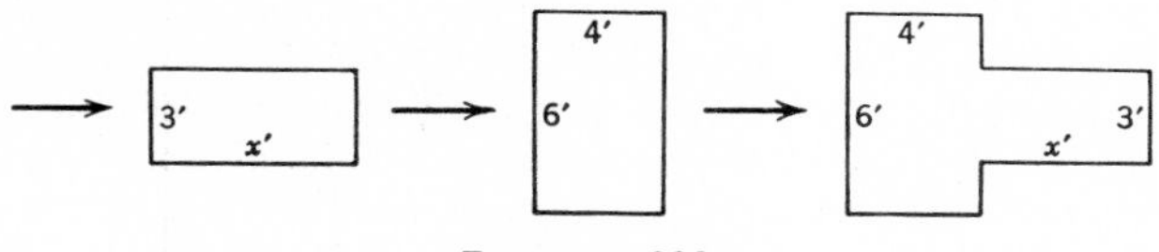

PROBLEM 983

984. If at the trailing edge of a smooth flat plate 10 ft long, covered by a laminar boundary layer, the shearing stress is 0.00001 lb/sqft, what is the total drag force on one side of the plate? What is the shearing stress at a point halfway between leading and trailing edges? What are the drag forces on the front and rear halves of one side of the plate?

985. A smooth flat plate 2.5 ft long is immersed in water at 68°F flowing at an undisturbed velocity of 2 fps. Calculate the shear stress at the center of the plate, assuming a laminar boundary layer over the whole plate.

986. A fluid stream of uniform velocity 10 fps approaches a flat plate 100 ft long which is parallel to the oncoming flow. The Reynolds number calculated with the full length of the plate is 400,000. How much space (sqft) is occupied by the boundary layer?

987. A boundary layer on a flat plate immersed in water increases in thickness from 0.50 ft to 0.51 ft in a distance of 1 ft. The velocity of the undisturbed flow is 25 fps. The flow in the boundary layer is not laminar. If there is a (fictitious) linear velocity distribution in the boundary layer, calculate the mean local shearing stress on this section of the plate.

988. The velocity profile through the boundary layer at the downstream end of a flat plate is found to conform to the equation $v/V_o = (y/\delta)^{\frac{1}{8}}$ in which V_o is 20 fps and δ is 1 ft. Calculate the drag force exerted on (one side of) this plate if the fluid density is 2.0 slugs/cuft.

989. The velocity distributions upstream and downstream from the transition region of a flat plate boundary layer are as indicated. Calculate the total drag force on the plate length AB.

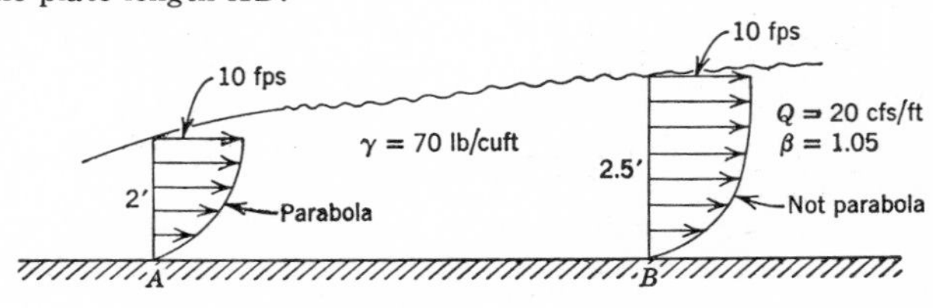

Problem 989

990. Show that for a laminar boundary layer on a flat plate the kinetic energy lost between free stream and any point in the boundary layer is 31.5 percent of the total kinetic energy of a portion of the free stream containing the flowrate in the boundary layer at the above point. How may this loss of kinetic energy be accounted for?

991. Accurate values of C_f from Fig. 255 are 0.00185, 0.00181, and 0.00199 for Reynolds numbers 514,000, 537,000, and 560,000, respectively, which seem to indicate that the local wall shear is larger at the beginning of transition than at the end of the laminar boundary layer. From these figures what percentage increase in mean wall shear is indicated for the reaches implied by the Reynolds numbers?

992. A smooth flat plate 20 ft long moves through a fluid of kinematic viscosity 0.0001 sqft/sec and s.g. 0.80. Where $\mathbf{R}_x$ is 537,000 the shear stress on the plate is 0.067 psf. Estimate the total drag (per foot of width) on one side of the plate.

993. Calculate the drag of a smooth sphere of 12-in. diameter in a stream of standard sea level air (Appendix IV) at Reynolds numbers of 1, 10, 100, and 1000.

994. Calculate the drag of a smooth sphere of 20-in. diameter when placed in an air stream (59°F and 14.7 psia) if the velocity is (*a*) 20 fps, and (*b*) 28 fps. At what velocity will the sphere attain the same drag which it had at a velocity of 20 fps?

995. Estimate the drag on a model of an N.P.L. airship hull of 6-in. diameter which is to be tested in an air stream (14.7 psia and 59°F) at 60 mph.

996. A sphere of 10-in. diameter is tested in a wind tunnel with standard sea level air (Appendix IV) at 80 mph. At what speed must a 2-in. sphere be towed in water (68°F) for these spheres to have the same drag coefficients? What are the drag forces on these two spheres?

997. A steel sphere (s.g. 7.82) of 2-in. diameter is released in a large tank of oil (s.g. 0.82, viscosity 0.020 lb-sec/sqft). Calculate the terminal velocity of this sphere.

998. A cylindrical chimney 3 ft in diameter and 75 ft high is exposed to a 35-mph wind (59°F and 14.7 psia); estimate the bending moment at the bottom of the chimney. Neglect end effects.

999. The drag force exerted on an object (having a volume of 1 cuft and s.g. of 2.80) is 100 lb when moving at 20 fps through oil (s.g. 0.90). If this object is allowed to fall through the same oil, is its terminal velocity larger or smaller than 20 fps?

1000. A standard marine torpedo is 21 in. in diameter and about 24 ft long. Make an engineering estimate of the horsepower required to drive this torpedo at 50 mph through fresh water at 68°F. Assume hemispherical nose, cylindrical body, and flat tail. C_D for a solid hemisphere (flat side downstream) is about 0.35.

1001. This thin smooth wing lands at a speed of 120 mph in air of specific weight 0.0765 lb/cuft and kinematic viscosity 0.00015 sqft/sec. A braking parachute is released to slow it down. Calculate the approximate diameter of the parachute required to produce an extra drag equal to the wing drag at this speed. Assume flow about the wing two-dimensional.

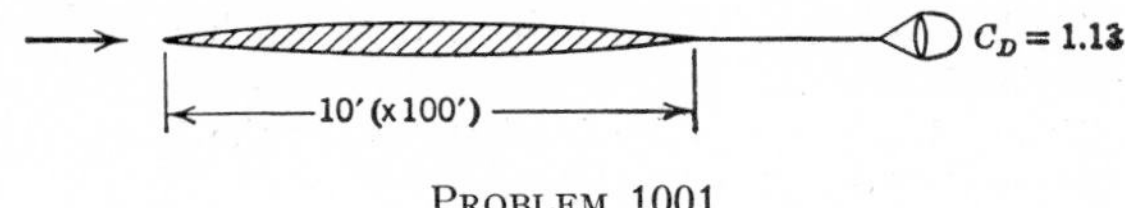

PROBLEM 1001

1002. A thin circular disk is placed (normal to the flow) in an air stream of velocity 500 fps, pressure 20 psia, and density 0.0025 slug/cuft. If the pressure on the upstream side of the disk may be assumed to vary elliptically from stagnation pressure at the center to p_x at the edges, estimate the magnitude of p_x if this pressure is also exerted over the downstream side of the disk.

1003. If Γ is the mean circulation about a wing per foot of span, calculate the circulation about the wing at midpoint and quarter-points of the span, assuming a semielliptical lift distribution.

1004. Derive a general expression for lift coefficient in terms of circulation.

1005. An airfoil of 5-ft chord and 30-ft span develops a lift of 3000 lb when moving through air of specific weight 0.0763 lb/cuft at a velocity of 100 mph. What is the mean circulation about the wing?

1006. If the mean velocity adjacent to the top of a wing of 6-ft chord is 90 mph and that adjacent to the bottom of the wing 70 mph when the wing moves through still air (0.075 lb/cuft) at 75 mph, estimate the lift per foot of span.

1007. A model wing of 5-in. chord and 3-ft span is tested in a wind tunnel (60°F and 14.5 psia) at 60 mph, and the lift and drag are found to be 9.00 and 0.460 lb, respectively, at an angle of attack of 6.7°. Assuming a semielliptical lift distribution, calculate (*a*) the lift and drag coefficients, (*b*) C_{D_i}, (*c*) C_{D_o}, (*d*) the corresponding angle of attack for an airfoil of infinite span, (*e*) the corresponding angle of attack for a foil of this type with aspect ratio 5, and (*f*) the lift and drag coefficients at this aspect ratio.

1008. An airfoil of infinite span has lift and drag coefficients of 1.31 and 0.062, respectively, at an angle of attack of 7.3°. Assuming semielliptical lift distribution, what will be the corresponding coeffiicents for an airfoil of the same profile but aspect ratio 6? What will be the corresponding angle of attack?

1009. From the data of Fig. 267, calculate the lift and drag coefficients for a Clark-Y airfoil of aspect ratio 8, and plot the polar diagram for this airfoil.

1010. The Clark-Y airfoil of Fig. 267 is to move at 180 mph through standard sea level air (Appendix IV). Determine the minimum drag, drag at optimum L/D, and drag at point of maximum lift. Calculate the lift at these points and the horsepower that must be expended to obtain these lifts.

1011. Using Fig. 255 and assuming a turbulent boundary layer, what approximate percentage of the total drag (at zero lift) can be attributed to skin friction for the Clark-Y airfoil of Fig. 267?

1012. If the pointed artillery projectile of Fig. 273 is of 12-in. diameter and is moving at 2000 fps through standard sea level air (Appendix IV), what drag force is exerted on it?

1013. What is the drag of the blunt-nosed projectile of Fig. 273 (if its diameter is 3 in.) when it travels at (*a*) 700 mph, and (*b*) 800 mph, through standard sea level air (Appendix IV).

1014. An airfoil travels through standard sea level air (Appendix IV) at 500 mph. Calculate the local pressure and velocity on the airfoil at which shock wave phenomena will be expected.

1015. The minimum pressure coefficient for an airfoil at a certain angle of attack in incompressible flow is -0.7. Predict the critical Mach number ($\mathbf{M}_o$) for this airfoil at this angle of attack.

1016. Plot lift coefficient against angle of attack for the Clark-Y airfoil of Fig. 267 for free stream Mach numbers of 0 and 0.4.

1017. A thin airfoil at a small angle of attack produces a certain lift force at a free stream Mach number of 0.90. At what Mach number(s) will the same airfoil at the same angle of attack in the same air stream produce the same lift?

appendix I

Symbols and Dimensions of Quantities

Symbol	Quantity	English Engineering Dimensions	Force-Length-Time Dimensions	Mass-Length-Time Dimensions
A	Area	ft^2	L^2	L^2
a	Wave velocity	ft/sec (fps)	L/T	L/T
a	Linear acceleration	ft/sec^2	L/T^2	L/T^2
B	Bottom width (open channel)	ft	L	L
b	Surface width (open channel)	ft	L	L
b	Span of airfoil	ft	L	L
C	Chezy coefficient	$ft^{\frac{1}{2}}/sec$	$L^{\frac{1}{2}}/T$	$L^{\frac{1}{2}}/T$
c_p	Specific heat at constant pressure	ft-lb/lb°R		
c_v	Specific heat at constant volume	ft-lb/lb°R		
D	Drag force	lb	F	ML/T^2
d	Diameter	ft	L	L
E	Modulus of elasticity	lb/ft^2	F/L^2	M/LT^2
E	Unit energy	ft-lb/lb	L	L
e	Size of roughness	ft	L	L
F	Force	lb	F	ML/T^2
G	Weight flowrate	lb/sec	F/T	ML/T^3
g	Gravitational acceleration	ft/sec^2	L/T^2	L/T^2
H	Total head	ft	L	L
H	Head on weir	ft	L	L
H	Enthalpy	ft-lb/lb	L	L
h	Head or height	ft	L	L
I	Moment of inertia	ft^4	L^4	L^4
I	Internal energy	ft-lb/lb	L	L
K	Coefficient of permeability	ft/sec (fps)	L/T	L/T
L	Lift force	lb	F	ML/T^2
l	Length	ft	L	L
M	Mass	slugs	FT^2/L	M
M	Moment	ft-lb	FL	ML^2/T^2

Symbol	Quantity	English Engineering Dimensions	Force-Length-Time Dimensions	Mass-Length-Time Dimensions
n	Manning's roughness coefficient	$ft^{\frac{1}{6}}$	$L^{\frac{1}{6}}$	$L^{\frac{1}{6}}$
P	Power	ft-lb/sec	FL/T	ML^2/T^3
P	Perimeter	ft	L	L
P	Weir height	ft	L	L
p	Pressure	lb/ft^2 (psf)	F/L^2	M/LT^2
Q	Flowrate	ft^3/sec (cfs)	L^3/T	L^3/T
q	Two-dimensional flowrate	ft^2/sec (cfs/ft)	L^2/T	L^2/T
R	Radius	ft	L	L
R	Engineering gas constant	ft-lb/lb°R		
R_h	Hydraulic radius	ft	L	L
r	Radial distance	ft	L	L
T	Torque	ft-lb	FL	ML^2/T^2
T	Absolute temperature	°R		
t	Temperature	°F		
t	Time	sec	T	T
u	Velocity	ft/sec (fps)	L/T	L/T
V	Velocity	ft/sec (fps)	L/T	L/T
V	Volume	ft^3 (cuft)	L^3	L^3
v	Velocity	ft/sec (fps)	L/T	L/T
W	Weight	lb	F	ML/T^2
w	Velocity	ft/sec (fps)	L/T	L/T
y	Depth (open channel)	ft	L	L
y	Distance from solid boundary	ft	L	L
z	Height above datum	ft	L	L
Γ	Circulation	ft^2/sec	L^2/T	L^2/T
γ	Specific weight	lb/ft^3	F/L^3	M/L^2T^2
δ	Boundary layer or film thickness	ft	L	L
ϵ	Eddy viscosity	$lb\text{-}sec/ft^2$	FT/L^2	M/LT
μ	Viscosity	$lb\text{-}sec/ft^2$	FT/L^2	M/LT
ν	Kinematic viscosity	ft^2/sec	L^2/T	L^2/T
ξ	Vorticity	sec^{-1}	$1/T$	$1/T$
ρ	Density	$slug/ft^3$	FT^2/L^4	M/L^3
σ	Surface tension	lb/ft	F/L	M/T^2
τ	Shear stress	lb/ft^2	F/L^2	M/LT^2
ϕ	Velocity potential	ft^2/sec	L^2/T	L^2/T
ψ	Streamfunction	ft^2/sec	L^2/T	L^2/T
ω	Angular velocity	sec^{-1}	$1/T$	$1/T$

SYMBOLS FOR DIMENSIONLESS QUANTITIES

Symbol	Quantity	Symbol	Quantity
C	Cauchy number	n	Polytropic exponent
C_c	Coefficient of contraction	**R**	Reynolds number
C_D	Drag coefficient	S	Slope of energy line
C_f	Frictional drag coefficient	S_o	Bottom slope (open channel)
C_L	Lift coefficient	S_c	Critical slope
C_p	Pressure coefficient	**W**	Weber number
C_v	Coefficient of velocity	Y	Expansion factor
C_w	Weir coefficient	α	Kinetic energy correction factor
E	Euler number		
F	Froude number	β	Momentum correction factor
f	Friction factor	η	Efficiency
K_L	Loss coefficient	κ	von Kármán's turbulence "constant"
k	Adiabatic exponent		
M	Mach number	Π	Dimensionless group

appendix **II**

Physical Properties of Water

Temperature, °F	Specific Weight,[a] γ, lb/ft^3	Density,[a] ρ, slug/ft^3	Modulus[b] of Elasticity,[c] $E/10^3$, psi	Viscosity,[a] $\mu \times 10^5$, lb-sec/ft^2	Kinematic Viscosity,[a] $\nu \times 10^5$, ft^2/sec	Surface[a] Tension,[d] σ, lb/ft	Kinematic Surface[a] Tension,[d] σ/ρ, ft^3/sec^2	Vapor Pressure,[e] p_v, psia	Vapor Pressure Head,[f] p_v/γ, ft
32	62.42	1.940	287	3.746	1.931	0.00518	0.00267	0.09	0.19
40	62.43	1.940	296	3.229	1.664	0.00514	0.00265	0.12	0.28
50	62.41	1.940	305	2.735	1.410	0.00509	0.00262	0.18	0.41
60	62.37	1.938	313	2.359	1.217	0.00504	0.00260	0.26	0.59
70	62.30	1.936	319	2.050	1.059	0.00498	0.00257	0.36	0.84
80	62.22	1.934	324	1.799	0.930	0.00492	0.00254	0.51	1.17
90	62.11	1.931	328	1.595	0.826	0.00486	0.00251	0.70	1.62
100	62.00	1.927	331	1.424	0.739	0.00480	0.00249	0.95	2.21
110	61.86	1.923	332	1.284	0.667	0.00473	0.00246	1.27	2.97
120	61.71	1.918	332	1.168	0.609	0.00467	0.00243	1.69	3.95
130	61.55	1.913	331	1.069	0.558	0.00460	0.00240	2.22	5.20
140	61.38	1.908	330	0.981	0.514	0.00454	0.00238	2.89	6.77
150	61.20	1.902	328	0.905	0.476	0.00447	0.00235	3.72	8.75
160	61.00	1.896	326	0.838	0.442	0.00441	0.00232	4.74	11.17
170	60.80	1.890	322	0.780	0.413	0.00434	0.00229	5.99	14.18
180	60.58	1.883	318	0.726	0.385	0.00427	0.00226	7.51	17.85
190	60.36	1.876	313	0.678	0.362	0.00420	0.00223	9.34	22.40
200	60.12	1.868	308	0.637	0.341	0.00413	0.00221	11.52	27.63
212	59.83	1.860	300	0.593	0.319	0.00404	0.00217	14.70	35.35

[a] From "Hydraulic Models," *A.S.C.E. Manual of Engineering Practice*, No. 25, A.S.C.E., 1942.
[b] Approximate values averaged from many sources.
[c] At atmospheric pressure.
[d] In contact with air.
[e] From J. H. Keenan and F. G. Keyes, *Thermodynamic Properties of Steam*, John Wiley and Sons, 1936.
[f] Computed from p_v and γ.

appendix III

Wave Velocities

The velocity (or celerity) of small waves through fluids may be predicted by application of the continuity and impulse-momentum principles. For the incompressible (inelastic) fluid as in a rigid (inelastic) body, such waves would be expected to travel at infinite speeds as disturbances are transmitted instantaneously from point to point. For a fluid assumed incompressible, such as a liquid with a free surface, a small wave appears as a slight localized rise in the liquid surface; this type of wave is called a *gravity wave.* For a compressible fluid (liquid or gas), a small wave is featured by a slight localized rise in pressure and density of the fluid; such a wave is called a *wave of compression.* These waves are depicted in Fig. 279 moving at velocity a on or through fluids at rest. To a stationary observer the fluid motion is unsteady as the wave passes, but if the observer moves with the wave at velocity a he sees a steady flow which may be easily analyzed with elementary principles. Taking V to be the velocity at the centers of the waves and applying the continuity principle (equations 55 and 50),

Gravity wave:

$$ay = V(y + dy)$$

Compression wave:

$$Aa\rho = AV(\rho + d\rho)$$

Applying the impulse-momentum principle (equation 102),

Gravity wave:

$$\frac{\gamma}{2} y^2 - \frac{\gamma}{2}(y + dy)^2 = ay\frac{\gamma}{g}(V - a)$$

Gravity wave

Wave of compression

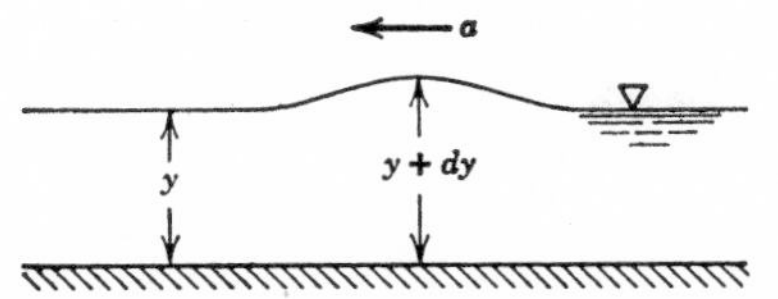

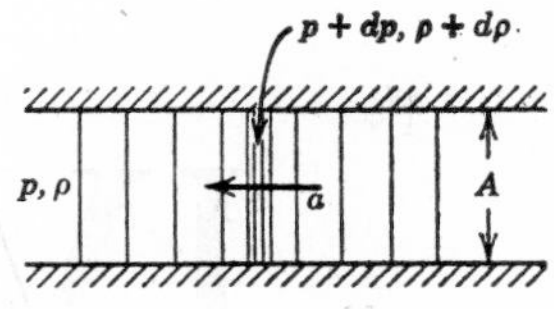

As seen by a stationary observer

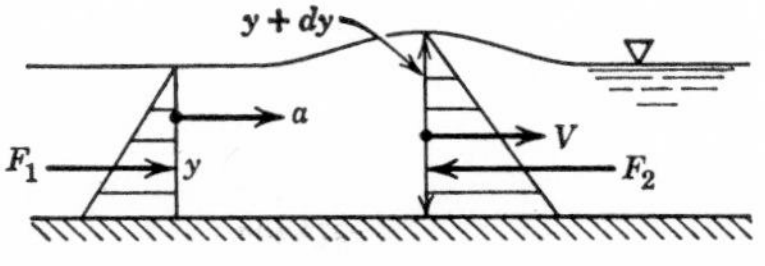

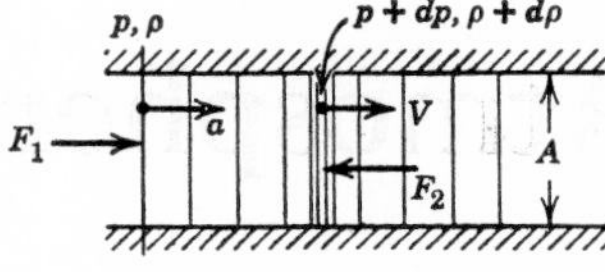

As seen by an observer moving with the wave

FIG. 279. Motion of small waves.

Compression wave:

$$pA - A(p + dp) = Aa\rho(V - a)$$

Simultaneous solution of the appropriate equations with the elimination of V yields

Gravity wave:

$$a^2 = g(y + dy)$$

Compression wave:

$$a^2 = \frac{dp}{d\rho}\left(\frac{\rho + d\rho}{\rho}\right)$$

As dy and $d\rho$ approach zero (the basic condition for a vanishingly small wave) these equations reduce to

Gravity wave:

$$a = \sqrt{gy}$$

Compression wave:

$$a = \sqrt{dp/d\rho}$$

appendix **IV**

The ICAO Standard Atmosphere[1]

Altitude, ft	Temperature, °F	Pressure, psia	Specific Weight, lb/cuft	Density, slug/cuft	Viscosity × 10^7, lb-sec/sqft
0	59.00	14.696	0.07648	0.002377	3.737
5,000	41.17	12.243	0.06587	0.002048	3.637
10,000	23.36	10.108	0.05643	0.001756	3.534
15,000	5.55	8.297	0.04807	0.001496	3.430
20,000	−12.26	6.759	0.04070	0.001267	3.325
25,000	−30.05	5.461	0.03422	0.001066	3.217
30,000	−47.83	4.373	0.02858	0.000891	3.107
35,000	−65.61	3.468	0.02367	0.000738	2.995
40,000	−69.70	2.730	0.01882	0.000587	2.969
45,000	−69.70	2.149	0.01481	0.000462	2.969
50,000	−69.70	1.690	0.01165	0.000364	2.969
55,000	−69.70	1.331	0.00917	0.000287	2.969
60,000	−69.70	1.049	0.00722	0.000226	2.969
65,000	−69.70	0.826	0.00568	0.000178	2.969
70,000	−69.70	0.650	0.00447	0.000140	2.969
75,000	−69.70	0.512	0.00352	0.000110	2.969
80,000	−69.70	0.404	0.00277	0.000087	2.969
85,000	−65.37	0.318	0.00216	0.000068	2.997
90,000	−57.20	0.252	0.00168	0.000053	3.048
95,000	−49.05	0.200	0.00131	0.000041	3.099
100,000	−40.89	0.160	0.00102	0.000032	3.150

[1] Data from R. A. Minzner, W. S. Ripley, and T. P. Condron, U.S. Extension of the ICAO Standard Atmosphere, U.S. Dept. of Commerce, Weather Bureau, 1958. This document also contains: "tentative" data between altitudes of 105,000 and 246,000 ft, and "speculative" data between altitudes of 246,000 and 985,000 ft.

appendix V

Basic Operations with Partial Derivatives

If F is some function of x and y, written $F = F(x, y)$, the partial derivative $\partial F/\partial x$ is obtained by differentiating the function in respect to x, *considering x to be the only variable;* to obtain $\partial F/\partial y$ the differentiation is performed in respect to y, considering y to be the only variable.

Example:

$$F = 2x^2 + 3xy + 4y^2 + C$$

$$\frac{\partial F}{\partial x} = 4x + 3y + 0 + 0$$

$$\frac{\partial F}{\partial y} = 0 + 3x + 8y + 0$$

The total differential of F may be calculated from,[1] in fact is defined by,

$$dF = \left(\frac{\partial F}{\partial x}\right) dx + \left(\frac{\partial F}{\partial y}\right) dy$$

Continuing the example:

$$dF = (4x + 3y)\, dx + (3x + 8y)\, dy$$

[1] Rigorous proof of this may be found in any text on higher mathematics. See, for example, H. W. Reddick and F. H. Miller, *Advanced Mathematics for Engineers*, Art. 66, 3rd edition, John Wiley and Sons, 1955.

Since the differential may be integrated to obtain F,

$$F = \int \left(\frac{\partial F}{\partial x}\right) dx + \int \left(\frac{\partial F}{\partial y}\right) dy + C$$

Again continuing the example:

$$F = \int (4x + 3y)\, dx + \int (3x + 8y)\, dy + C$$

Performing the integration depends upon recognition of the differentials of certain expressions, which often requires considerable insight and experience. Writing the foregoing expression for F as

$$F = 4\int x\, dx + 3\int y\, dx + 3\int x\, dy + 8\int y\, dy + C$$

the first and fourth integrals are routine. The sum of the second and third integrals may be written $3\int d(xy)$ since $y\, dx + x\, dy$ is recognized as the differential of xy. The integration may then be performed to yield

$$F = 2x^2 + 3xy + 4y^2 + C$$

With integration the opposite of differentiation, this result is, of course, to be expected.

A more difficult example (in polar coordinates) is to find F when

$$\frac{\partial F}{\partial r} = -\frac{\sin \theta}{r^2} \quad \text{and} \quad \frac{\partial F}{r\, \partial \theta} = \frac{\cos \theta}{r^2}$$

Here

$$dF = \left(\frac{\partial F}{r\, \partial \theta}\right) r\, d\theta + \left(\frac{\partial F}{\partial r}\right) dr$$

and the integral is

$$F = \int \frac{\cos \theta}{r^2}\, r\, d\theta - \int \frac{\sin \theta}{r^2}\, dr + C$$

However, $\cos\theta\, d\theta/r - \sin\theta\, dr/r^2$ may be recognized as $d[(\sin\theta)/r]$, so the solution is

$$F = \int d\left(\frac{\sin\theta}{r}\right) + C = \frac{\sin\theta}{r} + C$$

appendix **VI**

Properties of Areas and Volumes

	Sketch	Area or Volume	Location of Centroid	I or I_c
Rectangle	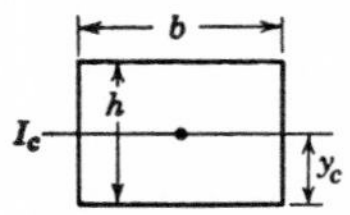	bh	$y_c = \frac{h}{2}$	$I_c = \frac{bh^3}{12}$
Triangle	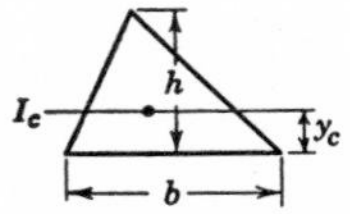	$\frac{bh}{2}$	$y_c = \frac{h}{3}$	$I_c = \frac{bh^3}{36}$
Circle	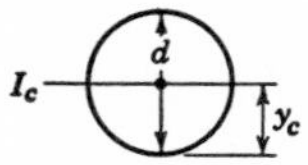	$\frac{\pi d^2}{4}$	$y_c = \frac{d}{2}$	$I_c = \frac{\pi d^4}{64}$
Semicircle	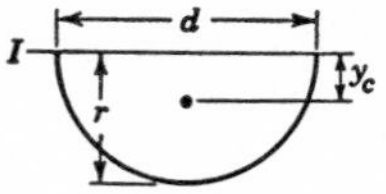	$\frac{\pi d^2}{8}$	$y_c = \frac{4r}{3\pi}$	$I = \frac{\pi d^4}{128}$
Ellipse	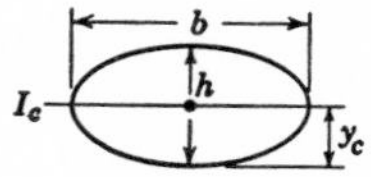	$\frac{\pi bh}{4}$	$y_c = \frac{h}{2}$	$I_c = \frac{\pi bh^3}{64}$
Semiellipse	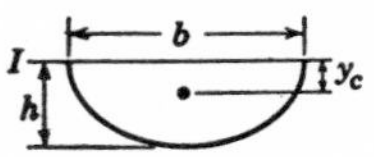	$\frac{\pi bh}{4}$	$y_c = \frac{4h}{3\pi}$	$I = \frac{\pi bh^3}{18}$
Parabola	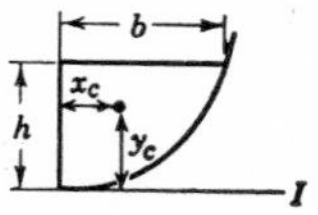	$\frac{2}{3}bh$	$y_c = \frac{3h}{5}$ $x_c = \frac{3b}{8}$	$I = \frac{2bh^3}{7}$
Cylinder	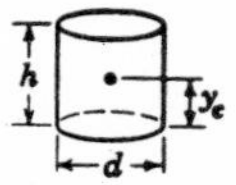	$\frac{\pi d^2 h}{4}$	$y_c = \frac{h}{2}$	

	Sketch	Area or Volume	Location of Centroid	I or I_c
Cone	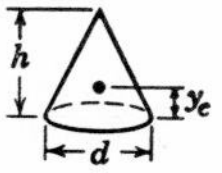	$\frac{1}{3}\left(\frac{\pi d^2 h}{4}\right)$	$y_c = \frac{h}{4}$	
Paraboloid of revolution	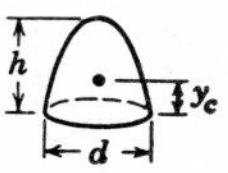	$\frac{1}{2}\left(\frac{\pi d^2 h}{4}\right)$	$y_c = \frac{h}{3}$	
Sphere	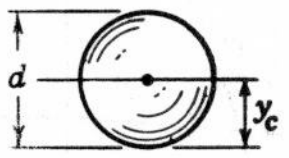	$\frac{\pi d^3}{6}$	$y_c = \frac{d}{2}$	
Hemisphere	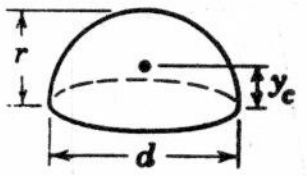	$\frac{\pi d^3}{12}$	$y_c = \frac{3r}{8}$	

appendix VII

Cavitation

The phenomenon of cavitation has become of great importance in the design of high-speed hydraulic machinery such as turbines, pumps, marine propellers, in the overflow and underflow structures of high dams, and in the high-speed motion of underwater bodies (submarines, torpedoes, missiles); it also may be of critical significance in pipe-line design and in certain problems of fluid metering.

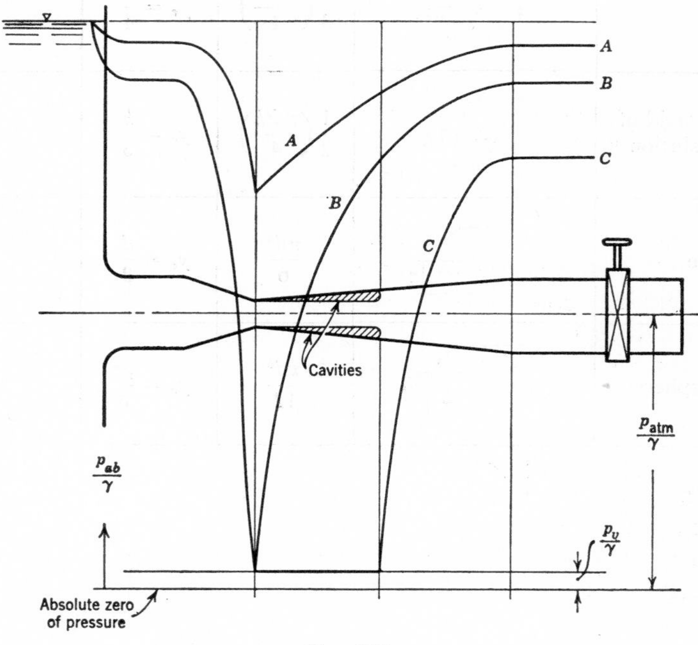

FIG. 280

Cavitation may be expected in a flowing liquid[1] wherever the local pressure falls to the vapor pressure of the liquid. Local vaporization of the liquid will then result, causing a hole, or cavity, in the flow, sometimes accompanied by erosion (pitting) of the solid boundary surfaces, losses of efficiency of machines, and vibration problems.

The nature of cavitation may be most easily observed by study of

[1] Cavitation is not possible in a gas because of the latter's capacity for expansion.

the ideal flow of a liquid through a constriction in a passage (Fig. 280). With the valve partially open, the variation of pressure head through passage and constriction is given by hydraulic grade line A, the point of lowest pressure occurring at the minimum area, where the velocity is highest. Increase of valve opening (causing larger flowrate) produces hydraulic grade line B, for which the absolute pressure in the throat of the constriction falls to the vapor pressure of the liquid, causing the *inception* of cavitation. Further opening of the valve *does not increase the flowrate* but serves to extend the zone of vapor pressure downstream from the throat of the constriction; here the live stream of liquid separates from the boundary walls, producing a cavity in which the mean pressure is the vapor pressure of the liquid. The cavity contains a swirling mass of droplets and vapor and, although appearing steady to the naked eye, actually forms and reforms many times a second. The formation and disappearance of a single cavity are shown schematically in Fig. 281, and the disappearance of the cavity is the clue to the destructive action caused by cavitation. The low-pressure cavity is swept swiftly downstream into a region of high pressure where it collapses suddenly, the surrounding liquid rushing in to fill the void. At the point of disappearance of the cavity the inrushing liquid comes together, momentarily raising the local pressure within the liquid to a very high value. If the point of collapse of the cavity is in contact with the boundary wall, the wall receives a blow (Fig. 282) as from a tiny hammer, and its surface may be stressed locally beyond its elastic limit, resulting eventually in fatigue and destruction of the wall material.

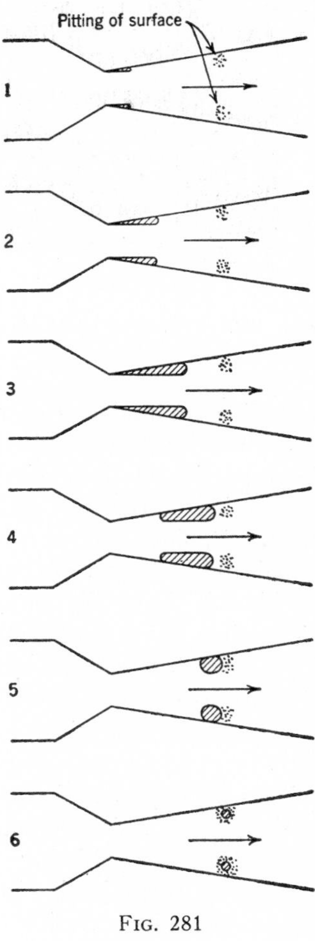

FIG. 281

Another form of cavitation is the *steady-state cavity* frequently observed in the tip vortices (Art. 120) of marine propellers or sur-

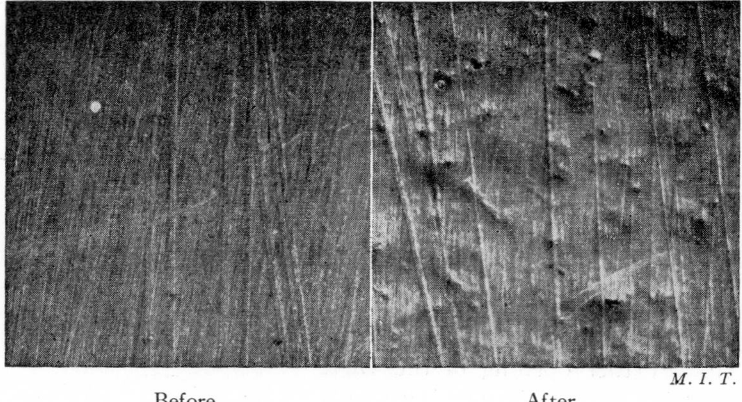

FIG. 282. Pitting of brass plate after 5 hours' exposure to cavitation (magnification 10×).

rounding an underwater missile (Fig. 283). Although the downstream ends of such cavities exhibit certain unsteady phenomena, the large portion of the cavity is steady with its outer boundary (A) a streamline of the flowfield. Such a streamline is a *free streamline* (Art. 28), since the pressure along it will be that in the cavity and equal to the vapor pressure of the liquid. Here the engineering problem is the prediction of cavity form and location of the separation points, the latter of which are critically dependent upon boundary layer growth (Art. 118) and upon the fine details of vaporization in the flow as the pressure falls to the vapor pressure of the liquid.

One of the objectives in the design of hydraulic machinery and structures is the prevention of cavitation, which the designer accomplishes by improved forms of boundary surfaces and the setting of limits beyond which the machine or structure should not be operated. In underwater ballistics, where there is little hope of preventing cavitation, the objective is a thorough understanding of the phenomenon so that its inclusion may lead to a design for reliable performance.

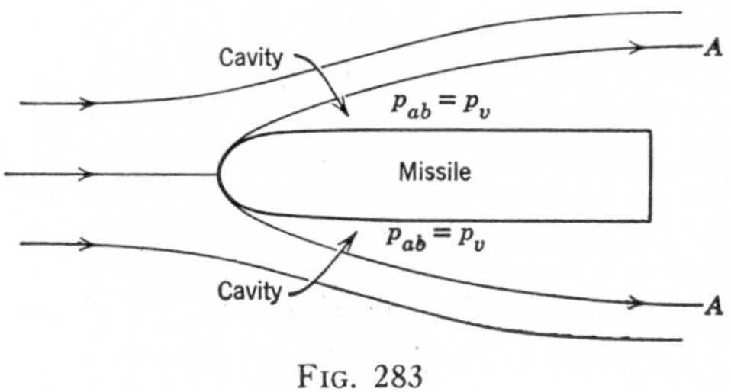

FIG. 283

REFERENCES

S. W. Barnaby, "Formation of Cavities in Water by Screw Propellers at High Speed," *Trans. Inst. Nav. Architects*, 1897.

O. Reynolds, "Experiments Showing the Boiling of Water in an Open Tube at Ordinary Temperatures," *Papers on Mechanical and Physical Subjects*, Vol. II, Cambridge University Press, 1901.

Lord Rayleigh, "Pressure Developed in a Liquid During the Collapse of a Spherical Cavity," *Phil. Mag.*, Vol. 34, 1917.

C. A. Parsons and S. S. Cook, "Investigation into the Causes of Corrosion or Erosion of Propellers," *Engineering* (*London*), Vol. 107, p. 515, April 18, 1919.

E. Englesson, "Turbines and the Cavitation Problem," *Canadian Eng.*, 1926.

W. Spannhake, "Cavitation and Its Influence on Hydraulic Turbine Design," *N.E.L.A.*, *Pub.* 222, June 1932.

L. P. Smith, "Cavitation on Marine Propellers," *Trans. A.S.M.E.*, July 1935.

S. L. Kerr, "Determination of the Relative Resistance to Cavitation Erosion by the Vibratory Method," *Trans. A.S.M.E.*, July 1937.

J. M. Mousson, "Pitting Resistance of Metals under Cavitation Conditions," *Trans. A.S.M.E.*, July 1937.

H. A. Thomas and E. P. Schuleen, "Cavitation in Outlet Conduits of High Dams," *Trans. A.S.C.E.*, Vol. 107, 1942.

J. K. Vennard, J. C. Harrold, J. E. Warnock, and G. H. Hickox, "Cavitation in Hydraulic Structures: A Symposium," *Trans. A.S.C.E.*, Vol. 112, 1947.

R. T. Knapp and A. Hollander, "Laboratory Investigations on the Mechanism of Cavitation," *Trans. A.S.M.E.*, Vol. 70, 1948.

M. S. Plesset, "The Dynamics of Cavitation Bubbles," *Trans. A.S.M.E.*, Vol. 71, 1949.

P. Eisenberg, "On the Mechanism and Prevention of Cavitation," *David Taylor Model Basin Rept.* 712, July 1950.

F. R. Gilmore, "The Growth or Collapse of a Spherical Bubble in a Viscous Compressible Liquid," *Preprints of Papers, Heat Transfer and Fluid Mechanics Institute*, Stanford University Press, 1952.

R. T. Knapp, "Cavitation Mechanics and Its Relation to the Design of Hydraulic Equipment," *Proc. Inst. Mech. Engrs.*, Vol. 166, 1952.

R. T. Knapp, "Recent Investigations of the Mechanics of Cavitation and Cavitation Damage," *Trans. A.S.M.E.*, Vol. 77, 1955.

Symposium, *Cavitation in Hydrodynamics*, National Physical Laboratory (England), 1956.

D. Gilbarg, "Free-Streamline Theory and Steady-State Cavitation"; M. S. Plesset, "Physical Effects in Cavitation and Boiling"; *Naval Hydrodynamics*, Nat. Acad. Sci.-Nat. Res. Council Publ. 515, 1957.

R. T. Knapp, "Accelerated Field Tests of Cavitation Intensity," *Trans. A.S.M.E.*, Vol. 80, 1958.

R. T. Knapp, "Cavitation and Nuclei," *Trans. A.S.M.E.*, Vol. 80, 1958.

F. Numachi, M. Yamabe, R. Oba, "Cavitation Effect on the Discharge Coefficient of the Sharp-Edged Orifice"; J. W. Holl, "The Inception of Cavitation on Isolated Surface Irregularities"; W. H. Wheeler, "Indentation of Metals by Cavitation," *Trans. A.S.M.E.* (Series D), Vol. 82, 1960.

appendix **VIII**

The Expansion Factor,[1] Y

$\frac{A_2}{A_1}$	k \ $\frac{p_2}{p_1}$	0.95	0.90	0.85	0.80	0.75
0	1.40	0.973	0.945	0.916	0.886	0.856
	1.30	0.971	0.941	0.910	0.878	0.846
	1.20	0.968	0.936	0.903	0.869	0.834
0.2	1.40	0.971	0.942	0.912	0.881	0.850
	1.30	0.969	0.938	0.906	0.873	0.839
	1.20	0.967	0.933	0.899	0.863	0.827
0.3	1.40	0.969	0.938	0.907	0.875	0.842
	1.30	0.967	0.934	0.900	0.866	0.831
	1.20	0.965	0.929	0.862	0.856	0.819
0.4	1.40	0.966	0.932	0.899	0.864	0.830
	1.30	0.964	0.928	0.891	0.855	0.818
	1.20	0.961	0.922	0.886	0.844	0.805
0.5	1.40	0.962	0.923	0.886	0.848	0.811
	1.30	0.959	0.918	0.878	0.839	0.799
	1.20	0.955	0.912	0.869	0.827	0.786
0.6	1.40	0.954	0.910	0.867	0.825	0.785
	1.30	0.951	0.904	0.858	0.814	0.772
	1.20	0.947	0.896	0.848	0.801	0.757

$$Y = \sqrt{\frac{1 - \left(\frac{A_2}{A_1}\right)^2}{1 - \left(\frac{A_2}{A_1}\right)^2\left(\frac{p_2}{p_1}\right)^{\frac{2}{k}}} \; \frac{\frac{k}{k-1}\left(\frac{p_2}{p_1}\right)^{\frac{2}{k}}\left[1 - \left(\frac{p_2}{p_1}\right)^{\frac{k-1}{k}}\right]}{1 - \frac{p_2}{p_1}}}$$

[1] The tabulated values were computed by J. P. Robb and R. E. Royer on the Bendix G-15 Digital Computer at the University of Wyoming.

appendix **IX**

Dimensional Analysis of Certain Problems

PIPE FRICTION (see Art. 64)

$$\tau_o = F(V, d, \rho, \mu, e)$$

$$\tau_o = CV^a d^b \rho^c \mu^d e^e$$

$$\frac{M}{T^2L} = \left(\frac{L}{T}\right)^a (L)^b \left(\frac{M}{L^3}\right)^c \left(\frac{M}{LT}\right)^d (L)^e$$

$$\begin{cases} M: & 1 = c + d \\ L: & -1 = a + b - 3c - d + e \\ T: & 2 = a + d \end{cases}$$

$$a = 2 - d, \quad c = 1 - d, \quad b = -d - e$$

$$\tau_o = CV^{2-d} d^{-d-e} \rho^{1-d} \mu^d e^e$$

$$\tau_o = C(\mu/Vd\rho)^d (e/d)^e \rho V^2$$

$$\tau_o = F'(Vd\rho/\mu, e/d)\rho V^2$$

WEIR FLOW (see Art. 105)

$$q = F(H, P, \mu, \sigma, \rho, g)$$

$$q = CH^a P^b \mu^c \sigma^d \rho^e g^f$$

$$\frac{L^2}{T} = (L)^a (L)^b \left(\frac{M}{LT}\right)^c \left(\frac{M}{T^2}\right)^d \left(\frac{M}{L^3}\right)^e \left(\frac{L}{T^2}\right)^f$$

$$\begin{cases} M: & 0 = c + d + e \\ L: & 2 = a + b - c - 3e + f \\ T: & 1 = c + 2d + 2f \end{cases}$$

$$e = -c - d, \quad f = \tfrac{1}{2} - c/2 - d,$$

$$a = \tfrac{3}{2} - b - 3c/2 - 2d$$

$$q = CH^{\frac{3}{2}-b-3c/2-2d}P^b\mu^c\sigma^d\rho^{-c-d}g^{\frac{1}{2}-c/2-d}$$

$$q = C\sqrt{g}\,H^{\frac{3}{2}}(P/H)^b(\mu/\rho\sqrt{g}\,H^{\frac{3}{2}})^c(\sigma/\rho g H^2)^d$$

$$q/\sqrt{g}\,H^{\frac{3}{2}} = F'(P/H,\ \rho\sqrt{g}\,H^{\frac{3}{2}}/\mu,\ \rho g H^2/\sigma)$$

DRAG OF IMMERSED OBJECTS (see Art. 116)

$$D = F(A, \rho, \mu, V_o, E)$$

$$D = CA^a\rho^b\mu^cV_o^{\ d}E^e$$

$$\frac{ML}{T^2} = (L^2)^a\left(\frac{M}{L^3}\right)^b\left(\frac{M}{LT}\right)^c\left(\frac{L}{T}\right)^d\left(\frac{M}{T^2L}\right)^e$$

$$\begin{cases} M: & 1 = b + c + e \\ L: & 1 = 2a - 3b - c + d - e \\ T: & 2 = c + d + 2e \end{cases}$$

$$b = 1 - c - e, \quad d = 2 - c - 2e, \quad a = 1 - c/2$$

$$D = CA^{1-c/2}\rho^{1-c-e}\mu^cV_o^{\ 2-c-2e}E^e$$

$$D = CA\rho V_o^{\ 2}(\mu/\sqrt{A}\,\rho V_o)^c(E/\rho V_o^{\ 2})^e$$

$$D = F'(\sqrt{A}\,\rho V_o/\mu,\ V_o^{\ 2}\rho/E)A\rho V_o^{\ 2}$$

appendix X

Mathematical Details of Turbulent Pipe Flow[1]

(*a*) To obtain the turbulent velocity profile (see Art. 68),

$$\tau = \tau_o\left(1 - \frac{y}{R}\right) = \rho\kappa^2 \frac{(dv/dy)^4}{(d^2v/dy^2)^2}$$

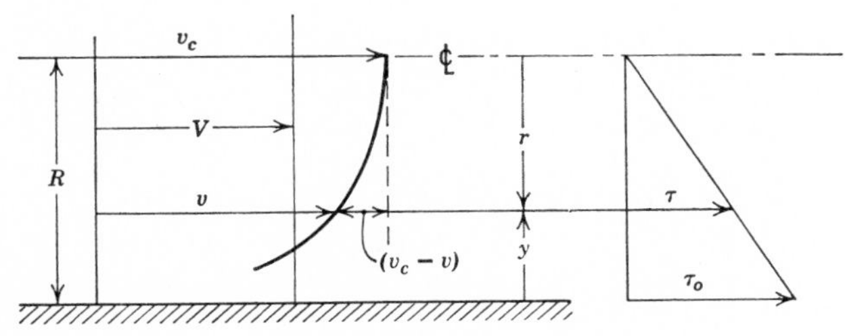

FIG. 284

Invert the equation, take the square root of both sides, substitute $v_\star$ for $\sqrt{\tau_o/\rho}$, and insert minus sign, since d^2v/dy^2 is seen (from Fig. 284) to be negative.

$$\frac{d^2v/dy^2}{(dv/dy)^2} = -\frac{\kappa}{v_\star} \frac{1}{\sqrt{1 - y/R}}$$

Let $x = dv/dy$ so that $dx/dy = d^2v/dy^2$. Rearrange the equation

[1] A considerably more comprehensive treatment is given by B. A. Bakhmeteff in *The Mechanics of Turbulent Flow*, Princeton University Press, 1936.

with this substitution to give

$$\frac{dx}{x^2} = -\frac{\kappa}{v_\star}\frac{dy}{\sqrt{1 - y/R}}$$

and integrate to obtain

$$\frac{1}{x} = \frac{2\kappa R}{v_\star}\sqrt{1 - y/R} + C_1$$

To obtain C_1 assume that, as $y \to 0$, $dv/dy(=x) \to \infty$. Note, however, that, although this is reasonable and convenient, it is not exact; toward the wall ($y \to 0$) the velocity gradient is indeed high but cannot be infinite. Substituting $x = \infty$ and $y = 0$ yields $C_1 = -2\kappa R/v_\star$. Including this value for C_1 and writing x as dv/dy, the equation then becomes

$$dv = -\frac{v_\star}{2\kappa R}\frac{dy}{1 - \sqrt{1 - y/R}} \tag{336}$$

which may be integrated to yield

$$v = \frac{v_\star}{\kappa}[\sqrt{1 - y/R} + \ln(1 - \sqrt{1 - y/R})] + C_2$$

For $v = v_c$, $y = R$, giving $C_2 = v_c$. Substituting v_c for C_2 and rearranging,

$$\frac{v_c - v}{v_\star} = -\frac{1}{\kappa}[\sqrt{1 - y/R} + \ln(1 - \sqrt{1 - y/R})]$$

(*b*) To obtain the velocity ratio V/v_c (see Art. 68),

$$(v_c - V)\pi R^2 = \int_0^R (v_c - v)2\pi r\, dr$$

Substitute $-2.5\, v_\star \ln(y/R)$ for $(v_c - v)$, $(R - y)$ for r, and $-dy$ for dr, and cancel π:

$$(v_c - V)R^2 = 5v_\star \int_R^0 (R - y)\ln\frac{y}{R}\,dy$$

Performing the indicated integration,

$$(v_c - V)R^2 = 5v_\star\left[Ry\ln\frac{y}{R} - Ry - \frac{y^2}{4}\left(2\ln\frac{y}{R} - 1\right)\right]_R^0$$

Substituting the lower limit, [ ] $= -3R^2/4$. Substituting the upper limit, [ ] is indeterminate. However, by substituting a small quantity Δ which is then allowed to approach zero it will be found that as, $\Delta \rightarrow 0$, [ ] $\rightarrow 0$. Introducing these values to the foregoing equation,

$$v_c - V = 15v_\star/4 = 3.75v_\star$$

or, by rearrangement and substitution of $V\sqrt{f/8}$ for $v_\star$,

$$\frac{V}{v_c} = \frac{1}{1 + 3.75\sqrt{f/8}}$$

There is better agreement with experimental results when 4.07 is substituted for 3.75. The discrepancy between the derived and experimental values results from the imperfections of equation 336; at the centerline of the pipe the velocity gradient is obviously zero, yet this result is not obtained from the equation.

Index